# PRECALCULUS MATHEMATICS
# A FUNCTIONAL APPROACH

## SECOND EDITION

# PRECALCULUS MATHEMATICS

# A FUNCTIONAL APPROACH

SECOND EDITION

## JAMES F. CONNELLY
## ROBERT A. FRATANGELO

MONROE COMMUNITY COLLEGE, ROCHESTER, NEW YORK

MACMILLAN PUBLISHING CO., INC.
NEW YORK
COLLIER MACMILLAN PUBLISHERS
LONDON

Macmillan Publishing Co., Inc.
866 Third Avenue, New York, New York 10022

Collier Macmillan Canada, Ltd.

Library of Congress Cataloging in Publication Data

Connelly, James Francis (date)
    Precalculus mathematics.

    Includes index.
    1. Mathematics—1961-   I. Fratangelo,
Robert A., joint author.   II. Title.
QA39.2.C66   1980        512′.1        78-26929
ISBN 0-02-324400-3

Printing:    2 3 4 5 6 7 8      Year:  0 1 2 3 4 5 6

TO OUR WIVES GINGER AND JO

# PREFACE
## TO THE SECOND EDITION

The basic goal of this text remains unchanged—to present, in a readable fashion, those topics in mathematics necessary for the successful study of calculus.

In an attempt to accomplish this goal, we have supplemented the illustrative examples of the first edition, included more applied problems, and made a special effort to expand and grade the exercises. In so doing, it is our intention that different instructors will have the freedom of choice to emphasize different points and make assignments according to the needs of the students.

Throughout the text, we have placed increased emphasis on the development of students' curve sketching ability. The importance of this skill cannot be overstated for the student planning on entering a calculus sequence.

There have been many changes in the text, and whereas they are impossible to completely enumerate, we feel that the following are especially significant.

1. In Chapter 1 the section on the Cartesian Coordinate System has been rewritten to incorporate the concept of a graph and then placed before the section on the Distance Formula.
2. In Chapter 5, the algebra of functions now directly precedes the discussion of inverse functions in the hope that the student will be better prepared for the composition viewpoint of inverse.
3. In Chapter 6 the section dealing with logarithmic functions has been separated into two distinct sections. This has been done to emphasize:
   a) the graphic properties of logarithmic functions,
   and
   b) the solutions of equations involving logarithmic functions.
4. In Chapter 10, the section on translation of axes has been rewritten and the exercises have been expanded.

We wish to thank the instructors and students who were generous enough to take the time to comment on the original text material. Their observations have been reflected in many of the changes. We would also like to acknowledge the valuable assistance of our production supervisor, John Travis, and the editorial staff at Macmillan, our Series Editor, Calvin A. Lathan, and the members of the Mathematics Department at Monroe Community College, especially Professors Thomas B. Dellaquila and Robert A. Gullo. We are grateful to Pamela Dretto for her usual excellent job of manuscript typing.

We again acknowledge our debt to our wives Ginger and Jo. Their support and encouragement have been immeasurable.

J.F.C.
R.A.F.

# CONTENTS

# PRECALCULUS MATHEMATICS

# A FUNCTIONAL APPROACH

**SECOND EDITION**

# CHAPTER 1
# SETS
# AND
# NUMBERS

## INTRODUCTION TO SETS

Since many mathematical concepts can be described most easily in terms of sets, it is advantageous to familiarize ourselves with a few basic set ideas. **Sets** are usually considered to be *collections of objects*. We can, for example, describe the set of all buildings on campus, or the set of all classrooms in a given building, or the set of all students in a particular classroom at a given time. In terms of our number system, we often refer to "the set of all positive integers" or "the set of all rational numbers" or "the set of all real numbers."

Geometrically, we can describe various curves in terms of sets. A *circle* is the set of all points equidistant from a given fixed point $C$ (Figure 1.1). A *parabola* is the set of all points equidistant from a fixed line (called the **directrix**) and a fixed point (called the **focus**) (Figure 1.2). The sets we have described are all **well defined.** That is, we can readily determine whether or not any element belongs to a particular set. The set of all good instructors is not well defined because the term "good" is not clear in this context.

The objects that belong to a given set are called the **elements** or the **members** of the set and are generally designated by lowercase letters, such as $x, y, a, b$. When we use capital

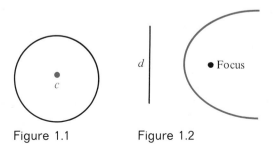

Figure 1.1         Figure 1.2

letters such as $A$ and $B$ to represent the set itself, the notation $a \in A$ is read "$a$ belongs to or is an element of the set $A$." Similarly, $a \notin A$ is read "$a$ does not belong to or is not an element of $A$." If $N$ is the set of all positive integers, then $2 \in N$, $5 \in N$, but $\frac{2}{3} \notin N$.

A particular set may be described in any one of three ways. The first method is simply a statement that indicates the elements of the set.

### Example 1.1

$A$ is the set of all vowels in the English language.

$B$ is the set of positive integers between 1 and 7.

$C$ is the set of even numbers greater than 4 and less than 2.

$D$ is the set of all positive prime numbers.

We may also describe a set by listing its members. When employing this means of description, the list of elements is enclosed in braces $\{\ \}$. The set of the first five positive integers is $\{1, 2, 3, 4, 5\}$. Note that in this listing method, each element of the set should appear only once, but the order of the elements is immaterial. Thus $\{1, 1, 2, 3, 3, 3, 4\}$ would be written as $\{1, 2, 3, 4\}$ and may also be written as $\{2, 1, 3, 4\}$.

If the number of elements in the set should be infinite, we can still use the braces as long as a pattern is indicated. The set of all natural numbers $N$ can be written as $\{1, 2, 3, \ldots\}$, where the three dots indicate that the elements of the set continue without end.

We can use the listing method to describe the sets in Example 1.1 as follows:

$$A = \{a, e, i, o, u\}$$
$$B = \{2, 3, 4, 5, 6\}$$
$$C = \{\ \}$$
$$D = \{2, 3, 5, 7, \ldots\}$$

When a set contains no elements, such as $C$ here, we refer to it as the **empty set** or **null set** and we use the symbol $\varnothing$. $C = \{\ \} = \varnothing$. The set of all elements under consideration is called the **universal set** and is denoted U.

Finally, sets may be described by means of a rule. $\{x \mid x$ is a rational number$\}$ is read "the set of all $x$ such that $x$ is a rational number."

### Example 1.2

"$\{x \mid x$ is a positive integer between 0 and 4$\}$" describes a set containing the numbers 1, 2, and 3.

"$\{x \mid x$ is a prime number between 0 and 4$\}$" describes a set containing the numbers 2 and 3.

"$\{x \mid x$ is a positive odd integer$\}$" describes a set containing an infinite number of elements.

$\{x \mid x$ is a positive odd integer$\} = \{1, 3, 5, 7, \ldots\}$.

The following list includes special sets of numbers that will be discussed throughout the text.

**Natural Numbers.** The natural numbers $N$ are the set of positive whole numbers 1, 2, 3, . . .:

$$N = \{1, 2, 3, \ldots\}$$

**Integers.** The integers $J$ are the positive and negative whole numbers and zero:

$$J = \{\ldots, -2, -1, 0, 1, 2, \ldots\}$$

**Rational Numbers.** The rational numbers $Q$ are numbers that can be expressed in the form $p/q$, where $p$ and $q$ are integers and $q \neq 0$:

$$Q = \{p/q \mid p \in J, \quad q \in J, \quad q \neq 0\}$$

**Irrational Numbers.** The irrational numbers $H$ are numbers that are not rational (i.e., cannot be expressed in the form $p/q$, where $p$ and $q$ are integers).

**Real Numbers.** The real numbers $\mathbb{R}$ are the numbers that are either rational or irrational:

$$\mathbb{R} = \{x \mid x \in Q \quad \text{or} \quad x \in H\}$$

## EXERCISES FOR SECTION 1

1. Use the concept of sets to determine which of the following are well defined.
   (a) The set of the three greatest living football players.
   (b) The set of all months of the year beginning with J.
   (c) The set of all living Americans.
   (d) The set of all living great Americans.
   (e) The set of all rivers in the world.

2. List each of the members of the following sets.
   (a) $\{x \mid x + 1 = 2\}$
   (b) $\{x \mid x$ is an integer greater than 2 and less than 9$\}$
   (c) $\{x \mid x$ is both even and odd$\}$
   (d) $\{x \mid (x - 1)(x + 3) = 0\}$
   (e) $\{x \mid x$ is a positive integral factor of 45$\}$

3. Which of the following are empty sets?
   (a) All integers ending in 8 that are perfect squares.
   (b) All integers ending in 5 that are perfect squares.
   (c) $\{x \mid 2x = x\}$
   (d) $\{x \mid 2x + 1 = 2(x + 1)\}$
   (e) The set of all even prime numbers.

4. Describe each of the following sets using the notation $\{x \mid x$ has some property and is defined by some rule$\}$.
   (a) $\{1, 3, 5, 7, \ldots\}$     (d) $\{\ \}$
   (b) $\{3, 4, 5, 6\}$     (e) $\{1, 2\}$
   (c) $\{3, 4, 5, 6, \ldots\}$     (f) $\{1, 2, 3, \ldots\}$

SECTION **2**
## SET RELATIONSHIPS

Consider the set $A = \{1, 2, 3\}$, $B = \{a, b, c\}$, $C = \{3, 1, 2\}$. We readily observe that the sets $A$ and $C$ contain the same elements. It would seem reasonable then to conclude that $A = C$. Formally, we have

**Definition 2.1** Two sets $A$ and $B$ are said to be **equal** ($A = B$) if and only if they have the same elements.

### Example 2.1
If $H$ is the set of all equilateral triangles, and $D$ is the set of all equiangular triangles, then $H = D$.

### Example 2.2
If

$$E = \left\{ x \left| \frac{x - 1}{x - 2} = 0 \right. \right\}$$

$$F = \{x \mid x - 1 = 0\}$$

then $E = F$.

We also observe that the sets $A$ and $B$ or $B$ and $C$ have the same number of elements. They are obviously not equal, but they are said to be **equivalent;** that is, they have the same number of elements or are the same size. For finite sets, equivalence is a simple matter. We count the number of elements. If the sets are infinite, the determination is more complex. For example, are there more odd or more even integers? Are there more prime numbers than numbers divisible by 5?

To answer some of these questions let us look at the idea of one-to-one correspondence. Let $A$ and $B$ be any two sets. A **one-to-one correspondence** exists between the two sets if each element of $A$ can be paired with one and only one element of $B$, and conversely.

### Example 2.3

$$A = \{1, 2, 3, 4, 5\}$$
$$\updownarrow \ \updownarrow \ \updownarrow \ \updownarrow \ \updownarrow$$
$$B = \{a, b, c, d, e\}$$

$A$ and $B$ are in one-to-one correspondence. This is demonstrated by the pairing that is shown. There are other ways to pair the elements, but it is only necessary to demonstrate one such correspondence.

### Example 2.4
Let

$$O = \{1, 3, 5, 7, \ldots\}$$
$$N = \{1, 2, 3, 4, \ldots\}$$
$$E = \{2, 4, 6, 8, \ldots\}$$

$O$ and $N$ can be put in one-to-one correspondence. For every number of the form $n$, $n \in N$, there exists a corresponding number of the form $2n - 1$, $2n - 1 \in O$. $N$ and $E$ can be put in one-to-one correspondence, since for any number of the form $n$, $n \in N$, there exists a corresponding number of the form $2n$, $2n \in E$.

At first glance it would seem that the set $N$ should have twice as many elements as the set $E$. But, since we can place the sets in one-to-one correspondence, they must have the same number of elements. This seeming inconsistency arises from the fact that both sets are infinite (see Exercise 6).

Not all sets are equal or can be put into one-to-one correspondence. But a given set may be contained in another set. For example, every element in $A = \{1, 2, 3\}$ also belongs to the set $B = \{1, 2, 3, 4, 5\}$. In this case we say that $A$ is a proper subset of $B$.

**Definition 2.2**  $A$ is said to be a **proper subset** of $B$, denoted $A \subset B$, if and only if every element in $A$ is also in $B$, and in addition $A \neq B$.

> **Example 2.5**
> If $A = \{2, 4, 6\}$ and $B = \{1, 2, 3, 4, 5, 6\}$, then $A \subset B$. If $O = \{1, 3, 5, \ldots\}$ and $N = \{1, 2, 3, \ldots\}$, then $O \subset N$.

Suppose that $B$ is the set of all students in a given classroom and $A$ is the set of all students in this classroom who are 20 years old. We know that every element or person in $A$ is also an element or person in $B$. We do not know if $A = B$. In this situation we say that $A$ is a subset of $B$.

**Definition 2.3**  $A$ is said to be a **subset** of $B$, denoted $A \subseteq B$, if and only if every element in $A$ is also in $B$.

We observe from the definition that any set $A$ is a subset of itself. $A \subseteq A$. This applies to the empty set as well. $\varnothing \subseteq \varnothing$. We agree that $\varnothing$ is a subset of any set. Further, $\varnothing$ is a proper subset of every set except itself.

## EXERCISES FOR SECTION 2

**1.** All equal sets are equivalent, but equivalent sets are not necessarily equal. Make up an example to show this.

**2.** $A = \{1, 2, 3, 4, 5\}$, $B = \{2, 3, 4\}$, $C = \{2, 4, 5\}$. Which of the following are true?
(a) $A \subset B$      (d) $A \subset C$      (g) $B \subset A$
(b) $B \subset C$      (e) $C \subseteq A$      (h) $C \subseteq B$
(c) $C \subset C$      (f) $\varnothing \subset B$      (i) $B = C$

**3.** If $C = \{0, 3, 6\}$, we say that $6 \in C$ or $\{6\} \subset C$, but we cannot say that $6 \subset C$, since 6 is not a set. Which of the following are true?
(a) $3 \in C$      (d) $\varnothing \subset C$      (g) $0 = \varnothing$
(b) $3 \subset C$      (e) $0 \in \{0\}$      (h) $0 \subset \varnothing$
(c) $\varnothing \in C$      (f) $\{\varnothing\} \subset C$      (i) $0 \in \varnothing$

4. For each of the following sets, determine whether $A$ and $B$ are equivalent and, if so, whether $A = B$.
   (a) $A = \{2, 4, 6\}$, $B = \{6, 4, 2\}$.
      Assume that $x \in \{-1, 0, 1\}$.
   (b) $A = \{-1, 1\}$, $B = \{x \,|\, x^2 = 1\}$.
   (c) $A = \{x \,|\, x^2 - x = 0\}$, $B = \{0\}$.
   (d) $A = \{x \,|\, (x + 1)^2 = 0\}$, $B = \{-1\}$.

5. Give two examples of an empty set.

6. A set is said to be **infinite** if it can be put in one-to-one correspondence with a proper subset of itself. Thus the set of all positive integers is an infinite set because we can demonstrate a one-to-one correspondence between the positive integers and the positive even integers (a proper subset of the integers) as follows:

$$N = \{1, 2, 3, \ldots, n, \ldots\}$$
$$\updownarrow \ \updownarrow \ \updownarrow \qquad \updownarrow$$
$$E = \{2, 4, 6, \ldots, 2n, \ldots\}$$

Show that each of the following sets is infinite.
   (a) $E = \{2, 4, 6, \ldots\}$
   (b) $O = \{1, 3, 5, \ldots\}$
   (c) $A = \{3, 6, 9, \ldots\}$
   (d) $J = \{x \,|\, x$ is an integer$\}$

7. Which of the following sets are infinite?
   (a) $\{x \,|\, x \in N,\ x > 2\}$
   (b) The set of all odd natural numbers.
   (c) $\{x \,|\, x \in N,\ (x - 3)(x + 2) = 0\}$
   (d) $\{1, 2, 3, \ldots, 999\}$
   (e) $\{4, 8, 12, \ldots, 4n, \ldots\}$, $n \in N$

# SECTION 3
# SET OPERATIONS

Suppose that we have a set $N = \{1, 2, 3, \ldots\}$. We know that $O = \{1, 3, 5, \ldots\}$ and $E = \{2, 4, 6, \ldots\}$ are both proper subsets of $N$. $O \subset N$ and $E \subset N$. We notice that the set of elements in $N$, but not in $O$, is $E$. The set $E$ is called the complement of $O$ relative to $N$ and is written $O'$.

**Definition 3.1**  Let $U$ be a given universal set and $A$ any set such that $A \subseteq U$. The **complement** of $A$ (written $A'$) is the set of all elements in $U$ that do not belong to $A$.

**Example 3.1**
If $U = \{1, 2, 3, 4, 5\}$ and $A = \{1, 3, 5\}$, then $A' = \{2, 4\}$.

**Example 3.2**
If $A = U$, then $A' = \varnothing$. If $A = \varnothing$, then $A' = U$.

There are two other set operations that are used to produce new sets, the operations of union and intersection.

**Definition 3.2** The **union** of the sets $A$ and $B$, written $A \cup B$, is the set of all elements that belong to $A$ or $B$ or both.

**Definition 3.3** The **intersection** of the sets $A$ and $B$, written $A \cap B$, is the set of all elements that $A$ and $B$ have in common.

Symbolically we have

$$A \cup B = \{x \mid x \in A \text{ or } x \in B\}$$
$$A \cap B = \{x \mid x \in A \text{ and } x \in B\}$$

**Example 3.3**
Let $U = \{1, 2, 3, 4, 5\}$, $A = \{1, 3, 5\}$, $B = \{2, 4\}$, $C = \{2, 3, 5\}$. Then

$$A' = \{2, 4\} = B$$
$$A \cup B = U$$
$$B \cup C = \{2, 3, 4, 5\}$$
$$A \cap C = \{3, 5\}$$
$$A \cap B = \emptyset$$

In this case $A$ and $B$ are said to be disjoint since they have no common elements.

## EXERCISES FOR SECTION 3

**1.** Let $U = \{1, 2, 3, \ldots, 10\}$, $A = \{1, 2, 3, 4, 5\}$, $B = \{2, 4, 6, 8, 10\}$, and $C = \{1, 3, 5, 7, 9\}$. List each of the following.
  (a) $B \cup C$     (e) $B' \cap C'$     (i) $A \cap C$
  (b) $B \cap C$     (f) $(B \cup C)'$     (j) $A \cup A'$
  (c) $B'$     (g) $A \cup B$     (k) $C \cap C'$
  (d) $C'$     (h) $A \cup C$     (l) $(A \cap B) \cup C'$

**2.** Under what conditions would each of the following be true?
  (a) $A \cup \emptyset = \emptyset$     (d) $A \cap B = A$
  (b) $A \cup \emptyset = U$     (e) $A \cap U = A$
  (c) $A \cup B = A$     (f) $A \cup B = A \cap B$

**3.** The set $A - B$ is called the **complement** of $B$ relative to $A$ and is defined by $\{x \mid x \in A \text{ and } x \notin B\}$ or $A - B = A \cap B'$. Let $U = \{1, 2, 3, 4, 5, 6, 7, 8, 9, 10\}$, $A = \{1, 2, 3, 4, 5\}$, and $B = \{5, 6, 7, 8, 9, 10\}$. List each of the following sets.
  (a) $A - B$     (c) $U - A$
  (b) $B - A$     (d) $U - (A - B)$

**4.** Let $J$ be the set of all integers, $N$ the set of all positive integers, $N^-$ the set of all negative integers, $Q$ the set of all rational numbers, $H$ the set of all irrational numbers, and $\mathbb{R}$ the set of all real numbers. Find each of the following sets. (Assume that the universal set is $\mathbb{R}$.)

(a) $H \cup Q$        (f) $N \cup N^-$
(b) $J \cap H$        (g) $\mathbb{R} \cap H$
(c) $J \cap Q$        (h) $\mathbb{R} \cup J$
(d) $J \cup Q$        (i) $N \cap Q$
(e) $N \cap N^-$      (j) $H'$

# SECTION 4
## REAL NUMBERS

Let us suppose that we start with an arbitrary line $L$ and an arbitrary segment on $L$ with an assigned length of $l$. We can construct a number scale on $L$ by selecting a point, which we call **zero** (0), and then marking off successive segments of length $l$ to the right and left of 0:

The points to the right of 0 are the positive integers and the points to the left of 0 are the negative integers. The set composed of positive integers, negative integers, and zero constitutes the **set of integers.** This set is often denoted by the letter $J$, where $J = \{\ldots, -3, -2, -1, 0, 1, 2, 3, \ldots\}$.

The integers account for some of the points on $L$, but obviously not all of them. To fill in a part of what is missing, we will consider the rational numbers.

**Definition 4.1** A **rational number** is any number that can be expressed in the form $p/q$, where $p$ and $q$ are integers and $q \neq 0$. If we let $Q$ denote the set of rational numbers, we have $Q = \{p/q \mid p \in J, q \in J, \text{ and } q \neq 0\}$.

Every rational number can be expressed in decimal form simply by dividing the denominator into the numerator. In addition, it can be shown that these decimal representations must eventually repeat themselves. For example, $\frac{1}{4} = 0.2500\ldots, \frac{1}{3} = 0.333\ldots,$ $\frac{2}{7} = 0.285714285714\ldots.$ In the case of $\frac{1}{4}$, the decimal $0.25000\ldots$ is a terminating decimal; that is, after a certain point, all the digits are 0. For $\frac{1}{3}$ and $\frac{2}{7}$, the decimals are repeating but nonterminating. At this point we can define the rational numbers as the set of all repeating decimals, terminating or nonterminating.

Let us consider what happens if we begin with a repeating decimal and wish to express it in the form $p/q$. If the decimal is terminating, the process is quite simple.

### Example 4.1
Express $0.25000\ldots$ as a fraction.

**Solution.** We abbreviate the decimal and write $0.25000\ldots$ as $0.25$. Then $0.25 = \frac{25}{100} = \frac{1}{4}$. $\bullet^*$

---
*The symbol $\bullet$ designates end of proof and solution.

If the decimal is nonterminating, the repetition must occur over some sequence of digits. We indicate this by means of a bar drawn over the digits that repeat. Thus 2.141414... would be written as $2.\overline{14}$ and 0.1345345... as $0.1\overline{345}$. To convert these types of expressions to fractions, we follow the procedure outlined in the following examples.

### Example 4.2
Express $1.\overline{4}$ as a rational number in the form $p/q$.

Solution. Let $x = 1.\overline{4}$. Then $10x = 14.\overline{4}$. If we subtract $x$ from $10x$ we have

$$
\begin{array}{r}
10x = 14.\overline{4} \\
- \quad x = \phantom{0}1.\overline{4} \\
\hline
9x = 13 \quad \text{or} \quad x = \tfrac{13}{9} \quad \bullet
\end{array}
$$

### Example 4.3
Express $3.2\overline{17}$ as a rational number in the form $p/q$.

Solution. Let $x = 3.2\overline{17}$. $10x = 32.\overline{17}$ and $1{,}000x = 3{,}217.\overline{17}$. When we subtract we obtain

$$
\begin{array}{r}
1{,}000x = 3{,}217.\overline{17} \\
- \quad 10x = \phantom{3{,}217.}32.\overline{17} \\
\hline
990x = 3{,}185 \\
x = \dfrac{3{,}185}{990} = \dfrac{637}{198} \quad \bullet
\end{array}
$$

### Example 4.4
Show that $0.\overline{9} = 1$.

Solution. Let $x = 0.\overline{9}$. Then $10x = 9.\overline{9}$. Therefore,

$$
\begin{array}{r}
10x = 9.\overline{9} \\
- \quad x = 0.\overline{9} \\
\hline
9x = 9 \quad x = 1 \quad \bullet
\end{array}
$$

In addition to the set of repeating decimals, there also exists a set of decimals that do not repeat. For example, the decimal 0.01002000300004... is nonrepeating. Other examples of nonrepeating decimals include $\pi = 3.14159...$, $\sqrt{2} = 1.41421...$, and $e = 2.71828...$. Numbers of this type are called **irrational numbers** and are denoted by $H$. The set $\mathbb{R}$ of real numbers is the union of the sets $Q$ and $H$:

$$\mathbb{R} = Q \cup H = \{x \mid x \in Q \text{ or } x \in H\}$$

Geometrically, the set of real numbers is represented as the set of all points on a line which we call the **real number line** or **real number axis.** This is possible because every point on the number line corresponds to one and only one real number; and, conversely, every real number corresponds to one and only one point.

Assuming familiarity with addition, $+$, and multiplication, $\cdot$, we now list some axioms governing the operations of addition and multiplication.

**Closure for Addition.** $a + b$ is a real number.

**Commutative Law of Addition.** $a + b = b + a$.

**Associative Law of Addition.** $a + (b + c) = (a + b) + c$.

**Identity.** There is a unique real number 0 such that $a + 0 = a$ for all $a$.

**Inverse.** For each $a$ there is a unique real number, $-a$, such that $a + (-a) = 0$.

**Closure for Multiplication.** $a \cdot b$ is a real number.

**Commutative Law of Multiplication.** $a \cdot b = b \cdot a$.

**Associative Law of Multiplication.** $a \cdot (b \cdot c) = (a \cdot b) \cdot c$.

**Identity.** There is a unique real number 1 such that $a \cdot 1 = a$ for all $a$.

**Inverse.** For each $a$ other than 0, there is a unique number, $1/a$, such that $a \cdot (1/a) = 1$.

**Distributive Law.** $a \cdot (b + c) = a \cdot b + a \cdot c$.

We will now define subtraction in terms of addition and division in terms of multiplication in order to make use of these eleven axioms for all four operations.

**Definition 4.2** Let $a$ and $b$ be any real numbers. Then the **difference** of $a$ and $b$, written $a - b$, is the real number $a + (-b)$.

**Definition 4.3** Let $a$ and $b$ be any real numbers, $b \neq 0$. Then the **quotient** of $a$ and $b$, written $a/b$ or $a \div b$, is the real number $a \cdot (1/b)$.

Note that if $a/b = c$, then $b \cdot c = a$ provided that $c$ exists and has only one possible value.

**Remark 4.1**

Definitions 4.2 and 4.3 enable us to handle problems that involve differences and quotients by converting them into sums and products, respectively.

In any initial study of the real numbers, the multiplicative and divisibility properties of zero generally cause the most trouble. To alleviate future problems, let us consider some of the **properties of zero:**

1. $a \cdot 0 = 0$.
2. If $a \cdot b = 0$, then either $a = 0$ or $b = 0$.
3. If $a \neq 0$, then $0/a = 0$.
4. If $a \neq 0$, then $a/0$ is undefined.
5. $0/0$ is indeterminate.

Property 2 is a consequence of property 1 and is especially useful in solving equations.

**Example 4.5**

Solve $x^2 - 3x + 2 = 0$.

**Solution.** By factoring, we have $(x - 1)(x - 2) = 0$. Property 2 states that $x - 1 = 0$ or $x - 2 = 0$. Hence $x = 1$ and $x = 2$ are solutions of the equation. ●

Now let us briefly outline the validity of the remaining properties. For property 3 we have $0/a = 0$ or $0 \cdot a = 0$. The result follows from property 1.

For property 4, let us suppose that $a/0 = c$. Then $a = c \cdot 0 = 0$, which is impossible since $a \neq 0$.

In a similar fashion, suppose that $0/0 = c$. Then $0 = c \cdot 0$ and we have an equation that is satisfied by any real number $c$. Since we are not able to "determine" a unique value for $c$, the expression $0/0$ is indeterminate. This accounts for our fifth property. Note that we may never divide by zero.

### Example 4.6

The expression $\dfrac{x-1}{x-2}$ is equal to 0 for $x = 1$ and is undefined for $x = 2$. The expression $\dfrac{2x-6}{x-3}$ is indeterminate for $x = 3$.

## EXERCISES FOR SECTION 4

1. Express the given rational numbers in terms of a decimal expansion.
   (a) $\frac{6}{7}$          (e) $\frac{3}{8}$
   (b) $\frac{1}{3}$          (f) $\frac{1}{11}$
   (c) $-\frac{1}{2}$         (g) $\frac{4}{10}$
   (d) $4\frac{1}{4}$         (h) $2\frac{3}{5}$

2. Express each of the following numbers as the ratio of integers.
   (a) $-2$               (e) $1.\overline{7}$
   (b) $1.3$              (f) $3.1\overline{23}$
   (c) $\frac{2}{11}$     (g) $3.\overline{9}$
   (d) $0.\overline{45}$  (h) $1.\overline{131}$

3. (a) Find a rational number between 0 and $\frac{1}{2}$.
   (b) How many rational numbers exist between 0 and $\frac{1}{2}$?
   (c) Is there a smallest positive rational number? Explain.

4. We can show, in the following manner, that $\sqrt{2}$ is not a rational number. Suppose that $\sqrt{2}$ is a rational number of the form $p/q$, where $p/q$ is in its lowest terms. If $\sqrt{2} = p/q$, then $2 = p^2/q^2$ or $2q^2 = p^2$. $2q^2$ is even and, therefore, $p^2$ is even and $p$ is even. Since $p$ is even, it can be expressed in the form $2n$, where $n$ is a positive integer. Now $2q^2 = p^2 = (2n)^2$. Therefore, $q^2 = 2n^2$ and $q$ must be even. But $p$ and $q$ cannot both be even, since $p/q$ was in lowest terms. Hence our supposition that $\sqrt{2}$ is rational must be false.
   (a) Show that $\sqrt{3}$ is not rational.
   (b) Show that $\sqrt{5}$ is not rational.
   (c) Show that $1 + \sqrt{2}$ is not rational.
   (d) Show that $2 - \sqrt{3}$ is not rational.

5. For what real values of $x$ are the following expressions equal to zero?
   (a) $\dfrac{0}{x}$          (b) $\dfrac{x}{x+1}$          (c) $\dfrac{x}{\sqrt{x+2}}$          (d) $\dfrac{4}{x^2-4}$

   (e) $\dfrac{x^2+1}{x-1}$    (f) $\dfrac{1}{x+1}$          (g) $\dfrac{x+1}{\sqrt{x}}$

**6.** For what real values of $x$ are the following expressions undefined?

(a) $\dfrac{x}{x-1}$         (b) $\dfrac{x-1}{x}$         (c) $\dfrac{2}{x^2-4}$         (d) $\dfrac{2}{x^2+4}$

(e) $\dfrac{3x+1}{x^2+2x+1}$     (f) $\dfrac{3x+1}{x^2+x-2}$     (g) $\dfrac{x-1}{x^2-1}$     (h) $\dfrac{x^2}{x^3}$

**7.** For what real values of $x$ are the following expressions indeterminate?

(a) $\dfrac{x+1}{2x+2}$     (b) $\dfrac{x+1}{2x+1}$     (c) $\dfrac{x}{x}$     (d) $\dfrac{x^2}{x^3}$

(e) $\dfrac{x-3}{x^2-9}$     (f) $\dfrac{x-1}{x^4-1}$     (g) $\dfrac{4x+1}{4x+1}$     (h) $\dfrac{0}{x^2}$

**8.** For what real values of $x$ are the following expressions equal to zero?, Undefined?, Indeterminate?

(a) $\dfrac{2}{x-1}-\dfrac{1}{x}$         (b) $\dfrac{x}{x+1}+\dfrac{2}{x}$

(c) $\dfrac{4}{x-3}-\dfrac{16}{x^2-9}$      (d) $\dfrac{x^2+2x-3}{(x^2-1)(x-1)}$

(e) $\dfrac{1}{x}+\dfrac{2}{x-1}-\dfrac{3}{x+2}$

# SECTION 5
## THE ORDER RELATION

Let $a$ be any point on the real number line. Any point $b$ to the right of $a$ represents a larger number and any point $c$ to the left of $a$ represents a smaller number:

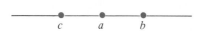

This concept of size or order is indicated by writing $a < b$ or $b > a$ and $a > c$ or $c < a$.

**Definition 5.1**   If $a$ and $b$ are any two real numbers, we say $a$ is **less than** $b$ ($a < b$) or $b$ is **greater than** $a$ ($b > a$) if and only if $b - a$ is a positive number.

> **Example 5.1**
> $5 < 7$ or $7 > 5$ since $7 - 5 = 2$, and 2 is a positive number.
> $-6 < 3$ since $3 - (-6) = 9$.
> $-5 < -3$ since $-3 - (-5) = 2$.
> $-1 > -5$ since $-1 - (-5) = 4$.

We now list some axioms which govern the order relation

5.1. **Trichotomy.** For any two real numbers $a$ and $b$, one and only one of the following relations is true:

$$a < b \quad \text{or} \quad a = b \quad \text{or} \quad a > b$$

This property is a consequence of the fact that every real number is positive, zero, or negative. If $b - a$ is positive, then $a < b$. If $b - a = 0$, then $a = b$. If $b - a$ is negative, then $-(b - a)$ is positive and $b < a$ or $a > b$.

5.2. **Transitivity.** If $a$, $b$, and $c$ are real numbers, where $a < b$ and $b < c$, then $a < c$.

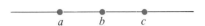

For example, since $2 < 3$ and $3 < 5$, $2 < 5$. If $2 < x$, $x < y$, then $2 < y$. The proof of this property depends upon the definition of order. If $a < b$, then $b - a$ is positive and if $b < c$, then $c - b$ is positive. The sum $(b - a) + (c - b) = c - a$ is positive. Therefore, $a < c$.

5.3. If $a < b$, then $a + c < b + c$ and $a - c < b - c$. We can add or subtract the same quantity to both sides of an inequality without changing the sense. Thus, if $x + 3 < 5$, then $x + 3 - 3 < 5 - 3$ or $x < 2$.

5.4. If $a < b$ and $c > 0$, then $ac < bc$ and $a/c < b/c$. We can multiply or divide both sides of an inequality by the same positive quantity without changing the sense.

5.5. If $a < b$ and $c < 0$, then $ac > bc$ and $a/c > b/c$. Multiplication or division of both sides of an inequality by a negative number reverses the sense of the inequality.

### Example 5.2

Let us consider a special case of axiom 5.5. Prove that if $a < b$, then $-a > -b$.

**Solution.** We have from axiom 5.5 that if $a < b$ and $c < 0$, then $ac > bc$. Let $c = -1$. Then $a(-1) > b(-1)$ or $-a > -b$. •

In finding the solutions for inequalities, we will have occasion to use the following subsets of the real number line $\mathbb{R}$. Each subset is called an **interval** and the various types are illustrated as follows:

1. $(a, b) = \{x \mid a < x < b\}$.

The statement $a < x < b$ exhibits the "betweenness" property of real numbers and is called a **compound inequality.** It is a combination of the statements $a < x$ *and* $x < b$. $\{x \mid a < x < b\} = \{x \mid a < x\} \cap \{x \mid x < b\}$.

2. $[a, b) = \{x \mid a \leq x < b\}$.

3. $(a, b] = \{x \mid a < x \leq b\}$.

4. $[a, b] = \{x \mid a \leq x \leq b\}$.

Each of the above subsets of $\mathbb{R}$ is called a **bounded interval.** $(a, b)$ is an **open interval** since neither endpoint is included in the interval; similarly, $[a, b]$ is a **closed interval** since both endpoints are included. $(a, b]$ and $[a, b)$ are said to be **half-opened** or **half-closed** since one endpoint is included in the interval but the other is not. The line graphs for the four bounded intervals may also be represented by line segments whose endpoints are circles. We will use an open circle if the endpoint is not included and a closed circle if the endpoint is a part of the interval. Thus,

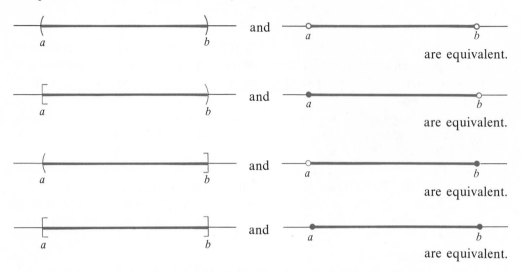

are equivalent.

are equivalent.

are equivalent.

are equivalent.

The choice of which type of diagram to use is often a matter of convenience.

We describe the unbounded intervals as follows:

5. $(a, \infty) = \{x \mid x > a\}$.

6. $[a, \infty) = \{x \mid x \geq a\}$.

7. $(-\infty, a) = \{x \mid x < a\}$.

8. $(-\infty, a] = \{x \mid x \leq a\}$.

We attach no meaning to the notation $[a, \infty]$, $(a, \infty]$, $[-\infty, a]$, $[-\infty, a)$, or $[-\infty, +\infty]$.

**Example 5.3**

Express each of the following in interval notation if possible and then graph the solution set.

(a) $1 \leq x < 4$.

This describes the set of values in the interval from 1 to 4 including 1 but not including 4. Hence $1 \leq x < 4 = [1, 4)$.

(b) $(2, 4] \cup (4, 5] = (2, 5]$.

(c) $(2, 4] \cap (4, 5) = \emptyset$.

(d) $(1, 3) \cup (4, 6]$ cannot be expressed using a single interval.

(e) $(-\infty, 2) \cap [0, 7] = [0, 2)$.

(f) $\{x \mid x > 1 \text{ or } x < 3\} = (1, \infty) \cup (-\infty, 3) = (-\infty, \infty) = R$.

## EXERCISES FOR SECTION 5

1. Which of the following inequalities are true?
   (a) $-5 < -2$
   (b) $-[-1(3 - 2)] > 4$
   (c) $-\frac{6}{2} < -\frac{12}{4}$
   (d) $\sqrt{2} < 1$
   (e) $\frac{7}{16} > \frac{8}{17}$
   (f) $2 < 4 < 1$
   (g) $6 > 3 > -1$
   (h) $\frac{22}{7} < \pi$

2. Simplify and sketch each of the following intervals on the real number line.
   (a) $(-3, 1]$
   (b) $(1, 4] \cap [2, 4]$
   (c) $(2, 3) \cup [1, 5]$
   (d) $(1, 5] \cap [5, 7]$
   (e) $(1, 5] \cap (5, 7]$
   (f) $(-\infty, 1] \cup (0, \infty)$
   (g) $(0, 1) \cup [2, 4)$
   (h) $[0, \infty) \cap (-\infty, 2)$
   (i) $[2, 7] \cup (3, 5)$
   (j) $[2, 7] \cap (3, 5)$

3. Describe the following by means of interval notation.

(a)

(b)

(c)

(d)

(e)

(f)

(g)

(h)

(i)

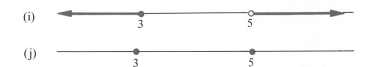

(j)

4. Find an example which shows that the following statements are not always true.
   (a) If $a < b$, then $a^2 < b^2$.
   (b) If $a < b$ and $c < d$, then $ac < bd$.
   (c) If $a < 3$, then $a^2 < 9$.

5. Show that if $a \neq 0$, $a^2 > 0$.

6. If $0 < a < 1$, show that $a^2 < a$.

7. Show that if $a$ is any real number greater than 0, the sum of $a$ and its reciprocal is greater than or equal to 2.

8. Prove the following.
   (a) If $a < b$ and $c < d$, then $a + c < b + d$.
   (b) If $0 < a < b$ and $0 < c < d$, then $ac < bd$.

9. Prove the following.
   (a) If $a > 0$ and $b > 0$, then $ab > 0$.
   (b) If $a > 0$ and $b < 0$, then $ab < 0$.

# SECTION 6
## LINEAR INEQUALITIES IN ONE VARIABLE

A **linear inequality in one variable** is any inequality whose variable is first degree. The solution set is the set of all real numbers that satisfy the inequality. Graphically, these sets will be subsets of the real number line. We can determine the solution sets algebraically in a manner that does not differ significantly from the methods used for linear equations. The following examples illustrate some of the methods used to solve linear inequalities and the graphs of the solution sets.

### Example 6.1
Solve and graph $3x + 1 < 7$.

### Solution

$$3x + 1 - 1 < 7 - 1$$
$$3x < 6$$
$$x < 2$$
$$(-\infty, 2)$$

### Example 6.2
Solve and graph $x + 5 \leq 3x - 7$.

Solution

$$x + 5 - 3x - 5 \le 3x - 7 - 3x - 5$$
$$-2x \le -12$$
$$-\frac{2x}{-2} \ge -\frac{12}{-2}$$
$$x \ge 6$$

$[6, \infty)$ •

6

### Example 6.3
Solve and graph $x + 2 > x - 3$.

Solution

$$x + 2 - x + 3 > x - 3 - x + 3$$
$$5 > 0$$

Since $5 > 0$ is always true, the solution set is all real numbers.   •

*Note:* Examples 6.1 and 6.2 are called **conditional inequalities** since they are true for some real values of $x$; that is, the solution sets of Examples 6.1 and 6.2 are **proper subsets of R.** In Example 6.3 the solution set consists of all real values of $x$. For this reason the inequality is called an **absolute inequality.**

### Example 6.4
Solve and graph $1 < 2x - 3 \le 4$.

Solution

$$3 + 1 < 2x - 3 + 3 \le 4 + 3$$
$$4 < 2x \le 7$$
$$2 < x \le \tfrac{7}{2}$$

$(2, \tfrac{7}{2}]$ •

2          $\frac{7}{2}$

### Example 6.5
Solve and graph $2 < 2x + 4 < -4$.

Solution.   We may observe that it is impossible for $2x + 4$ to be greater than 2 and less than $-4$. Therefore, $\{x \mid 2 < 2x + 4 < -4\} = \varnothing$. If we do not make this initial observation, we proceed as follows.

$$2 < 2x + 4 < -4$$
$$2 - 4 < 2x + 4 - 4 < -4 - 4$$
$$-2 < 2x < -8$$
$$-1 < x < -4$$

which means that $-1 < x$ *and* $x < -4$.

The intersection of the graphs shown below indicate the solution set is ∅.   •

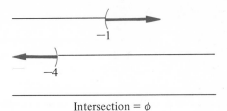

Intersection = $\phi$

### Example 6.6
Solve and graph $\{x\,|\,2x + 1 < -3\} \cup \{x\,|\,x + 2 > 5\}$.

### Solution

$\{x\,|\,2x < -4\} \cup \{x\,|\,x > 3\}$
$\{x\,|\,x < -2\} \cup \{x\,|\,x > 3\}$
$(-\infty, -2) \cup (3, \infty)$   •

### Example 6.7
If $1 < x < 3$, express $3x + 2$ in terms of an inequality.

**Solution.**   Axiom 5.4 allows us to multiply each member of the compound inequality $1 < x < 3$ by 3 to obtain $3 < 3x < 9$. We now use axiom 5.3 to add 2 to each member of $3 < 3x < 9$ to obtain $5 < 3x + 2 < 11$.   •

## EXERCISES FOR SECTION 6

1. Solve each of the following inequalities and graph their solution sets on the real number line.
   (a) $x - 2 > 1$
   (b) $x - 2 \leq 4$
   (c) $x + 1 \leq 2x$
   (d) $x + 3 > 5x + 7$
   (e) $4 < -2x + 1 \leq 6$
   (f) $2(x + 1) \geq 2x + 2$
   (g) $2(x + 1) > 2x + 2$
   (h) $x < 2x + 1 < 3x - 6$
   (i) $\{x\,|\,2x + 3 < 9\} \cup \{x\,|\,1 - x < -10\}$
   (j) $\{x\,|\,2x + 17 < 3\} \cup \{x\,|\,4 > x + 1\}$

2. Each of the following intervals represent graphical solutions for linear inequalities. Use interval notation to represent the given sets.

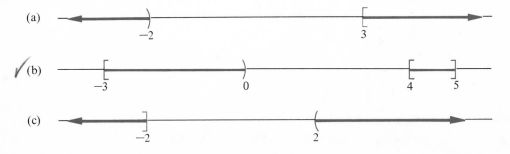

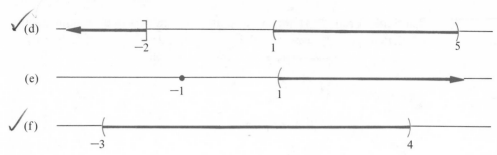

**3.** If $0 < x < 2$, express the following in terms of inequalities.

(a) $2x + 3$     (b) $3x - 1$     (c) $2 - x$     (d) $1 - 2x$

**4.** If $0 < x < 1$ and $2 < y < 4$, express the following in terms of inequalities. Use the results of Exercise 8 in Section 5.

(a) $x + y$         (d) $x^2$
(b) $x - y$         (e) $x^2 + y$
(c) $xy$             (f) $x^2 + y^2$

**5.** (a) Explain why the following statement is not correct. If $-2 \le x \le 5$, then $4 \le x^2 \le 25$.

(b) What is the correct statement. Hint: Examine the graph of the interval $-2 \le x \le 5$ and observe maximum and minimum values of these numbers when we square.

(c) If $-3 \le x \le 2$, express $x^2$ in terms of an inequality.

(d) If $-3 \le x \le -2$, express $x^2$ in terms of an inequality.

(e) If $2 \le x \le 3$, express $x^2$ in terms of an inequality.

(f) If $-1 < x < 1$, express $x^2 + 1$, $2 - x^2$ and $x^2 + 2x - 3$ in terms of inequalities.

# SECTION 7
# ABSOLUTE VALUE

Associated with each real number is a nonnegative number, called its **absolute value.** To be precise we formulate the following definition

**Definition 7.1**   The absolute value of the real number $x$, denoted $|x|$ is

$$|x| = \begin{cases} x, \text{ if } x \ge 0 \\ -x, \text{ if } x < 0 \end{cases} \tag{7.1}$$

### Example 7.1
Evaluate $|6|$, $|-5|$, and $|0|$.

Solution.   If $x = 6$, then it falls in the first case of (7.1) and thus $|x| = |6| = 6$. If $x = -5$, then it falls in the second case of (7.1) and we have $|x| = |-5| = -(-5) = 5$. If $x = 0$, then it falls in the first case of (7.1) and we have $|x| = |0| = 0$.   ●

### Example 7.2
Express $|2x - 1|$ without absolute value symbols.

Solution. From (7.1) we know that when $2x - 1 \geq 0$ or equivalently $x \geq \frac{1}{2}$, that $|2x - 1| = 2x - 1$. We also know that when $2x - 1 < 0$ or $x < \frac{1}{2}$ that $|2x - 1| = -(2x - 1) = 1 - 2x$. Thus we have

$$|2x - 1| = \begin{cases} 2x - 1 \text{ if } x \geq \dfrac{1}{2} \\ 1 - 2x \text{ if } x < \dfrac{1}{2} \end{cases} \bullet$$

**Example 7.3**
Express $|x + 1| - |2x|$ without absolute value symbols.

Solution. We note that $|x + 1| = 0$ for $x = -1$ and $|2x| = 0$ for $x = 0$. When $x \geq 0$ then $2x \geq 0$ and $x + 1 \geq 1$. From (7.1) we have $|2x| = 2x$ and $|x + 1| = x + 1$. Thus

$$|x + 1| - |2x| = (x + 1) - 2x = 1 - x \text{ when } x \geq 0.$$

When $-1 \leq x < 0$, then $2x < 0$ and $0 \leq x + 1 < 1$. From (7.1) we have $|2x| = -(2x)$ and $|x + 1| = x + 1$, thus

$$|x + 1| - |2x| = (x + 1) - (-2x) = 3x + 1 \text{ when } -1 \leq x < 0.$$

When $x < -1$, then $2x < -2$ and $x + 1 < 0$. From (7.1) we have $|2x| = -2x$ and $|x + 1| = -(x + 1)$ thus

$$|x + 1| - |2x| = -(x + 1) - (-2x) = x - 1 \text{ when } x < -1$$

Therefore

$$|x + 1| - |2x| = \begin{cases} 1 - x \text{ if } x \geq 0 \\ 3x + 1 \text{ if } -1 \leq x < 0 \\ x - 1 \text{ if } x < -1 \end{cases} \bullet$$

We have established a graphic interpretation of the set of real numbers by introducing the real number line. The development of the real number line gives us two distinct properties for real numbers, order and distance. We have investigated some of the order properties. Now let us consider the concept of **distance.**

Every real number represents the directed distance from the point 0 to the point that represents the real number. Thus the point 5 is 5 units from 0 and $-3$ is $-3$ units from 0 (Figure 7.1).

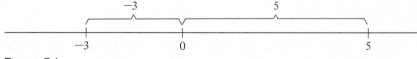

Figure 7.1

If we consider only the distance from 0 to any point on the real number line, it makes no difference in which direction we move. In the case of undirected distance, the length of the segment from 0 to 5 is the same as the length of the segment from 5 to 0. In addition, the length of the segment from $-5$ to $0$ is the same as that from $0$ to $5$ (Figure 7.2).

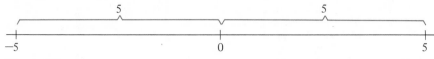

Figure 7.2

Now, we can define the absolute value of a real number as "the undirected distance from the point that represents the real number to the origin." This statement is equivalent to definition 7.1.

We note from the definition that $|x|$ is never a negative number, and we also note that $-x$ is not necessarily a negative number.

**Example 7.4**
Find $x$ if $|x| = 5$.

Solution.   The absolute value of $x$ is a point 5 units from 0. The point 5 units to the right is $+5$ and the point 5 units to the left is $-5$. Thus $|x| = 5$ implies that $x = +5$, or $x = -5$ (Figure 7.3). Alternatively, we have from the definition that $x = 5$ if $x \geq 0$ or $-x = 5$ if $x < 0$. Thus $|x|$ implies that $x = 5$ or $x = -5$.   ●

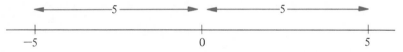

Figure 7.3

**Example 7.5**
Solve for $x$ if $|x - 2| = 3$. Graph the solution set and note the geometric significance.

Solution.   $|x - 2| = 3$ means that $(x - 2) = 3$ or $-(x - 2) = 3$. Thus we have $x = 5$ or $x = -1$. The graph of this set is shown in Figure 7.4. Geometrically, the solutions of $|x - 2| = 3$ are the endpoints of an interval whose midpoint is at $(-1 + 5)/2 = 2$. The values $-1$ and 5 are obtained by moving 3 units in either direction from the midpoint, $(-1 + 5)/2 = 2$ (Figure 7.5).   ●

Figure 7.4

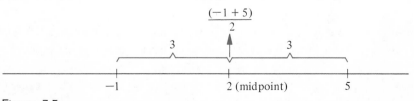

Figure 7.5

**Example 7.6**
Solve for $x$ when $|x - 3| = 1$.

Solution.   $|x - 3| = 1$ describes the points that are 1 unit from 3. Thus $x = 2$ or $x = 4$. *Note:* Use of the definition will yield the same results.   ●

**Example 7.7**
Solve for $x$ if $|2x - 3| = 7$.

Solution.   $|2x - 3| = 7$ is equivalent to $(2x - 3) = 7$ or $-(2x - 3) = 7$. Thus $x = 5$ or $x = -2$. The graph of this solution set is shown in Figure 7.6. In this case the solutions are the endpoints of an interval that has its midpoint at $\frac{3}{2}$, and a half-width of $\frac{7}{2}$. Explain how we might obtain this information from the original expression.   ●

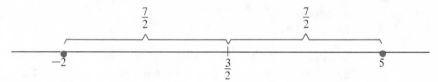

Figure 7.6

**Example 7.8**
Use absolute value to describe the points that are 2 units from the point $a = 3$.

Solution.   The points we are looking for are the endpoints of an interval whose midpoint is at $a = 3$. The half-width of the interval is 2. Thus $|x - 3| = 2$.   ●

The absolute value has the following properties:

1. $|-a| = |a|.$ The points $a$ and $-a$ are the same distance from the origin.

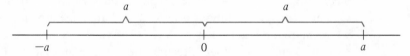

2. $|a - b| = |b - a|$. The undirected distance between the points $a$ and $b$ on the real number line is given by either $|a - b|$ or $|b - a|$.

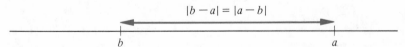

3. $|a| > 0$ if $a \neq 0$. Any number except 0 is some positive distance from the origin.
4. $|ab| = |a| \cdot |b|$. The absolute value of a product is equal to the product of the absolute values.
5. $|a/b| = |a|/|b|$, $b \neq 0$. The absolute value of a quotient is equal to the quotient of the absolute values.

6. $|a| + |b| \geq |a + b|$ (Triangle Inequality). *Note:* The absolute value of the sum of two real numbers is *not* equal to the sum of the absolute values; it is less than or equal to the sum of the absolute values.

7. $|a|^2 = a^2$.

The absolute value is often useful in the study of inequalities. The inequality $|x| \leq p$ is satisfied by a number $x$ if the distance from the origin to the point which represents the graph of $x$ is less than or equal to $p$. Figure 7.7 indicates that $x$ must be $p$, or $-p$ or else lie between these numbers. Thus the solution set for $|x| \leq p, p > 0$ is $\{x \mid -p \leq x \leq p\}$. We state this as

$$\{x \mid |x| \leq p, p > 0\} = \{x \mid -p \leq x \leq p\} \tag{7.2}$$

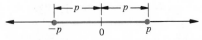

Figure 7.7

### Example 7.9
Find the solution set for $|2x - 4| \leq 8$ and then graph it.

Solution.   We use (7.2) and the indicated axioms of inequalities to solve as follows:

$-8 \leq 2x - 4 \leq 8$     by using (7.2)
$-4 \leq 2x \leq 12$     by adding 4
$-2 \leq x \leq 6$     by dividing by 2.

Thus the solution set is $\{x \mid -2 \leq x \leq 6\}$, and its graph is shown in Figure 7.8.  ●

Figure 7.8

The inequality $|x| \geq p$ is satisfied by a number $x$ if the distance from the origin to the point which represents the graph of $x$ is greater than or equal to $p$. Figure 7.9 indicates that $x$ must be $p$ or lie to the right of $p$ or else $x$ must be $-p$ or lie to the left of $-p$. From this we can formulate the following statement.

$$\{x \mid |x| \geq p, p > 0\} = \{x \mid x \leq -p\} \cup \{x \mid x \geq p\} \tag{7.3}$$

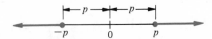

Figure 7.9

### Example 7.10
Find the solution set for $|2x + 1| \geq 5$ and then graph it.

**Solution.** From 7.3 we are told that the solution set will be the union of the solution sets for the inequalities $2x + 1 \leq -5$ and $2x + 1 \geq 5$. We solve as follows:

$$2x + 1 \leq -5 \qquad \text{and} \qquad 2x + 1 \geq 5$$
$$2x \leq -6 \qquad\qquad\qquad 2x \geq 4$$
$$x \leq -3 \qquad\qquad\qquad x \geq 2$$

Thus the solution set for $|2x + 1| \geq 5$ is $\{x \,|\, x \leq -3\} \cup \{x \,|\, x \geq 2\}$ or $\{x \,|\, x \leq -3 \text{ or } x \geq 2\}$. The graph of this set is shown in Figure 7.10. ●

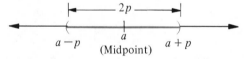

Figure 7.10

The interval $a - p < x < a + p$ shown in Figure 7.11 has a midpoint or center at $a$ and a length of $2p$ units. Since $a - p < x < a + p$ or $-p < x - a < p$ is equivalent to $|x - a| < p$ this suggests a method for expressing intervals such as $-4 < x < 2$ in the form of $|x - a| < p$.

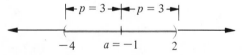

Figure 7.11

**Example 7.11**
Rewrite $-4 < x < 2$ in the form $|x - a| < p$.

**Solution.** The midpoint of this interval can be found by averaging the endpoints. Thus $a = (-4 + 2)/2 = -1$. The length $2p$ of this interval is $2 - (-4) = 6$. Therefore $2p = 6$ or $p = 3$. Thus we can rewrite $-4 < x < 2$ as $|x - (-1)| < 3$ or $|x + 1| < 3$. The graph of this interval is shown in Figure 7.12. ●

$$\leftarrow p = 3 \rightarrow\leftarrow p = 3 \rightarrow$$

$$-4 \qquad a = -1 \qquad 2$$

Figure 7.12

With the aid of 7.2 and 7.3 and axioms 5.4 and 5.5 we can expand our knowledge of absolute value to include the solution sets for $|1/x| \leq p, \; p > 0$ and $|1/x| \geq p, \; p > 0$.

Since $|x| \geq 0$ for all $x$, we use axiom 5.4 to note that $|1/x| \leq p$ may be rewritten as $1 \leq p|x|$. Since this is equivalent to $|x| \geq 1/p$ we have by 7.3

$$\left\{ x \,\Big|\, \left|\frac{1}{x}\right| \leq p, p > 0 \right\} = \left\{ x \,\Big|\, |x| \geq \frac{1}{p} \right\} = \left\{ x \,\Big|\, x \leq -\frac{1}{p} \right\} \cup \left\{ x \,\Big|\, x \geq \frac{1}{p} \right\} \qquad (7.4)$$

A similar examination of $|1/x| \geq p, \; p > 0$ will show that

$$\left\{ x \,\Big|\, \left|\frac{1}{x}\right| \geq p, p > 0 \right\} = \left\{ x \,\Big|\, |x| \leq \frac{1}{p}, x \neq 0 \right\} = \left\{ x \,\Big|\, -\frac{1}{p} \leq x \leq \frac{1}{p}, x \neq 0 \right\} \qquad (7.5)$$

**Example 7.12**

Find the solution set for $|1/(x - 2)| \geq 3$ and then graph it.

**Solution.** From 7.5 we note that $|1/(x - 2)| \geq 3$ is equivalent to $|x - 2| \leq \frac{1}{3}$, $x - 2 \neq 0$. We solve this inequality as follows:

$$|x - 2| \leq \frac{1}{3}, \ x \neq 2$$

$$-\frac{1}{3} \leq x - 2 \leq \frac{1}{3}, \ x \neq 2$$

$$\frac{5}{3} \leq x \leq \frac{7}{3}, \ x \neq 2$$

Thus the solution set for $|1/(x - 2)| \geq 3$ is $\{x \mid \frac{5}{3} \leq x \leq \frac{7}{3}, \ x \neq 2\}$. The graph of this set is shown in Figure 7.13.   •

$$\frac{5}{3} \qquad\qquad \frac{7}{3}$$

Figure 7.13

## EXERCISES FOR SECTION 7

1. Solve each of the following for $x$ and graph the solution sets on the real number line.

(a) $|x| = 6$
(b) $|x - 6| = 2$
(c) $|x + 2| = -2$
(d) $|2x + 1| = 1$

(e) $|2x - 3| = 0$
(f) $|x| = 2x$
(g) $|2x + 1| = |-3|$
(h) $|1 - x| = 2$

(i) $\left|\dfrac{x^2 - 3x}{x}\right| = 4$
(j) $|x^2 + 4| = 5$
(k) $|x^2 + 4| = 1$

2. For each of the following, write equivalent statements that do not involve absolute value.

(a) $|x + 1| < 5$
(b) $|2x - 3| \geq 6$
(c) $|x - 3| < 0.1$
(d) $|x - 1| < \epsilon, \ \epsilon > 0$
(e) $|2y + 3| < \delta, \ \delta > 0$

(f) $|x - a| < b$
(g) $|x - a| \geq b$
(h) $|x - a| < -2$
(i) $|x^2 + 2x - 8| < 2$
(j) $\left|\dfrac{1}{x} - 4\right| < 1$

3. Solve each of the following for $x$, and graph the solution sets on the real number line.

(a) $|x + 1| < 2$
(b) $|x| \geq 3$
(c) $|2x + 1| < 5$
(d) $|x + 1| < -3$
(e) $\left|\dfrac{x - 2}{3}\right| < 2$

(f) $|x - 2| \leq 0$
(g) $|2 - x| < 1$
(h) $|x - 2| < 1$
(i) $|x + 1| \geq -2$
(j) $|2x - 3| > 0$

4. Express the following intervals as inequalities that involve absolute value.

(a) $2 < x < 10$
(b) $-8 < x < 6$
(c) $-2 < x < 2$
(d) $4 - a < x < 4 + a$
(e) $-2.0 < x < -1.90$

**5.** What can be said about the value of $|x|$ if
  (a) $2 < x < 6$?
  (b) $-8 < x < 6$?
  (c) $-1 < x < 1$?
  (d) $-1 < x < 0$?
  (e) $-5 < x < -2$?

**6.** If $|x - 1| < 3$, express each of the following in terms of inequalities.
  (a) $|x|$      (b) $|x + 1|$      (c) $|x + 3|$      (d) $|x - 3|$

**7.** Solve each of the following for $x$ and graph the solution sets on the real number line.

  (a) $\left| \dfrac{1}{x} \right| \le 4$   (d) $\left| \dfrac{2}{x + 1} \right| < 1$   (g) $\left| \dfrac{x}{x - 1} - 1 \right| < \dfrac{1}{2}$

  (b) $\left| \dfrac{1}{x - 1} \right| \le 2$   (e) $\left| \dfrac{3}{1 - x} \right| \ge 3$

  (c) $\left| \dfrac{1}{2x - 1} \right| \ge \dfrac{1}{2}$   (f) $\left| \dfrac{-1}{2x + 3} \right| \ge \dfrac{1}{3}$   (h) $\left| \dfrac{2x}{3x - 1} - \dfrac{2}{3} \right| < 5$

**8.** Express the following without absolute value symbols.
  (a) $|x + 2| - |2x|$
  (b) $|x + 4| - |3x|$
  (c) $|4x| - 2|x - 3|$

**9.** Solve the following:
  (a) $|x + 2| - |2x| = 1$
  (b) $|x + 4| - |3x| = -12$
  (c) $|4x| - 2|x - 3| = 4$

**10.** Solve the following
  (a) $|x + 2| - |2x| < -2$
  (b) $|x + 4| - |3x| < 0$
  (c) $|4x| - 2|x - 3| < 6$

**11.** Show that $|a \cdot b| = |a| \cdot |b|$.

**12.** Show that $|a + b| \le |a| + |b|$.

# SECTION **8**
# EXPONENTS

A positive integral exponent is simply a shorthand method used to denote the product of a number of identical factors.

**Definition 8.1**   If $n$ is a positive integer and $a$ is any real number, then

$$a^n = \underbrace{a \cdot a \cdot \ldots \cdot a}_{n \text{ times}}$$

The integer $n$ is called the **exponent** and $a$ is called the **base.**

The basic laws governing the uses of exponents are a direct consequence of Definition 8.1 and are given in the following theorem.

**Theorem 8.1** Suppose that $m$ and $n$ are positive integers and $a$ and $b$ are real numbers. Then

1. $a^m \cdot a^n = a^{m+n}$.
2. $(a^m)^n = a^{m \cdot n}$.
3. $(ab)^n = a^n \cdot b^n$.
4. $(a/b)^n = a^n/b^n$, $b \neq 0$.
5. $a^m/a^n = \begin{cases} a^{m-n} & \text{if } m > n \\ 1 & \text{if } m = n, \ a \neq 0. \\ 1/a^{n-m} & \text{if } m < n \end{cases}$

**Example 8.1**

$$a^2 \cdot a^3 = a^5$$

$$(a^2)^3 = a^6$$

$$(ab)^4 = a^4 \cdot b^4$$

$$a^6/a^2 = a^4$$

$$a^3/a^6 = 1/a^3$$

The following definition enables us to include zero and the negative integers in the laws of exponents. The definition is chosen so that the properties in Theorem 8.1 can still be used.

**Definition 8.2** If $a$ is any real number and $a \neq 0$, then $a^0 = 1$ and $a^{-n} = 1/a^n$, where $n$ is a positive integer.

**Example 8.2**

$$(2a)^0 = 1$$

$$2a^0 = 2$$

$$3^{-2} = 1/3^2 = 1/9$$

$$(1/2)^{-2} = 1/(1/2)^2 = 4$$

$$a^4/a^6 = a^{4-6} = a^{-2} = 1/a^2$$

At this point we have only the use of exponents which are integers. We would like to expand the use of exponents to include such expressions as $2^{1/2}$, $8^{-1/3}$, $a^{1/4}$, and $b^{-2/3}$. Further, we would like to include rational exponents in such a way that the properties listed in Theorem 8.1 are still valid.

**Definition 8.3** If $a$ is any real number and $n$ is a positive integer, then the real number $a^{1/n}$ (written $\sqrt[n]{a}$) is called the **principal $n$th root** of $a$. If the principal root is $r$, then $r$ must satisfy the equation $r^n = a$. If $a$ is positive and $n$ is an even integer, there will always be two real numbers that satisfy the equation $r^n = a$. One of these will be positive and the other negative. We *always* select the positive number as our principal $n^{th}$ root.

A number may have more than one $n^{th}$ root, but only the positive root is called the principal root. For example, $(2)^2 = 4$ and $(-2)^2 = 4$. Thus both 2 and $-2$ are square roots of 4, but $\sqrt{4} = 2$.

In the taking of square roots, there is one instance where we must exercise particular care. To illustrate this situation, let us consider the equation $x^2 = 4$. If we take the square root of both sides of this equation, we would usually write $x = 2$. Although this is true, there is more to the solution set than $x = 2$. We know that $x^2 = 4$ is equivalent to $x^2 - 4 = 0$. Factoring, we have $(x - 2)(x + 2) = 0$ or $x = 2$ and $x = -2$.

The reason for not obtaining both solutions in our initial attempt lies in the interpretation of taking the square root of $x^2$. We note that $\sqrt{(3)^2} = \sqrt{9} = 3$ and $\sqrt{(-3)^2} = \sqrt{9} = 3$. This example suggests that $\sqrt{x^2} = x$ if $x \geq 0$ and $-x$ if $x < 0$. But, this is precisely our definition of absolute value, and hence we may write $\sqrt{x^2} = |x|$. In fact, we must write $\sqrt{x^2} = |x|$, not $\sqrt{x^2} = x$. The latter is true only if $x$ is a positive real number or zero.

Now, we use that for $x^2 = 4$, the correct statement after taking the square root of both sides is $|x| = 2$. Using our definition of absolute value, we have $x = 2$ and $x = -2$. Situations that involve inequalities are handled in a similar fashion.

**Example 8.3**
Solve for $x$ if $x^2 - 9 < 0$.

Solution.  $x^2 - 9 < 0$ is equivalent to $x^2 < 9$. Taking the square root of both sides of the inequality yields $|x| < 3$. Thus the solution set for $x^2 - 9 < 0$ is $|x| < 3$ or, equivalently, $-3 < x < 3$.  ●

**Example 8.4**
For which values of $x$ does $\sqrt{x^2 + 2x + 1} = x + 1$?

Solution.  $\sqrt{x^2 + 2x + 1} = \sqrt{(x + 1)^2}$. Now, $\sqrt{(x + 1)^2} = x + 1$ if and only if $x + 1 \geq 0$. Thus $x \geq -1$ is the set of values we seek. Note that in general $\sqrt{x^2 + 2x + 1} = |x + 1|$, not $x + 1$. It is only with the restriction $x \geq -1$ that we may write $\sqrt{x^2 + 2x + 1} = x + 1$.  ●

Now let us consider two other possibilities that arise from Definition 8.3. If $a$ is a negative real number and $n$ is an odd positive integer, there is one real $n$th root of $a$ and it is always negative. But, if $a$ is a negative real number and $n$ is an even positive integer, there is *no real* $n$th root of $a$. This is easy to see since no real number raised to an even power yields a negative result. Thus, while $\sqrt[3]{-27} = -3$ and $\sqrt[5]{-32} = -2$, expressions such as $\sqrt{-9}$, $\sqrt[4]{-16}$, and $\sqrt{x}$, $x < 0$, are not defined in the set of real numbers. In Chapter 9 we shall extend our number system to include numbers whose even powers are negative. For now, we summarize our results as follows:

1. If $a > 0$ and $n$ is any positive integer, $a^{1/n}$ is positive.
2. If $a < 0$ and $n$ is a positive even integer, $a^{1/n}$ is nonreal.
3. If $a < 0$ and $n$ is a positive odd integer, $a^{1/n}$ is negative.

### Example 8.5

$$9^{1/2} = \sqrt{9} = 3 \qquad \text{since } 3^2 = 9$$
$$(-27)^{1/3} = \sqrt[3]{-27} = -3 \qquad \text{since } (-3)^3 = -27$$
$$(-1)^{1/2} \text{ is nonreal}$$
$$\sqrt{(-3)^2} = \sqrt{9} = 3$$

Statement 2 of the summary says that we may take only even roots of positive quantities if we expect to obtain a real result. This observation is illustrated in Examples 8.6 and 8.7.

### Example 8.6

For what real values of $x$ is $\sqrt{2x - 4}$ defined?

Solution.  In order to take the square root of $2x - 4$, we must have values that are positive or zero. Thus we have the inequality $2x - 4 \geq 0$, from which we have $x \geq 2$.  •

### Example 8.7

For what real values of $x$ is $\dfrac{1}{\sqrt[4]{x - 1}}$ defined?

Solution.  We know that the expression is not defined when $x = 1$, since division by zero is impossible. In addition, since we are attempting to take an even root, we must have $x - 1 > 0$. Thus our solution set is given by $x > 1$.  •

### Example 8.8

For what real values of $x$ is $\sqrt{x^2 - 16}$ defined?

Solution.  In order to take the square root, $x^2 - 16 \geq 0$. We note that $x^2 - 16 \geq 0$ is equivalent to $x^2 \geq 16$. Taking the square root of both sides yields $|x| \geq 4$. Thus our solution set is $x \leq -4$ or $x \geq 4$.  •

To further generalize our laws of exponents, we have the following definition.

**Definition 8.4**  If $m/n$ is a positive rational number in lowest terms and if $a$ is any real number, then $a^{m/n} = (a^m)^{1/n} = (a^{1/n})^m$ provided that $a^{1/n}$ is real. If $m/n$ is negative and $a \neq 0$, then $a^{m/n} = 1/a^{-m/n}$.

The following examples illustrate the rules of exponents using fractional values.

### Example 8.9

$$8^{2/3} = (8^{1/3})^2 = (2)^2 = 4$$
$$(-3)^{3/2} = (-3^3)^{1/2} = (-27)^{1/2} \text{ which is nonreal}$$
$$(-32)^{-3/5} = \frac{1}{(-32)^{3/5}} = \frac{1}{(-32^{1/5})^3} = \frac{1}{(-2)^3} = -\frac{1}{8}$$

### Example 8.10

$(-4)^{1/2}(-4)^{1/2} = (-4)^{1/2+1/2} = -4$ if we use property 1 of Theorem 8.1. But $(-4)^{1/2}(-4)^{1/2} = [(-4)(-4)]^{1/2} = 16^{1/2} = 4$ if we use property 3 of Theorem 8.1. Thus we conclude that $-4 = 4$. What is wrong with this example?

### Example 8.11

Find $x$ if $3^{4x} = \frac{1}{9}$.

Solution.  Since $1/9 = 1/3^2 = 3^{-2}$, we have $3^{4x} = 1/9 = 3^{-2}$. Equating exponents, using the principle that if $a^x = a^y$, $a \neq 0$, then $x = y$, we have $4x = -2$ or $x = -1/2$.  ●

Having considered the workings of integral and rational exponents, it is natural to ask if the properties of exponents listed in Theorem 8.1 can be extended to include the set of all real numbers. By employing methods presently beyond our means, the following theorem can be proved. Note that the exponents can be any real numbers.

**Theorem 8.2**  Let $a$ and $b$ be any positive real numbers. For all real numbers $m$ and $n$,

1. $a^m \cdot a^n = a^{m+n}$.
2. $(a^m)^n = a^{m \cdot n}$.
3. $(ab)^n = a^n \cdot b^n$.
4. $(a/b)^n = a^n/b^n$, $b \neq 0$.
5. $a^m/a^n = a^{m-n}$, $a \neq 0$.

## EXERCISES FOR SECTION 8

**1.** Write each of the following as rational numbers, if possible.
   (a) $8^{2/3}$
   (h) $(49)^{-3/2}$
   (b) $32^{-1/5}$
   (i) $(0.008)^{-1/3}$
   (c) $(\frac{1}{4})^{5/2}$
   (j) $42^{2/3} \div 4^{1/6}$
   (d) $(5^{1/2})^4$
   (k) $[(\frac{1}{3})^{1/2}]^{-2}$
   (e) $(16)^{3/2}$
   (l) $[(\frac{1}{8})^2]^{-1/3}$
   (f) $27^{4/3}$
   (m) $4^{1/4} \cdot 8^{1/2}$
   (g) $(\frac{16}{9})^{3/2}$
   (n) $\sqrt[3]{\sqrt{27}}$

**2.** Write the following without negative exponents.
   (a) $(x^{1/3})^{-3/4}$
   (f) $(x^{-2}y^1)^{-4}$
   (b) $5x^{-5} \cdot 3x^2$
   (c) $8^{-2/3} + 3x^0 + (\frac{1}{81})^{-1/4}$
   (g) $\dfrac{32x^{-5/2}y^{1/2}}{8^{4/3}x^{-3/2}y^{-2}}$
   (d) $x^{1/2} \cdot y^{-1/2}$
   (e) $(x^{1/3})^{-1/2}$
   (h) $(x^6y^3)^{-1/3}$
   (i) $\dfrac{x + 3x^{-2} - x^{-3/2}}{x^{3/2}}$

3. Use the method of Example 8.3 to solve each of the following. Graph the solution sets on the real number line.
   (a) $16 - x^2 \geq 0$
   (b) $25 - x^2 \leq 0$
   (c) $9 - x^2 \geq 0$
   (d) $x^2 - 4 \leq 0$
   (e) $4x^2 - 1 > 0$
   (f) $x^2 + 9 \leq 0$
   (g) $\dfrac{1}{x^2} \leq 4$
   (h) $\dfrac{1}{x^2} \geq 4 \quad (x \neq 0)$
   (i) $\dfrac{1}{(x - 2)^2} < 1$

4. Simplify each of the following.
   (a) $(x^2 + 8x + 16)^{1/2}$
   (b) $(x^4 + 2x^2 + 1)^{1/2}$
   (c) $\dfrac{\sqrt{x^2 + 2x + 1}}{x + 1}, \; x < -1$
   (d) $x\sqrt{1 - \dfrac{2}{x} + \dfrac{1}{x^2}}, \; x > 0$
   (e) $\dfrac{(4x^2 + 4x + 1)^{1/2}}{2x + 1}$
   (f) $(x^2 + 8x + 16)^{1/2}, \; x \geq -4$
   (g) $x\sqrt{1 - \dfrac{2}{x} + \dfrac{1}{x^2}}, \; x > 1$
   (h) $\dfrac{(4x^2 + 4x + 1)^{1/2}}{2x + 1}, \; x > -\frac{1}{2}$

5. For what real values of $x$ are the following equations valid?
   (a) $\sqrt{(x - 3)^2} = x - 3$?
   (b) $\sqrt{(2 - x)^2} = x - 2$?
   (c) $|4 - x| = x - 4$?
   (d) $\sqrt{x^2 - 8x + 16} = x - 4$?
   (e) $\sqrt{x^2 + 8x + 16} = x - 4$?
   (f) $|x - 1| = 1 - x$

6. If $x \geq 2$, simplify the following.
   (a) $\sqrt{(x - 2)^2} + \sqrt{(2 - x)^2}$
   (b) $\sqrt{(x - 2)^2} - \sqrt{(2 - x)^2}$

7. If $x \leq 2$, simplify the following.
   (a) $\sqrt{(x - 2)^2} + \sqrt{(2 - x)^2}$
   (b) $\sqrt{(x - 2)^2} - \sqrt{(2 - x)^2}$

8. For what real values of $x$ are the following expressions defined?
   (a) $\sqrt{3x - 5}$
   (b) $\sqrt{3 - x}$
   (c) $\sqrt{x - 2}$
   (d) $\dfrac{1}{\sqrt{x - 2}}$
   (e) $\dfrac{1}{\sqrt{x^2 + 2}}$
   (f) $\sqrt{x^2 + 4x + 4}$
   (g) $\sqrt[4]{2x + 6}$
   (h) $\sqrt[6]{5x - 10}$
   (i) $\sqrt[n]{x + 1}$, $n$ a positive even integer
   (j) $\sqrt[3]{x}$

9. Find $x$ in each of the following using the technique of Example 8.11.
   (a) $2^x = 16$
   (b) $2^x = \frac{1}{16}$
   (c) $10^x = 1$
   (d) $10^x = 0.001$
   (e) $10^x = 100$
   (f) $2^x = \frac{1}{2}$
   (g) $4^x = \frac{1}{2}$
   (h) $3^x = 27$
   (i) $(\frac{1}{3})^x = 27$
   (j) $(\frac{9}{16})^x = \frac{4}{3}$
   (k) $2^{x^2} = 16$
   (l) $2^{x^2} = \frac{1}{16}$
   (m) $2^{x^2 + 2x - 1} = 4$

10. (a) Show that $(x^2 + 2x + 1)^{1/2} + (x^2 - 2x + 1)^{1/2} \neq 2x$ by finding a counterexample.
    (b) Simplify $(x^2 + 2x + 1)^{1/2} + (x^2 - 2x + 1)^{1/2}$.

11. Show that $(a/b)^{-n} = (b/a)^n$ if $a \neq 0$ and $b \neq 0$.

SECTION **9**

# THE CARTESIAN COORDINATE SYSTEM

When we use set notation to describe the set having members 1 and 3 either $\{1, 3\}$ or $\{3, 1\}$ is acceptable since the order of the pairing is immaterial. We now wish to consider "ordered pairs" of real numbers. Any two real numbers form a **pair** and when order is designated we call it an **ordered pair of real numbers.** If the first real number is represented by $x$ and the second real number is represented by $y$, then the ordered pair is denoted by the symbol $(x, y)$. For the ordered pair $(x, y)$ the number $x$ is called the **first component** and the number $y$ is called the **second component.** The ordered pair $(1, 2)$ having 1 as the first component and 2 as the second component is different from the ordered pair $(2, 1)$ which has 2 for its first component and 1 for its second component.

Now that we have established the meaning of an ordered pair we wish to define another set operation that will produce ordered pairs.

**Definition 9.1** The **Cartesian product** of two sets $A$ and $B$, denoted by $A \times B$ (read "$A$ cross $B$"), is the set of all possible ordered pairs $(x, y)$ for which the first component is an element of $A$ and the second component is an element of $B$. In terms of set notation we write

$$A \times B = \{(x, y) \mid x \in A \text{ and } y \in B\}$$

**Example 9.1**
If $A = \{1, 3\}$ and $B = \{2, 4, 5\}$, then

$$A \times B = \{(1, 2), (1, 4), (1, 5), (3, 2), (3, 4), (3, 5)\}$$

and

$$B \times A = \{(2, 1), (2, 3), (4, 1), (4, 3), (5, 1), (5, 3)\}.$$

In this text we are concerned with the Cartesian product of the set $\mathbb{R}$ of real numbers with itself. This set which can be described as

$$\mathbb{R} \times \mathbb{R} = \{(x, y) \mid x \in \mathbb{R} \text{ and } y \in \mathbb{R}\}$$

is an infinite set of ordered pairs of real numbers.

Geometrically the set $\mathbb{R} \times \mathbb{R}$ is represented by a plane called the **real** plane. This can be done by first selecting a horizontal line in the plane extending indefinitely to the left and to the right. This line is called the **x-axis.** Next a vertical line extending indefinitely up and down is chosen in plane. This line is called the **y-axis.** The point of intersection of the $x$-axis and the $y$-axis is called the **origin.** We consider the positive direction on the $x$-axis to be to the right of the origin and the positive direction on the $y$-axis to be above the origin. See Figure 9.1.

We can now associate ordered pairs of real numbers $(x, y)$ in $\mathbb{R} \times \mathbb{R}$ with a point $P$ in the plane. The distance of $P$ from the $y$-axis is called the **x-coordinate** or **abscissa** of $P$ and the distance of $P$ from the $x$-axis is called the **y-coordinate** or **ordinate** of $P$. The $x$- and

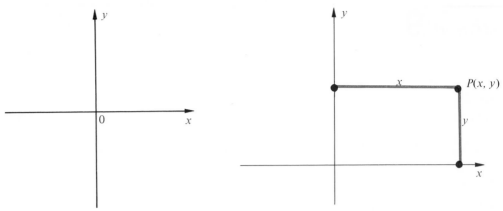

Figure 9.1                              Figure 9.2

$y$-coordinates of $P$ are called the **rectangular Cartesian coordinate** of the point. There is a one-to-one correspondence between the points in the plane and the Cartesian coordinate. This one-to-one correspondence is called a **Cartesian coordinate system.** See Figure 9.2. In this system to locate a point $(x, y)$ is to **plot** the point. The point associated with a given ordered pair $(x, y)$ is called the **graph** of the ordered pair.

**Example 9.2**
Plot the points $(-2, 0)$, $(1, 3)$, $(-4, 2)$, $(0, 1)$, $(-2, -2)$, and $(4, -1)$.

Solution.   Figure 9.3 shows the Cartesian coordinate system with the given points plotted.   ●

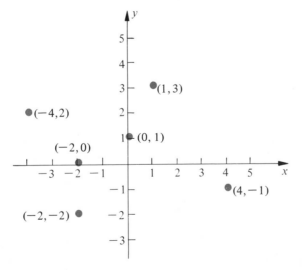

Figure 9.3

We note that those points in the plane which do not lie on either the $x$- or the $y$-axis are divided into four parts. Each of these parts represent a subset of the plane and is called a **quadrant** (Figure 9.4). These quadrants can be defined as shown on the following page.

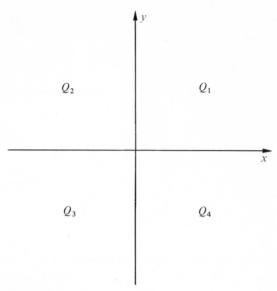

Figure 9.4

$$Q_1 = \{(x,y)\,|\,x > 0 \text{ and } y > 0\} \qquad Q_3 = \{(x,y)\,|\,x < 0 \text{ and } y < 0\}$$
$$Q_2 = \{(x,y)\,|\,x < 0 \text{ and } y > 0\} \qquad Q_4 = \{(x,y)\,|\,x > 0 \text{ and } y < 0\}$$

If $A$ is a subset of $\mathbb{R} \times \mathbb{R}$, then $A$ is a set of ordered pairs of real numbers. We say the **graph** of $A$ is the set of all points $(x, y)$ in the Cartesian coordinate system for which $(x, y)$ is an ordered pair in $A$.

**Example 9.3**
If $A = \{(-2, -1), (2, 0), (3, 1)\}$, the graph of $A$ consists of the 3 points shown in Figure 9.5. ●

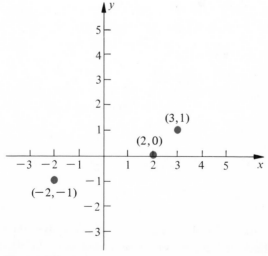

Figure 9.5

**Example 9.4**

If $A = \{1, 3\}$ and $B = \{2, 4, 5\}$, then graph $A \times B$ and $B \times A$.

Solution.  We use the results of Example 9.1 to show the graphs of $A \times B$ and $B \times A$ in Figures 9.6 and 9.7 respectively.  ●

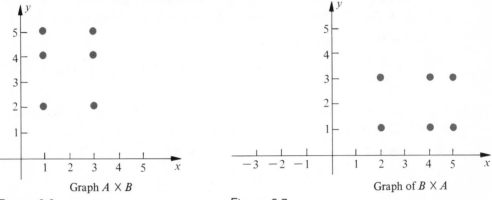

Graph $A \times B$                              Graph of $B \times A$

Figure 9.6                              Figure 9.7

Equations in two variables, $x$ and $y$, where $x \in \mathbb{R}$ and $y \in \mathbb{R}$ have as solutions ordered pairs of real numbers. For example, the ordered pair $(2, 3)$ is said to be a solution of the equation

$$3x + 4y = 18$$

since substitution of 2 for $x$ and 3 for $y$ yields the following true statement

$$3(2) + 4(3) = 18.$$

Whereas, the ordered pair $(3, 2)$ is not a solution of the equation

$$3x + 4y = 18$$

since substitution of 3 for $x$ and 2 for $y$ yields the following false statement.

$$3(3) + 4(2) = 18$$

The set of all ordered pairs that satisfy the equation is called the **solution set** of the equation. We will denote the solution set for the above equation as

$$S = \{(x, y) \mid 3x + 4y = 18\}$$

This solution set is an infinite subset of $\mathbb{R} \times \mathbb{R}$.

The **graph of such equations** as $3x + 4y = 18$ is the graph of its solution set.

**Example 9.5**

If $A = \{-2, -1, 0, 1, 2, 3\}$, find the solution set for $y = x + 1$ when $(x, y) \in A \times A$ and then show the graph of this set in the real plane.

Solution.  Since $(x, y) \in A \times A$ we need only to replace $x$ in $y = x + 1$ with each element in $A$ to determine the associated values for $y$. We do this and find that when

$$x = -2, y = -1$$
$$x = -1, y = 0$$

$x = 0, y = 1$
$x = 1, y = 2$
$x = 2, y = 3$
$x = 3, y = 4$

We note that when we replace $x$ by 3 we obtain a value of 4 for $y$ that is not in $A$. Thus the entire solution set $S$, for $y = x + 1$ when $(x, y) \in A \times A$ is

$$S = \{(-2, -1), (-1, 0), (0, 1), (1, 2), (2, 3)\}$$

We show the graph of the equation $y = x + 1$ where $(x, y) \in A \times A$ in Figure 9.8. ●

The points in Figure 9.8 appear to lie in a straight line. In fact, if we consider the solution set for $y = x + 1$ when $(x, y) \in \mathbb{R} \times \mathbb{R}$ we see that its solution set

$$S = \{(x, y) \mid y = x + 1 \text{ and } (x, y) \in \mathbb{R} \times \mathbb{R}\}$$

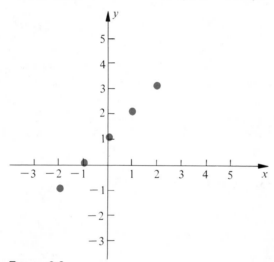

Figure 9.8

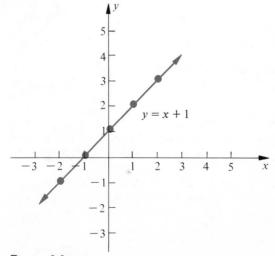

Figure 9.9

is an infinite set and its graph will be a straight line. This will be shown in Section 13. See Figure 9.9. We say that the line in Figure 9.9 is the **graph** of $y = x + 1$. Since this line continues indefinitely in both directions, we indicate this with arrows and we call the portion of the line shown a **sketch of the graph** of the line $y = x + 1$.

## EXERCISES FOR SECTION 9

1. Let $A = \{2, 3, 4\}$ and $B = \{-1, 0, 1, 2\}$.
   (a) Find $A \times A$, $B \times B$, $A \times B$, $B \times A$.
   (b) Find $C$ if $C = \{(x, y) | (x, y) \in A \times B, y > 0\}$.
   (c) Find $D$ if $D = \{(x, y) | (x, y) \in A \times B, x = y\}$.
   (d) Find $E$ if $E = \{(x, y) | (x, y) \in B \times A, x = y\}$.
   (e) Find $F$ if $F = \{(x, y) | (x, y) \in A \times B, x > y\}$.
   (f) Find $G$ if $G = \{(x, y) | (x, y) \in A \times B, x \le y\}$.
   (g) Does $A \times B = B \times A$.

2. If $(x, y) \in A \times A$ where $A = \{-3, -2, -1, 0, 1, 2\}$ then find the graphs of the following equations.
   (a) $y = x - 1$        (d) $y = x - 2$
   (b) $y = x$              (e) $y = -x^2$
   (c) $y = x + 2$         (f) $y = x^2$

3. Plot each set of points. Join the successive points of each set by straight lines and identify the figure formed.
   (a) $(0, 0)$, $(8, -2)$, $(9, 2)$, $(1, 4)$
   (b) $(0, 0)$, $(-3, 5)$, $(5, 3)$
   (c) $(0, 0)$, $(14, 4)$, $(5, 7)$, $(-2, 5)$
   (d) $(3, 9)$, $(12, 6)$, $(9, -3)$, $(0, 0)$
   (e) $(-4, 3)$, $(0, -3)$, $(4, 4)$

4. For which quadrant or quadrants are the following statements true?
   (a) $x$ and $y$ have the same sign         (d) $x < 0$
   (b) $y > 0$                                          (e) $x > 0$    or    $y > 0$
   (c) $y > 0$    and    $x < 0$               (f) $x > 0$    or    $y < 0$

5. If $X$ is the set of all real numbers $x$ such that $-1 \le x \le 1$ and $Y$ is the set of all real numbers $y$ such that $-1 \le y \le 1$, graph $X \times Y$.

6. If $X$ is the set of all real numbers $x$ such that $0 \le x \le 2$ and $Y$ is the set of all real numbers $y$ such that $y \ge 2$, graph $X \times Y$.

## SECTION 10
# THE DISTANCE FORMULA

The distance between any two points on a number line is the absolute value of the difference between the coordinates of the points (Figure 10.1):

$$\overline{AC} = |5 - (-3)| = |-3 - 5| = 8$$
$$\overline{BC} = |5 - 1| = |+1 - 5| = 4$$

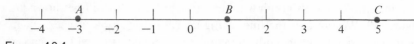

Figure 10.1

If two points in the coordinate plane lie on the same horizontal or vertical line, the distance between them is determined in the same way.

Let us consider the pairs of points in Figure 10.2. The distances are found as follows:

$$\overline{AB} = |3 - (-2)| = |3 + 2| = |5| = 5$$
$$\overline{CD} = |2 - (-1)| = |2 + 1| = |3| = 3$$

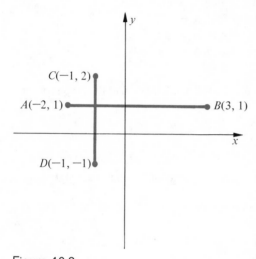

Figure 10.2

To find the distance between any two points in the coordinate plane we make use of the **Pythagorean Theorem:** *The square of the length c of the hypotenuse of a right triangle equals the sum of the squares of the length of the other two sides a and b.* $c^2 = a^2 + b^2.$

**Theorem 10.1** Let $P_1(x_1, y_1)$ and $P_2(x_2, y_2)$ be any two points in the plane. The distance $d$ between $P_1$ and $P_2$ is given by the formula

$$d = \overline{P_1P_2} = \sqrt{(x_2 - x_1)^2 + (y_2 - y_1)^2} \tag{10.1}$$

Proof. Let $Q$ be the point of intersection between the line through $P_1$ parallel to the $x$-axis and the line through $P_2$ parallel to the $y$-axis (Figure 10.3). The coordinates of $Q$ are $(x_2, y_1)$. $\overline{P_1Q} = |x_2 - x_1|$ and $\overline{QP_2} = |y_2 - y_1|$. Since triangle $P_1QP_2$ is a right triangle, we can use the Pythagorean Theorem. Thus $(\overline{P_1P_2})^2 = (\overline{P_1Q})^2 + (\overline{QP_2})^2$. Since $(P_1Q)^2 = |x_2 - x_1|^2 = (x_2 - x_1)^2$ and $(\overline{QP_2})^2 = |y_2 - y_1|^2 = (y_2 - y_1)^2$, we have

$$(\overline{P_1P_2})^2 = (x_2 - x_1)^2 + (y_2 - y_1)^2$$

or

$$\overline{P_1P_2} = \sqrt{(x_2 - x_1)^2 + (y_2 - y_1)^2} \quad \bullet$$

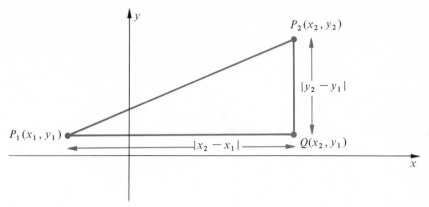

Figure 10.3

**Example 10.1**
Find the distance between $P_1(3, -2)$ and $P_2(1, 4)$.

Solution

$$\overline{P_1 P_2} = \sqrt{(1 - 3)^2 + (4 - (-2))^2}$$
$$= \sqrt{(-2)^2 + (6)^2} = \sqrt{4 + 36} = \sqrt{40} = 2\sqrt{10} \quad \bullet$$

**Example 10.2**
Find the distance $r$ from the origin to any point $(x, y)$.

Solution

$$r = \sqrt{(x - 0)^2 + (y - 0)^2}$$
$$= \sqrt{x^2 + y^2} \quad \bullet$$

**Example 10.3**
Use the distance formula to show that the points $P(3, 2)$, $Q(6, 3)$, and $R(12, 5)$ lie on a straight line.

Solution

$$\overline{PQ} = \sqrt{(6 - 3)^2 + (3 - 2)^2} = \sqrt{3^2 + 1^2} = \sqrt{10}$$
$$\overline{QR} = \sqrt{(12 - 6)^2 + (5 - 3)^2} = \sqrt{40} = 2\sqrt{10}$$
$$\overline{PR} = \sqrt{(12 - 3)^2 + (5 - 2)^2} = \sqrt{9^2 + 3^2} = \sqrt{90} = 3\sqrt{10}$$

Since $\overline{PQ} + \overline{QR} = \overline{PR}$, the points are collinear. $\quad \bullet$

**Example 10.4**
Find the equation of a circle whose center is at $(0, 0)$ and whose radius is equal to 1.

Solution  In Figure 10.4 we see that the radius $r$ is equal to 1. We use the result of Example 10.3 to obtain $\sqrt{x^2 + y^2} = r = 1$ or $x^2 + y^2 = 1$. This is the desired equation. $\quad \bullet$

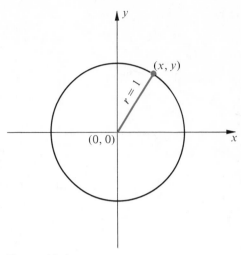

Figure 10.4

## EXERCISES FOR SECTION 10

1. Use the distance formula to find the length of the line segment joining the given pairs of points.
   (a) $(3, 1), (6, 5)$     (c) $(1, 2), (2, 3)$     (e) $(-3, 4), (-3, 5)$
   (b) $(4, -2), (7, -2)$     (d) $(-2, 3), (-1, 5)$     (f) $(2, 7), (-2, -4)$

2. (a) Find a formula that gives the coordinates of the point that is midway between the points $(x_1, y_1)$ and $(x_2, y_2)$. This point is called the **midpoint.**
   (b) Find the coordinates of the midpoint using the result of Exercise 2(a) and the length of the line segment joining the points with the given coordinates.
      (i) $(4, 1), (4, -6)$     (iii) $(3, 1), (5, 7)$     (v) $(a, b), (2a, 3b)$
      (ii) $(2, 3), (6, -3)$     (iv) $(a, b), (x, y)$

3. Show that the points $(-4, 3), (0, -3)$, and $(4, 4)$ represent the vertices of an isosceles triangle.

4. Show that the following points are collinear by using the distance formula.
   (a) $(3, 0), (0, -4), (9, 8)$     (b) $(1, 1), (0, -2), (2, 4)$     (c) $(-2, 3), (1, 6), (3, 8)$

5. Show that the points $(10, -7), (11, -8), (10, -9)$, and $(9, -8)$ represent the vertices of a square.

6. Show that the points $(0, 0), (8, 0)$, and $(0, 6)$ represent the vertices of a right triangle.

7. The point $(x, 3)$ is 4 units from the point $(2, 7)$. Find $x$.

8. Find the point on the $y$-axis that is equidistant from the points $(1, 3)$ and $(5, 4)$.

9. If the point $C$ $(x, 3)$ is the vertex of the right triangle formed by the points $A(-4, -2)$, $B(7, -3)$ and $C(x, 3)$ find the value(s) of $x$.

10. Find the length of the diagonal of a square whose area is 8.

11. Find the equation of a circle whose center is at $(0, 0)$ and whose radius is $r$ units.

12. Find the relationship between $x$ and $y$ if the point $(x, y)$ is always 4 units from the point $(2, 3)$.

13. Find the relationship between $x$ and $y$ if the point $(x, y)$ is always equidistant from the points $(4, 0)$ and $(-4, y)$.

# CHAPTER 2
# RELATIONS
# AND
# FUNCTIONS

**RELATIONS**

In Chapter 1 we defined the Cartesian product of two sets $A$ and $B$ as $A \times B = \{(x, y) \mid x \in A, y \in B\}$. We noted that the elements of such a set were ordered pairs of the form $(x, y)$, where the first component $x$ was selected from set $A$ and the second component $y$ was selected from set $B$. We used the term "ordered pair" to indicate that elements such as $(1, 2)$ and $(2, 1)$ were distinct. We then examined the special Cartesian product $\mathbb{R} \times \mathbb{R}$, where $\mathbb{R} \times \mathbb{R} = \{(x, y) \mid x \in \mathbb{R}, y \in \mathbb{R}\}$. The geometric or graphic interpretation of $\mathbb{R} \times \mathbb{R}$ was the entire plane. Therefore, any subset of $\mathbb{R} \times \mathbb{R}$ has for its graph a set of points in the plane. In the future $\mathbb{R} \times \mathbb{R}$ will be our universal set unless otherwise stated. Sets such as

$$A = \{(x, y) \mid y = 2x + 1\}$$
$$B = \{(x, y) \mid y > x\}$$

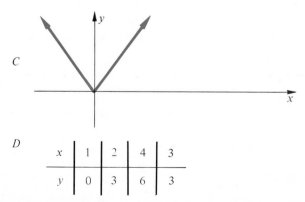

$C$

| $x$ | 1 | 2 | 4 | 3 |
|-----|---|---|---|---|
| $y$ | 0 | 3 | 6 | 3 |

$D$

are all subsets of $\mathbb{R} \times \mathbb{R}$.

Many different means are used to define the $x$ and $y$ components of the ordered pairs which belong to the relation.

The means may be an equation or inequality involving two variables, a statement, graph, or chart. The "relationship" between $x$ and $y$ leads us to the following definition.

**Definition 11.1** Any subset of $\mathbb{R} \times \mathbb{R}$ is called a **relation.** Alternatively, we may say that any set of ordered pairs is a relation in $\mathbb{R}$.

**Definition 11.2** The set of all first components (usually $x$) is called the **domain** of the relation. The set of all second components (usually $y$) is called the **range** of the relation.

### Example 11.1
What is the domain of the relation $A = \{(0, 1), (2, 1), (2, 2), (3, 2)\}$? The range?

Solution. The set of all first components is the domain of this relation. Thus the domain is $\{0, 2, 3\}$. The set of all second components is the range of this relation. Thus the range is $\{1, 2\}$. ●

### Example 11.2
Let $A = \{-1, 0, 1, 2, 3\}$. What is the domain of the relation $r = \{(x, y) \mid y = x + 1, x \in A, y \in A\}$. The range?

Solution. It is understood that the relation $r$ is a subset of $A \times A$ or, more specifically, $r = \{(-1, 0), (0, 1), (1, 2), (2, 3)\}$. (*Note:* The replacement of $x$ by 3 does not give us a $y$ value in $A$.) Thus the domain of $r$ is $\{-1, 0, 1, 2\}$ and the range of $r$ is $\{0, 1, 2, 3\}$. Figures 11.1 and 11.2 give us a graphic representation of $A \times A$ and the relation $r$. ●

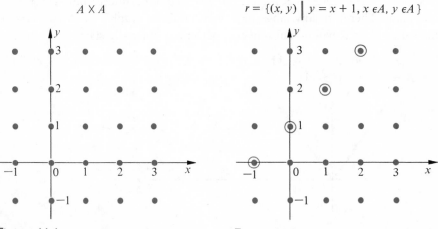

Figure 11.1                                   Figure 11.2

### Example 11.3
Let $A = \{1, 2, 3\}$. What is the domain of the relation $r = \{(x, y) \mid y > x, x \in A, y \in A\}$? The range?

Solution.  Here the universal set is $A \times A$ and $r = \{(x, y)|y > x\} = \{(1, 2),$ $(1, 3), (2, 3)\}$. Thus the domain of $r$ is $\{1, 2\}$ and the range of $r$ is $\{2, 3\}$. Figure 11.3 illustrates the graph of $r$.  ●

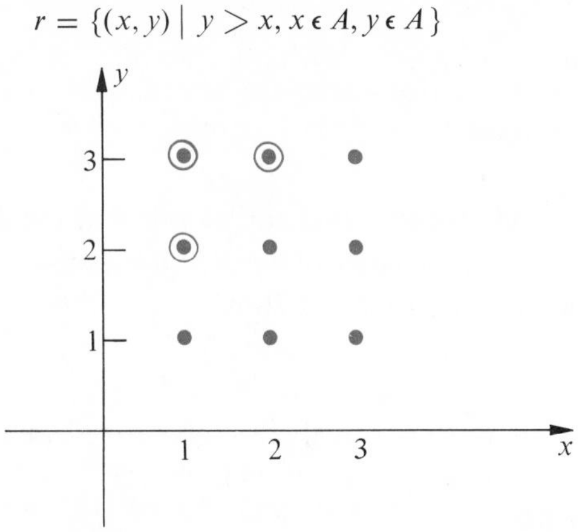

$$r = \{(x, y) \mid y > x, x \in A, y \in A\}$$

Figure 11.3

In each of the previous examples the relations had a finite number of elements (ordered pairs). Thus it was quite simple to determine the domain and range. We now wish to examine the domain and range of some relations that contain an infinite number of elements.

### Example 11.4
Find the domain and range of the relation $r = \{(x, y)|y = 3x - 2\}$.

Solution.  Since $r$ is a subset of $\mathbb{R} \times \mathbb{R}$, we know that the domain of $r$ must be a subset of $\mathbb{R}$ or possibly $\mathbb{R}$ itself. To determine the domain of $r$, we look for those values of $x$ that will *not* yield a real value for $y$. If such values exist, they must be excluded from the domain. The excluded values usually occur when we have an indicated division by zero or when the relation contains an even root. These cases will be treated in the following examples. In this example there are no excluded values. Thus the domain is $\{x|x \in \mathbb{R}\}$.

We determine the range of $r$ in exactly the same manner. We look for possible values of $y$ that will not yield real values for $x$. But any real value of $y$ gives a real value for $x$. We therefore conclude that the range of $r$ is $\{y|y \in \mathbb{R}\}$.  ●

### Example 11.5
Find the domain and range of the relation $r = \{(x, y)|y = 1/(x - 1)\}$.

Solution.  Inspection of the equation $y = 1/(x - 1)$ indicates that if $x = 1$, then $y = \frac{1}{0}$ is undefined. Therefore, $x = 1$ must be excluded from the domain of $r$. The domain of $r$ is $\{x|x \in \mathbb{R}, x \neq 1\}$. We now solve for $x$ in terms of $y$ in the given equation as follows:

$$y = \frac{1}{x - 1}$$

$$xy - y = 1$$

$$xy = 1 + y \quad or \quad x = \frac{1 + y}{y}$$

Inspection of the equation $x = \dfrac{1 + y}{y}$ indicates that if $y = 0$, then $x$ is undefined. Thus the range of $r$ is $\{y \mid y \in \mathbb{R}, y \neq 0.\}$   •

The method used for finding domain and range in Example 11.5 can be generalized as follows. We can determine the domain and range of a relation by solving for $y$ in terms of $x$ and $x$ in terms of $y$ and then examining each equation for possible excluded values.

### Remark 11.1

There are relations where it is not possible to solve for $y$ in terms of $x$ and $x$ in terms of $y$. One such example is $x^5 + 4xy^3 - 3y^5 = 2$. We shall not attempt to deal with these situations.

### Example 11.6

Find the domain and range of the relation $r = \{(x, y) \mid x^2 + y^2 = 25\}$.

**Solution.**   To find the domain of $r$ we must solve for $y$ in terms of $x$ as follows:

$$x^2 + y^2 = 25$$
$$y^2 = 25 - x^2$$
$$|y| = \sqrt{25 - x^2}$$
$$y = \pm\sqrt{25 - x^2}$$

From this equation we see that $25 - x^2 \geq 0$ if $y$ is to be a real number. (Why?) Thus the domain of $r$ is $\{x \mid 25 - x^2 \geq 0\} = \{x \mid -5 \leq x \leq 5\}$, since the solution set of $25 - x^2 \geq 0$ is $-5 \leq x \leq 5$.

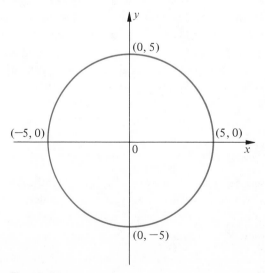

Figure 11.4

To find the range of $r$ we must solve for $x$ in terms of $y$ as follows:

$$x^2 + y^2 = 25$$
$$x^2 = 25 - y^2$$
$$|x| = \sqrt{25 - y^2}$$
$$x = \pm\sqrt{25 - y^2}$$

From the equation we see that $25 - y^2 \geq 0$ if $x$ is to be a real number. Thus the range of $r$ is $\{y\,|\,25 - y^2 \geq 0\} = \{y\,|\,-5 \leq y \leq 5\}$. The graph of this relation is the *circle* shown in Figure 11.4. ●

### Example 11.7

Find the domain and range of the relation $r = \{(x, y)\,|\,y = x/\sqrt{4 - x^2}\}$.

**Solution.** Inspection of the equation $y = x/\sqrt{4 - x^2}$ indicates that $y$ will be nonreal when $4 - x^2 \leq 0$. Thus the solution set for $4 - x^2 > 0$ will be the domain of $r$. To find this solution set we proceed as follows.

$$4 - x^2 > 0$$
$$4 > x^2 \qquad \text{(add } x^2 \text{ to both sides)}$$
$$2 > |x| \qquad \text{(take the square root of both sides)}$$

Since $2 > |x|$ is equivalent to $-2 < x < 2$, the domain of $r$ is $\{x\,|\,-2 < x < 2\}$. To find the range of $r$ we now solve for $x$ in terms of $y$ as follows.

$$y = \frac{x}{\sqrt{4 - x^2}}$$

$$y^2 = \frac{x^2}{4 - x^2} \qquad \text{(squaring both sides)}$$

$$4y^2 - y^2 x^2 = x^2$$
$$4y^2 = x^2 + y^2 x^2$$
$$4y^2 = x^2(1 + y^2)$$
$$x^2 = \frac{4y^2}{1 + y^2}$$
$$x = \pm\sqrt{\frac{4y^2}{1 + y^2}}$$

Inspection of the equation $x = \pm\sqrt{4y^2/(1 + y^2)}$ indicates that $x$ is a real number when $4y^2/(1 + y^2) \geq 0$. Since $4y^2 \geq 0$ and $1 + y^2 > 0$ we have $4y^2/(1 + y^2) \geq 0$ for all real values of $y$. Therefore the range is $\mathbb{R}$. ●

We mentioned earlier that the "means" for describing the relationship between the elements in the domain and the elements in the range of a given relation do not necessarily have to be an equation or inequality. The means might well be a graph or chart, as in the following example.

**Example 11.8**

The graph in Figure 11.5 determines a relation with a domain of $\{x \mid -4 \leq x \leq 10\}$ and a range of $\{y \mid -5 \leq y \leq 6\}$.

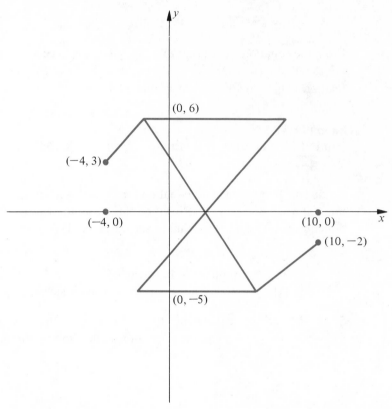

Figure 11.5

# EXERCISES FOR SECTION 11

*Do* all odd numbers

Find the domain and range of each relation in exercises 1-19.

**1.** $A = \{(1, 2), (1, 3), (2, 4), (2, 5), (3, 4)\}$

**2.** $A = \{(1, 1), (1, 2), (1, 3), (1, 4)\}$

**3.** $A = \{(1, 2), (2, 2), (3, 2), (6, 2)\}$

**4.** $A = \{(x, y) \mid x + y = 1\}$

**5.** $A = \{(x, y) \mid x^2 + y^2 = 36\}$

**6.** $A = \{(x, y) \mid x^2 + y^2 = -1\}$

**7.** $A = \left\{(x, y) \mid y = \dfrac{1}{(x - 3)}\right\}$

**11.** $r = \{(x, y) \mid y^2 = 4 - x\}$

**12.** $r = \left\{(x, y) \mid y = \dfrac{2x + 1}{5 - x}\right\}$

**13.** $r = \left\{(x, y) \mid y = \dfrac{4x}{x - 2}\right\}$

**14.** $r = \{(x, y) \mid y = x^2 - 9\}$

**15.** $r = \{(x, y) \mid y = 4 - x^2\}$

**16.** $r = \{(x, y) \mid 9x^2 - 4y^2 = 36\}$

**17.** $r = \{(x, y) \mid y^2 - 4x^2 = 4\}$

**8.** $A = \{(x, y) \mid y = \sqrt{x^2}\}$

**9.** $A = \left\{(x, y) \mid x + 3 = \dfrac{2}{y - 1}\right\}$

**10.** $r = \{(x, y) \mid y^2 = x + 1\}$

**18.** $r = \left\{(x, y) \mid y = \dfrac{1}{(x - 2)^2}\right\}$

**19.** $r = \left\{(x, y) \mid x = \dfrac{1}{(y - 2)^2}\right\}$

Assume that each of the following relations $r$ is a subset of $A \times A$, where $A = \{-1, 0, 1, 2, 3\}$. Find the domain and range of each relation. Then graph each relation.

**20.** $r = \{(x, y) \mid y > x\}$      **23.** $r = \{(x, y) \mid y = |x|\}$

**21.** $r = \{(x, y) \mid y = x^2\}$      **24.** $r = \{(x, y) \mid x = 1\}$

**22.** $r = \{(x, y) \mid x = y^2\}$      **25.** $r = \{(x, y) \mid y = 3\}$

Determine the domain and the range of each of the following relations. Each of these relations is determined by the given graph.

**26.**

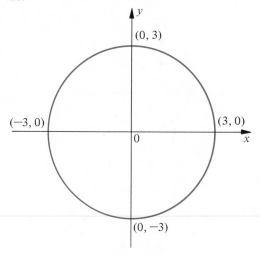

**27.**

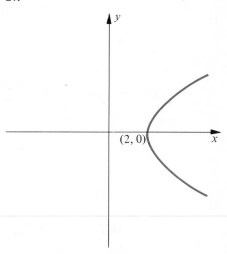

**28.**

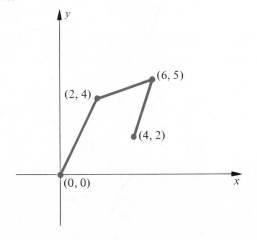

**29.**

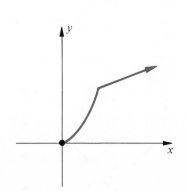

**30.**

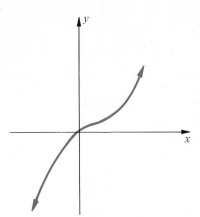

**31.**

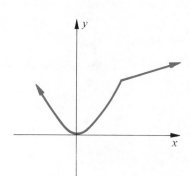

**32.**

**33.**

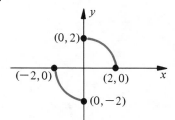

**34.**

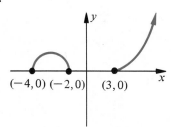

**35.**

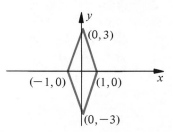

**36.**

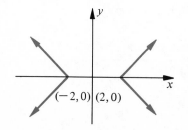

**37.**

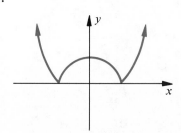

**Example 12.8**

If $f(x) = x^2 + 3x$, find $\dfrac{f(1 + h) - f(1)}{h}$, $h \neq 0$.

Solution. We first find $f(1 + h)$ and $f(1)$ by noting that since

$f(x) = x^2 + 3x$
$f(1) = (1)^2 + 3(1) = 4$ **(A)**

and

$f(1 + h) = (1 + h)^2 + 3(1 + h)$

or

$f(1 + h) = 1 + 2h + h^2 + 3 + 3h = 4 + 5h + h^2$ **(B)**

We now use the results of (A) and (B) as follows.

$$\frac{f(1 + h) - f(1)}{h} = \frac{(4 + 5h + h^2) - (4)}{h}$$

or

$$\frac{f(1 + h) - f(1)}{h} = \frac{5h + h^2}{h} = 5 + h$$

**Example 12.9**

The function $f : x \to x^2 + 1$ can be expressed as a subset of $\mathbb{R} \times \mathbb{R}$ as follows: $f = \{(x, y) \mid y = x^2 + 1\}$. What is the domain and range of $f$?

Solution. Inspection of the equation $y = x^2 + 1$ indicates that $y$ will be defined for all real values of $x$. Thus the domain of $f$ is $\mathbb{R}$.

We determine the range by inspection as follows. Since $x^2 \geq 0$ for all real $x$, $x^2 + 1 \geq 1$ and hence the range is given by $\{y \mid y \geq 1\}$. This same result is obtained algebraically by solving for $x$ in terms of $y$ and then determining the appropriate values for $y$ as follows.

$y = x^2 + 1$
$x^2 = y - 1$
$x = \pm\sqrt{y - 1}$

Now $x$ will be defined when $y - 1 \geq 0$ or $y \geq 1$. Thus the range is $\{y \mid y \geq 1\}$. ●

## EXERCISES FOR SECTION 12

Determine whether or not each of the following relations is a function in $\mathbb{R} \times \mathbb{R}$.

**1.** $A = \{(2, 4), (4, 6), (6, 8), (7, 8)\}$

**2.** $A = \{(2, 1), (3, 1), (5, 1), (-1, 1)\}$

**6.** $A = \{(x, y) \mid y = \sqrt{x}\}$

**7.** $A = \{(x, y) \mid y = \pm\sqrt{x}\}$

**3.** $A = \{(1, -2), (1, 0), (2, 1), (3, 2)\}$

**4.** $A = \{(3, 1), (3, 3), (3, 6), (3, 9), (3, 11)\}$

**5.** $A = \{(x, y) \mid y = 1/x\}$

**8.** $A = \{(x, y) \mid y = \sqrt[3]{x}\}$

**9.** $A = \{(x, y) \mid y^2 = x^3\}$

**10.** Exercises 26–37 in Section 11.

Find the domain and the range of each of the following functions.

**11.** $f = \left\{(x, y) \mid y = \dfrac{1}{x - 1}\right\}$

**12.** $f = \{(x, y) \mid y = x^2 - 1\}$

**13.** $f = \{(x, y) \mid y = \sqrt{x + 5}\}$

**14.** $f = \{(x, y) \mid y = \sqrt{x^2 - 1}\}$

**15.** $f = \left\{(x, y) \mid y = 4 - \dfrac{2}{x}\right\}$

**16.** $f = \{(x, y) \mid y = \sqrt[4]{x - 2}\}$

**17.** $f = \{(x, y) \mid y = \sqrt[3]{x - 2}\}$

**18.** $f = \{(x, y) \mid y - 1 = (x + 2)^3\}$

**19.** $f = \{(x, y) \mid y + 2 = 8x^3\}$

**20.** $f = \{(x, y) \mid y = \sqrt{1 - x^2}\}$

**21.** $f = \{(x, y) \mid y = \sqrt{x^2 + 1}\}$

**22.** If $f(x) = 2x - \dfrac{1}{x}$, find the following.

(a) $f(2)$        (d) $f(h)$          (g) $f(1) + f(-1)$

(b) $f(1)$        (e) $f(x + h)$

(c) $f(-\tfrac{1}{2})$      (f) $[f(1)]^2$

**23.** What element(s) in the domain of each of the following functions is (are) associated with the element 4 in the range?

(a) $y = x - 4$      (d) $y = 2$

(b) $y = x + 4$      (e) $y = 3x^2$

(c) $y = 3x$

**24.** Consider the functions $f$ and $g$ defined by the graphs.

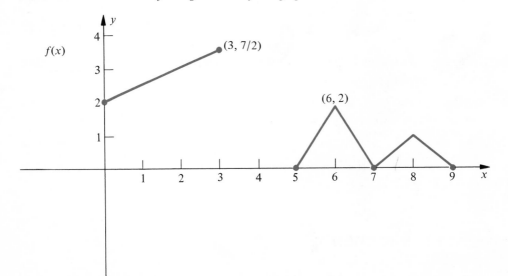

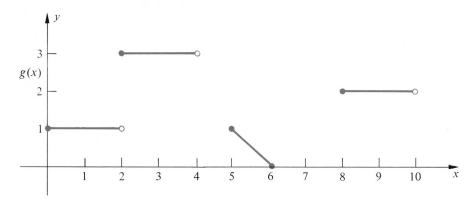

(a) Determine the domain and the range of $f$ and $g$.
(b) Find $f(0)$, $f(3)$, $f(5)$, $f(8)$, $f(9)$, and $f[f(6)]$.
(c) Find $g(0)$, $g(3)$, $g(5)$, $g(8)$, $g(9)$, and $g[g(6)]$.

25. If $f(x) = x^2 - 4$, find all values of $x$ for which $f(x) = 0$, $f[f(x)] = 0$.

26. If $f(x) = x^2$, find all values of $x$ for which $f(x + 1) = f(2x)$.

27. Suppose that $f(x^m) = mx^{m-1}$, where $m$ is a nonnegative integer.
    *Example: $f(x^4) = 4x^{4-1} = 4x^3$.*
    (a) Find $f(x^3)$, $f(x^5)$, and $f(x^{10})$.
    (b) Does $f(x^3 \cdot x^5) = f(x^3) \cdot f(x^5)$?

28. Suppose that $g(x^m) = \dfrac{1}{m + 1} x^{m+1}$, where $m$ is a natural number.

    *Example: $g(x^2) = \dfrac{1}{2 + 1} x^{(2+1)} = \dfrac{1}{3} x^3$.*
    (a) Find $g(x^3)$, $g(x^4)$, and $g(x^{10})$.
    (b) Does $g(x^3 \cdot x^4) = g(x^3) \cdot g(x^4)$?

29. Express the area $(A)$ of a circle as a function of the radius $(r)$ of a circle, and specify an appropriate domain.

30. Express the area $(A)$ of a triangle as a function of its base $(b)$ when the height of the triangle is 6 inches, and specify an appropriate domain.

31. Express the perimeter $(P)$ of a rectangle as a function of its width $(w)$ when the length of the rectangle is 10 inches, and specify an appropriate domain.

32. Express the width $(w)$ of a rectangle as a function of its perimeter $(P)$ when the length of the rectangle is 10 inches, and specify an appropriate domain.

33. If $f(x) = x^2 + x$, find the following.

    (a) $f(3 + h) - f(3)$
    (b) $\dfrac{f(3 + h) - f(3)}{h}$
    (c) $\dfrac{f(x + h) - f(x)}{h}$

34. If $f(x) = 2x + 1$, find $\dfrac{f(x + h) - f(x)}{h}$, $h \neq 0$.

✓**35.** Same as exercise 34 when $f(x) = x^2 - 2x$.

**36.** Same as exercise 34 when $f(x) = \dfrac{1}{x}$.

✓**37.** Suppose $f(x) = x^4 + 3x^2$.
    (a) Find $f(h)$ and $f(-h)$.
    (b) How are $f(h)$ and $f(-h)$ related?

**38.** Same as exercise 37 when $f(x) = 2x^3 - x$.

✓**39.** If $f(x) = 2x - 3$ and if $|f(x) - f(2)| < 0.1$ when $|x - 2| < L$, find $L > 0$.

**40.** Same as exercise 39 when $f(x) = 2 - 3x$.

# SECTION 13
## LINEAR EQUATIONS IN TWO VARIABLES: LINEAR FUNCTIONS

The equation $Ax + By + C = 0$, $A,B,C \in \mathbb{R}$, $A$ and $B$ both not zero, is called a **linear** or **first-degree equation in two unknowns (variables)**, $x$ and $y$. The solution set for $Ax + By + C = 0$ is the set of all ordered pairs of real numbers $(x, y)$ that satisfy the given equation. If we assume that $B \neq 0$, we may solve $Ax + By + C = 0$ for $y$ as follows:

$$Ax + By + C = 0$$
$$By = -Ax - C$$
$$y = -\left(\frac{A}{B}\right)x - \frac{C}{B}$$

or equivalently, $y = mx + b$, where $m = -A/B$ and $b = -C/B$. Thus the solution set for $y = mx + b$ is $\{(x,y) \mid y = f(x) = mx + b\}$. The graph of this set is always a straight line whose domain is $\{x \mid x \in \mathbb{R}\}$. $f = \{(x,y) \mid y = mx + b\}$ is called a **linear function** and is simply written $y = mx + b$. [*Note:* We determine the graph or trace of a given function by plotting a number of the points in the geometric (Cartesian) plane that are associated with the elements (ordered pairs) of the function and then sketching the remaining curve.]

    We know that any two distinct points will determine a straight line. So, if we wish to graph a linear equation, we need two such points. Usually the easiest way to find these points is to set $x = 0$ and solve for $y$ in the given equation and then set $y = 0$ and solve for $x$ in the given equation to obtain the respective points $(0, y)$ and $(x, 0)$.

    The $x$ component of the point $(x, 0)$ is called the **$x$-intercept** and the $y$ component of the point $(0, y)$ is called the **$y$-intercept**.

> **Example 13.1**
> Graph the linear function $\{(x, y) \mid x + 2y = 1\}$.
>
> **Solution.** Here we set $x = 0$ to obtain $0 + 2y = 1$ or $y = \frac{1}{2}$. Thus the $y$-intercept is at $(0, \frac{1}{2})$. We then set $y = 0$ to obtain $x + 2(0) = 1$ or $x = 1$. Thus the $x$-intercept is at $(1, 0)$. The graph of this function is shown in Figure 13.1. *Note:* The range of this function is $\{y \mid y \in \mathbb{R}\}$.   ●

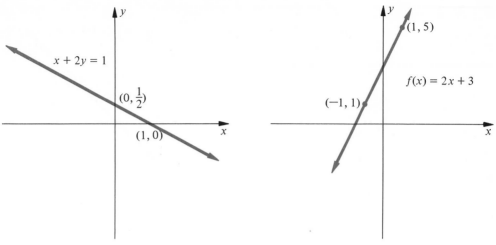

Figure 13.1                                      Figure 13.2

When describing the linear function $\{(x, y) \mid Ax + By + C = 0,\ B \neq 0\}$ or its equivalent form $\{(x, y) \mid y = f(x) = mx + b\}$, we will simply use its rule (correspondence), $f(x) = mx + b$ or $y = mx + b$.

**Example 13.2**
Graph the linear function $f(x) = 2x + 3$.

Solution. We need only two distinct points to determine the graph $f$. So we arbitrarily choose $f(1) = 2(1) + 3 = 5$ and $f(-1) = 2(-1) + 3 = 1$, which gives us the points $(1, 5)$ and $(-1, 1)$. The graph of this function is shown in Figure 13.2. [*Note:* This function could also have been graphed by finding the respective $x$- and $y$-intercepts. The reader should verify that they are at $(-\frac{3}{2}, 0)$ and $(0, 3)$, respectively.] ●

In general, the $x$-intercept for $y = f(x) = mx + b$ is $(-b/m, 0)$. Using functional notation, this means that $f(-b/m) = m(-b/m) + b = 0$. We say that $-b/m$ is a **zero** of this function. In general, zeros of a function are the values of $x$ for which $f(x) = 0$. In Example 13.2 the zero of $f(x) = 2x + 3$ is $-\frac{3}{2}$. If $A = 0$, then the linear function is $\{(x, y) \mid By + C = 0\} = \{(x, y) \mid y = -C/B\}$ or $y = f(x) = D$, where $D = -C/B$ is a real number. This particular linear equation is called a **constant function.** Its domain is $\{x \mid x \in \mathbb{R}\}$, and its range is $\{D\}$. The graph of this type of function is a straight line parallel to the $x$-axis.

**Example 13.3**
Graph $y = f(x) = 2$.

Solution. We need two distinct points. If we arbitrarily let $x = 1$, $f(1) = 2$ gives us the point $(1, 2)$. Similarly, if we let $x = -3$, $f(-3) = 2$ gives another point $(-3, 2)$. The graph of $f$ is shown in Figure 13.3. [*Note:* The graph of $f(x) = 2$ is a straight line parallel to the axis and $f(x) = 2$ has no zero.] ●

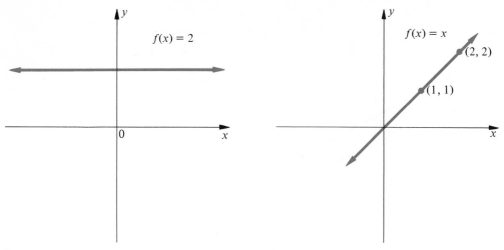

Figure 13.3                                   Figure 13.4

**Example 13.4**
Graph $f(x) = x$.

Solution.   We need two distinct points. We arbitrarily compute $f(1) = 1$ and $f(2) = 2$, to obtain the point $(1, 1)$ and $(2, 2)$, respectively. The graph of $f$ is shown in Figure 13.4. This function is called the **identity function.**  ●

Consider $\{(x, y) \mid Ax + By + C = 0,\ A \neq 0,\ B = 0\}$ or $\{(x, y) \mid Ax + C = 0\} = \{(x, y) \mid x = -C/A\}$. This subset of $\mathbb{R} \times \mathbb{R}$ is *not* a function since any two distinct ordered pairs such as $(-C/A, 1)$ and $(-C/A, 2)$ have the same first component, $x = -C/A$. The graph of this relation ($x = $ constant) is a straight line parallel to the $y$-axis.

**Example 13.5**
Graph $\{(x, y) \mid x - 1 = 0\} = \{(x, y) \mid x = 1\}$.

Solution.   We see that the ordered pairs $(1, 0)$ and $(1, 2)$ are elements of this set. More specifically, the domain is $\{1\}$ and the range $\{y \mid y \in \mathbb{R}\}$. The graph of this relation is shown in Figure 13.5. [*Note:* The graph of $\{(x, y) \mid x = 1\}$ is a straight line parallel to the $y$-axis and one unit to the right of it.]  ●

**Example 13.6**
Graph $\{(x, y) \mid y = 0\}$.

Solution.   Again, we only need two distinct points. Since $y = 0$, we arbitrarily select $(2, 0)$ and $(3, 0)$. The graph of $f$ is the $x$-axis and is shown in Figure 13.6.  ●

**Summary**
The graph of the solution set for $Ax + By + C = 0$ is always a straight line. If $B \neq 0$ and $A = 0$, then $By + C = 0$ is a straight line parallel to the $x$-axis and $y = -C/B$

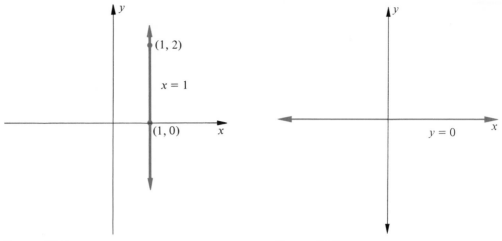

Figure 13.5                                    Figure 13.6

is a linear function. If $B = 0$ and $A \neq 0$, then $Ax + C = 0$ is a straight line parallel to the $y$-axis and $x = -C/A$ is *not* a linear function, but simply a *relation*.

### Remark 13.1

Geometrically the $x$-intercept for $f(x) = mx + b$ is the intersection of the graph of $f$ and the $x$-axis. When $f(x) = mx$, the $x$- and $y$-intercepts coincide and the graph passes through the origin. In this case the graph of $f(x) = mx$ can be determined by the origin and one additional point.

## EXERCISES FOR SECTION 13

1. Find the $x$- and $y$-intercepts and then graph each of the following.
   (a) $y = x + 2$         (d) $y = 3x - 2$
   (b) $y = 3 - x$         (e) $y = 4$
   (c) $y = 2x$            (f) $x = 3$

2. Graph each of the following.
   (a) $y = 2x + 1$                    (c) $y = 2x + 1, x \geq 2, y \leq 10$
   (b) $y = 2x + 1, x \geq 2$          (d) $y = 2x + 1, y \leq 0$

3. Suppose that $f(x) = 2x$. Find each of the following.
   (a) $f(2)$              (e) Does $f(2 + 3) = f(2) + f(3)$?
   (b) $f(3)$              (f) Does $f(2 \cdot 3) = f(2) \cdot f(3)$?
   (c) $f(2 + 3)$          (g) Does $f(2 \cdot 3) = 2 \cdot f(3)$?
   (d) $f(2) + f(3)$

4. Suppose that $f(x) = 3x + 2$
   (a) Show that $f(3x) \neq 3f(x)$
   (b) Show that $f(3x) = 3f(x) - 4$

5. Suppose that $f(x) = 5x$
      Show that $f(3x) = 3f(x)$

**6.** Suppose that $f(x) = 2x - 1$
   (a) Show that $f(3x + 1) \neq f(3x) + f(1)$
   (b) Show that $f(3x + 1) = f(3x) + 2f(1)$

**7.** Suppose that $f(x) = 2x$
   Show that $f(4x + 3) = f(4x) + f(3)$.

**8.** The velocity $V$ of an object thrown upward is a linear function of the time $t$ and is given by the equation $V = 160 - 32t$
   (a) Find $V$ when $t = 1$.
   (b) Find $V$ when $t = 3$.
   (c) Find $t$ when $V = 0$.
   (d) Sketch the graph of the function for $0 \leq t \leq 5$.

**9.** The pressure $P$ of a fixed volume of gas is a linear function of the temperature $T$ and is given by the equation $P = \dfrac{T}{4} + 80$.

   (a) Find $P$ when $T = 40$.
   (b) Find $T$ when $P = 100$.
   (c) Sketch the graph of this function for $40 \leq T \leq 80$.

**10.** Suppose that $f(x) = mx, m \in \mathbb{R}$. Show that $f(a + b) = f(a) + f(b)$, and $f(ab) = af(b)$.

**11.** Suppose that $f(x) = mx + c$. Does $f(a + b) = f(a) + f(b)$? Does $f(ab) = af(b)$?

# SECTION 14
## SYSTEMS OF LINEAR EQUATIONS

Since $\{(x, y)|Ax + By + C = 0\}$ is a relation in $R$ (that is, a subset of $\mathbb{R} \times \mathbb{R}$), we can investigate sets formed from the intersection of such sets. The elements of the set $A = \{(x, y)|x + y - 1 = 0\} \cap \{(x, y)|x + 2y - 3 = 0\}$ are all the ordered pairs $(x, y)$ that satisfy both the equation $x + y - 1 = 0$ and $x + 2y - 3 = 0$. We also say that the set $A$ is the solution set of the **system of linear equations**

$$x + y - 1 = 0$$
$$x + 2y - 3 = 0$$

The general form of such a system involving two variables is

$$a_1x + b_1y + c_1 = 0$$
$$a_2x + b_2y + c_2 = 0$$

The solution set for this system is

$$\{(x, y)|a_1x + b_1y + c_1 = 0\} \cap \{(x, y)|a_2x + b_2y + c_2 = 0\}$$

   The techniques used in solving a system of two linear equations in two variables employ methods that reduce the original system to a single equation in one unknown. One such method is **substitution.** We shall use this method to solve one of the equations for one variable in terms of the other and then substitute the result into the second equation. This procedure yields a single equation in one variable. The following examples illustrate this method.

**Example 14.1**
Find the solution set for the system

$$3x - y + 6 = 0$$
$$2x + 3y - 7 = 0$$

Solution.   We first solve for $y$ in the equation $3x - y + 6 = 0$ to obtain $y = 3x + 6$. Next we substitute this result in the second equation to obtain

$$2x + 3(3x + 6) - 7 = 0$$

or

$$11x + 11 = 0$$

or

$$x = -1$$

We now substitute the value $x = -1$ in the first equation $y = 3x + 6$ to obtain $y = -3 + 6 = 3$. Thus the solution set for this system is the ordered pair $(-1, 3)$. In set terminology we have $\{(x, y) | 3x - y + 6 = 0\} \cap \{(x, y) | 2x + 3y - 7 = 0\} = \{(-1, 3)\}$. We say that this system is composed of **consistent equations.**   • *Lines intersect*

**Example 14.2**
Find the solution set for the system

$$x + y - 1 = 0$$
$$x + y - 2 = 0$$

Solution.   We first solve for $y$ in the first equation to obtain $y = -x + 1$. Next we substitute this result into the second equation to obtain

$$x + (-x + 1) - 2 = 0$$

or

$$-1 = 0$$

Since $-1 \neq 0$, we must conclude that there are no ordered pairs that satisfy this system. In set terminology we have

$$\{(x, y) | x + y - 1 = 0\} \cap \{(x, y) | x + y - 2 = 0\} = \emptyset$$

We say that this system is composed of **inconsistent equations.**   •

**Example 14.3**
Find the solution set for the system

$$x + y + 1 = 0$$
$$2x + 2y + 2 = 0$$

Solution. We first solve for $x$ in the first equation to obtain $x = -y - 1$, and then we substitute this result into the second equation to obtain

$$2(-y - 1) + 2y + 2 = 0$$

or

$$0 = 0$$

Since $0 = 0$ is always true, we can conclude that any ordered pair that satisfies the equation $x + y + 1 = 0$ also satisfies the equation $2x + 2y + 2 = 0$. Thus the solution set for this system is $\{(x, y) \mid x + y + 1 = 0\}$ or $\{(x, y) \mid 2x + 2y + 2 = 0\}$. We say that this system is composed of **dependent equations.** ●

The three previous examples lead us to the conclusion that one and only one of the following is true for any given system of linear equations in two unknowns.

*Case 1.* The solution set contains one and only one ordered pair. See Example 14.1.
*Case 2.* The solution set is the empty (null) set. See Example 14.2.
*Case 3.* The solution set contains all those ordered pairs contained in either one of the given equations. See Example 14.3.

Since the graphs of $a_1 x + b_1 y + c_1 = 0$ and $a_2 x + b_2 y + c_2 = 0$ are straight lines, we can give a geometric interpretation to the possible solution sets for a given system.

1. The graphs of the two lines intersect at a unique point.
2. The graphs of the two lines are parallel and distinct (that is, there is no point of intersection).
3. The graphs of the two lines are the *same.*

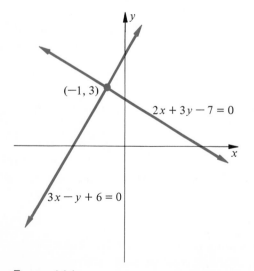

Figure 14.1

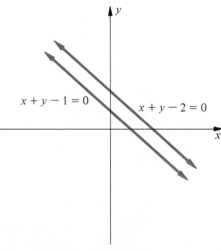

Figure 14.2

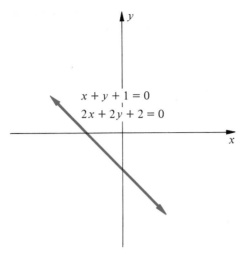

$$x + y + 1 = 0$$
$$2x + 2y + 2 = 0$$

Figure 14.3

The graphs of the systems of equations given in Examples 14.1, 14.2, and 14.3, respectively, illustrate these geometric possibilities and are shown in Figures 14.1, 14.2, and 14.3.

A second method for solving a system of two linear equations in two variables is the method of **elimination.** In this method we multiply each of the original equations by numbers that will make the coefficients of one of the variables numerically equal. We then subtract the two new equations to obtain one equation in one variable. We illustrate this method with the following example.

**Example 14.4**
Find the solution set for the system

$$3x - 5y = 10$$
$$4x - 3y = 6$$

Solution.   Here we seek to eliminate the variable $y$. To do this we will multiply the first equation by 3 and the second equation by 5 to obtain the equivalent system

$$9x - 15y = 30$$
$$20x - 15y = 30$$

We now subtract one equation from the other in this new system to obtain the equation

$$-11x = 0$$

or

$$x = 0$$

We now can solve for $y$ by substituting the value $x = 0$ into either of the equations in the original system. We find that $3(0) - 5y = 10$ or $y = -2$ when we substitute $x = 0$ into the first equation of the original system. It should be noted that we might also solve for $y$ by multiplying the first and second equations of the original system by 4 and 3, respectively, and then subtracting. In any case the solution set is $\{(0, -2)\}$. ●

The method of elimination can be extended to solving three linear equations in three variables. In this situation we first eliminate one variable from any two of the equations and then eliminate the same variable from any other pair of equations. The two results then form a system of two equations in two variables and we repeat the process.

**Example 14.5**
Find the solution set for the system

$$3x + 2y - z = -1$$
$$-2x + y - 2z = -1$$
$$x + y - z = 0$$

Solution. We first eliminate the variable $z$ by subtracting the third equation from the first equation to obtain

$2x + y = -1$ **(A)**

Next we eliminate the variable $z$ by multiplying the first equation by 2 and then subtracting the second equation from it to obtain

$8x + 3y = -1$ **(B)**

We now multiply (A) by 3 and subtract it from (B) to obtain

$2x = 2$    or    $x = 1$

We find that $y = -3$ by substituting $x = 1$ into **(B)**. Finally, we find that $z = -2$ by substituting the value $x = 1$ and $y = -3$ into the first equation of the original system. We conclude that the solution set for this system consists of the **ordered triple** $(1, -3, -2)$. ●

## EXERCISES FOR SECTION 14

Find the solution sets for each of the following systems.

**1.** $x - y = 0$
$\phantom{x}x + 3 = 0$

**2.** $\phantom{x}x + y - 2 = 0$
$2x - 2y + 6 = 0$

**3.** $x - y + 6 = 0$
$\phantom{x}x - y + 5 = 0$

**4.** $\phantom{x}x + 2y - 1 = 0$
$2x + 4y - 2 = 0$

**5.** $\frac{1}{4}x - \frac{1}{3}y = -\frac{3}{12}$
$\frac{1}{10}x + \frac{2}{5}y = \frac{2}{5}$

**6.** $2y - x = 1$
$2x + y = 8$

**7.** $2x - y + 2z = 6$
$\phantom{2x - }y + z = 4$
$\phantom{2x - y + 2}z = 1$

**8.** $2x - y + 2z = -8$
$\phantom{2}x + 2y - 3z = 9$
$3x - y - 4z = 3$

**9.** Find the linear equation $f(x) = ax + b$ when $f(2) = 1$ and $f(4) = -2$.

**10.** Find the values of $a$ and $b$ so that the graph $ax + by + 2 = 0$ passes through the points $(1, 2)$ and $(2, 0)$.

11. The equation of a circle is of the form $x^2 + y^2 + Ax + By + C = 0$. Find the equation of the circle that contains the points $(5, 3)$, $(6, 2)$, and $(3, -1)$.

12. The equation of a parabola whose axis is parallel to the $x$-axis is of the form $y^2 + Ay + Bx + C = 0$. Find the equation of the parabola that contains the points $(-1, 2)$, $(-3, 6)$, and $(-3, -2)$.

13. Suppose the temperature $T°$ above the surface of the Earth is assumed to be a linear function of the height $h$ (feet) above the surface of the Earth. If the temperature on the surface of the Earth is $72°$, and the temperature at 3200 feet is $62°$, find the temperature at 6400 feet.

14. Find the linear relationship between Fahrenheit temperature, $F$, and centigrade temperature, $C$, of any body. (*Hint:* Let $C = aF + b$ and recall that $F = 32$ when $C = 0$ and that $F = 212$ when $C = 100$.) Express $F$ as a linear equation of $C$.

15. Find the coordinates of the vertices of the triangle formed by the lines $3x + 4y = 5$, $6x + y = 17$, and $x - y = -3$.

## SECTION 15
## QUADRATIC FUNCTIONS

The general quadratic function is

$$f(x) = ax^2 + bx + c \qquad (15.1)$$

where $a$, $b$, and $c$ are constant real numbers and $a \neq 0$. The graph of the function $f = \{(x, y) \mid y = ax^2 + bx + c\}$ is the same as the graph of the equation $y = ax^2 + bx + c$. The graph of $f$ is called a **parabola.**

### Example 15.1
Sketch the graph of $f = \{(x, y) \mid y = x^2 - 2x - 3\}$.

Solution. Since the graph of $f$ is the same as the graph of the equation $y = x^2 - 2x - 3$ we can find a "sufficient" number of points on the graph by substituting values of $x$ into the equation $y = x^2 - 2x - 3$. For example, when $x = 1$ then $y = -4$. Thus the point $(1, -4)$ is on the graph. Table 15.1 shows the points that were used to sketch the graph shown in Figure 15.1. ●

### Table 15.1

| $x$ | $-2$ | $-1$ | $0$ | $1$ | $2$ | $3$ | $4$ |
|-----|------|------|-----|-----|-----|-----|-----|
| $y$ | $5$  | $0$  | $-3$ | $-4$ | $-3$ | $0$ | $5$ |

We now wish to develop some properties of the graph of the quadratic function that will enable us to sketch its graph while minimizing the use of table values such as in Example 15.1.

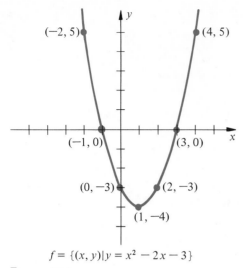

$$f = \{(x, y)|y = x^2 - 2x - 3\}$$

Figure 15.1

We can find the $y$-intercept for $y = ax^2 + bx + c$ by setting $x = 0$ to obtain $y = c$. To find its $x$-intercept we set $y = 0$ and we have

$$0 = ax^2 + bx + c \tag{15.2}$$

This equation is quadratic in the variable $x$. The solutions (if any) to (15.2) are the $x$-intercepts or **the real zeros of the function.**

We can solve (15.2) by using the following theorem.

**Theorem 15.1**  The solutions to the quadratic equation $ax^2 + bx + c = 0$, $a \neq 0$ are given by the formula

$$x = \frac{-b \pm \sqrt{b^2 - 4ac}}{2a} \tag{15.3}$$

*← Write on note card*

where (15.3) is called the **quadratic formula** and $b^2 - 4ac$ is called the **discriminant** of the quadratic equation.

We will defer the derivation of this formula until we have completed the discussion of the properties of the graph of the quadratic function.

Since $a \neq 0$, the quadratic equation $y = ax^2 + bx + c$ can be written as

$$y = a\left[x^2 + \frac{b}{a}x + \frac{c}{a}\right]$$

Now if we add and then subtract the term $\left(\dfrac{b}{2a}\right)^2$ inside the brackets we obtain

$$y = a\left[\left(x^2 + \frac{b}{a}x + \frac{b^2}{4a^2}\right) + \left(\frac{c}{a} - \frac{b^2}{4a^2}\right)\right]$$
$$y = a\left[\left(x + \frac{b}{2a}\right)^2 + \left(\frac{4ac - b^2}{4a^2}\right)\right]$$

or

$$y = a\left(x + \frac{b}{2a}\right)^2 - \left(\frac{b^2 - 4ac}{4a}\right) \tag{15.4}$$

From 15.4 we see that if $a > 0$, then the first term on the right equals 0 when $x = -\dfrac{b}{2a}$ and otherwise this term is positive. Thus the **lowest** point on the graph is

$$\left(-\frac{b}{2a}, -\frac{b^2 - 4ac}{4a}\right) \quad \text{or} \quad \left(-\frac{b}{2a}, \frac{4ac - b^2}{4a}\right) \tag{15.5}$$

Similarly, if $a < 0$, then the first term on the right equals 0 when $x = -\dfrac{b}{2a}$, and otherwise this term is negative. The **highest** point on the graph is

$$\left(-\frac{b}{2a}, -\frac{b^2 - 4ac}{4a}\right) \quad \text{or} \quad \left(-\frac{b}{2a}, \frac{4ac - b^2}{4a}\right) \tag{15.6}$$

The high or low point for $y = ax^2 + bx + c$ is called the **vertex** of the parabola.

The preceding discussions lead us to the following properties.

**Properties of the graph of** $f = \{(x, y) \mid y = ax^2 + bx + c\}$

1. The $x$-intercepts of the graph are the real solutions to (15.2) and can be found by using the quadratic formula

$$x = \frac{-b \pm \sqrt{b^2 - 4ac}}{2a}$$

2. The vertex of the parabola is located at $\left(-\dfrac{b}{2a}, \dfrac{4ac - b^2}{4a}\right)$

3. If $a > 0$, then the parabola opens upward and the vertex is a **minimum point.** If $a < 0$, then the parabola opens downward and the vertex is a **maximum point.**

### Example 15.2

Locate the $x$-intercepts, the vertex, and draw the graph of $f = \{(x, y) \mid y = -x^2 + x + 2\}$.

**Solution.** To find the $x$-intercepts we set $y = 0$ to obtain the quadratic equation

$$-x^2 + x + 2 = 0$$

We can now apply the quadratic formula using $a = -1$, $b = 1$, and $c = 2$ to obtain

$$x = \frac{-1 \pm \sqrt{(1)^2 - 4(-1)(2)}}{2(-1)} = \frac{-1 \pm \sqrt{9}}{-2}$$

Thus

$$x = \frac{-1 + 3}{-2} = -1 \quad \text{and} \quad x = \frac{-1 - 3}{-2} = 2$$

The $x$-intercepts are $-1$ and $2$. The vertex has $x = -\dfrac{b}{2a} = \dfrac{-1}{2(-1)} = \dfrac{1}{2}$ for

its $x$-coordinate and $\dfrac{4ac - b^2}{4a} = \dfrac{4(-1)(2) - 1^2}{4(-1)} = \dfrac{9}{4}$ for its $y$-coordinate.

We see that $a = -1 < 0$; thus the graph of $f$ is a parabola that opens *downward*. We show a sketch of this parabola in Figure 15.2. From the graph, we note that the range is $\{y \mid y \le \frac{9}{4}\}$.  •

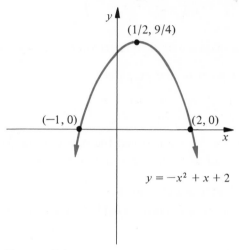

Figure 15.2

### Example 15.3

Locate the vertex and the $x$-intercepts (if any) for $f = \{(x, y) \mid y = x^2 + x + 1\}$, and sketch the graph of $f$.

**Solution.** The $x$-intercepts will be the real roots of the quadratic equation $x^2 + x + 1 = 0$. Since its discriminant is $b^2 - 4ac = 1^2 - 4(1)(1) < 0$, we have *no real* roots. Thus we have *no x*-intercepts. Geometrically, this means

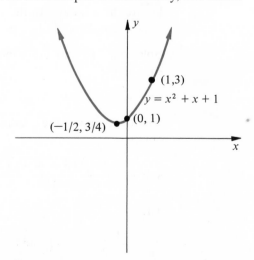

**Table 15.2**

| $x$ | 0 | 1 |
|-----|---|---|
| $y$ | 1 | 3 |

Figure 15.3

that the parabola neither cuts nor touches the $x$-axis. The vertex is located at the point $\left(-\frac{b}{2a}, \frac{4ac - b^2}{4a}\right) = \left(-\frac{1}{2}, \frac{3}{4}\right)$. Since $a = 1 > 0$, the parabola breaks upward. Thus the range is $\left\{y \,|\, y \geq \frac{3}{4}\right\}$. We draw this parabola by using the vertex and the values from Table 15.2; see Figure 15.3. ●

**Proof of Theorem 15.1**  We can reexpress $ax^2 + bx + c = 0$, $a \neq 0$, as

$$x^2 + \left(\frac{b}{a}\right)x + \frac{c}{a} = 0$$

by dividing by $a$ or

$$x^2 + \left(\frac{b}{a}\right)x = -\frac{c}{a}$$

We then add $\left[\frac{1}{2}\left(\frac{b}{a}\right)\right]^2 = \frac{b^2}{4a^2}$ to both sides to obtain

$$x^2 + \left(\frac{b}{a}\right)x + \frac{b^2}{4a^2} = \frac{b^2}{4a^2} - \frac{c}{a}$$

Since the left-hand member is now a perfect square, we have

$$\left[x + \left(\frac{b}{2a}\right)\right]^2 = \frac{b^2 - 4ac}{4a^2}$$

We next extract the square root to obtain

$$x + \left(\frac{b}{2a}\right) = \pm\sqrt{\frac{b^2 - 4ac}{4a^2}}$$

or

$$x = \frac{-b \pm \sqrt{b^2 - 4ac}}{2a} \qquad ●$$

From (15.3) we observe that the real zeros of the quadratic functions are the real solutions to the quadratic equation. We also note $x$ will be a real number only if the discriminant $b^2 - 4ac$ is nonnegative. Thus we have the following possibilities for the real zeros of the function.

*Case 1.* If $b^2 - 4ac > 0$, then we have the two distinct real zeros

$$x = \frac{-b + \sqrt{b^2 - 4ac}}{2a} \quad \text{and} \quad x = \frac{-b - \sqrt{b^2 - 4ac}}{2a}$$

*Case 2.* If $b^2 - 4ac = 0$, then we have *one* real zero, $x = -b/2a$.
*Case 3.* If $b^2 - 4ac < 0$, then we have no real zeros.

### Example 15.4

Locate the $x$-intercepts, the vertex, and then sketch the graph of $f = \{(x, y) \,|\, y = x^2 - 4x + 4\}$.

**Solution.** We can find the $x$-intercepts by setting $y = 0$ and then solving for $x$ as follows.

$$x^2 - 4x + 4 = 0$$
$$(x - 2)(x - 2) = 0$$

Since 2 is the solution to this equation we say 2 is the $x$-intercept. We can find the $y$-intercept by setting $x = 0$ and obtaining $y = 4$. We can locate the coordinates of the vertex by noting that when $a = 1$, $b = -4$, we have $x = -b/2a = 4/2 = 2$. Since 2 is also the $x$-intercept we note that the vertex is located at $(2, 0)$. Finally, since $a > 0$ we have the parabola opening upward. We use the above information to sketch the graph shown in Figure 15.4.

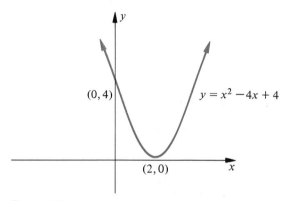

Figure 15.4

**Remark 15.1**

It is much easier to calculate the $y$ coordinate of the vertex by simply substituting the value of the $x$ coordinate into the original equation. This eliminates memorizing the formula for the $y$ coordinate.

**Remark 15.2**

The algebraic manipulation we used to develop Formula (15.4) and to prove Theorem 15.1 is called **completing the square.** This manipulation is extremely useful in mathematics and is further discussed in Chapter 10.

The question of finding the maximum or minimum value for a function $y = f(x)$ frequently arises in applications. We have observed that a quadratic function will have a maximum or minimum value equal to the value of the $y$ coordinate of the vertex when $a < 0$ or $a > 0$ respectively. Thus we can solve any problem concerning a maximum or minimum value if the function describing the quantity in question is quadratic.

**Example 15.5**

If the population of a bacteria culture at time $t$ minutes is $P = f(t) = 200 + 4t - 0.04t^2$, find (a) the time $t$ when the culture will be at its maximum size; (b) the maximum size of the culture.

Solution. We observe that $P = 200 + 4t - 0.04t^2$ is a quadratic function of $t$ where $a = -0.04$, $b = 4$ and $c = 200$. Since $a < 0$, the vertex of $P$ will be a maximum point. We calculate the coordinates of this vertex as follows.

$$t = -\frac{b}{2a} = -\frac{(4)}{2(-0.04)} = 50$$

when $t = 50$ we have

$$P = f(50) = 200 + 4(50) = (0.04)(50)^2$$

or

$$P = 200 + 200 - 100 = 300$$

Thus the culture reaches a maximum value of 300 after 50 minutes. ●

# EXERCISES FOR SECTION 15
*Odd Numbers*

In exercises 1 through 10, solve for $x$ by using the quadratic formula (15.3).

**1.** $4x^2 - 12x + 7 = 0$    **6.** $x^2 + 4x + 3 = 0$

**2.** $4 - 6x + x^2 = 0$    **7.** $x^2 - 2x - 5 = 0$

**3.** $x^2 + x + 1 = 0$    **8.** $x^2 + 3x + 1 = 0$

**4.** $12x - 9x^2 = 5$    **9.** $y = \dfrac{(x-1)^2}{x^2}$

**5.** $x^2 + 2x - 11 = 0$    **10.** $y^2 = \dfrac{x^2}{x-2}$

In exercises 11 through 20, find the $x$-intercepts (if any), the $y$-intercepts, the coordinates of the vertex, and then sketch the graph of the given quadratic functions. Use the graph to state the range of each.

**11.** $y = x^2 + 2x + 1$    **16.** $y = x^2 + 4x + 1$

**12.** $y = 2 - 3x - x^2$    **17.** $y = (x-1)^2 + 2$

**13.** $y = 4 - x^2$    **18.** $y = (x+2)^2 - 3$

**14.** $y = x^2 + x + 1$    **19.** $y - 3 = -2(x-1)^2$

**15.** $y = 15 + 2x - x^2$    **20.** $y = 2x^2 - 6x + 7$

**21.** Find the equation of the quadratic function $f(x) = ax^2 + bx + c$ if $f(0) = 4$, $f(1) = 0$, and $f(3) = 0$.

**22.** Find the equation of the quadratic function $f(x) = ax^2 + bx + c$ if $f(2) = 3$, $f(1) = 2$, and $f(0) = 3$.

**23.** (a) Graph $y = -x^2 - 1$.
   (b) How many times does the graph of $y = -x^2 - 1$ cut or touch the $x$-axis?
   (c) How many times does the graph of $y$ in part (a) cut or touch the $x$-axis if we add 1 to all values of $y$?
   (d) How many times does the graph of $y$ in part (a) cut or touch the $x$-axis if we add 3 to all values of $y$?

**24.** Find the value of $b$ so that the graph of $y = 8x^2 + bx + 8$ will have its vertex on the line $y = 4$.

**25.** Find the dimensions of the rectangle with a perimeter of 20 inches and a maximum area.

**26.** If an object is thrown vertically upward with a velocity of 32 ft/sec, its height $y$ after $t$ seconds is given by the equation $y = 32t - 16t^2$. At what time $t$ will the object be at its highest point, and how high will it be? (*Hint:* Examine the vertex of $y = 32t - 16t^2$.)

**27.** Show that a rectangle with a fixed perimeter and a maximum area is a square.

**28.** A company's profits in producing $x$ units is $P = f(x) = 3000 + 40x - 0.05x^2$. How many units should be produced to maximize the profits?

**29.** A company has found that its total cost to produce $x$ units is $C = f(x) = 1500 - 80x + 0.04x^2$. How many units should be produced to minimize the cost?

**30.** A manufacturer offers to deliver 100 lamps to a dealer at a cost of $20 per lamp. He offers to reduce the price per lamp on the entire order by $0.05 for each additional lamp over 100. What is the manufacturer's largest possible gross revenue?

**31.** A rectangular pen is to be made out of 120 feet of fencing. If one side of the pen is the side of a barn, what should the lengths of the other three sides be to assure the largest possible area?

**32.** A wire 40 inches long is cut into two pieces. One piece is bent into the shape of a circle; the other, into the shape of a square. Where should the wire be cut in order that the sum of the areas of the circle and the square be a minimum?

**33.** A rifle bullet is fired directly upward from the ground. Its distance $S$ (in feet) from the ground is the quadratic function $S = 1280t - 16t^2$ where $t$ is time in seconds. How long does it take the bullet to reach its maximum height and what is the maximum height?

SECTION **16**
## SYSTEMS OF NONLINEAR EQUATIONS

Any system of equations containing a nonlinear equation is said to be a nonlinear system of equations. The solution set of a nonlinear system is the intersection of the solution sets of its equations. One nonlinear system that we will frequently encounter is the system that contains one linear and one quadratic equation in two variables. The most efficient method of solution for such systems is that of substitution. We illustrate this method with the following examples.

**Example 16.1**
Solve the system

$$x - 2y = 10$$
$$y = x^2 + 2x - 15$$

Solution. $x - 2y = 10$ can be rewritten as $2y = x - 10$ or $y = \frac{1}{2}x - 5$. We can then substitute $y = \frac{1}{2}x - 5$ into the equation $y = x^2 + 2x - 15$ to obtain

$$\frac{1}{2}x - 5 = x^2 + 2x - 15$$
$$x^2 + \frac{3}{2}x - 10 = 0$$

$$2x^2 + 3x - 20 = 0$$
$$(2x - 5)(x + 4) = 0$$

Thus $x = \frac{5}{2}$ or $x = -4$. We now substitute these values of $x$ into $y = \frac{1}{2}x - 5$ to obtain their respective $y$-values. When $x = \frac{5}{2}$, then $y = \frac{1}{2}(\frac{5}{2}) - 5$ or $y = \frac{-15}{4}$. When $x = -4$, then $y = \frac{1}{2}(-4) - 5$ or $y = -7$. Thus the solution set for the system is $\{(\frac{5}{2}, \frac{-15}{4}), (-4, -7)\}$. ●

### Example 16.2
Solve the system

$$y^2 = 9 - x$$
$$y = x - 7$$

Solution.   Here we can substitute $y = x - 7$ into the equation $y^2 = 9 - x$ to obtain

$$(x - 7)^2 = 9 - x$$
$$x^2 - 14x + 49 = 9 - x$$
$$x^2 - 13x + 40 = 0$$
$$(x - 8)(x - 5) = 0$$

Thus $x = 8$ or $x = 5$. We now substitute these values of $x$ into the equation $y = x - 7$ to obtain the respective values of $y$. When $x = 8$, then $y = 8 - 7$ or $y = 1$. When $x = 5$, then $y = 5 - 7$ or $y = -2$. Thus the solution set for this system is $\{(8, 1), (5, -2)\}$. ●

### Example 16.3
Solve the system

$$y = x^2 - 4x + 7$$
$$y = 3x - 8$$

Solution.   Here we can substitute $y = 3x - 8$ into the equation $y = x^2 - 4x + 7$ to obtain

$$3x - 8 = x^2 - 4x + 7$$
$$x^2 - 7x + 15 = 0$$

Since the discriminant of $x^2 - 7x + 15$ is $(-7)^2 - 4(1)(15) = -11 < 0$, there are *no* real values for $x$. Thus the solution set for this system is $\emptyset$. ●

### Example 16.4
Solve the system

$$2x^2 - y^2 = 14$$
$$x - y = 1$$

Solution.   $x - y = 1$ can be rewritten as $y = x - 1$ and then substituted into the equation $2x^2 - y^2 = 14$ to obtain

$$2x^2 - (x - 1)^2 = 14$$
$$2x^2 - x^2 + 2x - 1 = 14$$

$$x^2 + 2x - 15 = 0$$
$$(x + 5)(x - 3) = 0$$

Thus $x = -5$ or $x = 3$. When we substitute these values into $y = x - 1$, we obtain the solution set $\{(-5, -6), (3, 2)\}$.   ●

## EXERCISES FOR SECTION 16

Odd Numbers

Find the solution set for each of the following systems of equations.

**1.** $y = x^2$
    $y = x$

**2.** $y = 9 - x^2$
    $y = 5$

**3.** $y = x^2 - x - 6$
    $y = 2x - 8$

**4.** $y = x^2 - 2x + 4$
    $y = x$

**5.**    $y = x^2 - 2x + 1$
    $y + x = 3$

**6.** $y = 6x - x^2$
    $y = x^2 - 2x$

**7.**    $x = y^2$
    $x^2 = -8y$

**8.** $x = 2y^2 - 4$
    $x = y^2$

**9.** $y = x^2 + 2x - 3$
    $y = 1 - x^2$

**10.** $y = x^2 + 2x - 3$
    $y = x^2 + x - 2$

**11.** $x^2 + y^2 = 1$
    $y - x = 0$

**12.** $x^2 + y^2 = 9$
    $x^2 - y^2 = 9$

**13.**    $xy = 1$
    $x + y = 2$

**14.** $y = x - x^3$
    $y = x^3 - x^2 - 2x$

**15.** $y^2 = x$
    $y = 2x$

**16.** $x^2 - y^2 = 1$
    $y + x = 0$

**17.** $x^2 - y^2 = 1$
    $x^2 + y^2 = 9$

**18.** $3x = 8 - 2y - y^2$
    $y = -\frac{3}{5}x$

**19.** $y^2 = x^3$
    $x = 4$

**20.** $x + y = 1$
    $-xy = 12$

**21.** Find the dimensions of a rectangle if its diagonal is 17 inches and its perimeter is 46 inches.

**22.** The circle $x^2 + y^2 = a^2$ and the straight line $y = mx + b$ will intersect at two points, be tangent, or not intersect. Find the value of $b$ in terms of $a$ and $m$ so that the straight line will be tangent to the circle.

**23.** The sum of two numbers is 16 and their product is 63. Find the numbers.

**24.** A rectangle has a perimeter of 40 inches and an area of 96 square inches. Find its dimensions.

**25.** The diagonal of a rectangle is 85 inches. If the length of the rectangle is decreased by 7 inches and the width is increased by 11 inches, the length of the diagonal remains the same. Find the dimensions of the rectangle.

**26.** At a constant temperature the pressure $P(\text{lb/in}^2)$ and the volume $V(\text{in}^3)$ are related by the formula $PV = K$. The product of the pressure and the volume of a certain gas is 48 in-lb. If the temperature remains constant as the pressure is decreased by $4\,\text{lb}/\text{in}^2$, the volume is increased by $2\,\text{in}^3$. What are the original pressure and volume of this gas?

27. A rectangular piece of sheet metal has an area of 300 in². A 3 in square is cut from each corner and open box is formed by turning the ends and sides. If the volume of the box is 378 in³, what are the dimensions of the piece of tin?

28. The bottom of a 13 foot ladder that is leaning against a wall is 5 ft from the wall. If the bottom of the ladder is pulled out 2 ft farther, find how far the top end moves down the wall.

SECTION **17**
## SOME SPECIAL FUNCTIONS AND THEIR GRAPHS

The study of the graphs of functions plays a vital role in both mathematics and the natural sciences. The mathematician uses the graph to provide a geometric visualization for many types of problems. The scientist uses the graphs in an attempt to discover a natural law or pattern for data obtained during experimentation.

Earlier in this chapter we saw that the graphs of the linear and the quadratic functions were lines and parabolas, respectively. We now wish to investigate the graphs of some special functions that are formed from pieces or parts of the linear and quadratic functions. These special functions might well be called **"piecewise" functions.** The graphs of most of these piecewise functions will be completely connected, as were the graphs of the linear and quadratic functions. Functions whose graphs are completely connected are said to be **continuous functions.** Intuitively, a function is continuous if its graph can be traced out by a pencil point without ever having to lift the point.

(*Note:* a detailed study of this concept is left for the branch of mathematics known as calculus)

We now introduce some examples that typify piecewise functions.

**Example 17.1**
Graph the function

$$y = f(x) = \begin{cases} x, & x \geq 0 \\ -x, & x < 0 \end{cases}$$

Solution. The domain of $f$ is $\{x \mid x \in \mathbb{R}\}$. The function can be described as

$$f = \{(x, y) \mid y = x, x \geq 0\} \cup \{(x, y) \mid y = -x, x < 0\}$$

The graphs of $\{(x, y) \mid y = x, x \geq 0\}$ and $\{(x, y) \mid y = -x, x < 0\}$ are parts of the straight lines $y = x$ and $y = -x$, respectively. The graph of $f$ is obtained by combining the two parts of these straight lines and is shown in Figure 17.1 on page 74. We note that the range of $f$ is $\{y \mid y \geq 0\}$, and that $f$ is continuous. Another description for this function is $y = f(x) = |x|$. ●

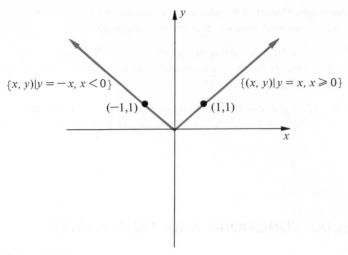

$\{x, y)|y = -x, x < 0\}$

$\{(x, y)|y = x, x \geqslant 0\}$

$(-1,1)$    $(1,1)$

Figure 17.1

**Example 17.2**
Graph the function

$$y = f(x) = \begin{cases} x^2 - 4x + 7, & x \geq 2 \\ x + 1, & x < 2 \end{cases}$$

Solution. The domain of $f$ is $\{x \mid x \in \mathbb{R}\}$. The function can be described as

$$f = \{(x, y)|y = x^2 - 4x + 7, x \geq 2\} \cup \{(x, y)|y = x + 1, x < 2\}$$

The graph of $\{(x, y)|y = x^2 - 4x + 7, x \geq 2\}$ is a part of the parabola whose vertex is $(2, 3)$ and which opens upward. The graph of $\{(x, y)|y = x + 1, x < 2\}$ is a part of the straight line whose $x$ and $y$ intercepts are $(-1, 0)$ and $(0, 1)$, respectively. Thus the graph of $f$ will be the combination of the two parts. This graph is shown in Figure 17.2. We see that $f$ is a continuous function whose range is $\{y \mid y \in \mathbb{R}\}$.  ●

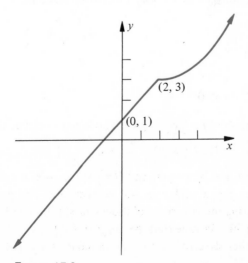

$(2, 3)$

$(0, 1)$

Figure 17.2

**Example 17.3**

Graph the function

$$y = f(x) = \begin{cases} \frac{2}{3}x + 2, & x \le 3 \\ 4, & 3 < x \le 6 \\ -2x + 16, & 6 < x \le 8 \end{cases}$$

Solution. The domain of $f$ is $\{x \mid x \le 8\}$. The function can be described as

$$f = \{(x, y) \mid y = \tfrac{2}{3}x + 2, x \le 3\} \cup \{(x, y) \mid y = 4, 3 < x \le 6\} \cup$$
$$\{(x, y) \mid y = -2x + 16, 6 < x \le 8\}$$

The graphs of the three sets are parts of the straight lines $y = \frac{2}{3}x + 2, y = 4$, and $y = -2x + 16$, respectively. Thus the graph of $f$ is obtained by combining these three parts of straight lines. This graph is shown in Figure 17.3. We see that $f$ is a continuous function whose range $\{y \mid y \le 4\}$. ●

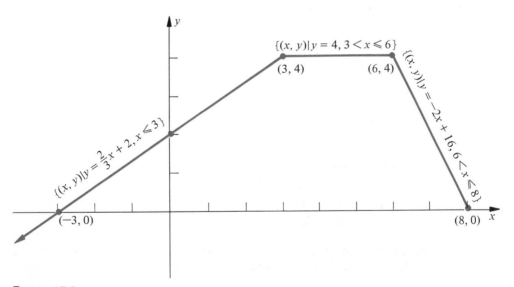

Figure 17.3

**Example 17.4**

Graph the function

$$y = f(x) = \begin{cases} 1, & x > 0 \\ -1, & x < 0 \end{cases}$$

Solution. The domain of $f$ is $\{x \mid x \in \mathbb{R}, x \ne 0\}$. The function can be described as

$$f = \{(x, y) \mid y = 1, x > 0\} \cup \{(x, y) \mid y = -1, x < 0\}$$

The graphs of $\{(x, y) \mid y = 1, x > 0\}$ and $\{(x, y) \mid y = -1, x < 0\}$ are parts of the straight lines $y = 1$ and $y = -1$, respectively. The graph of $f$ is obtained by combining these two parts and is shown in Figure 17.4. The range of $f$ is $\{1, -1\}$ and $f$ is *not* a continuous function. Another description of $f$ is $y = f(x) = \dfrac{|x|}{x}$. ●

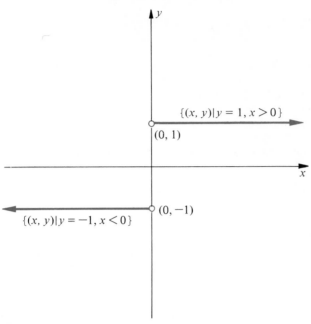

Figure 17.4

**Example 17.5**

Graph the function

$$y = f(x) = \begin{cases} -2, & 0 \le x < 2 \\ 2, & 2 \le x < 4 \\ 4, & 4 \le x < 6 \end{cases}$$

Solution.   The domain of $f$ is $\{x \mid 0 \le x < 6\}$ and the range of $f$ is $\{-2, 2, 4\}$. The function can be described as

$$f = \{(x, y) \mid y = -2, 0 \le x < 2\} \cup \{(x, y) \mid y = 2, 2 \le x < 4\} \cup$$
$$\{(x, y) \mid y = 4, 4 \le x < 6\}$$

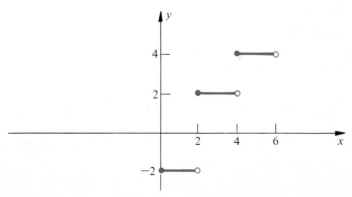

Figure 17.5

The graphs of the three sets are parts of the straight lines $y = -2, y = 2$, and $y = 4$, respectively. The graph of $f$ is obtained by combining these three parts and is shown in Figure 17.5.  ●

**Example 17.6**
Graph the function $y = |x + 1| - |2x|$.

Solution.   We use the results of example 7.3 to note that

$$y = |x + 1| - |2x| = \begin{cases} 1 - x & \text{if } x \geq 0 \\ 3x + 1 & \text{if } -1 < x < 0 \\ x - 1 & \text{if } x \leq -1 \end{cases}$$

Thus the function can be described as

$$\{(x, y) \mid y = 1 - x, x \geq 0\} \cup \{(x, y) \mid y = 3x + 1, -1 < x < 0\} \cup \{(x, y) \mid y = x - 1, x \leq -1\}.$$

The graphs of the three sets are parts of the straight lines $y = 1 - x, y = 3x + 1$, and $y = x - 1$, respectively. Thus the graph of the function is obtained by combining these three parts of straight lines. This graph is shown in Figure 17.6. From the graph we note that the range of the function is $\{y \mid y \leq 1\}$.  ●

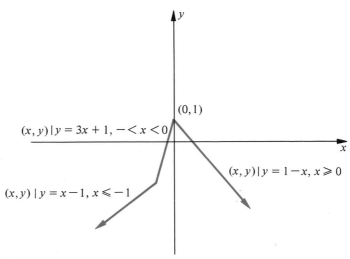

Figure 17.6

## EXERCISES FOR SECTION 17

# 16 → Homework to Hand in

Graph each of the following functions and then determine their range.

**1.** $y = f(x) = \begin{cases} x + 1, & x \geq -1 \\ -x - 1, & x < -1 \end{cases}$

**2.** $y = f(x) = \begin{cases} 3 - x^2, & x > 0 \\ x^2 + 3, & x \leq 0 \end{cases}$

**3.** $y = f(x) = \begin{cases} x + 4, & x \geq -1 \\ (\frac{1}{2})x + \frac{1}{2}, & x < -1 \end{cases}$

**4.** $y = g(x) = \begin{cases} 0, & 0 < x < 3 \\ 2, & 3 \leq x < 6 \\ 4, & 6 \leq x < 9 \end{cases}$

odd Numbers

**5.** $y = g(x) = \begin{cases} 3, & x > 0 \\ -3, & x < 0 \end{cases}$

**6.** $y = g(x) = \begin{cases} 6, & x > 3 \\ x^2 - 2x + 3, & 0 \le x \le 3 \\ 3, & x < 0 \end{cases}$

**7.** $y = f(x) = \begin{cases} 3x + 1, & x \ge 0 \\ x + 1, & -\frac{1}{2} \le x < 0 \\ -3x - 1, & x < -\frac{1}{2} \end{cases}$

**8.** $y = h(x) = \begin{cases} 1, & x \ge 0 \\ 2x + 1, & -1 \le x < 0 \\ -1, & x < -1 \end{cases}$

**9.** $y = g(x) = \begin{cases} 4 - x^2, & -2 \le x \le 2 \\ x^2 - 4, & x < -2 \text{ or } x > 2 \end{cases}$

**10.** $y = f(x) = \begin{cases} x^2 + x - 4, & x > 2 \\ -x^2 + x + 4, & 0 \le x \le 2 \\ -x^2 - x + 4, & -2 \le x < 0 \\ x^2 - x - 4, & x < -2 \end{cases}$

**11.** $y = f(x) = \begin{cases} x^2 - 6x + 5, & x \ne 1 \text{ or } 5 \\ 4, & x = 1 \text{ or } 5 \end{cases}$

**12.** $y = f(x) = \begin{cases} \dfrac{x^2 - 4}{x + 2}, & x \ne -2 \\ 0, & x = -2 \end{cases}$

**13.** $y = g(x) = \begin{cases} -2, & -2 \le x < -1 \\ -1, & -1 \le x < 0 \\ 0, & 0 \le x < 1 \\ 1, & 1 \le x < 2 \\ 2, & 2 \le x < 3 \end{cases}$

**14.** $y = f(x) = \begin{cases} 4, & x \ge 2 \\ x^2, & -1 \le x < 2 \\ 1, & x < -1 \end{cases}$

**15.** $y = \dfrac{|x - 1|}{x - 1}$

**16.** $y = \dfrac{|x + 3|}{x + 3}$

**17.** $y = \dfrac{\sqrt{x^2 - 2x + 1}}{x - 1}$

**18.** $y = \dfrac{x - 2}{|x - 2|}$

Use the graphs of each of the following functions to determine which functions are continuous at $x = 1$.

**19.** $f(x) = \left\{ \dfrac{x^2 - 1}{x - 1} \right.$;   $g(x) = \begin{cases} x + 1, & x \ne 1, \\ 3, & x = 1 \end{cases}$;   $h(x) = \begin{cases} x + 1, & x \ne 1 \\ 2, & x = 1 \end{cases}$

**20.** $f(x) = \left\{ \dfrac{x^2 - 6x + 5}{x - 1} \right.$;   $g(x) = \begin{cases} x - 5, & x \ne 1 \\ -4, & x = 1 \end{cases}$;   $h(x) = \begin{cases} x - 5, & x \ne 1 \\ 0, & x = 1 \end{cases}$

**21.** Use the definition of absolute value to determine that
  (a) $y = |2x + 1| + |x|$ is equivalent to Exercise 7.
  (b) $y = |x + 1| - |x|$ is equivalent to Exercise 8.
  (c) $y = |4 - x^2|$ is equivalent to Exercise 9.
  (d) $y = |4 - x^2| + |x|$ is equivalent to Exercise 10.

**22.** The *step function* denoted as $[x]$ is defined to be

$$f(x) = [x] = n \text{ where } n \in J \text{ such that } n \le x.$$

Thus to find $[x]$ we simply "round off" the expression for $x$ to the next lowest whole number. For example $\left[\dfrac{1}{2}\right] = 0$, $[3.1] = 3$, $[-4] = -4$ and $[-2.9] = -3$.

Use the above definition to graph the following:

  (a) $f(x) = [x]$, $-3 \le x \le 2$        (b) $f(x) = [x]$, $x \ge 0$
  (c) $f(x) = [x] + 2$, $-1 \le x \le 2$     (d) $f(x) = [x] + x$, $0 \le x < 4$
  (e) $f(x) = [x] - x$, $-2 \le x \le 2$

Graph each of the following functions and then determine their range.

**23.** $y = x|x|$                    **24.** $y = (x + 2)|x + 2|$

**25.** $y = \dfrac{x^2|x - 1|}{x - 1}$            **26.** $y = \dfrac{x^2|x + 4|}{x + 4}$

**27.** $y = \dfrac{x^3}{|x|}$    **28.** $y = \dfrac{(x-2)^3}{|x-2|}$

**29.** $y = |x-2| + |3x+6|$    **30.** $y = |3x+1| + |2x|$

**31.** $y = |x-2| - |3x+6|$    **32.** $y = |3x+1| - |2x|$

**33.** A faucet on a tank lets water enter at the rate of 2 gallons per minute and the tank has a drain that lets the water out at a rate of 5 gallons per minute. Suppose the tank is empty when the faucet is opened. Six minutes later the drain is also opened. If $t$ is the time (in minutes) elapsed from the start and if $G$ is the number of gallons of water in the tank at time $t$, express $G$ as a function of $t$ for $t \geq 0$ and then sketch the graph of $G$.

**34.** Use the graph of $G$, in exercise 29, to determine when the tank will be empty again.

**35.** A delivery company will deliver packages for \$1.25 up to 1 pound, and \$.25 for each additional pound or fraction thereof. If $C$ is the delivery charge for $x$ lbs, express $C$ as a function of $x$ and sketch its graph.

**36.** It is found that the yield of a certain crop is a function of the time $t$ (days) between planting and harvesting. If the yield is given by

$$H(t) = \begin{cases} (t-20)(60-t) & \text{if } 20 \leq t \leq 60 \\ 0 & \text{if } t < 20 \text{ or } t > 60 \end{cases}$$

Graph $H(t)$ and then determine the maximum yield and the harvest day $t$ that it will occur.

# SECTION 18
# SOME SPECIAL RELATIONS AND THEIR GRAPHS

We will now examine some special types of relations in $\mathbb{R}$ that involve the absolute value of one or both variables. Recall that

$$|x| = \begin{cases} x & \text{if } x \geq 0 \\ -x & \text{if } x < 0 \end{cases}$$

### Example 18.1
Graph $S = \{(x, y) \mid |x| + |y| = 1\}$.

Solution. Since $|x| + |y| = 1$ involves the absolute values of both variables, we will examine it in each of the four quadrants as follows:

In *quadrant 1,* both $x$ and $y$ are positive. Thus $|x| + |y| = 1$ can be expressed as $x + y = 1$.

In *quadrant 2,* $x$ is negative and $y$ is positive. Thus $|x| + |y| = 1$ can be expressed as $-x + y = 1$.

In *quadrant 3,* both $x$ and $y$ are negative. Thus $|x| + |y| = 1$ can be expressed as $-x - y = 1$.

In *quadrant 4,* $x$ is positive and $y$ is negative. Thus $|x| + |y| = 1$ can be expressed as $x - y = 1$.

We see that in each quadrant we obtain a linear equation with a restricted domain and range. The graph of $S$ will be the combined graphs of

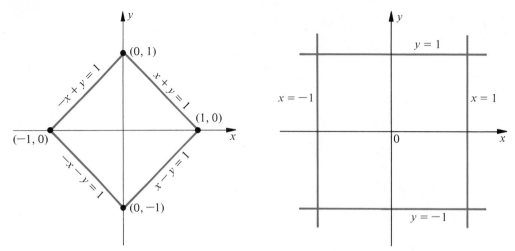

Figure 18.1                  Figure 18.2

these straight lines shown in Figure 18.1. From the figure we see that the domain and the range of $S$ are $\{x \mid -1 \leq x \leq 1\}$ and $\{y \mid -1 \leq y \leq 1\}$, respectively.   ●

**Example 18.2**
Graph $S = \{(x, y) \mid |x| = 1 \text{ or } |y| = 1\}$.

**Solution.** Since $|x| = 1$ is equivalent to $x = 1$ or $x = -1$, and $|y| = 1$ is equivalent to $y = 1$ or $y = -1$, the graph of $S$ will be the combined graphs of these straight lines. The graph of $S$ is shown in Figure 18.2.   ●

**Example 18.3**
Graph $S = \{(x, y) \mid |y| = 9 - x^2\}$.

**Solution.** $|y| = 9 - x^2$ can be reexpressed as $y = 9 - x^2$ if $y \geq 0$ or as $-y = 9 - x^2$ if $y < 0$. In each case we have a quadratic function with a restricted domain and range. If we denote $S_1 = \{(x, y) \mid y = 9 - x^2, y \geq 0\}$ and $S_2 = \{(x, y) \mid -y = 9 - x^2, y < 0\}$, the graph of $S = S_1 \cup S_2$ is the union of the graphs of $S_1$ and $S_2$, respectively. The graphs of $S_1$, $S_2$, and $S$ are shown in Figure 18.3.   ●

**Example 18.4**
Graph $S = \{(x, y) \mid x^2 - y^2 = 0\}$.

**Solution.** We can factor $x^2 - y^2$ to obtain $(x + y)(x - y) = 0$. Setting each factor equal to zero we have $x + y = 0$ or $x - y = 0$. Thus the graph of $S$ will simply be the union of the graphs of the straight lines $x + y = 0$ and $x - y = 0$. The graph of $S$ is shown in Figure 18.4.   ●

**Example 18.5**
Graph $S = \{(x, y) \mid (y - x - 2)(y - x^2 - 2) = 0\}$.

# CHAPTER 3
# INEQUALITIES

## QUADRATIC INEQUALITIES IN ONE VARIABLE

We are interested in finding the solution sets for the following quadratic inequalities in one real variable:

$$ax^2 + bx + c > 0$$
$$ax^2 + bx + c \geq 0$$
$$ax^2 + bx + c < 0$$
$$ax^2 + bx + c \leq 0$$

where $a, b, c \in \mathbb{R}$ and the variable $x \in \mathbb{R}, a \neq 0$. We know that we can find the real roots or solutions to $ax^2 + bx + c = 0$ by applying the quadratic formula (15.3). Suppose that

$$r_1 = \frac{-b + \sqrt{b^2 - 4ac}}{2a} \quad \text{and} \quad r_2 = \frac{-b - \sqrt{b^2 - 4ac}}{2a}$$

are the roots or solutions to $ax^2 + bx + c = 0$. These roots will be real and distinct if $b^2 - 4ac > 0$. (*Why?*)

The sum of these two roots can be expressed as follows:

$$r_1 + r_2 = \frac{-b + \sqrt{b^2 - 4ac}}{2a} + \frac{-b - \sqrt{b^2 - 4ac}}{2a}$$

$$= \frac{-2b}{2a} = \frac{-b}{a} \tag{19.1}$$

The product of the two roots can be expressed as

$$r_1 r_2 = \left( \frac{-b + \sqrt{b^2 - 4ac}}{2a} \right) \cdot \left( \frac{-b - \sqrt{b^2 - 4ac}}{2a} \right)$$

$$= \frac{(-b)^2 - (\sqrt{b^2 - 4ac})^2}{(2a)^2}$$

$$= \frac{b^2 - (b^2 - 4ac)}{4a^2}$$

$$= \frac{4ac}{4a^2} = \frac{c}{a} \tag{19.2}$$

We now use the results (19.1) and (19.2) as follows:

$$ax^2 + bx + c = a\left( x^2 + \frac{b}{a}x + \frac{c}{a} \right)$$

$$= a[x^2 - (r_1 + r_2)x + r_1 r_2)]$$

$$= a(x - r_1)(x - r_2) \tag{19.3}$$

Result (19.3) gives us a method for factoring *all* quadratic expressions.

### Example 19.1
Factor the quadratic expression $4x^2 - 8x - 4$.

Solution. The roots or solution to $4x^2 - 8x - 4 = 0$ are

$$r_1 = \frac{8 + \sqrt{64 - 4(4)(-4)}}{8} \quad \text{or} \quad r_1 = \frac{8 + \sqrt{128}}{8} = 1 + \sqrt{2}$$

and

$$r_2 = \frac{8 - \sqrt{64 - 4(4)(-4)}}{8} \quad \text{or} \quad r_2 = \frac{8 - \sqrt{128}}{8} = 1 - \sqrt{2}$$

Thus $4x^2 - 8x - 4 = 4[x - (1 + \sqrt{2})][x - (1 - \sqrt{2})]$.    ●

Now, suppose that we wish to find the solution set for $ax^2 + bx + c > 0$, $a > 0$. Result (19.3) enables us to express $ax^2 + bx + c > 0$ as $a(x - r_1)(x - r_2) > 0$, where $r_1$ and $r_2$ are the real solutions to $ax^2 + bx + c = 0$.

We now assume that $r_1 < r_2$ and draw the number line in Figure 19.1 to represent this. The points $r_1$ and $r_2$ divide the number line into three regions: $x < r_1$, $r_1 < x < r_2$, and $x > r_2$. We see that the solution set for $ax^2 + bx + c > 0$ or $a(x - r_1)(x - r_2) > 0$ will be those regions of the number line for which the product $(x - r_1)(x - r_2) > 0$. Our task is to examine this product in each of these regions.

In region 1 we have $x < r_1$ and $x < r_2$. Thus $x - r_1 < 0$ and $x - r_2 < 0$. Therefore, $(x - r_1)(x - r_2) > 0$ and region 1 is part of the solution set.

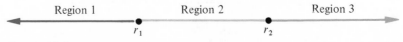

Region 1        Region 2        Region 3

$r_1$        $r_2$

Figure 19.1

In region 2 we have $r_1 < x < r_2$ or $r_1 < x$ and $x < r_2$, which means that $x - r_1 > 0$ and $x - r_2 < 0$. Thus $(x - r_1)(x - r_2) < 0$ and region 2 is *not* part of the solution set.

In region 3 we have $x > r_2$ and $x > r_1$, which means that $x - r_2 > 0$ and $x - r_1 > 0$. Thus $(x - r_1)(x - r_2) > 0$ and region 3 is also part of the solution set.

We see that regions 1 and 3 are the solution set for $ax^2 + bx + c > 0$, where $a > 0$ and $r_1 \neq r_2$. Therefore, $\{x \mid ax^2 + bx + c > 0, a > 0\} = \{x \mid x < r_1 \text{ or } x > r_2\}$. Since the product $(x - r_1)(x - r_2) < 0$ in region 2, we can conclude that $\{x \mid ax^2 + bx + c < 0, a > 0\} = \{x \mid r_1 < x < r_2\}$.

The case where $r_1 = r_2$ can be handled in a similar fashion. We summarize all of these results with the following theorems.

**Theorem 19.1**　If $ax^2 + bx + c = 0, a > 0$, has real and distinct roots $r_1$ and $r_2$, where $r_1 < r_2$, the solution set for

1. $ax^2 + bx + c > 0$ is $x < r_1$ or $x > r_2$.
2. $ax^2 + bx + c \geq 0$ is $x \leq r_1$ or $x \geq r_2$.
3. $ax^2 + bx + c < 0$ is $r_1 < x < r_2$.
4. $ax^2 + bx + c \leq 0$ is $r_1 \leq x \leq r_2$.

**Theorem 19.2**　If $ax^2 + bx + c = 0, a > 0$, has real and equal roots $r_1 = r_2 = r$, the solution set for

1. $ax^2 + bx + c > 0$ is $x < r$ or $x > r$.
2. $ax^2 + bx + c \geq 0$ is $x \leq r$ or $x \geq r$ (that is, all real numbers).
3. $ax^2 + bx + c < 0$ is the empty set, $\emptyset$.
4. $ax^2 + bx + c \leq 0$ is $x = r$.

The only other possibility for the roots of $ax^2 + bx + c = 0$ is that $r_1$ and $r_2$ are not real roots. For this situation we have the following theorem.

**Theorem 19.3**　If $ax^2 + bx + c = 0, a > 0$, has nonreal solutions, the real solution set for

$$ax^2 + bx + c \geq 0 \quad \text{is all real } x$$
$$ax^2 + bx + c \leq 0 \quad \text{is the empty set } \emptyset$$

The proof of Theorem 19.3 is left as an exercise.

### Example 19.2
Find the solution set for $x^2 - 4x - 5 < 0$, and graph it.

**Solution.**　We first find the real roots of $x^2 - 4x - 5 = 0$. $x^2 - 4x - 5 = (x - 5)(x + 1) = 0$. Thus the roots are $-1$ and $5$. We now apply Theorem 19.1, where $a = 1 > 0, r_1 = -1$, and $r_2 = 5$. The solution set for $x^2 - 4x - 5 < 0$ is $-1 < x < 5$. Therefore, $\{x \mid x^2 - 4x - 5 < 0\} = (-1, 5)$. The graph of this set is shown in Figure 19.2.　●

$\qquad\qquad\quad -1 \qquad\qquad\qquad\qquad\qquad\qquad 5$

Figure 19.2

*[handwritten in left margin: Write on note cards]*

**Example 19.3**

Find the solution set for $3x^2 + 13x + 4 > 0$, and graph the result.

**Solution.** We first find the real roots of $3x^2 + 13x + 4 = 0$. We use the quadratic formula to obtain the roots $r_1 = -4$ and $r_2 = -\frac{1}{3}$. Use Theorem 19.1 and note that $a = 3 > 0$, $r_1 = -4$, and $r_2 = -\frac{1}{3}$. We conclude that the solution set for $3x^2 + 13x + 4 > 0$ is $x < -4$ or $x > -\frac{1}{3}$; that is, $\{x \mid 3x^2 + 13x + 4 > 0\} = (-\infty, -4) \cup (-\frac{1}{3}, \infty)$. The graph is shown in Figure 19.3. ●

Figure 19.3

**Example 19.4**

Find the solution set for $x^2 - 4x + 4 \geq 0$.

**Solution.** We first find the real roots of $x^2 - 4x + 4 = 0$. We can rewrite $x^2 - 4x + 4$ as $(x - 2)^2$. Thus $x^2 - 4x + 4 = 0$ or $(x - 2)^2 = 0$ has 2 as a double root. We now apply Theorem 19.2, where $a = 1 > 0$ and $r_1 = r_2 = 2$ to conclude that the solution set for $x^2 - 4x + 4 \geq 0$ is all real numbers. Thus $\{x \mid x^2 - 4x + 4 \geq 0\} = (-\infty, \infty)$. The graph of this solution set is shown in Figure 19.4. ●

Figure 19.4

**Example 19.5**

Find the solution set for $x^2 + x + 1 < 0$.

**Solution.** We must find the roots of $x^2 + x + 1 = 0$. Since the discriminant $b^2 - 4ac = 1^2 - 4(1)(1) = -3 < 0$, we note that the roots of $x^2 + x + 1 = 0$ are nonreal. We now apply Theorem 19.3, where $a = 1 > 0$, to conclude that the solution set for $x^2 + x + 1 < 0$ is the empty set $\varnothing$. $\{x \mid x^2 + x + 1 < 0\} = \varnothing$. ●

**Example 19.6**

Find the solution set for $-x^2 + 4x - 4 \leq 0$.

**Solution.** We note that if we multiply $-x^2 + 4x - 4 \leq 0$ by $-1$ we obtain the equivalent inequality, $x^2 - 4x + 4 \geq 0$. Since this is the inequality of Example 19.4, we conclude that the solution set for $-x^2 + 4x - 4 \leq 0$ is the set of all real numbers. ●

**Remark 19.1**

When we encounter an inequality where $a < 0$, we can multiply or divide by $-1$ to obtain an equivalent inequality where $a > 0$. We can then apply the appropriate theorem to solve the equivalent inequality. See Example 19.6.

## EXERCISES FOR SECTION 19

Express each of the following quadratic expressions in the factored form $a(x - r_1)(x - r_2)$, where $r_1$ and $r_2$ are the roots of the associated quadratic equation $ax^2 + bx + c = 0$.

**1.** $2x^2 - 5x - 3$      **4.** $3x^2 + 8x + 5$

**2.** $2x^2 - 3x - 3$      **5.** $4x^2 + 4x - 5$

**3.** $x^2 - 12x + 31$

Use Theorem 19.1, 19.2, or 19.3 to find the solution sets for each of the following, and graph each solution set.

**6.** $x^2 - 6x + 5 \geq 0$      **15.** $x^2 - 9 > 0$

**7.** $-x^2 + 2x - 4 \leq 0$      **16.** $x \leq 2x^2$

**8.** $2x^2 - 5x - 3 > 0$      **17.** $x^2 + 16 > 0$

**9.** $2x^2 - 5x - 3 \geq 0$      **18.** $9 - 6x - 8x^2 \leq 0$

**10.** $x^2 + 2x + 2 > 0$      **19.** $25 - 4x^2 > 0$

**11.** $x^2 - 6x + 9 > 0$      **20.** $16 - x^2 \leq 0$

**12.** $3x^2 + 2x + 6 < 0$      **21.** $4x^2 - 4x > -1$

**13.** $2x^2 > 7x - 6$      **22.** $(x - 3)^2 \leq 4$

**14.** $\dfrac{x^2}{4} + \dfrac{x}{4} \leq 3$      **23.** $(2x - 1)^2 \geq 9$

Use formulas 19.1 and 19.2 to solve exercises 24 through 26.

**24.** Find a quadratic equation having $\dfrac{-2 \pm \sqrt{5}}{4}$ for its roots.

**25.** Find the roots of a quadratic equation if the sum of its roots is $-\frac{1}{6}$ and their product is $-\frac{1}{3}$.

**26.** Find the roots of a quadratic equation if the sum of its roots is $-\frac{9}{4}$ and their product is $\frac{9}{8}$.

**27.** The width of a rectangle is 2 m more than the length l. If the area of the rectangle is to be greater than 40m², find l.

**28.** The height of a triangle is 4 in. less than the base $b$. If the area of the triangle is to be greater than 20 in², find $b$.

**29.** (a) Prove Theorem 19.2.
(b) Prove Theorem 19.3.

## SECTION 20
# A GRAPHIC SOLUTION FOR INEQUALITIES IN ONE VARIABLE

We now wish to introduce a graphic method for solving inequalities in one variable. Consider the algebraic expression $x - 3$, where $x$ is a real number. The real number $x - 3$ has the following possibilities:

$$x - 3 > 0 \quad \text{or} \quad x - 3 = 0 \quad \text{or} \quad x - 3 < 0$$

In the language of sets we have $R = \{x \mid x - 3 > 0\} \cup \{x \mid x - 3 = 0\} \cup \{x \mid x - 3 < 0\}$. A graph of $R$ is given in Figure 20.1. The negative signs represent the region of the number line where $x - 3 < 0$. The positive signs represent the region of the number line where $x - 3 > 0$. The point $x = 3$ represents the solution to $x - 3 = 0$. We will refer to the graph in Figure 20.1 as the **sign graph** for $x - 3$. We use the following examples to introduce the graphic method for solving inequalities in one variable.

$$(x - 3) \qquad - \quad - \quad - \quad - \quad - \quad - \quad - \quad 0 \; + \; + \; + \; + \; + \; + \; + \; +$$
$$x = 3$$

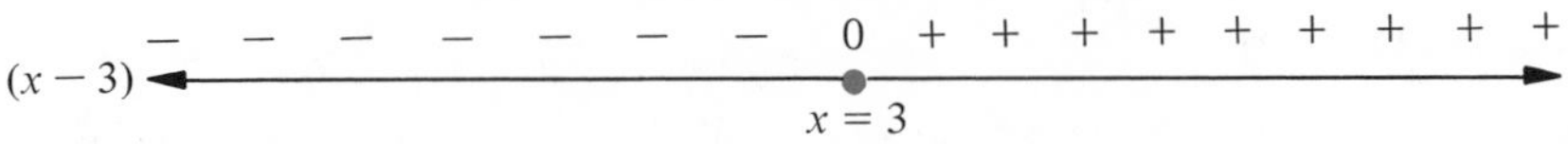

Figure 20.1

### Example 20.1

Find the solution set for $(x + 3)(x - 1)(x - 3) > 0$, and graph this set.

Solution.  We first construct a sign graph for each of the three factors $(x + 3), (x - 1), (x - 3)$ in Figure 20.2. The solution set will be the regions of the number line where the product $(x + 3)(x - 1)(x - 3)$ is positive, that is, regions where any two of the three factors or none of three factors are negative. In Figure 20.2 we have shaded these regions and we conclude that the solution set for $(x + 3)(x - 1)(x - 1)(x - 3) > 0$ is $\{x \mid -3 < x < 1$ or $x > 3\} = (-3, 1) \cup (3, \infty)$. The graph of this solution set is shown on the number line in Figure 20.2.  •

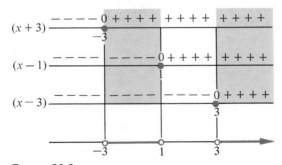

Figure 20.2

### Example 20.2

Find the solution set for $(x + 3)(x - 1)(x - 3)(x - 4) \leq 0$, and graph this set.

Solution.  We first construct a sign graph for each of the four factors $(x + 3), (x - 1), (x - 3)$, and $(x - 4)$ in Figure 20.3. The solution set will be the regions of the number line where the product $(x + 3)(x - 1)(x - 3)(x - 4)$ will be negative or zero, that is, regions where any one or any three of the factors are negative and the points where the factors are equal to zero. In Figure 20.3 we have shaded these regions and we conclude that the solution set for $(x + 3)(x - 1)(x - 3)(x - 4) \leq 0$ is $\{x \mid -3 \leq x \leq 1$ or $3 \leq x \leq 4\} = [-3, 1] \cup [3, 4]$. The graph of this solution set is shown on the number line in Figure 20.3.  •

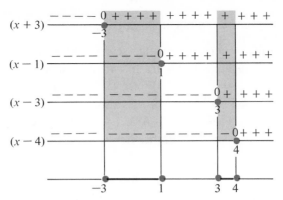

Figure 20.3

The methods of Examples 20.1 and 20.2 lead us to the following theorem.

**Theorem 20.1**   If $a(x - r_1)(x - r_2) \cdots (x - r_n) = 0$, $a > 0$, has $n$ real and distinct solutions, then

1. $a(x - r_1)(x - r_2) \cdots (x - r_n) < 0$ if and only if an odd number of the factors $(x - r_1)$, $(x - r_2), \ldots, (x - r_n)$ are negative.
2. $a(x - r_1)(x - r_2) \cdots (x - r_n) > 0$ if and only if an even number of the factors $(x - r_1)$, $(x - r_2), \ldots, (x - r_n)$ are negative.

The graphic method can also be used to solve fractional inequalities in one variable whose numerators and denominators can be expressed as products of factors of the form $x - r$. Consider the following examples.

**Example 20.3**

Find the solution set for $\dfrac{(x + 3)(x - 1)}{x(x + 2)} > 0$, and graph this set.

Solution.   We first construct a sign graph for each of the four factors $(x + 3)$, $(x - 1)$, $x$, and $(x + 2)$. The solution set will be the regions of the

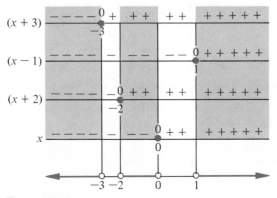

Figure 20.4

number line where the quotient $\dfrac{(x + 3)(x - 1)}{x(x + 2)}$ will be positive, that is, regions where any even number of the factors in *both* the numerator and the denominator are negative. In Figure 20.4 we have shaded these regions and we conclude that the solution set for $\dfrac{(x + 3)(x - 1)}{x(x + 2)} > 0$ is $\{x \mid x < -3$ or $-2 < x < 0 \text{ or } x > 1\} = \{-\infty, -3) \cup (-2, 0) \cup (1, \infty)$. The graph of this solution set is shown on the number line in Figure 20.4.   ●

## Example 20.4

Find the solution set for $\dfrac{x}{x - 2} < 5$, and graph this set.

**Solution.** We must first reexpress $\dfrac{x}{x - 2} < 5$ as follows:

$$\frac{x}{x - 2} - 5 < 0$$

$$\frac{x}{x - 2} - \frac{5(x - 2)}{x - 2} < 0$$

$$\frac{x - 5x + 10}{x - 2} < 0$$

$$\frac{-4x + 10}{x - 2} < 0$$

$$\frac{4x - 10}{x - 2} > 0$$

This last inequality was obtained by multiplying $\dfrac{-4x + 10}{x - 2} < 0$ by $-1$. Since $\dfrac{x}{x - 2} < 5$ is equivalent to $\dfrac{4x - 10}{x - 2} > 0$ we can now apply the graphic method to $\dfrac{4x - 10}{x - 2} > 0$ to obtain the solution set for $\dfrac{x}{x - 2} < 5$. We construct a sign graph for each of the factors $4x - 10$ and $x - 2$ in Figure 20.5.

The solution set will be the regions of the number line where the quotient $\dfrac{4x - 10}{x - 2}$ will be positive. That is, regions where both the factors will

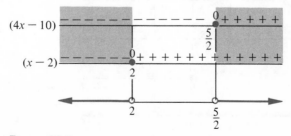

Figure 20.5

be negative or both the factors will be positive. In Figure 20.5 we have shaded these regions and we conclude that the solution set for $\dfrac{x}{x-2} < 5$ is $\{x \mid x < 2 \text{ or } x > \frac{5}{2}\} = (-\infty, 2) \cup (\frac{5}{2}, \infty)$. The graph of this solution set is shown on the number line in Figure 20.5.   •

### Example 20.5

Consider the following *erroneous* solution for $\dfrac{x}{x-2} < 5$. We multiply both sides of the inequality by $x - 2$. Thus $x < 5(x - 2)$ or $x < 5x - 10$. This implies that $10 < 4x$ or $x > \frac{5}{2}$. We conclude that $\left\{x \mid \dfrac{x}{x-2} < 5\right\} = (\frac{5}{2}, \infty)$. Explain why our method did not give us the right result.

### Example 20.6

Find the solution set for $\dfrac{1}{x-2} \geq \dfrac{3}{x}$, and graph this set.

Solution.   Again, we use the properties of inequalities to reexpress $\dfrac{1}{x-2} \geq \dfrac{3}{x}$ in an equivalent form as follows:

$$\frac{1}{x-2} \geq \frac{3}{x}$$

$$\frac{1}{x-2} - \frac{3}{x} \geq 0$$

$$\frac{x - 3(x-2)}{x(x-2)} \geq 0$$

$$\frac{-2x + 6}{x(x-2)} \geq 0$$

$$\frac{2x - 6}{x(x-2)} \leq 0$$

We now apply the sign graph method to $\dfrac{2x-6}{x(x-2)} \leq 0$. We construct a sign graph in Figure 20.6. The solution set will be the regions of the number line

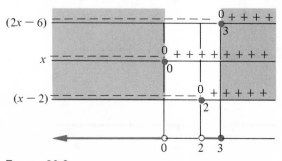

Figure 20.6

where the quotient $\dfrac{2x-6}{x(x-2)}$ will be negative or zero. That is, regions where there will be an odd number of negative factors. In Figure 20.6 we have shaded these regions and we conclude that the solution set for $\dfrac{1}{x-2} \geq \dfrac{3}{x}$ is $\{x \mid x < 0 \text{ or } 2 < x \leq 3\} = (-\infty, 0) \cup (2, 3]$. [*Note:* The values of $x = 0$ and $x = 2$ have been excluded since each makes the expression $\dfrac{2x-6}{x(x-2)}$ undefined]. The graph of this solution set is shown on the number line in Figure 20.6. •

## Example 20.7

Find the domain and the range for $\left\{(x, y) \mid y = \dfrac{x^2 - 4}{9 - x^2}\right\}$.

Solution. Inspection of the equation $y = \dfrac{x^2 - 4}{9 - x^2}$ indicates that $y$ will not be defined when $9 - x^2 = 0$ or equivalently when $x = \pm 3$. Thus the domain is $\{x \mid x \neq \pm 3\}$. To find the range we solve the given equation for $x$ as follows.

$$y = \frac{x^2 - 4}{9 - x^2}$$
$$9y - yx^2 = x^2 - 4$$
$$9y + 4 = x^2 + yx^2$$
$$9y + 4 = x^2(1 + y)$$
$$\frac{9y + 4}{1 + y} = x^2$$
$$x = \pm\sqrt{\frac{9y + 4}{1 + y}}$$

From this latter equation we see that $x$ will be defined for $\dfrac{9y + 4}{1 + y} \geq 0$. We

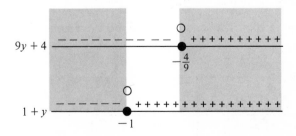

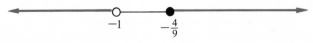

Figure 20.7

now apply the sign graph method to solve $\dfrac{9y + 4}{1 + y} \geq 0$ by constructing the sign graphs shown in Figure 20.7. The solution set will be the regions of the number line where the quotient $\dfrac{9y + 4}{1 + y}$ will be positive or zero. In Figure 20.7 we have shaded these regions and we conclude that the range is $\{y \mid y < -1 \text{ or } y \geq -\frac{4}{9}\}$. •

## EXERCISES FOR SECTION 20

**1.** Apply the sign graph method (if possible) to Exercises 6–19 in Section 19.

Use the sign graph method to solve the following, and graph each solution set.

**2.** $(x - 1)(x + 2)(x - 3) < 0$

**3.** $4(x + 2)(x + 3)(x + 4)(x + 6) \geq 0$

**4.** $(2x - 3)(3x + 7)(x - 4) > 0$

**5.** $x^3 - 9x \geq 0$

**6.** $x^5 - 4x^3 \leq 0$

**7.** $x(x^2 - 5x + 4) > 0$

**8.** $(2 - x)(2x^2 - 10x + 12) \leq 0$

**9.** $\dfrac{(x - 5)(x + 3)}{x^2 - 4} \geq 0$

**10.** $\dfrac{x^2 - 16}{9 - x^2} > 0$

**11.** $\dfrac{18 - 5x}{x - 3} > 0$

**12.** $\dfrac{3}{4x + 1} > \dfrac{2}{x - 5}$

**13.** $\dfrac{1}{x - 1} \geq \dfrac{2}{x}$

**14.** $\dfrac{4}{x} < 2$

**15.** $\dfrac{x - 5}{x(x - 2)} < 0$

**16.** $\dfrac{-1}{x + 1} \leq 0$

**17.** $\dfrac{4}{x - 3} - \dfrac{16}{x^2 - 9} > 1$

In exercises 18 through 21, find the domain for the given relations.

**18.** $\left\{ (x, y) \mid y = \dfrac{x^2}{x^2 - 4} \right\}$

**20.** $\left\{ (x, y) \mid y^2 = \dfrac{x + 1}{x - 1} \right\}$

**19.** $\{ (x, y) \mid y = \sqrt{x(x^2 - 1)} \}$

**21.** $\left\{ (x, y) \mid y^2 = \dfrac{x^2 - 4}{1 - x^2} \right\}$

**22.** Show that if $x, y \in R$ and if $x + y = 8$, then $xy \leq 16$.

**23.** Show that for $x \in R$, $x^2 + 4 \geq 4x$.

**24.** Show that the sum of any positive number and its reciprocal is greater than or equal to 2.

**25.** Determine which of the following are equivalent inequalities (that is, have the same solution set).

    (a) $(x + 1)^2(x - 3)(x - 1) > 0$     (c) $(x - 3)(x - 1) \geq 0$

    (b) $(x - 3)(x - 1) > 0$     (d) $(x + 1)^2(x - 3)(x - 1) \geq 0$

**26.** Determine which of the following are equivalent inequalities.

(a) $(x + 3)(x - 2)(x + 6) > 0$      (c) $\dfrac{(x + 3)^2(x - 2)}{x + 6} > 0$      (e) $\dfrac{x + 3}{(x - 2)(x + 6)} > 0$

(b) $\dfrac{(x + 3)(x - 2)}{x + 6} > 0$      (d) $\dfrac{(x + 3)(x - 2)}{x + 6} \geq 0$

**27.** When is the surface area of a cube, whose edge is $e$ units, larger than the volume of the cube?

# SECTION 21
## INEQUALITIES IN TWO VARIABLES

We have seen that the graphs of $y = ax + b$ and $y = ax^2 + bx + c$ are straight lines and parabolas, respectively. Now we wish to consider the graphs of inequalities in two variables such as $y > ax + b$ and $y \geq ax^2 + bx + c$. The graph of a linear function such as $y = ax + b$ divides the plane into three parts: the set of points above the line, the line itself, and the set of points below the line. The points above and below the line are called *half-planes*.

Consider $S = \{(x. y) \mid y > 2x + 1\}$. It may seem obvious that the graph of the set $S$ is the half-plane that lies above the line since $y$ is greater than $2x + 1$. If, in fact, we check a few points above the line we see that the inequality $y > 2x + 1$ is indeed satisfied. For example, if we check the point $(0, 3)$ we see that $y > 2x + 1$ is satisfied since $3 > 2(0) + 1$. Similarly, the point $(1, 4)$ satisfies the inequality since $4 > 2(1) + 1$. This process of checking points is never ending since we have an infinite number of ordered pairs to deal with. We would like to conclude that the graph of the solution set for the inequality can be determined by checking just a single point. But, can this be done? The answer is yes, as we will now show.

Consider again, $S = \{(x, y) \mid y > 2x + 1\}$. Let $P$ be any point $(x_0, y_0)$ on the straight line $y = 2x + 1$ and let $Q$ be any point $(x_0, y_1)$ directly above $P$ (Figure 21.1). Now, if $P$ lies on the line $y = 2x + 1$, its coordinates must satisfy the equation. Thus, we have

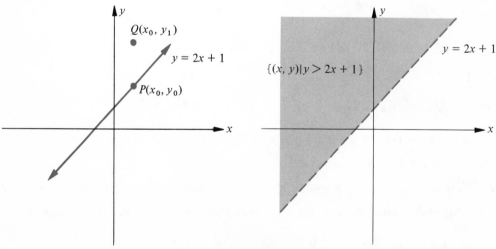

Figure 21.1                                                            Figure 21.2

$y_0 = 2x_0 + 1$. Since the point $Q(x_0, y_1)$ was chosen above $P(x_0, y_0)$, we know that $y_1 > y_0$. Now, if $y_1 > y_0$ and $y_0 = 2x_0 + 1$, then $y_1 > 2x_0 + 1$. We conclude that since $Q(x_0, y_1)$ was any point above the line, $Q$ satisfies the inequality and, therefore, any point above the line will satisfy the inequality. The graph of $S$ is the half-plane above the line $y = 2x + 1$. [*Note:* The line $y = 2x + 1$ is not part of the set $S$. We indicate this graphically by means of a dashed line (Figure 21.2).]

This suggests a general procedure to be followed when graphing inequalities of this type.

*Step 1.* Graph the equation.

*Step 2.* Check a point in each one of the regions. If the point satisfies the inequality, that region is part or possibly the entire graph of the solution set.

### Example 21.1
Graph $S = \{(x, y) \mid y < -2x + 4\}$.

**Solution.** The corresponding equation is $y = -2x + 4$. The graph of this equation is a straight line that divides the plane into two regions shown in Figure 21.3. We know choose any point not on the line, say $(0, 0)$. Substituting the coordinates of this point into our original inequality we obtain $0 < 2(0) + 4$, which is true. Since $(0, 0)$ is a point in the half-plane below the line, we conclude that the graph of $S$ includes this region. If we check a point above the line such as $(4, 1)$, we have $4 < -2(1) + 4$ or $4 < 2$, which is not true. Therefore, the half-plane below the line represents the graph of $S$. The line itself is not part of this graph (Figure 21.3). •

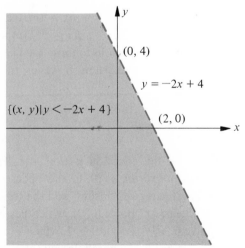

Figure 21.3

### Example 21.2
Graph $S = \{(x, y) \mid (y - 2x + 1)^2 > 0\}$.

**Solution.** We first graph the corresponding equation $(y - 2x + 1)^2 = 0$. To do this we reexpress $(y - 2x + 1)^2 = 0$ as $y - 2x + 1 = 0$ or

$y = 2x - 1$ by extracting the square root of both sides of the equation. We see that the graph of $(y - 2x + 1)^2 = 0$ is the same as the graph of the straight line $y = 2x - 1$. The point $(2, 0)$ lies below the line and satisfies the inequality since $[0 - 2(2) + 1]^2 > 0$ or $9 > 0$. Therefore, the region below the line is part of the graph of $S$. If we select a point above the line, say $(0, 0)$, we see that these coordinates also satisfy the original inequality since $[0 - 2(0) + 1]^2 > 0$ or $1 > 0$. Thus the half-plane above the line is also part of the graph of $S$. The line $y = 2x - 1$ is not part of the graph of $S$ (Figure 21.4). ●

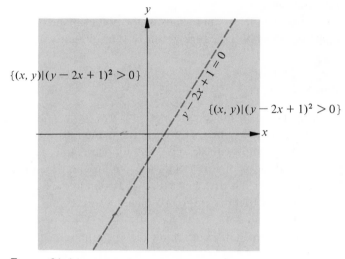

Figure 21.4

Inequalities such as $y > ax^2 + bx + c$ and $y < ax^2 + bx + c$ have solution sets $S = \{(x, y) \mid y > ax^2 + bx + c\}$ and $R = \{(x, y) \mid y < ax^2 + bx + c\}$, respectively. The graphs of these sets can be obtained by procedures analogous to those used in graphing the solution sets of the linear inequalities in two variables. Here, however, the term "half-plane" has no meaning.

### Example 21.3
Graph $S = \{(x, y) \mid y > x^2 + 1\}$.

Solution. We first must graph the equation $y = x^2 + 1$. The graph of $y = x^2 + 1$ is a parabola with its vertex at $(0, 1)$ and opens upward. The parabola divides the plane into the region "inside" and the region "outside" the parabola. If we take the "easiest" point $(0, 0)$, which lies in the region outside the parabola, we have $0 > 0^2 + 1$ or $0 > 1$, which is *not true*. Thus we can conclude that any point inside the parabola will satisfy the equality, and this region (excluding the parabola itself) is the graph of $S$. This graph is shown in Figure 21.5. ●

The procedure for graphing the solution sets of inequalities in two variables can be extended to higher-degree inequalities as well as those involving the absolute value symbol. We are restricted only by our ability to graph the associated equations.

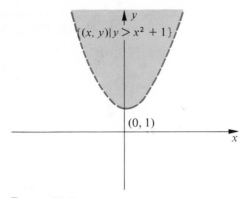

Figure 21.5

**Example 21.4**

Graph $S = \{(x, y) \mid |x| + |y| < 1\}$.

Solution.   We first must graph the associated equation $|x| + |y| = 1$ (see Example 18.1). The graph of $|x| + |y| = 1$ divides the plane into two regions, the region "inside" and the region "outside" (Figure 21.6). We now test an arbitrary point in either of these regions, say $(2, 1)$, which belongs to the outside region. We see that these coordinates do *not* satisfy the original inequality, since $|2| + |1| < 1$ or $3 < 1$ is *not true*. If we test a point in the region inside, such as $(0, 0)$, we see that these coordinates do satisfy the original inequality, since $|0| + |0| < 1$ or $0 < 1$ is true. We conclude, therefore, that the region inside, excluding the graph of the equality, is the graph of $S$. This graph is shown in Figure 21.6.   ●

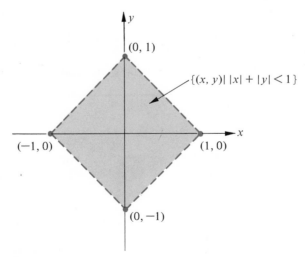

Figure 21.6

**Example 21.5**

Graph $S = \{(x, y) \mid x^2 + y^2 \geq 1\}$.

Solution.   We first graph the associated equation $x^2 + y^2 = 1$. The graph of $x^2 + y^2 = 1$ is a circle whose center is at $(0, 0)$ and whose radius is 1 (Figure

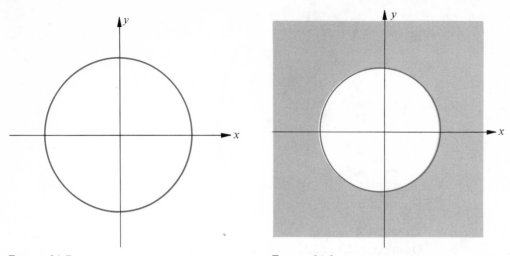

Figure 21.7                                             Figure 21.8

21.7) See example 10.4. The circle divides the plane into two regions, the region inside the circle and the region outside the circle. We now test an arbitrary point in either region, say $(0, 0)$, which belongs to the inside region. We see that these coordinates do *not* satisfy the original inequality, since $0^2 + 0^2 \geq 1$ or $0 \geq 1$ is *not* true. We can conclude that the outside region and the circle itself constitute the graph of $S$. This graph is shown in Figure 21.8. ●

**Example 21.6**

Graph $S = \{(x, y) \,|\, |y| \leq 9 - x^2\}$.

**Solution.** We first must graph the associated equation $|y| = 9 - x^2$ (see Example 18.3). This graph is shown in Figure 21.9. The graph of $|y| = 9 - x^2$ divides the plane into two regions, the region inside and the region

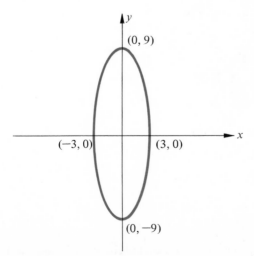

Figure 21.9

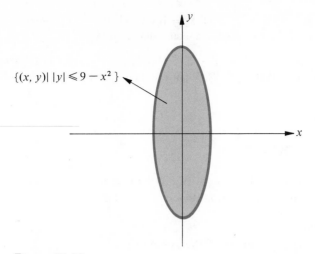

$\{(x, y)|\ |y| \leqslant 9 - x^2\}$

Figure 21.10

outside. We now test an arbitrary point in either region, say $(0, 0)$, in the inside region. The coordinates of this point satisfy the inequality since $0 \leq 9 - 0^2$ or $0 \leq 9$ is true. We can conclude that the inside region as well as the graph of the equality constitute the graph of $S$. This graph is shown in Figure 21.10 in final form. •

## EXERCISES FOR SECTION 21

Graph the solution set for each of the following.

**1.** $x + 2y + 4 \leq 0$          **14.** $x + |y| > 6$

**2.** $3x + 2y + 6 \geq 0$          **15.** $x - |x| < y$

**3.** $x^2 + y^2 > 0$          **16.** $x + |x| \geq y$

**4.** $y > x^2 - 4$          **17.** $x^2 + y^2 \leq 16$

**5.** $y < x^2 - 4$          **18.** $y^2 > 0$

**6.** $(x - 2y - 3)^2 < 0$          **19.** $y \leq |(x - 2)(x + 4)|$

**7.** $(x - y + 2)^2 \leq 0$          **20.** $|x| < 1$

**8.** $y \leq -x^2 + 2x + 3$          **21.** $|x - 2| > 6$

**9.** $x^2 < y$          **22.** $y < 0$

**10.** $y^2 > x$          **23.** $y \geq |x + 1| - |2x|$

**11.** $0 \geq x + y$          **24.** $|y + 2| \leq 2x + 1$

**12.** $x^2 + 6x + 4 > y$          **25.** $|x + 2y| \leq 1$

**13.** $|x| + |y| < 3$          **26.** $|x - 3y| \geq 6$

**27.** Use the graphs of each of the relations in exercises 1 through 26 to determine their domain and range.

**28.** Suppose that $S_1 = \{(x, y)\,|\,|x| + |y| < 1\}$, $S_2 = \{(x, y)\,|\,x^2 + y^2 < 1\}$. Use their graphs to show that $S_1 \subset S_2$. What are their respective domain and range?

**29.** Suppose that $R_1 = \{(x, y)\,|\,|x - 1| + |y - 2| < 1\}$ and $R_2 = \{(x, y)\,|\,(x - 1)^2 + (y - 2)^2 < 1\}$. Use their graphs to show that $R_1 \subset R_2$. What are their respective domain and range?

SECTION **22**
## SYSTEMS OF INEQUALITIES

The graph of the solution set of a system of inequalities in two variables consists of the intersection of the graphs of the inequalities in the system.

**Example 22.1**
Graph the solution set of the system

$$y - x \geq 0$$
$$y + x \geq 0$$

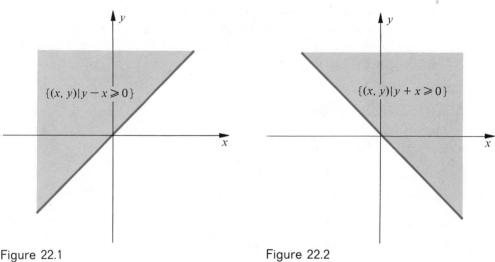

Figure 22.1                                   Figure 22.2

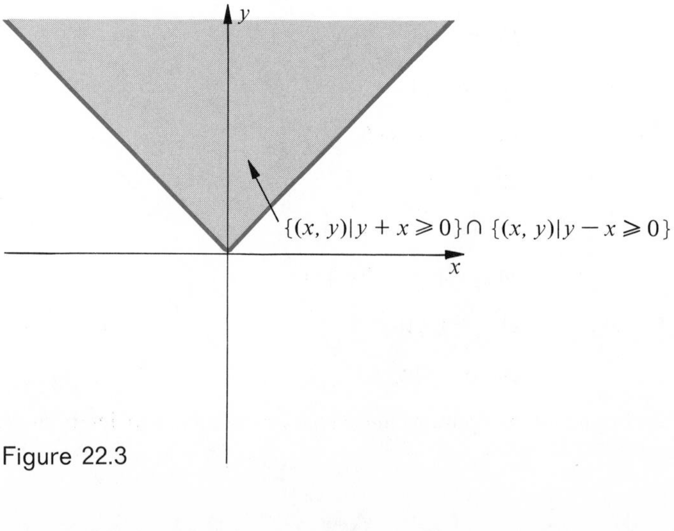

Figure 22.3

Solution.   We must first graph $\{(x, y) \mid y - x \geq 0\}$ and then $\{(x, y) \mid y + x \geq 0\}$. We use the methods of Section 21 to obtain their respective graphs, which are shown in Figures 22.1 and 22.2. We then graph $\{(x, y) \mid y - x \geq 0\} \cap \{(x, y) \mid y + x \geq 0\}$ in Figure 22.3.   ●

**Example 22.2**
Graph the solution set of the system

$$y < x - 2$$
$$y \geq x^2 - 3x - 4$$

Solution.   We must graph $\{(x, y) \mid y < x - 2\}$ and then graph $\{(x, y) \mid y \geq x^2 - 3x - 4\}$. We again apply the methods of Section 21 to obtain their respective graphs shown in Figures 22.4 and 22.5. We then graph $\{(x, y) \mid y < x - 2\} \cap \{(x, y) \mid y \geq x^2 - 3x - 4\}$, which is shown in Figure 22.6.   ●

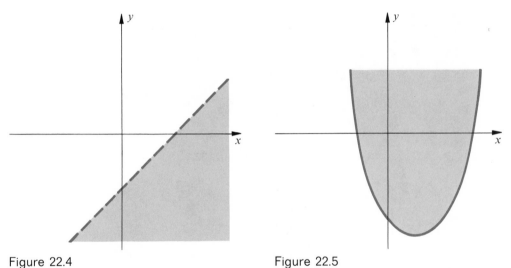

Figure 22.4                                         Figure 22.5

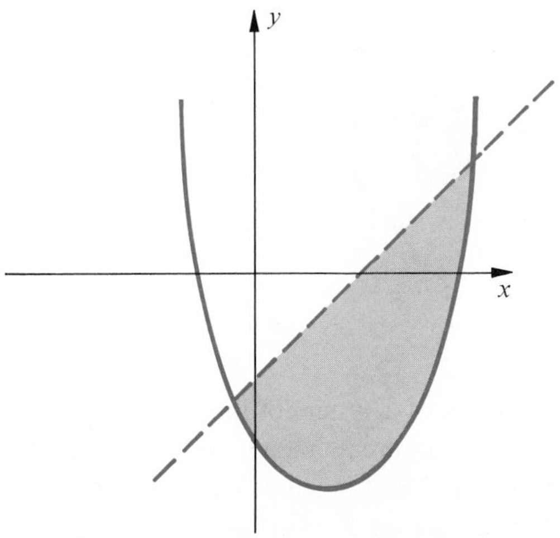

Figure 22.6

**Example 22.3**

Graph the solution of the system

$$|x| < 1$$
$$|y| < 1$$

Solution.  We must graph $\{(x, y)\,|\,|x| < 1\}$ and then graph $\{(x, y)\,|\,|y| < 1\}$. We show these respective graphs in Figures 22.7 and 22.8. We then graph $\{(x, y)\,|\,|x| < 1\} \cap \{(x, y)\,|\,|y| < 1\}$ in Figure 22.9.  ●

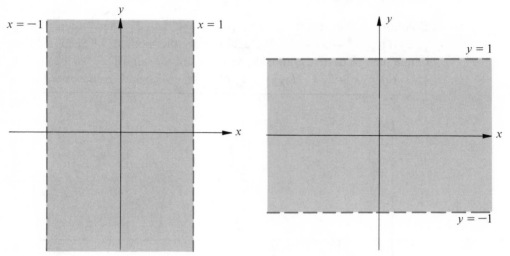

Figure 22.7                                   Figure 22.8

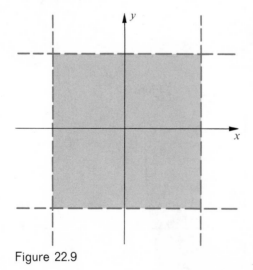

Figure 22.9

      In some problems the solution set of a system of linear inequalities can be used to determine the maximum and the minimum values of expressions of the form $F = ax + by + c$ where $a$, $b$, and $c \in \mathbb{R}$. The branch of mathematics that deals with this type of problem is called linear programming.

      It can be shown that if $F = ax + by + c$ is defined on a domain described by a system of linear inequalities, then any maximum or minimum values (that exist) will occur at the vertices of the polygon formed by the graph of the system of linear inequalities.

If the domain of $F$ consists of a *convex* polygon and its interior, then $F$ has a maximum value and a minimum value. Roughly speaking, a convex polygon has no indentations.

**Example 22.4**

Find the maximum and minimum values for $F = 20x + 8y$ if its domain is described by the following system of linear inequalities.

$$x + 2y \leq 8$$
$$x + 2y \geq 2$$
$$4 \leq x \leq 6$$

Solution.   We apply the methods of Section 21 to obtain the graphs of $x + 2y \leq 8$, $x + 2y \geq 2$, and $4 \leq x \leq 6$ shown respectively in Figures 22.10, 22.11, and 22.12. The intersection of these three graphs shown in Figure 22.13 represents the graph of the given system. We use the indicated ordered pairs at the vertices of the polygon in Figure 22.13 to find the values for $F$ shown in Table 22.1 on page 104. From this table we see that the maximum value of $f$ is 128 and the minimum value of $F$ is 72.   ●

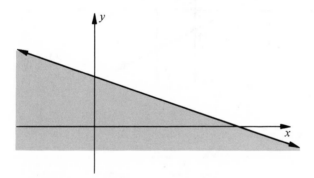

Figure 22.10

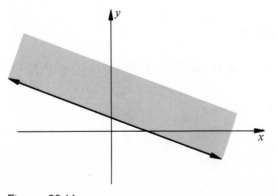

Figure 22.11

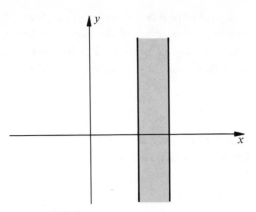

Figure 22.12

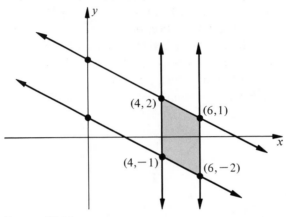

Figure 22.13

**Table 22.1**

| Vertices<br>(x, y) | (4, −1) | (4, 2) | (6, 1) | (6, −2) |
|---|---|---|---|---|
| F = 20x + 8y | 72 | 96 | 128 | 104 |

## EXERCISES FOR SECTION 22

In exercises 1 through 15 graph the solution sets for each of the given systems, and state their domain and range.

**1.** $y < 2x + 3$
$\quad y > 2x - 1$

**2.** $y \geq x - 4$
$\quad y < -x + 4$

**3.** $x + 2y + 3 < 0$
$\quad x^2 + 6x + 4 > y$

**4.** $x^2 + 4 < y$
$\quad x^2 - 4 > y$

**5.** $\quad y > 2$
$\quad y + 3x < 6$

**6.** $\quad x = 2$
$\quad y + x < 3$

7. $y \geq 3$
   $x \leq 2$
   $y < x$

11. $4x + 2y \geq 8$
    $x + y \geq 5$
    $2x + 7y \geq 20$
    $x \geq 1$
    $y \geq 0$

8. $|x - 1| < 1$
   $|y - 2| < 1$

12. $y \leq 6x - x^2$
    $y \geq x^2 - 2x$

9.   $y > x^2$
     $y < x + 6$
   $x + y - 6 < 0$

13. $x \geq y^2$
    $x^2 \leq -8y$

10. $|x| \geq 1$
    $|y| \geq 1$

14. $x \geq 2y^2 - 4$
    $x \leq y^2$

15. $y \geq x^2 + 2x - 3$
    $y \leq 1 - x^2$

16. Find the maximum and minimum values for $F = 5x + 10y$ if its domain is

$$2y - x \leq 10$$
$$2y - x \geq 4$$
$$1 \leq x \leq 3$$

17. Repeat exercise 16 for $F$ when its domain is

$$x + y \leq 16$$
$$0 \leq y \leq 12$$
$$1 \leq x \leq 6$$

18. Find the maximum and minimum values for $F = 100 + 2x - y$ when its domain is

$$x + y \geq 25$$
$$x + y \leq 80$$
$$0 \leq x \leq 35$$
$$0 \leq y \leq 60$$

19. Repeat exercise 18 for $F$ when its domain is

$$3x + 2y \geq 6$$
$$6x + 8y \leq 48$$
$$x \geq 0$$
$$y \geq 0$$

20. Determine which of the following relations are subsets of one another by comparing their respective graphs.
    (a) $S_1 = \{(x, y) \,|\, |x| + |y| < 1\}$
        $S_2 = \{(x, y) \,|\, x^2 + y^2 < 1\}$
        $S_3 = \{(x, y) \,|\, |x| < 1 \text{ and } |y| < 1\}$
    (b) $S_1 = \{(x, y) \,|\, |x - 1| + |y - 2| < 1\}$
        $S_2 = \{(x, y) \,|\, (x - 1)^2 + (y - 2)^2 < 1\}$
        $S_3 = \{(x, y) \,|\, |x - 1| < 1 \text{ and } |y - 2| < 1\}$
    (*Hint:* Refer to Exercises 28 and 29 in Section 21.)

21. Show that the graph of the system

$$y - x \geq 0$$
$$y + x \geq 0$$

shown in Example 22.1 is the same as the graph of $S = \{(x, y) \,|\, y \geq |x|\}$.

# CHAPTER 4
# POLYNOMIALS

## INTRODUCTION

An expression containing algebraic symbols (constants and variables) that is generated by a finite number of the algebraic operations of addition, subtraction, multiplication, extracting roots, and raising to powers is called an **algebraic expression.** We wish to consider a special type of algebraic expression called a **polynomial.**

**Definition 23.1** A **polynomial** is an algebraic expression whose variables have only nonnegative integers for exponents.

We will generally be concerned with polynomials in one variable. If this variable is $x$, the general $n$th-degree polynomial in $x$ is given by

$$P(x) = a_0 x^n + a_1 x^{n-1} + a_2 x^{n-2} + \ldots + a_{n-1} x + a_n, \quad \text{where } a_0 \neq 0$$

Each $a_i$ is a constant and these constants are called the **coefficients** of the polynomial. If the coefficients of $P(x)$ are rational, real, or complex numbers, then $P(x)$ is said to be a polynomial over the rational numbers, the real numbers, or the complex numbers. For the purposes of our discussion, we will consider only polynomials over the real numbers. The degree of a polynomial is the same as that of the term having the highest degree and a nonzero coefficient. Thus $P(x) = 6x^5 - x^3 + 2x^2 - 6$ is a fifth-degree polynomial in $x$ since the degree of $6x^5$ is five. If $P(x) = c, c \neq 0$, then $P(x)$ is a polynomial of degree zero, since $c = cx^0$. If $P(x) = 0$, we refer to $P(x)$ as the zero polynomial, but we do not assign a degree to it.

**107**

### Example 23.1

If $P(x) = x^2 + 3x + 2$, $P(x)$ is a second-degree polynomial. Hence $a_0 = 1$, $a_1 = 3$, and $a_2 = 2$. $P(x) = x^3 + x + x^{1/2}$ is not a polynomial since the expression contains an exponent that is not a positive integer.

**Definition 23.2** Two polynomials are **equal** if and only if the coefficients of corresponding powers are equal.

### Example 23.2

If $x^3 + 2x^2 - 3 = a_0 x^3 + a_1 x^2 + a_2 x + a_3$, find the values of $a_0$, $a_1$, $a_2$ and $a_3$.

Solution. We use definition (23.2) to equate the coefficients of like powers to obtain $a_0 = 1$, $a_1 = 2$, $a_2 = 0$, $a_3 = -3$. We note that $a_2 = 0$ since there was no term involving $x$ in the expression $x^3 + 2x^2 - 3$.  ●

### Example 23.3

If $P(x) = A(x + 1) + B(x - 2)$ and $Q(x) = 8 - x$, find the values of $A$ and $B$ such that $P(x) = Q(x)$.

Solution. If $P(x) = Q(x)$ we have

$$A(x + 1) + B(x - 2) = 8 - x$$
$$Ax + A + Bx - 2B = 8 - x$$
$$(A + B)x + (A - 2B) = -x + 8$$

We now use definition (23.2) and this latter equation to equate coefficients of like powers. Thus $A + B = -1$ and $A - 2B = 8$. To find the values of $A$ and $B$; we now solve the system of equations $\begin{cases} A + B = -1 \\ A - 2B = 8 \end{cases}$ to find that $A = 2$ and $B = -3$.  ●

### Example 23.4

Find the values of $A$ and $B$ such that $\dfrac{7x + 21}{(2x + 3)(x + 5)} = \dfrac{A}{2x + 3} + \dfrac{B}{x + 5}$.

Solution. We first multiply both sides of the given equation by $(2x + 3)(x + 5)$ to obtain

$$7x + 21 = A(x + 5) + B(2x + 3)$$
$$7x + 21 = Ax + 5A + 2Bx + 3B$$
$$7x + 21 = (A + 2B)x + (5A + 3B)$$

We again equate coefficients of like powers to obtain the system $\begin{cases} A + 2B = 7 \\ 5A + 3B = 21 \end{cases}$ whose solution is $A = 3$ and $B = 2$.  ●

Polynomials consisting of one, two, or three terms are called **monomials, binomials, and trinomials,** respectively. Thus $6x^2$ is a monomial, $x^3 + x$ is a binomial, and $3x^4 - x^2 + 6$ is a trinomial.

The degree of a monomial is determined by the exponent of the variable or variables that appear. If we have more than one variable in a single term, the degree of the monomial is the sum of the exponents of the variables.

### Example 23.5

$12x^3y^2z$ is a sixth-degree monomial in $xyz$ with a coefficient of 12. It is also third degree in $x$, second degree in $y$, and first degree in $z$.

### Example 23.6

$P(x) = 6x^4 + x^2 + 1$ is a fourth-degree polynomial in $x$. $Q(x, y) = x^3y^2 + 6x^2y^2 + xy$ is fifth degree in $xy$. $R(y) = y^3 + 2y - 6$ is third degree in $y$.

The symbols $P(x)$, $Q(x)$, etc., are used to denote polynomials in one variable. $P(x, y)$, $Q(x, y)$, etc., are used to denote polynomials in two variables. The notation extends easily to cover any situation we may encounter.

If $P(x)$ is a polynomial in $x$ and $a$ is any real number, $P(a)$ denotes the number we obtain by substituting $a$ for $x$ in $P(x)$. Thus, if $P(x) = x^3 - 2x + 1$, then $P(2) = 2^3 - 2(2) + 1 = 5$. When $x = 2$, $P(x) = 5$. We now have a pair of numbers of the general form $(x, P(x))$. If we think of $P(x)$ as $y$ or $f(x)$, we note that the polynomial notation $P(x)$, $Q(x)$, etc., is equivalent to the functional notation $f(x)$, $g(x)$, etc.

### Example 23.7

$P(x) = 2x + 1$ is a first-degree polynomial in $x$. It is also a linear function, since $P(x) = f(x) = 2x + 1$. $Q(x) = x^2 + x - 1$ is a polynomial of degree *two* in $x$. We see that $Q(x)$ is also a quadratic function.

## EXERCISES FOR SECTION 23

In Exercises 1–5, give the degree of each polynomial. If the algebraic expression is not a polynomial, so state.

**1.** $4x^3 + 3x^2 + 6$

**2.** $x^5 + x^2 - 3$

**3.** $x^4 + 3x^3 + 2x^2 + \dfrac{1}{x}$

**4.** $x^5 - 5x^4$

**5.** $x^2y^3 + x^2 + xy - 1$

**6.** If $P(x) = x^3 - x^2 + x - 1$, find
  (a) $P(1)$      (c) $P(1 + 2)$
  (b) $P(0)$      (d) $P(1) + P(2)$

(e) $P(1) \cdot P(2)$        (h) $3P(2) + 2P(3)$

(f) $\dfrac{P(1)}{P(2)}$        (i) $[P(2)]^2$

(g) $2P(0)$        (j) $P(3) - P(2) - P(1)$

**7.** If $P(x) = x^2 + 1$, $Q(x) = x^4 - 3x^3 + 2x^2 - 5x + 4$, and $R(x) = x^5 - 3x^4 + 5x^3 + 7x^2 + 2x - 1$, find:

(a) $P(x) \cdot Q(x)$        (e) $P(x) \cdot R(x)$

(b) $Q(x) + R(x)$        (f) $[P(x)]^2$

(c) $R(x) - Q(x)$        (g) $3P(x) + 2Q(x)$

(d) $P(x)[Q(x) + R(x)]$        (h) $2R(x) - 3Q(x)$

**8.** If $P(x) = A(x + 6) + B(x - 3)$ and $Q(x) = 5x + 3$, find values for $A$ and $B$ such that $P(x) = Q(x)$.

**9.** If $P(x) = (A - 2)x^2 + (B + 3)x + 1$ and $Q(x) = 4x^2 + 0x + 1$, find values for $A$ and $B$ such that $P(x) = Q(x)$.

**10.** Find the values for $A$ and $B$ such that $\dfrac{x - 18}{(x + 2)(x - 3)} = \dfrac{A}{x + 2} + \dfrac{B}{x - 3}$.

**11.** Find the values for $A$ and $B$ such that $\dfrac{3x - 15}{x(x - 3)} = \dfrac{A}{x} + \dfrac{B}{x - 3}$.

**12.** Find the values for $A$, $B$, and $C$ such that

$$\frac{x^2 - 9x - 6}{x(x - 2)(x + 3)} = \frac{A}{x} + \frac{B}{x - 2} + \frac{C}{x + 3}$$

**13.** If $P(x)$ and $Q(x)$ are polynomials of degree $m$ and $n$, respectively ($m > n$), what is the degree of

(a) their product, $P(x) \cdot Q(x)$?

(b) their sum, $P(x) + Q(x)$?

(c) their difference, $P(x) - Q(x)$?

(d) their quotient, $\dfrac{P(x)}{Q(x)}$?

# SECTION 24
# OPERATIONS ON POLYNOMIALS: SYNTHETIC DIVISION

The addition, subtraction, and multiplication of polynomials should be familiar to us. In fact, these operations with polynomials always yield a polynomial. Thus $(x^3 - x^2 + 2x - 6) + (x^4 - 2x^3) = x^4 - x^3 - x^2 + 2x - 6$, and $(x^3 + 2)(x^4 + x + 1) = x^7 + 3x^4 + x^3 + 2x + 2$.

The division of polynomials, however, is not quite so simple. Just as the division of two integers may not yield an integer for a result, the division of two poynomials may not yield a polynomial for a result. If the division of integers were limited to those cases where the result is an integer, expressions such as $\frac{7}{3}$ or $\frac{13}{5}$ would not be defined, since there are no integers $p$ and $q$ such that $7 = 3p$ or $13 = 5q$. To avoid this problem, we define the division of integers in the following manner.

**Definition 24.1** Let $a/b$ be the ratio of any two integers. There exist integers $q$ and $r (0 \leq r < b)$ such that $a = q \cdot b + r$. The number $q$ is called the **quotient** and the number $r$ is the **remainder.** If $r = 0$, we say the division is **exact** and that $b$ is a **factor** of $a$.

**Example 24.1**

If $a = 37$ and $b = 13$, find $q$ and $r$ if $a$ is divided by $b$.

Solution. Using long division:

$$
\begin{array}{r}
2 \to q \\
13\overline{)37} \\
\underline{26} \\
11 \to r
\end{array}
$$

Thus $37 = 2 \cdot 13 + 11$. •

We encounter the same type of situations when we divide polynomials.

**Definition 24.2** If $P(x)$, $D(x)$, and $Q(x)$ are polynomials $[D(x) \neq 0]$ and if $P(x) = Q(x) \cdot D(x)$, then $D(x)$ divides $P(x)$ or $P(x)$ is exactly divisible by $D(x)$. We write this as $\dfrac{P(x)}{D(x)} = Q(x)$. $P(x)$ is the dividend, $D(x)$ is the divisor, and $Q(x)$ is the quotient.

**Example 24.2**

If $P(x) = x^3 - 3x^2 + 5x - 6$ and $D(x) = x - 2$, find $Q(x)$.

Solution

$$
\begin{array}{r}
x^2 - x + 3 \\
x - 2\overline{)x^3 - 3x^2 + 5x - 6} \\
\underline{x^3 - 2x^2} \\
-x^2 + 5x \\
\underline{-x^2 + 2x} \\
3x - 6 \\
3x - 6
\end{array}
$$

Thus, $Q(x) = x^2 - x + 3$, $R(x) = 0$, and $x^3 - 3x^2 + 5x - 6 = (x^2 - x + 3)(x - 2)$. The division is exact. •

But, in general, polynomial division is not always exact. For example, $(6x^3 - 14x + 25) \div (2x + 4)$ yields a quotient of $3x^2 - 6x + 5$ with a remainder of 5. We note that division of polynomials always yields a polynomial quotient and a polynomial remainder. This statement is the basis for the following property:

Let $P(x)$ and $D(x)$ be polynomials and $D(x) \neq 0$. Then there are unique polynomials $Q(x)$ and $R(x)$, where $R(x) = 0$ or the degree of $R(x)$ is less than the degree of $D(x)$, such that

$$P(x) = Q(x)D(x) + R(x)$$

$P(x)$ is the dividend, $D(x)$ is the divisor, $Q(x)$ is the quotient, and $R(x)$ is the remainder.

**Example 24.3**

Let $P(x) = 2x^5 + 4x^4 + x^2 - 1$ and $D(x) = x^2 + 1$. Find $Q(x)$ and $R(x)$.

Solution

$$
\require{enclose}
\begin{array}{r}
2x^3 + 4x^2 - 2x - 3 \\[-2pt]
x^2 + 1 \enclose{longdiv}{2x^5 + 4x^4 + 0x^3 + x^2 + 0x - 1} \\
\underline{2x^5 \qquad\quad + 2x^3} \\
4x^4 - 2x^3 + \ x^2 \\
\underline{4x^4 \qquad\quad + 4x^2} \\
-2x^3 - 3x^2 + 0x \\
\underline{-2x^3 \qquad\quad - 2x} \\
-3x^2 + 2x - 1 \\
\underline{-3x^2 \qquad\quad - 3} \\
2x + 2
\end{array}
$$

Therefore, $Q(x) = 2x^3 + 4x^2 - 2x - 3$ and $R(x) = 2x + 2$.   ●

If the divisor $D(x)$ of a polynomial is of the form $x - a$, the long-division process may be shortened by a method called **synthetic division.** This method is illustrated in the following examples.

**Example 24.4**

Divide $2x^4 - 3x^3 - 5x^2 + 3x + 9$ by $x - 2$.

Solution

$$
\require{enclose}
\begin{array}{r}
2x^3 + \ x^2 - 3x - 3 \\[-2pt]
x - 2 \enclose{longdiv}{2x^4 - 3x^3 - 5x^2 + 3x + 9} \\
\underline{2x^4 - 4x^3} \\
x^3 - 5x^2 \\
\underline{x^3 - 2x^2} \\
-3x^2 + 3x \\
\underline{-3x^2 + 6x} \\
-3x + 9 \\
\underline{-3x + 6} \\
3
\end{array}
$$

Here $Q(x) = 2x^3 + x^2 - 3x - 3$ and $R(x) = 3$.

If we examine the division carefully, we notice that the initial coefficient in the quotient is the same as the initial coefficient in the dividend. Each succeeding coefficient of the quotient results from multiplying the coefficient that precedes it by $-2$ and then subtracting that product from the corresponding coefficient of the dividend. Thus the coefficient of the $x^2$ term in the quotient is obtained by subtracting $2 \cdot (-2)$ from $-3$. Successive coefficients of the quotient and the remainder are determined in a similar fashion.

The process may be further simplified by multiplying by $+2$ and adding, instead of multiplying by $-2$ and then subtracting. In schematic form we have the following:

$$
\begin{array}{r}
\text{Coefficients of the dividend} \\
\overbrace{\begin{array}{rrrrr} 2 & -3 & -5 & +3 & +9 \end{array}} \;\underline{\;\big|\;2\;} \\
\begin{array}{rrrr} 4 & 2 & -6 & -6 \end{array} \\
\hline
\begin{array}{rrrr} 2 & 1 & -3 & -3 \end{array} \;|+3 \;\rightarrow\; \text{Remainder} \\
\underbrace{\qquad\qquad\qquad\qquad} \\
\text{Coefficients of the quotient}
\end{array}
$$

We write the number 2 at the upper right because we are dividing by $x - 2$. If we were dividing by $x + 2$, we would write the number $-2$ since $x + 2 = x - (-2)$.

### Example 24.5

Use synthetic division to find the quotient and remainder if $P(x) = 2x^4 - 3x^3 + 2x + 1$ and $D(x) = x + 4$.

**Solution.** The coefficient of the $x^2$ term in $P(x)$ is 0. Thus $P(x) = 2x^4 - 3x^3 + 0x^2 + 2x + 1$. Since $x + 4 = x - (-4)$, we use the factor $-4$ in our process.

$$
\begin{array}{rrrrr}
2 & -3 & 0 & 2 & +1 \quad\underline{\;\big|\;-4\;} \\
  & -8 & +44 & -176 & +696 \\
\hline
2 & -11 & +44 & -174 & |\;\;697
\end{array}
$$

Therefore, $Q(x) = 2x^3 - 11x^2 + 44x - 174$ and $R = 697$ or $2x^4 - 3x^3 + 2x + 1 = (x + 4)(2x^3 - 11x^2 + 44x - 174) + 697$. ●

### Example 24.6

Use synthetic division to find $Q(x)$ and $R(x)$ if $P(x) = x^5 - 1$ and $D(x) = x - 1$.

**Solution.** Since $P(x) = x^5 + 0x^4 + 0x^3 + 0x^2 + 0x - 1$, we have

$$
\begin{array}{rrrrrr}
1 & 0 & 0 & 0 & 0 & -1 \quad\underline{\;\big|\;1\;} \\
  & 1 & 1 & 1 & 1 & 1 \\
\hline
1 & 1 & 1 & 1 & 1 & |\;\;0
\end{array}
$$

Thus $Q(x) = x^4 + x^3 + x^2 + x + 1$ and $R = 0$. We note that $x^5 - 1 = (x - 1)(x^4 + x^3 + x^2 + x + 1)$ ●

In the preceding example, we note that $x - 1$ and $x^4 + x^3 + x^2 + x + 1$ are factors of $x^5 - 1$. However, we must be very careful when we discuss the factored or reduced form of a polynomial. The factorization depends upon the set of numbers that is to be used as coefficients. Thus the polynomial $x^2 - 4$ factors over the set of integers since $x^2 - 4 = (x - 2)(x + 2)$, and each of these factors has integral coefficients. $x^2 - 4$ also factors over the rational numbers and the real numbers, since the set of integers is a subset of both the rational and real numbers.

**Example 24.7**

$x^2 - 3$ does not factor over the set of rational numbers. However, $x^2 - 3 = (x - \sqrt{3}) \cdot (x + \sqrt{3})$, so $x^2 - 3$ does factor over the real numbers.

**Example 24.8**

$x^2 + 3$ does not factor over the set of rational numbers or the set of real numbers. It does factor over the set of complex numbers, but we do not wish to enter into that discussion at this time.

## EXERCISES FOR SECTION 24

1. For each of the following, find positive integers such that $a = q \cdot b + r(0 \le r < b)$.
    (a) $a = 317, b = 23$     (c) $a = 15, b = 5$
    (b) $a = 1257, b = 49$     (d) $a = 17, q = 21$

2. For each of the following, find $Q(x)$ and $R(x)$ if
    (a) $P(x) = x^3 + x^2 - 5x + 2, D(x) = x^2 - 2x$.
    (b) $P(x) = x^4 - 7x^3 + x^2 + x - 5, D(x) = x^2 - x - 1$.
    (c) $P(x) = 2x^4 - x^3 - x^2 - 7x - 1, D(x) = 3x^2 + 2x + 2$.
    (d) $P(x) = x^2 + 1, D(x) = x^3 + 2x + 3$.
    (e) $P(x) = x^8 + 6x^4 + 3x^3 + x - 1, D(x) = x^3$
    (f) $P(x) = x^6 - x^4 + x^3 - x + 1, D(x) = x^2 + 1$
    (g) $P(x) = x^5 + x^4 + x^3 + x^2 + x + 1, D(x) = x^3 + 1$
    (h) $P(x) = 2x^4 - 3x^3 + x^2 + x - 2, D(x) = x^2 - 3x + 2$
    (i) $P(x) = 2x^3 + x^5 - 3x - 2, D(x) = x^2 - 3x + 1$
    (j) $P(x) = x^4 - x^3 - 2x^2 - 3x - 3, D(x) = x^2 + x + 1$

3. One factor of $x^4 + x^2 + 1$ is $x^2 + x + 1$. Find the other factor.

4. One factor of $x^3 + x^2 + 3x - 5$ is $x^2 + 2x + 5$. Find the other factor.

5. Use synthetic division to determine the quotient and remainder in each of the following. Write the results in the form $P(x) = D(x) \cdot Q(x) + R(x)$.
    (a) $(x^3 + 3x^2 - 2x - 5) \div (x - 2)$     (g) $(x^4 + 16) \div (x - 2)$
    (b) $(3x^5 + x^3 - 4) \div (x + 2)$     (h) $(x^5 + 1) \div (x + 1)$

    (c) $(x^4 - 2x^3 - 4x^2 + 2x + 3) \div (x + 1)$     (i) $(2x^4 - 3x^3 - 5x^2 + 7x + 4) \div \left(x - \frac{1}{2}\right)$

    (d) $(2x^3 + x^2 - 2x - 1) \div (2x + 1)$     (j) $(2x^4 - 3x^3 - 5x^2 + 7x + 4) \div (2x - 1)$

    (e) $(4x^3 + 6x^2 - 2x + 3) \div (2x - 1)$     (k) $(3x^3 + 4x^2 + 6x + 8) \div \left(x + \frac{4}{3}\right)$

    (f) $(2x^5 - 7) \div (x - 1)$     (l) $(3x^3 + 4x^2 + 6x + 8) \div (3x + 4)$

6. Is $x^2 - 5$ factorable (a) over the rational numbers? (b) Over the real numbers? If so, what are the factors?

7. Write $P(x) = (x^2 - 4)(x^2 - 5)(x^2 + 9)$ in completely factored form
    (a) Over the rational numbers.
    (b) Over the real numbers.

**8.** Using the result that $x^3 - a^3 = (x - a)(x^2 + ax + a^2)$, factor each of the following:
(a) $x^3 - 8$      (d) $x^3 - 16$
(b) $y^6 - 64$      (e) $x^3 - (a + b)^3$
(c) $27x^3 - 1$

**9.** If $(x + a)^3 = x^3 + 3ax^2 + 3a^2x + a^3$, factor each of the following:
(a) $x^3 + 6x^2 + 12x + 8$
(b) $x^3 - 3x^2 + 3x - 1$
(c) $a^3x^3 + 3a^2bx^2 + 3ab^2x + b^3$

**10.** (a) Show that $(r^n - 1) = (r - 1)(r^{n-1} + r^{n-2} + \ldots + r + 1)$ by multiplying.
(b) Use the result in part (a) to show that the sum of the geometric sequence, $a, ar, ar^2, \ldots,$

$$ar^{n-1} (r \neq 1) \text{ is given by } \frac{a(r^n - 1)}{r - 1}$$

**11.** The basis for synthetic division can be found in the following argument. Suppose that $D(x) = x - c$, where $c$ is any constant. If $P(x)$ is of the form $a_0x^n + a_1x^{n-1} + \ldots + a_n$, then $Q(x)$ will necessarily be of the form $b_0x^{n-1} + b_1x^{n-2} + \ldots + b_{n-2}x + b_{n-1}$. Theorem 24.1 tells us that since $x - c$ is a polynomial of degree 1, $R(x)$ must be some constant $r$ (possibly 0) and that $P(x) = Q(x) \cdot (x - c) + r$. Substituting, we have

$$a_0x^n + a_1x^{n-1} + \ldots + a_n = (b_0x^{n-1} + b_1x^{n-2} + \ldots + b_{n-1})(x - c) + r$$
$$a_0x^n + a_1x^{n-1} + \ldots + a_n = b_0x^n - cb_0x^{n-1} + b_1x^{n-1} - cb_1x^{n-2}$$
$$+ b_2x^{n-2} + \ldots + b_{n-1}x - cb_{n-1} + r$$

Then

$$a_0x^n + a_1x^{n-1} + \ldots + a_n = b_0x^n + (b_1 - cb_0)x^{n-1} + (b_2 - cb_1)x^{n-2}$$
$$+ \ldots + (b_{n-1} - cb_{n-2})x + (r - cb_{n-1}).$$

By definition, equal polynomials have equal coefficients. This enables us to obtain the following relationships between the coefficients:

$$b_0 = a_0$$
$$b_1 - cb_0 = a_1 \quad \text{or} \quad b_1 = a_1 + cb_0$$
$$b_2 - cb_1 = a_2 \quad \text{or} \quad b_2 = a_2 + cb_1$$
$$b_{n-1} - cb_{n-2} = a_{n-1} \quad \text{or} \quad b_{n-1} = a_{n-1} + cb_{n-2}$$
$$r - cb_{n-1} = a_n \quad \text{or} \quad r = a_n + cb_{n-1}$$

These equations suggest the following method for computing the $b$'s of $Q(x)$.

| $a_0$ | $a_1$ | $a_2$ | $\cdots$ | $a_{n-1}$ | $a_n$ | $\underline{c}$ |
|---|---|---|---|---|---|---|
|  | $+cb_0$ | $cb_1$ |  | $cb_{n-2}$ | $cb_{n-1}$ |  |
| $b_0$ | $b_1$ | $b_2$ | $\cdots$ | $b_{n-1}$ | $r$ |  |

If we wish to divide $x^3 + 3x^2 - 6x + 1$ by $x - 2$ using the procedure described, we note that $Q(x) = b_0x^2 + b_1x + b_2$. Since $P(x) = D(x) \cdot Q(x) + r$, we have

1. $x^3 + 3x^2 - 6x + 1 = (x - 2)(b_0x^2 + b_1x + b_2) + r$.
2. $x^3 + 3x^2 - 6x + 1 = b_0x^3 + b_1x^2 - 2b_0x^2 + b_2x - 2b_1x - 2b_2 + r$.
3. $x^3 + 3x^2 - 6x + 1 = b_0x^3 + (b_1 - 2b_0)x^2 + (b_2 - 2b_1)x - 2b_2 + r$.

Equating coefficients, $b_0 = 1$, $b_1 - 2b_0 = 3$, $b_2 - 2b_1 = -6$, and $-2b_2 + r = 1$. Therefore, $b_0 = 1$, $b_1 = 5$, $b_2 = 4$, and $r = 9$. We now conclude that $Q(x) = x^2 + 5x + 4$ and $r = 9$.

For a better appreciation of the ease of synthetic division, obtain the quotient and remainder by the method outlined above if

(a) $x^3 - 3x^2 + 2x + 6$ is divided by $x - 2$.

(b) $x^3 + x^2 - 4$ is divided by $x + 1$.

(c) $x^4 - 3x^2 + 3x - 4$ is divided by $x - 1$.

SECTION **25**

## FACTORS AND ZEROS OF REAL POLYNOMIALS

With every polynomial $P(x)$ we may associate a polynomial function defined by $y = P(x)$. A number $a$ is called a **zero** of a polynomial $P(x)$ if $P(a) = 0$. The zeros of the polynomial function are the same as the roots of the polynomial equation $P(x) = 0$. We now wish to discuss some of the methods used to solve a polynomial equation. However, it might be beneficial to first ask if every polynomial equation has a solution. The answer is given in the following theorem.

**Theorem 25.1  Fundamental Theorem of Algebra.**  Every polynomial function of degree $n \geq 1$ has at least one zero (not necessarily real).

The proof of this theorem is omitted.

To continue our study of the properties of polynomial functions, we need the following theorems.

**Theorem 25.2  The Remainder Theorem.**  Let $P(x)$ be any real polynomial and $a$ any real number. If we divide $P(x)$ by $(x - a)$ until a remainder $R$ of degree 0 or equal to 0 is found, then $R = P(a)$.

**Proof.**  According to Theorem 24.1, the division of $P(x)$ by $x - a$ may be described as $P(x) = Q(x)(x - a) + R$, where $Q(x)$ is a polynomial of degree $n - 1$ and $R$ is a real number. Since $P(x) = Q(x)(x - a) + R$ is true for all real values of $x$, we may substitute $a$ for $x$ and obtain

$$P(a) = Q(a)(a - a) + R$$

Thus $P(a) = 0 \cdot Q(a) + R$ or $R = P(a)$.  ●

**Example 25.1**

Evaluate $P(2)$ when $P(x) = x^3 + 2x^2 - x + 5$.

**Solution.**  According to the remainder theorem, $R = P(2)$. $P(2) = 2^3 + 2 \cdot 2^2 - 2 + 5 = 19$. Thus $R = 19$. The same result is obtained by means of synthetic division.

$$
\begin{array}{rrrr|r}
1 & 2 & -1 & +5 & \underline{2} \\
  & 2 & 8 & 14 & \\
\hline
1 & 4 & 7 & | \; 19 & = R
\end{array}
$$
●

**Example 25.2**

Evaluate $P(-2)$ when $P(x) = x^3 + 6x^2 + 11x + 6$.

Solution.  Again, according to the remainder theorem, $R = P(-2) = (-2)^3 + 6(-2)^2 + 11(-2) + 6 = -8 + 24 - 22 + 6 = 0$.  Thus $R = 0$ and $x + 2$ is a factor of $x^3 + 6x^2 + 11x + 6$.  ●

**Theorem 25.3  Factor theorem.**  If $P(x)$ is a real polynomial and if $P(a) = 0$, then $x - a$ is a factor of $P(x)$.

Proof.  The remainder theorem tells us that $P(x) = (x - a) \cdot Q(x) + P(a)$. Since $P(a) = 0$ by assumption, we have that $P(x) = (x - a) \cdot Q(x)$. Thus $x - a$ is a factor of $P(x)$.  ●

The factor theorem is especially useful in finding the roots of a polynomial equation $P(x) = 0$. Each factor we can find yields a root of the equation. Thus from Example 25.2 we note that since $x + 2$ is a factor of $x^3 + 6x^2 + 11x + 6$, $-2$ is a root of $x^3 + 6x^2 + 11x + 6 = 0$. We use the factor theorem and synthetic division in an attempt to find other solutions of the polynomial equation. Synthetic division enables us to write $P(x)$ in the form $(x + 2) \cdot Q(x)$ as follows:

$$
\begin{array}{rrrr|r}
1 & +6 & +11 & +6 & \underline{-2} \\
  & -2 & -8 & -6 & \\
\hline
1 & +4 & +3 & \mid \ 0 &
\end{array}
$$

$P(x) = (x + 2)(x^2 + 4x + 3)$. Now $P(x) = 0$ is equivalent to $(x + 2)(x^2 + 4x + 3) = 0$. Thus

$$x + 2 = 0 \quad \text{or} \quad x^2 + 4x + 3 = 0$$
$$x = -2 \quad \text{or} \quad (x + 1)(x + 3) = 0$$

Therefore, the solutions of $P(x) = 0$ are $x = -1, -2$, and $-3$.

**Example 25.3**

Find all real roots of $x^3 - 4x^2 + 5x - 2 = 0$ if 1 is a root.

Solution.  We use synthetic division to reduce the polynomial as follows:

$$
\begin{array}{rrrr|r}
1 & -4 & +5 & -2 & \underline{1} \\
  & 1 & -3 & +2 & \\
\hline
1 & -3 & +2 & \mid \ 0 &
\end{array}
$$

Thus $x^3 - 4x^2 + 5x - 2 = (x - 1)(x^2 - 3x + 2) = (x - 1)(x - 1)(x - 2) = 0$. Setting each factor equal to 0, we have $x - 1 = 0$, $x - 1 = 0$, and $x - 2 = 0$. The solutions are seen to be $+1, +1$, and $+2$. In this case we say that $x = 1$ is a multiple root or root of multiplicity 2.  ●

**Example 25.4**

Find all real roots of $x^3 - 2x^2 + x - 2 = 0$ if 2 is a root.

Solution

```
1  −2  +1  −2  │ 2
        2   0   2
───────────────────
1   0   1  │ 0
```

Thus $x^3 - 2x^2 + x - 2 = (x - 2)(x^2 + 1) = 0$. The factor $x - 2$ yields the given root, $x = 2$. $x^2 + 1 = 0$ has nonreal solutions according to the quadratic formula. We conclude that the equation has only one real root, $x = 2$. ●

### Example 25.5

Find a third-degree polynomial equation whose real and distinct roots are $-1$, 2, and 4.

Solution. Since the zeros of the polynomial we seek are $-1$, 2, and 4, the factors must be $(x + 1)$, $(x - 2)$, and $(x - 4)$. Thus $P(x) = (x + 1)(x - 2)(x - 4)$ and the corresponding equation is $(x + 1)(x - 2)(x - 4) = 0$. A general solution for this problem is $a(x + 1)(x - 2)(x - 4) = 0$, where $a \in \mathbb{R}$ and $a \neq 0$. ●

## EXERCISES FOR SECTION 25

**1.** Determine the remainder for each of the following divisions.

(a) $(x^3 - 2x^2 + 4x + 6) \div (x - 2)$     (d) $(3x^3 - x^2 - 15x + 5) \div (x - 5)$

(b) $(x^4 - 8x^2 + 16) \div (x - 2)$        (e) $(x^3 + 3x^2 - 2x - 5) \div (x - 2)$

(c) $(x^4 - 8x^2 + 16) \div (x + 2)$        (f) $(4x^3 - 2x^2 + 6x - 8) \div (2x - 1)$

**2.** Evaluate $P(1)$ when $P(x) = x^6 - x^4 + x^2 + 2$.

**3.** Evaluate $P(3)$ when $P(x) = x^3 - 3x^2 + 2x + 1$.

**4.** Evaluate $P(-2)$ when $P(x) = x^4 - x^3 + 2x^2 - x + 3$.

**5.** (a) Find a fourth-degree equation having the roots 1, 2, $-3$, and $-3$.

(b) Find the general solution.

**6.** (a) Find a cubic equation having the roots 0, 1, and 3.

(b) Find the general solution.

**7.** (a) Find the equation with integral coefficients whose roots are 0, 1, $\frac{1}{2}$, and $\frac{2}{3}$.

(b) Find the general solution.

**8.** If $P(x) = x^3 - 3x^2 - 13x + 15$ use the factor theorem to determine which of the following are factors of $P(x)$.

(a) $x + 3$      (b) $x + 5$      (c) $x - 1$

**9.** If $P(x) = x^4 - 5x^3 + 6x^2 + 4x - 8$ use the factor theorem to determine which of the following are factors of $P(x)$.

(a) $x + 2$      (b) $x - 2$      (c) $x + 1$

**10.** Show that $(x + 1)$ is a factor of $x^4 - 5x^3 - 13x^2 + 53x + 60$.

11. Find $Q(x)$ if $x^4 - 10x^2 + 9 = (x + 3) \cdot Q(x)$.

12. Find the roots of $x^3 + 6x^2 + 11x + 6 = 0$, given that one root is $-2$.

13. (a) Show that $x^2 - 4$ is a factor of $x^4 + x^3 - 10x^2 - 4x + 24$.
    (b) Find all roots of $x^4 + x^3 - 10x^2 - 4x + 24 = 0$.

14. Two of the roots of $x^4 - 2x^3 - 7x^2 + 8x + 12 = 0$ are $-1$ and $+2$. Find the remaining roots.

15. $x^4 + 2x^3 - 7x^2 - 8x + 12 = 0$ has the roots 1 and $-2$. Find the quadratic equation whose roots are the remaining two roots of the given equation. Find these roots.

16. For what value of $p$ is $x^2 + 5x + p$ divisible by $x - p$?

17. Find a value for $p$ such that $x - 4$ is a factor of $x^4 + 2x^3 - 21x^2 + px + 40$.

18. Find a value for $p$ such that $x + 1$ is a factor of $x^3 - 9x^2 + 14x + p$.

19. Show that $x + a$ is a factor of $x^n - a^n$ when $n$ is any even positive integer.

20. Show that $x + a$ is not a factor of $x^n - a^n$ when $n$ is an odd positive integer.

21. Show that $x - a$ is a factor of $x^n - a^n$ if $n$ is any positive integer.

22. Show that $x - 1$ is a factor of $x^{64} - 1$.

23. Show that $x + 1$ is a factor of $17x^{99} + 32x^{66} - 15$.

24. Prove the converse of the Factor Theorem.

SECTION **26**

# GRAPHS OF POLYNOMIAL FUNCTIONS: THE RATIONAL ROOT THEOREM

The graphing of polynomial functions is usually left to a first course in calculus, where the curve-sketching techniques are more exact and elaborate. But we can still make some worthwhile observations here. We know that any real polynomial equation in factored form has roots that are easy to find. We simply set each factor equal to zero and solve for the root. Thus $\{x \mid (x - 1)(x - 2)(x + 3) = 0\} = \{1, 2, -3\}$.

The real roots of the polynomial equations $P(x) = 0$ correspond to the real zeros of the polynomial function $y = P(x)$. With the aid of the zeros of the polynomial function, the $y$-intercept, a few selected points, and the fact that all polynomial functions are continuous, we can usually sketch the curve.

**Example 26.1**
Find the zeros and graph $f(x) = (x + 2)(x - 1)(x - 4)$.

Solution. The zeros of $f$ occur at $x = -2$, $x = 1$, and $x = 4$. To graph the function, we would like to know the $y$-intercept and a few additional points. These are given in Table 26.1. The graph is given in Figure 26.1. ●

**Table 26.1**

| x | −3 | −1 | 0 | 2 | 3 | 5 |
|---|---|---|---|---|---|---|
| y = f(x) | −28 | 10 | 8 | −8 | −10 | 28 |

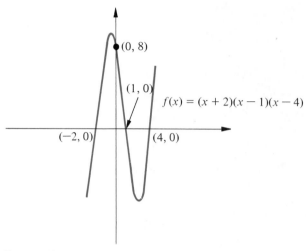

Figure 26.1

**Example 26.2**
Use the graph of the polynomial function in Example 26.1 to find the solution set for $(x + 2)(x - 1)(x - 4) > 0$.

**Solution.** We first note that $(x + 2)(x - 1)(x - 4) > 0$ is equivalent to $y > 0$. From the graph of $y = (x + 2)(x - 1)(x - 4)$ shown in Figure 26.1 we see that $y > 0$ when $-2 < x < 1$ or $x > 4$. Thus the desired solution set is $\{x \mid -2 < x < 1 \text{ or } x > 4\}$.

**Example 26.3**
Find the real zeros and graph $f(x) = (x^2 + 1)(x - 1)$.

**Solution.** $f(x) = 0$ has only one real zero and that occurs when $x = 1$. To graph the function we construct Table 26.2. The graph is shown in Figure 26.2. From the graph we note that $\{x \mid (x^2 + 1)(x - 1) \geq 0\} = [1, \infty)$ and $\{x \mid (x^2 + 1)(x - 1) < 0\} = (-\infty, 1)$.  ●

If we start with a real polynomial function of degree greater than 2 which is not in factored form, it may be difficult to find the real zeros. However, if we restrict our

**Table 26.2**

| x | −2 | −1 | 0 | 1 | 2 | 3 |
|---|---|---|---|---|---|---|
| f(x) | −15 | −4 | −1 | 0 | 5 | 20 |

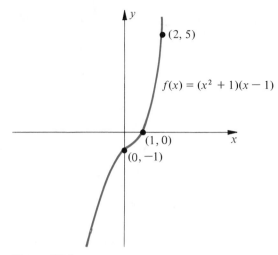

Figure 26.2

polynomials to those with integral coefficients, there is a general method which enables us to find the rational zeros of the polynomial function.

**Theorem 26.1    Rational Root Theorem.**    If $P(x) = a_0 x^n + a_1 x^{n-1} + \ldots + a_{n-1} x + a_n$, where the coefficients are integers and $p/q$ is a rational root in lowest terms of $P(x) = 0$, then $p$ is a divisor of $a_n$ and $q$ is a divisor of $a_0$.

**Remark 26.1**

The theorem says nothing about the existence of rational roots. It merely states how we can determine the possibilities if $P(x)$ is of the appropriate form.

**Proof.**    Since $p/q$ is a root of $P(x) = 0$, we have

$$a_0(p/q)^n + a_1(p/q)^{n-1} + \ldots + a_{n-1}(p/q) + a_n = 0$$

Multiplying by $q^n$, we obtain

$$a_0 p^n + a_1 p^{n-1} q + \ldots + a_{n-1} p q^{n-1} + a_n q^n = 0$$

Thus

$$a_0 p^n + a_1 p^{n-1} q + \ldots + a_{n-1} p q^{n-1} = -a_n q^n$$

Both sides of the equation are integers. Now $p$ is a factor of the left-hand side and therefore a factor of the right-hand side, $a_n q^n$. Since we assumed that $p/q$ was in lowest terms, $p$ cannot be a factor of $q^n$ and must therefore be a factor of $a_n$.

Similarly, if $a_1 p^{n-1} q + \ldots + a_{n-1} p q^{n-1} + a_n q^n = -a_0 p^n$, then $q$ is a factor of the left-hand side and also the right-hand side, $a_0 p^n \cdot q$ cannot be a factor of $p^n$ and must therefore be a factor of $a_0$.    ●

**Example 26.4**

List the possible rational roots of $P(x) = 3x^3 - 4x^2 + 5x - 2 = 0$.

Solution.   If $p/q$ is to be a rational root of $P(x) = 0$, $p$ must be a factor of $-2$ and $q$ must be a factor of 3. Therefore, $p = \pm 1, \pm 2$, and $q = \pm 1, \pm 3$. The possible rational roots are $\pm 1, \pm 2, \pm \frac{1}{3}, \pm \frac{2}{3}$.   ●

### Remark 26.2
In Theorem 26.1, if $a_0 = 1$, then the possible rational roots, $p/q$, are simply the divisors of $a_n$ (see Example 26.5).

### Example 26.5
Find all rational roots of $P(x) = x^4 - 4x^3 + 3x^2 + 4x - 4 = 0$.

Solution.   If $p/q$ is to be a rational root, $p$ must be a factor of $-4$ and $q$ must be a factor of 1. Therefore, $p = \pm 1, \pm 2, \pm 4$, and $q = \pm 1$. The possible rational roots are $\pm 1, \pm 2$, and $\pm 4$. To find the roots of the equation we use synthetic division:

$$
\begin{array}{rrrrr|r}
1 & -4 & 3 & 4 & -4 & \underline{2} \\
  & 2 & -4 & -2 & +4 & \\
\hline
1 & -2 & -1 & +2 & | & 0
\end{array}
$$

One of the roots of the equation is $x = 2$. To find the remaining roots we use the "reduced equation" $x^3 - 2x^2 - x + 2 = 0$. This "reduced equation" is a consequence of the fact that our division was even. Thus the coefficients 1, $-2, -1$, and 2 represent $x^3 - 2x^2 - x + 2$, which is a factor of our original equation.

　　　We again make use of the rational-root theorem to note that the possible rational roots are now $\pm 1$ and $\pm 2$:

$$
\begin{array}{rrrr|r}
1 & -2 & -1 & 2 & \underline{1} \\
  & 1 & -1 & -2 & \\
\hline
1 & -1 & -2 & | & 0
\end{array}
$$

We see that a second root of the equation is $x = 1$.

　　　The remaining roots of the equation now come from $x^2 - x - 2 = 0$. We could continue to use synthetic division, but it is somewhat easier to factor the reduced equation $x^2 - x - 2 = (x - 2)(x + 1) = 0$. Hence $x = -1$, and $x = 2$. The solutions for our original equation are $x = 1, -1, 2$, and 2. In this case 2 is a multiple root. *Note:* Since we know the roots of the equation, we also know the factors of the polynomial. Thus

$$(x^4 - 4x^3 + 3x^2 + 4x - 4) = (x - 1)(x + 1)(x - 2)(x - 2)$$   ●

### Example 26.6
Graph the polynomial function $y = x^4 - 4x^3 + 3x^2 + 4x - 4$.

Solution.   We know from Example 26.5 that the zeros of the function occur at $x = -1, 1$, and 2. The $y$-intercept is at $(0, -4)$. A few additional points are given in Table 26.3. The graph is shown in Figure 26.3.   ●

**Table 26.3**

| $x$ | $0$ | $\frac{3}{2}$ | $3$ |
|---|---|---|---|
| $y = f(x)$ | $-4$ | $\frac{5}{16}$ | $8$ |

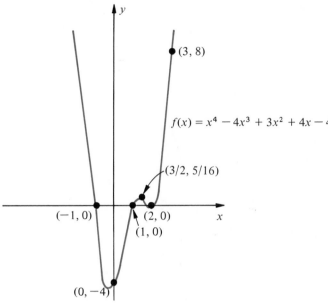

Figure 26.3

**Example 26.7**
Use the graph of the polynomial in Example 26.6 to find the solution set for $x^4 - 4x^3 + 3x^2 + 4x - 4 < 0$.

Solution. We first note that $x^4 - 4x^3 + 3x^2 + 4x - 4 < 0$ is equivalent to $y < 0$. From the graph of $y = x^4 - 4x^3 + 3x^2 + 4x - 4$ shown in Figure 26.3 we see that $y < 0$ when $-1 < x < 1$. Thus the desired solution set is $\{x | -1 < x < 1\}$.

**Example 26.8**
Find the rational roots of $\dfrac{x^3}{3} - x - \dfrac{2}{3} = 0$, and graph the function $y = \dfrac{x^3}{3} - x - \dfrac{2}{3}$.

Solution. The equation does not have integral coefficients, so we must first multiply by the common denominator 3. This yields the equation $x^3 - 3x - 2 = 0$. Our possible rational roots are $\pm 1$ and $\pm 2$. Synthetic division shows that $x = 2$ is one of the roots.

```
1   0   -3   -2  | 2
        2    4    2
─────────────────────
1   2    1   |  0
```

From the reduced equation $x^2 + 2x + 1 = 0$, we have the double root $x = -1$. Thus $x = 2$ and $x = -1$ are the only roots of the equation. The graph of the function is sketched using the zeros and the points in Table 26.4. See Figure 26.4.  ●

**Table 26.4**

| $x$ | $-2$ | $0$ | $1$ | $3$ |
|---|---|---|---|---|
| $y = f(x)$ | $-\frac{4}{3}$ | $-\frac{2}{3}$ | $-\frac{4}{3}$ | $\frac{16}{3}$ |

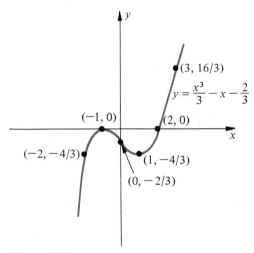

Figure 26.4

**Example 26.9**
Show that $\sqrt{3}$ is not rational.

Solution.   $x^2 - 3 = 0$ has $\sqrt{3}$ as a root. The possible rational roots of the equation are $\pm 1$ and $\pm 3$. Synthetic division or substitution will show that none of these are roots. Therefore, $\sqrt{3}$ is not rational.   ●

**EXERCISES FOR SECTION 26**

For Exercises 1–14 find all the rational roots when $P(x) = 0$.

**1.** $P(x) = x^3 - 2x^2 - x + 2$

**2.** $P(x) = 24x^3 - 2x^2 - 5x + 1$

**3.** $P(x) = 2x^3 - x^2 - 4x + 2$

**4.** $P(x) = 2x^3 - 7x^2 + 10x - 6$

**5.** $P(x) = 2x^4 + 5x^3 - x^2 + 5x - 3$

**6.** $P(x) = 6x^4 + 7x^3 - 27x^2 - 28x + 12$

**7.** $P(x) = 3x^4 - 40x^3 + 130x^2 - 120x + 27$

**8.** $P(x) = x^4 - 8x^3 + 22x^2 - 24x + 9$

**9.** $P(x) = x^4 - 6x^2 - 8x - 3$

**10.** $P(x) = x^3 + 2x^2 + 4x - 1$

**11.** $P(x) = x^3 - 2x^2 - 5x + 6$

**12.** $P(x) = 5x^3 + 12x^2 - 36x - 16$

**13.** $P(x) = 2x^4 + 5x^3 - 11x^2 - 20x + 12$

**14.** $P(x) = 2x^3 - 13x^2 + 27x - 18$

**15.** Suppose that $x^2 - 1$ is a factor of a quartic polynomial $P(x)$. Two of the roots of $P(x) = 0$ are 2 and 3. Find $P(x)$.

**16.** Suppose that $x^2 + 1$ is a factor of a cubic polynomial $P(x)$. If the graph of $P(x)$ crosses the $x$-axis at $(1, 0)$, find $P(x)$.

**17.** Suppose that the zeros of a cubic polynomial $P(x)$ are 1, 2, and $\frac{1}{2}$. If $P(0) = 4$, find $P(x)$.

**18.** The zeros of a quartic polynomial $P(x)$ are $-2$, 1, 2, and $-1$. The graph of $P(x)$ crosses the $y$-axis at $y = 2$. Find $P(x)$.

**19.** Let $P(x)$ be a real polynomial function of degree 4. What is the greatest number of times the graph of $P(x)$ can cross or touch the $x$-axis? The least number of times?

**20.** Let $P(x)$ be a real polynomial function of degree 8. Suppose that $P(x) = 0$ has a root of multiplicity 4. What is the greatest number of times the graph of $P(x)$ can cross or touch the $x$-axis? The least number of times?

**21.** Graph the given polynomial functions.
- (a) $P(x) = 2x^3 - x^2 - 4x + 2$
- (d) $P(x) = x^3 - 9x$
- (b) $P(x) = x^4 - 6x^2 - 8x - 3$
- (e) $P(x) = x^3 - 5x^2 + 8x - 4$
- (c) $P(x) = x^3 + 2x^2 - 8x$
- (f) $P(x) = (x - 1)^3$

**22.** Use the graphs found in Exercise 21 to solve the following inequalities for $x$.
- (a) $2x^3 - x^2 - 4x + 2 \leq 0$
- (d) $x^3 - 9x < 0$
- (b) $x^4 - 6x^2 - 8x - 3 < 0$
- (e) $x^3 - 5x^2 + 8x - 4 \geq 0$
- (c) $x^3 + 2x^2 - 8x < 0$
- (f) $(x - 1)^3 > 0$

**23.** Show that the following numbers are not rational.
- (a) $\sqrt{2}$
- (b) $\sqrt[3]{4}$
- (c) $\sqrt[4]{2}$

**24.** To show that $\sqrt{5} - \sqrt{3}$ is not a rational number, we begin with the polynomial $x - (\sqrt{5} - \sqrt{3})$, which has $\sqrt{5} - \sqrt{3}$ as a zero. In an attempt to obtain integral coefficients we multiply by $x + (\sqrt{5} - \sqrt{3})$:

$$[x - (\sqrt{5} - \sqrt{3})][x + (\sqrt{5} - \sqrt{3})] = x^2 - (8 - 2\sqrt{15}) = 0$$

Now $x^2 - 8 = -2\sqrt{15}$. Squaring both sides, $x^4 - 16x^2 + 64 = 4 \cdot 15$ or $x^4 - 16x^2 + 4 = 0$. One of the roots of $x^4 - 16x^2 + 4 = 0$ is $\sqrt{5} - \sqrt{3}$. Does the equation have any rational roots? The answer is no, since the only possible rational roots are $\pm 1$, $\pm 2$, and $\pm 4$, and none of these are roots. Therefore, $\sqrt{5} - \sqrt{3}$ is not rational.

For each of the following numbers, find a polynomial with integral coefficients having that number as a zero. Then prove that the number is not rational.

- (a) $1 + \sqrt{2}$
- (c) $3 - \sqrt{2}$
- (b) $\sqrt{3} - \sqrt{2}$
- (d) $1 - \sqrt[3]{2}$

**25.** (a) Reexpress the polynomial equation $96x^3 - 16x^2 - 6x + 1 = 0$ as $y^3 - 2y^2 - 9y + 18 = 0$ by first multiplying by 18 and then substituting $12x = y$.
(b) Find all the integral roots of the equation $y^3 - 2y^2 - 9y + 18 = 0$. Use this result to find all the rational roots of $96x^3 - 16x^2 - 6x + 1 = 0$.

**26.** (a) Reexpress the polynomial equation $2x^3 - 7x^2 + 10x - 6 = 0$ as $y^3 - 7y^2 + 20y - 24 = 0$ by first multiplying by 4 and then substituting $2x = y$.
(b) Find all the integral roots of the equation $y^3 - 7y^2 + 20y - 24 = 0$. Use this result to find all the rational roots of $2x^3 - 7x^2 + 10x - 6 = 0$.

27. The lengths of three sides of a rectangular box, measured in centimeters, are three consecutive integers. If the volume is 720 cm$^3$, find the lengths of the sides.

28. An open box is to be made from a square sheet of metal 18 in on each side by cutting out equal squares from the corners and folding up the sides. How long should the edge of the cutout square be if the volume of the box is 392 in$^3$?

29. Sketch the graph of the regions enclosed by the following.
    (a) $3y + x = 4$, $4x - 3y = 16$ and $y = x^3$.
    (b) $y - x = 6$, $x + 2y = 0$, and $y = x^3$.
    (c) $y = x^3 - 6x^2 + 8x$ and $y = x^2 - 4x$.
    (d) $y = 2x^3 - 3x^2 - 9x$ and $y = x^3 - 2x^2 - 3x$.

## SECTION 27
### APPROXIMATING THE REAL ZEROS OF POLYNOMIAL FUNCTIONS

For an equation of the type $x^3 + 2x - 1 = 0$, the possible rational roots are $\pm 1$. Synthetic division or the remainder theorem will show that neither of these values works. Therefore, the equation has no rational roots. But this is not the only possibility. The Fundamental Theorem of Algebra tells us that any polynomial equation of degree $n \geq 1$ has at least one root and may have as many as $n$ distinct roots. These roots may be complex, real, or rational. Now let us consider the problem of finding or "estimating" the real roots of a polynomial equation if they exist.

    The graph of $f(x) = x^3 + 2x - 1$ is shown in Figure 27.1. But where does the graph come from in the first place? We usually find the zeros of the function and then plot some points and connect the points with a smooth curve. In this case we may have such points as $(-2, -13)$, $(-1, -4)$, $(0, -1)$, $(1, 2)$, and $(2, 11)$. We notice that the points $(0, -1)$ and $(1, 2)$ lie on opposite sides of the $x$-axis and therefore assume that the curve crosses the axis somewhere between the two values for $x$. This is true, although we will not attempt to develop a formal proof. In general, we have the following theorem.

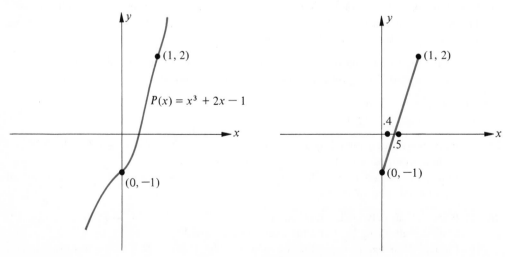

Figure 27.1                                 Figure 27.2

**Theorem 27.1**  If $P(x)$ is a real polynomial and $a$ and $b$ are distinct numbers such that $P(a)$ and $P(b)$ have different signs, there exists at least one number $r$ between $a$ and $b$ such that $P(r) = 0$.

In effect, the theorem tells us the interval in which we have a real root. Repeated use of the theorem allows us to compute an approximate value for the root to any degree of accuracy required. The process involved is illustrated in the following example.

**Example 27.1**
Find the real root of $P(x) = x^3 + 2x - 1$ to the nearest tenth.

Solution.  We have noted that $P(0) = -1$ and $P(1) = 2$. Therefore, the equation has a real root between 0 and 1. Although we cannot find the exact root, we can come closer and closer by repeated use of our theorem. By drawing a straight line between the points $(0, -1)$ and $(1, 2)$, and by estimating where the line crosses the $x$-axis, we obtain a linear approximation to the root in the tenth's place. This is shown in Figure 27.2. We note that the line crosses the axis between 0.4 and 0.5. On this basis, we use synthetic division to test 0.4.

$$
\begin{array}{rrrr|l}
1 & 0 & 2 & -1 & \underline{\;0.4\;} \\
 & 0.4 & 0.16 & 0.864 & \\
\hline
1 & 0.4 & 2.16 & \;|\; -0.136 = P(0.4) &
\end{array}
$$

The root must be between 0.4 and 1, since $P(1) = 2$ and $P(0.4) = -0.136$ have opposite signs. To narrow the interval, we test 0.5:

$$
\begin{array}{rrrr|l}
1 & 0 & 2 & -1 & \underline{\;0.5\;} \\
 & 0.5 & 0.25 & 1.125 & \\
\hline
1 & 0.5 & 2.25 & \;|\; 0.125 = P(0.5) &
\end{array}
$$

The real root we seek is now between 0.4 and 0.5. We again resort to a linear approximation to choose a starting point for our values between 0.4 and 0.5 (see Figure 27.3). The graph indicates that $r$ is between 0.45 and 0.46:

$$
\begin{array}{rrrr|l}
1 & 0 & 2 & -1 & \underline{\;0.45\;} \\
 & 0.45 & 0.2025 & 0.991125 & \\
\hline
1 & 0.45 & 2.2025 & \;|\; -0.008875 = P(0.45) &
\end{array}
$$

$$
\begin{array}{rrrr|l}
1 & 0 & 2 & -1 & \underline{\;0.46\;} \\
 & 0.46 & 0.2116 & 1.017336 & \\
\hline
1 & 0.46 & 2.2116 & \;|\; 0.017336 = P(0.46) &
\end{array}
$$

The real root is between 0.45 and 0.46; that is, $0.45 < r < 0.46$. Thus, to the nearest tenth, $r = 0.5$.  ●

If more accuracy is desired, the process must be repeated. In doing so, we would generate a decimal expansion for the real root of our given equation.

**Example 27.2**
Discuss the real roots of $P(x) = x^3 + 2x^2 - 2x - 1 = 0$.

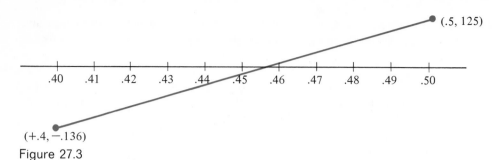

(.5, 125)

(+.4, −.136)

Figure 27.3

Solution. Using either synthetic division or the remainder theorem, we construct Table 27.1. From the table we see that the equation has one rational root, $x = 1$. In addition, the equation has a real root between $-3$ and $-2$ and another real root between $-1$ and $0$. Since the equation can have at most three roots, we have accounted for all possibilities. The graph of the function is shown in Figure 27.4. ●

**Table 27.1**

| $x$ | $-3$ | $-2$ | $-1$ | $0$ | $1$ | $2$ |
|---|---|---|---|---|---|---|
| $P(x)$ | $-4$ | $3$ | $2$ | $-1$ | $0$ | $11$ |

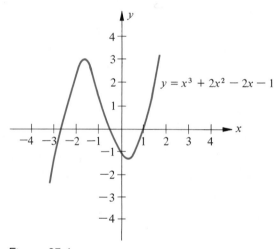

$$y = x^3 + 2x^2 - 2x - 1$$

Figure 27.4

The problem of locating the real zeros of a real polynomial can be simplified by the following theorem.

**Theorem 27.2** Suppose $P(x) = a_0 x^n + a_1 x^{n-1} + \ldots + a_n$ is a real polynomial and that $a_0 > 0$. Then:

1. If we divide $P(x)$ synthetically by $x - a$, where $a \geq 0$, and find that all the numbers obtained in the third row are nonnegative, then no real zero for $P(x)$ can be greater than $a$. We call $a$ an **upper bound** for the real zeros of $P(x)$.

2. If we divide $P(x)$ synthetically by $x - b$, where $b < 0$, and find that all the numbers

obtained in the third row are alternately positive and negative (or zero), then no real zero for $P(x)$ can be less than $b$. We call $b$ a **lower bound** for the real zeros of $P(x)$.

### Example 27.3

Show that 3 is an upper bound for the real zeros of $P(x) = x^4 + 3x^3 - 45x + 8$.

Solution.   We divide $P(x)$ synthetically by $x - 3$ to obtain

$$
\begin{array}{rrrrr|r}
1 & 3 & 0 & -45 & 8 & \underline{\;3\;} \\
  & 3 & 18 & 54 & 27 & \\
\hline
1 & 6 & 18 & 9 & 35 &
\end{array}
$$

Since all numbers in the third row of this synthetic division are positive an upper bound for the real zeros is 3. That is, no zero is greater than 3. The reader can show that 3 is the smallest integer upper bound.   •

### Example 27.4

Show that $-1$ is a lower bound for the real zeros of $P(x) = x^3 + 2x - 1$.

Solution.   We divide $P(x)$ synthetically by $x + 1$ to obtain

$$
\begin{array}{rrrr|r}
1 & 0 & 2 & -1 & \underline{\;-1\;} \\
  & -1 & 1 & -3 & \\
\hline
1 & -1 & 3 & -4 &
\end{array}
$$

Since all the numbers in the third row of this synthetic division are alternately positive and negative a lower bound for the real zeros is $-1$.   •

Another rule that is often helpful in finding the real zeros of a real polynomial is **Descartes' Rule of Signs.** This rule deals with the variation of sign in a polynomial. Before stating the rule, we note that a *variation* of sign occurs when two consecutive terms in a polynomial, with terms in order of decreasing degree, differ by sign. Thus, $x^3 - x^2 + 5x - 6$ has three variations of sign, while $x^4 + 6x^2 - 7x + 8$ has two variations of sign.

### Theorem 27.3   Descartes' Rule of Signs.   

If $P(x)$ is a real polynomial, the number of positive real zeros for $P(x)$ is either equal to the number of variations of sign of $P(x)$ or is this number less an even positive integer. The number of negative zeros of $P(x)$ is equal to the number of variations in sign of $P(-x)$ or is this number less an even positive integer.

### Example 27.5

Use Descartes' rule of signs to discuss the possible number of positive and negative real zeros of $P(x) = 2x^3 - 4x^2 + x - 2$.

Solution.   The number of positive zeros of $P(x) = 2x^3 - 4x^2 + x - 2$ is either three or one, since $P(x)$ has three variations in sign. We see that $P(-x) = 2(-x)^3 - 4(-x)^2 - x - 2$ or $P(-x) = -2x^3 - 4x^2 - x - 2$ has no variations in sign. Thus $P(x)$ has no negative real zeros.   •

## EXERCISES FOR SECTION 27

Find the real zeros of each of the following polynomials to the nearest tenth.

**1.** $P(x) = x^3 + 18x - 30, 1 \leq x \leq 2$

**6.** $P(x) = x^3 - 2x^2 - 2x - 7, 1 \leq x \leq 4$

**2.** $P(x) = x^3 + x - 3, 1 \leq x \leq 2$

**7.** $P(x) = x^3 - 3x^2 - 3x + 18, -5 \leq x \leq -1$

**3.** $P(x) = 2x^3 + 3x^2 - 9x - 7, 1 \leq x \leq 2$

**8.** $P(x) = x^3 + 6x^2 + 9x + 17, -6 \leq x \leq -4$

**4.** $P(x) = x^3 - 3x^2 - 3x + 18, -3 \leq x \leq -2$

**9.** $P(x) = x^4 - 12x^2 - 40x - 21, 4 \leq x \leq 5$

**5.** $P(x) = x^4 - 2x^2 - 1, 1 \leq x \leq 2$

**10.** Use the methods outlined in this section to find the values of the following numbers correct to two decimal places.
  (a) $\sqrt{2}$  (b) $\sqrt{5}$  (c) $\sqrt[3]{3}$  (d) $\sqrt[4]{13}$

**11.** Given that $x^4 - 2x^3 - 3x^2 + 4x + 2 = 0$:
  (a) Use Theorem 27.1 to show that the equation has four real roots.
  (b) Show that none of these roots are rational.
  (c) Find the smallest positive real root correct to the nearest hundredth.

**12.** The dimensions of a box are 1 foot by 3 feet by 4 feet. The volume is to be doubled by increasing each side by the same amount. Find how much each side should be increased, to the nearest tenth.

**13.** How many real roots do each of the following have?
  (a) $x^n - 1 = 0$, $n$ even
  (b) $x^n - 1 = 0$, $n$ odd
  (c) $x^n + 1 = 0$, $n$ odd

**14.** Use Descartes' Rule of Signs and Theorem 27.1 to find the number of positive roots of each of the following equations.
  (a) $x^4 - 1 = 0$  (d) $x^4 + x^2 - 5 = 0$
  (b) $x^5 + x^4 - 2x^3 - x - 1 = 0$  (e) $x^5 + x^3 - 3x^2 + x - 3 = 0$
  (c) $x^4 + x^2 - 3x - 1 = 0$

**15.** For each of the equations in Exercise 14 find the smallest positive root correct to the nearest tenth.

**16.** Show that $P(x) = x^6 + 2x^3 + 3x - 4$ has exactly two real zeros.

**17.** Show that $P(x) = x^4 - x^3 - x^2 - 1$ has exactly two real zeros.

**18.** Show that $P(x) = x^6 + 2x^4 + 3x^2 + 8$ has no real zeros.

**19.** Show that $P(x) = 3x^3 + 2x^2 + 5$ has exactly one real zero.

In exercises 20 through 25, find the smallest integer upper bound and the greatest integer lower bound for the real zeros of each.

**20.** $P(x) = x^5 - x^3 + x + 3$

**23.** $P(x) = x^4 - 4x^3 + 20x^2 - 64x + 64$

**21.** $P(x) = 7x^3 - 15x^2 - 16x + 10$

**24.** $P(x) = 4x^3 + 15x - 36$

**22.** $P(x) = 8x^3 + 9x^2 - 7x - 8$

**25.** $P(x) = 2x^4 + 5x^2 - 6x - 14.$

# CHAPTER 5
# GRAPHS OF RELATIONS AND FUNCTIONS

## ALGEBRAIC FUNCTIONS

We have examined the linear, quadratic, and polynomial functions and their associated graphs. We have also noted that the linear and quadratic functions are simply special cases of the class of functions known as the polynomial functions.

We frequently encounter functions such as $y = \sqrt{x + 2}$, $y = \dfrac{x}{x + 1}$, and $y = \dfrac{(x - 4)^2}{x^2}$ which are not polynomials. (Why?) Our primary interest in this type of function will be graphical. In order to include such functions as $y = \sqrt{x + 2}$, $y = \dfrac{x}{x + 1}$, and $y = \dfrac{(x - 4)^2}{x^2}$ in the graphic analysis of this chapter, we introduce a larger class of functions, known as algebraic functions.

**Definition 28.1** **Algebraic functions** are functions generated by a finite number of the algebraic operations of addition, subtraction, multiplication, division, and extraction of roots.

Since polynomial functions can be generated by a finite number of the algebraic operations of addition, subtraction, and multiplication, they are included in the class of functions known as the algebraic functions. We briefly examined the graphs of polynomial of third or higher degree in Section 25. Any further analysis of their graphs is best left for the branch of mathematics known as calculus.

Another type of algebraic function is the rational function.

**131**

**Definition 28.2** A **rational function** is the ratio of two polynomials. If we let $R(x)$ denote any rational function, then we have $R(x) = \dfrac{P(x)}{Q(x)}$ where $P(x)$ and $Q(x)$ are polynomials.

### Example 28.1

$$y = \frac{x^2}{x^3 + 2x + 1}, \qquad y = \frac{1}{x^2 - 4}, \qquad xy = 1 \qquad y = \frac{x}{x + 1}$$

are *all* rational functions.

### Example 28.2

The polynomial function $y = x^2 - 4x + 6$, is also a rational function since it can be expressed in the form $y = \dfrac{P(x)}{Q(x)}$, where $P(x) = x^2 - 4x + 6$ and $Q(x) = 1$. In general, all polynomial functions are rational functions.

## EXERCISES FOR SECTION 28

**1.** Determine which of the following define rational functions.

(a) $y = x^2 + 2x + 1$        (h) $y = \dfrac{x^2}{x^3}$

(b) $y = \dfrac{x}{(x + 1)(x + 1)}$        (i) $y = \dfrac{1}{\sqrt{x - 1}}$

(c) $y = \dfrac{x^3}{x^2 + 1}$        (j) $y = \dfrac{3\sqrt{x} + 1}{x + 2}$

(d) $y = \dfrac{x^2 + 1}{x^2 + \sqrt{3}}$        (k) $y = \dfrac{\sqrt{3x} + 1}{x + 2}$

(e) $y = \dfrac{x^3 + x^2 + 1}{\pi}$        (l) $y = \dfrac{1}{(x - 2)^3}$

(f) $y = \dfrac{4}{x^3}$        (m) $y = \dfrac{\sqrt[3]{x + 8}}{x - 1}$

(g) $y = \dfrac{2^x}{x^2}$        (n) $y = \dfrac{x^2 - 4}{x + 2}$

**2.** State the domain for each function in exercise 1.

**3.** State the kind of function and the domain of each of the following.

(a) $y = x^3 + \sqrt{3x} + 2$        (f) $y = x^{1/3} + x^{1/2}$

(b) $y = x^3 + \sqrt{3}x + 2$        (g) $y = x^2 + \sqrt{x} + 1$

(c) $y = \dfrac{x}{\sqrt{x + 2}}$        (h) $y = \sqrt{x} + 2^x$

(d) $y = \dfrac{x^{1/4}}{x + 1}$        (i) $y = \dfrac{x^2 + 4}{x^2 - 4}$

(e) $y = 2^x + 2$

# SECTION **29**
## TECHNIQUES OF GRAPHING:
## GRAPHS OF RATIONAL FUNCTIONS

The graphing of any function or relation can often be simplified if we know how to determine some of the geometric properties of the curve. The determination of some of these properties has been discussed previously. For example, we know how to find the domain and the range of many expressions. We are also able to determine the $x$- and $y$-intercepts. These observations will be helpful in our discussions concerning rational functions.

### Example 29.1

$y = \dfrac{1}{x - 1}$ is a rational function whose domain is $\{x \mid x \neq 1\}$. The range is $\{y \mid y \neq 0\}$. The $y$-intercept is at $(0, -1)$ and there is no $x$-intercept. (Why?)

We could sketch the graph of $y = \dfrac{1}{x - 1}$ using the information from Example 29.1 and a table of values. But we would rather consider some of the additional geometric properties that the graph of the function might have. For if the graph of a given function possesses some of these properties, we will find that the number of table values needed will be minimal.

We first consider the graph of the parabola $y = x^2$ shown in Figure 29.1. We notice that for every point $P$ with coordinates $(x, y)$ which satisfies the equation $y = x^2$, there is a point $Q$ with coordinates $(-x, y)$ which also satisfies the equation. In this case $Q$ is said to be the mirror image or the reflection of $P$ through the $y$-axis. Geometrically, the $y$-axis is the perpendicular bisector of the line segment $PQ$ and we say the $P$ and $Q$ are symmetric with respect to the $y$-axis. Since $P$ could be any point on the curve, we can say that the

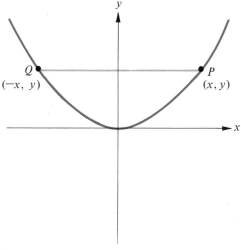

Figure 29.1

graph of $y = x^2$ is symmetric with respect to the $y$-axis; that is, the graph to the left of the $y$-axis is the reflection of the graph to the right of the $y$-axis. This suggests the following test for symmetry about the $y$-axis.

**Test for Symmetry with Respect to the $y$-Axis**
To test a relation for symmetry with respect to the $y$-axis, we replace $x$ by $-x$ in the given expression. If after simplification the resulting expression is the same as the original expression, we have symmetry with respect to the $y$-axis.

### Example 29.2
Determine whether or not $y = \dfrac{x^2}{1 - x^2}$ is symmetric with respect to the $y$-axis.

**Solution.** If we replace $x$ by $-x$ we have $y = \dfrac{(-x)^2}{1 - (-x)^2}$ or $y = \dfrac{x^2}{1 - x^2}$. Since the resulting expression is the same as the original expression, we conclude that the graph of $y = \dfrac{x^2}{1 - x^2}$ is symmetric with respect to the $y$-axis. ●

### Example 29.3
Determine whether or not $y = \dfrac{x}{x^2 - x + 2}$ is symmetric with respect to the $y$-axis.

**Solution.** If we replace $x$ by $-x$ in the given expression we have $y = \dfrac{-x}{(-x)^2 - (-x) + 2}$ or $y = \dfrac{-x}{x^2 + x + 2}$. Since the resulting expression is not the same as the original expression, we conclude that the graph of $y = \dfrac{x}{x^2 - x + 2}$ is *not* symmetric with respect to the $y$-axis. ●

We now wish to generalize the concept of symmetry with respect to a line $L$.

**Definition 29.1** A relation $R$ is **symmetric** with respect to a line $L$ if and only if for each point $P$ in $R$ there is a point $Q$ in $R$ such that the line $L$ is the perpendicular bisector of the segment $PQ$.

Although the definition of line symmetry allows us to consider an infinite number of symmetries, we are basically concerned with line symmetry with respect to the $y$-axis and line symmetry with respect to the $x$-axis.

Suppose that the graph of a relation $R$ is as shown in Figure 29.2. This graph will be symmetric with respect to the $x$-axis if for each point $P$ in $R$ there is a point $Q$ in $R$ such that the $x$-axis is the perpendicular bisector of the segment $PQ$. We see from the diagram that this will occur if the coordinates of $Q$ are $(x, -y)$. We use this observation to formulate the following test for symmetry with respect to the $x$-axis.

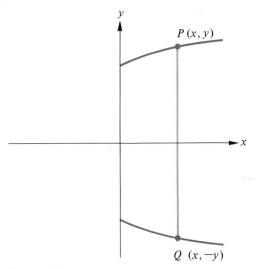

Figure 29.2

**Test for Symmetry with Respect to the *x*-Axis**

To test a relation for symmetry with respect to the *x*-axis, we replace $y$ by $-y$ in the given expression. If after simplification the resulting expression is the same as the original expression, we then have symmetry with respect to the *x*-axis.

**Example 29.4**

Determine whether or not $x^2 + y^2 - 2x + 1 = 0$ is symmetric with respect to the *x*-axis.

Solution.   If we replace $y$ by $-y$ in the given expression, we have $x^2 + (-y)^2 - 2x + 1 = 0$ or $x^2 + y^2 - 2x + 1 = 0$. Since the resulting expression is the same as the original expression, we conclude that the curve is symmetric with respect to the *x*-axis. Note that this relation is *not* symmetric with respect to the *y*-axis. (Why?)   ●

A curve is symmetric with respect to the *x*-axis when all $y$ terms appear to even powers and it is symmetric with respect to the *y*-axis when all $x$ terms appear to even powers.

In addition to line symmetry, there is also point symmetry.

**Definition 29.2**   A relation $R$ is **symmetric with respect to a point** $O$ if and only if for each point $P$ in $R$ there is a point $Q$ in $R$ such that the point $O$ is the midpoint of the line segment $PQ$.

Suppose that the graph of a relation $R$ is as shown in Figure 29.3. This graph will be symmetric with respect to the origin $(0, 0)$ if and only if for each point $P$ there is a point $Q$ such that $O$ is the midpoint of segment $PQ$. We see from the diagram that this will occur if the coordinates of $Q$ are $(-x, -y)$. We use this observation to formulate the following test.

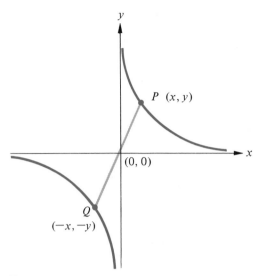

Figure 29.3

**Test for Symmetry with Respect to the Origin**

To test a relation for symmetry with respect to the origin, we replace $y$ by $-y$ and $x$ by $-x$ in the given expression. If after simplification the resulting expression is the same as the original expression, we then have symmetry with respect to the origin.

> **Example 29.5**
> Determine whether or not $x^2 + xy + y^2 = 1$ is symmetric with respect to the origin.
>
> Solution. If we replace $x$ by $-x$ and $y$ by $-y$ in the original expression, we have $(-x)^2 + (-x)(-y) + (-y)^2 = 1$ or $x^2 + xy + y^2 = 1$. Since the resulting expression is the same as the original expression, we conclude that there is symmetry with respect to the origin. ●

Our future discussions dealing with possible symmetries will be restricted to line symmetry with respect to the $x$- and the $y$-axis and point symmetry with respect to the origin.

> **Example 29.6**
> Find all possible symmetries for $y^2 = x^3$.
>
> Solution. $(-y)^2 = x^3$ is equivalent to $y^2 = x^3$. Therefore, we have symmetry with respect to the $x$-axis. The reader can verify that we do not have symmetry with respect to the $y$-axis or the origin. ●

**Summary of Tests for Symmetry**

Suppose that our given expression is $F(x, y) = 0$.

1. If $F(x, y) = F(-x, y) = 0$, the curve is symmetric with respect to the $y$-axis.

2. If $F(x, y) = F(x, -y) = 0$, the curve is symmetric with respect to the $x$-axis.
3. If $F(x, y) = F(-x, -y) = 0$, the curve is symmetric with respect to the origin.

Since rational functions are the ratio of two polynomials, we will frequently have expressions whose denominators may be 0 for some real value or values of $x$. These values must be excluded from the domain of the function since division by 0 is not defined, but they can be of use in our curve sketching.

Consider the rational function $y = \dfrac{1}{x}$. The domain of this function is $\{x \mid x \neq 0\}$.
While $y$ is not defined for $x = 0$, we can see that as $x$ approaches 0 from the right, $y$ gets very large in a positive sense (see Table 29.1). Table 29.2 shows us that as $x$ approaches 0 from the left, $y$ gets very large in a negative sense. The line $x = 0$ is called a **vertical asymptote** and represents a "guideline" for the graph of our function. Although we cannot give a precise definition of what we mean by the term "asymptote," intuitively it is a line that the curve approaches. We say that the distance between the curve and the line tends to 0 as (in this case) $x$ tends to 0.

**Table 29.1**

| $x$ | 2 | 1 | $\frac{1}{2}$ | $\frac{1}{4}$ | $\frac{1}{10}$ | $\frac{1}{100}$ | $\frac{1}{1,000}$ |
|---|---|---|---|---|---|---|---|
| $y$ | $\frac{1}{2}$ | 1 | 2 | 4 | 10 | 100 | 1,000 |

**Table 29.2**

| $x$ | $-2$ | $-1$ | $-\frac{1}{2}$ | $-\frac{1}{10}$ | $-\frac{1}{100}$ | $-\frac{1}{1,000}$ |
|---|---|---|---|---|---|---|
| $y$ | $-\frac{1}{2}$ | $-1$ | $-2$ | $-10$ | $-100$ | $-1,000$ |

If we solve for $x$ in terms of $y$, we obtain $x = \dfrac{1}{y}$. The range of this function is $\{y \mid y \neq 0\}$. Examination of the behavior of $x$ as $y$ nears 0 from both the left and right leads us to conclude that the line $y = 0$ is an asymptote. We call this line a **horizontal asymptote.** The graph of $y = \dfrac{1}{x}$ is shown in Figure 29.4 on page 138.

We now outline a procedure for finding horizontal and vertical asymptotes.

**Vertical Asymptotes**
1. Solve for $y$ in terms of $x$ (if possible).
2. The real values of $x$ for which the denominator is equal to 0 represent vertical asymptotes provided that the numerator is *not equal* to 0.

**Horizontal Asymptotes**
1. Solve for $x$ in terms of $y$ (if possible).
2. The real values of $y$ for which the denominator is equal to 0 represent the horizontal asymptotes provided that the numerator is *not equal* to 0.

When seeking horizontal asymptotes it frequently can be difficult or impossible to

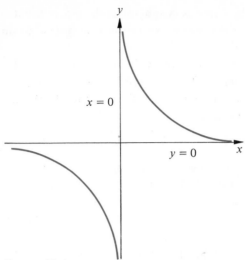

Figure 29.4

*solve* for $x$ in terms of $y$. For these situations we introduce another method for finding horizontal asymptotes.

If $y = l$ is a horizontal asymptotes for $y = f(x)$, then the values of $y = f(x)$ must approach the value $l$ when $|x|$ becomes very large.

**Example 29.7**

Find the horizontal asymptotes for $y = \dfrac{2x^3 + 3x^2 + x}{4x^3 - 2x - 3}$.

Solution. We first divide both the numerator and the denominator by $x^3$ (the highest power of $x$ in the given expression) to obtain

$$y = \frac{\dfrac{2x^3}{x^3} + \dfrac{3x^2}{x^3} + \dfrac{x}{x^3}}{\dfrac{4x^3}{x^3} - \dfrac{2x}{x^3} - \dfrac{3}{x^3}} \quad \text{or} \quad y = \frac{2 + \dfrac{3}{x} + \dfrac{1}{x^2}}{4 - \dfrac{2}{x^2} - \dfrac{3}{x^3}}$$

From this latter form of $y$ we can see that the expressions with $x$ in the denominator will approach the value of 0 when $|x|$ grows very large. Thus the values of $y$ will approach the value $\frac{2}{4} = \frac{1}{2}$ when $|x|$ grows very large. Therefore $y = \frac{1}{2}$ is a horizontal asymptote for $y$. ●

**Example 29.8**

Find the horizontal asymptotes for $y = \dfrac{x^2 + 1}{x^4 - x + 1}$.

Solution. We first divide the numerator and the denominator by $x^4$ (the highest power of $x$ in the given expression) to obtain

$$y = \frac{\dfrac{x^2}{x^4} + \dfrac{1}{x^4}}{\dfrac{x^4}{x^4} - \dfrac{x}{x^4} + \dfrac{1}{x^4}} \quad \text{or} \quad y = \frac{\dfrac{1}{x^2} + \dfrac{1}{x^4}}{1 - \dfrac{1}{x^3} + \dfrac{1}{x^4}}$$

From this latter form of $y$ we can see that the expressions with $x$ in the denominator will approach the value of 0 when $|x|$ grows very large. Thus the values of $y$ will approach the value $\frac{0}{1} = 0$ when $|x|$ grows very large. Therefore, $y = 0$ (the $x$-axis) is a horizontal asymptote.  •

In the examples that follow, we now have two choices for determining horizontal asymptotes. Since in most cases we must solve for $x$ in terms of $y$ in order to discuss the range, we often use the denominator obtained to determine the asymptotes we seek. This method may not always be the easiest, but it is often the most convenient. In any event, the method used in examples 29.7 and 29.8 might also be used in the following examples.

Now, let us consider a few examples that will illustrate the finding and use of some of the geometric properties that we have discussed.

### Example 29.9

Discuss and sketch the graph of $y = \dfrac{3}{x - 4}$.

### Solution

**Domain:** $\{x \mid x \neq 4\}$.

**Range:** We first solve for $x$ in terms of $y$ as follows:

$$y = \frac{3}{x - 4}$$

$$xy - 4y = 3$$

$$x = \frac{3 + 4y}{y}$$

The range is $\{y \mid y \neq 0\}$.

**Intercepts:** When $x = 0$, $y = -\frac{3}{4}$. When $y = 0$, $x$ is undefined. Thus we have no $x$-intercept and a $y$-intercept located at $(0, -\frac{3}{4})$.

**Symmetry:** There is no symmetry.

**Asymptotes:** We see that the denominator $x - 4$ will be equal to 0 when $x = 4$. Thus the line $x = 4$ is a vertical asymptote. Since $x = \dfrac{3 + 4y}{y}$ we see that the denominator will be 0 when $y = 0$. Thus the line $y = 0$ is a horizontal asymptote.

**Graphing Comments:** We first place our vertical and horizontal asymptotes in Figure 29.5 on page 140 and then plot our $y$-intercept. The domain tells us that the graph exists for all real $x$ except $x = 4$. Finally, we plot some values from Table 29.3 on page 140 and sketch the curve.  •

### Example 29.10

Discuss and sketch the graph of $y = \dfrac{x}{x - 1}$.

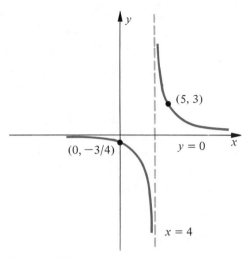

Figure 29.5

**Table 29.3**

| $x$ | $-1$ | 1 | 5 | 6 |
|---|---|---|---|---|
| $y$ | $-\frac{3}{5}$ | $-1$ | 3 | $\frac{3}{2}$ |

Solution

**Domain:** $\{x \,|\, x \neq 1\}$.

**Range:** We first solve for $x$ in terms of $y$ as follows:

$$y = \frac{x}{x - 1}$$
$$xy - y = x$$
$$xy - x = y$$
$$x(y - 1) = y$$
$$x = \frac{y}{y - 1}$$

The range is $\{y \,|\, y \neq 1\}$.

**Intercepts:** When $x = 0$, then $y = 0$. Thus both the $x$- and $y$-intercepts are located at $(0, 0)$.

**Symmetry:** There is no symmetry.

**Asymptotes:** We see that the denominator $x - 1$ will be equal to 0 when $x = 1$. Thus the line $x = 1$ is a vertical asymptote. Since $x = \dfrac{y}{y - 1}$ we see that the denominator $y - 1$ will be equal to 0 when $y = 1$. Thus the line $y = 1$ is a horizontal asymptote.

**Graphing Comments:** We first place our asymptotes in Figure 29.6 and then plot our intercept $(0, 0)$. The domain tells us that the graph exists for all real $x$ except $x = 1$. Finally, we plot the points from Table 29.4 and then sketch the curve.   ●

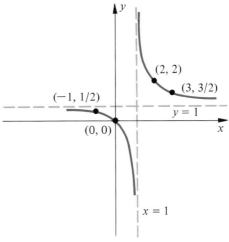

Figure 29.6

**Table 29.4**

| $x$ | $-1$ | $2$ | $3$ |
|---|---|---|---|
| $y$ | $\frac{1}{2}$ | $2$ | $\frac{3}{2}$ |

**Example 29.11**

Discuss and sketch the graph of $y = \dfrac{x^2}{x^2 - 4}$.

Solution
**Domain:** $\{x \,|\, x \neq 2, -2\}$.

**Range:** We first solve for $x$ in terms of $y$ as follows:

$$y = \frac{x^2}{x^2 - 4}$$

$$yx^2 - 4y = x^2$$

$$yx^2 - x^2 = 4y$$

$$x^2(y - 1) = 4y$$

$$x^2 = \frac{4y}{y - 1}$$

$$x = \pm\sqrt{\frac{4y}{y - 1}}$$

The range

$$\left\{ y \,\Big|\, \frac{4y}{y - 1} \geq 0 \right\} = \{y \,|\, y \leq 0 \text{ or } y > 1\}$$

**Intercepts:** When $x = 0$, then $y = 0$. Thus both the $x$- and $y$-intercepts are located at $(0, 0)$.

**Asymptotes:** We see that the denominator $x^2 - 4$ will be equal to 0 for the real values 2 and $-2$. Thus the lines $x = 2$ and $x = -2$ are vertical asymptotes. Since $x = \pm\sqrt{\dfrac{4y}{y - 1}}$ we see that the denominator $\sqrt{y - 1}$ will be equal to 0 when $y = 1$. Thus the line $y = 1$ is a horizontal asymptote.

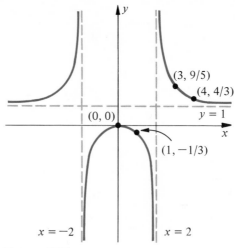

**Table 29.5**

| x | 1 | 3 | 4 |
|---|---|---|---|
| y | $-\frac{1}{3}$ | $\frac{9}{5}$ | $\frac{4}{3}$ |

Figure 29.7

**Symmetry:** If we replace $x$ by $-x$ in the original expression we have $y = \dfrac{(-x)^2}{(-x)^2 - 4}$ or $y = \dfrac{x^2}{x^2 - 4}$. Since the resulting expression is the same as the original, we have symmetry with respect to the $y$-axis.

**Graphing Comments:** We first place our asymptotes in Figure 29.7 and then plot our intercept at $(0, 0)$. The domain tells us that the graph exists for all real $x$ except $x = 2$ and $x = -2$. Next we plot the values from Table 29.5 and sketch the right portion of the curve. Finally we use the symmetry with respect to the $y$-axis to determine the left portion of the curve. ●

**Example 29.12**

Discuss and sketch the graph of $y = \dfrac{(x - 1)^2}{x^2}$.

Solution

**Domain:** $\{x \mid x \neq 0\}$.

**Range:** We first solve for $x$ in terms of $y$ as follows:

$$y = \frac{(x - 1)^2}{x^2}$$

$$yx^2 = x^2 - 2x + 1$$

$$yx^2 - x^2 + 2x - 1 = 0 \quad \text{or} \quad (y - 1)x^2 + 2x - 1 = 0$$

We now use the quadratic formula to obtain

$$x = \frac{-2 \pm \sqrt{4 - 4(y - 1)(-1)}}{2(y - 1)} \quad \text{or} \quad \frac{-2 \pm \sqrt{4y}}{2(y - 1)} = \frac{-1 \pm \sqrt{y}}{y - 1}$$

It appears that the range is $y \geq 0$, $y \neq 1$. But since $y = 1$ when $x = \frac{1}{2}$, the range is $\{y \mid y \geq 0\}$.

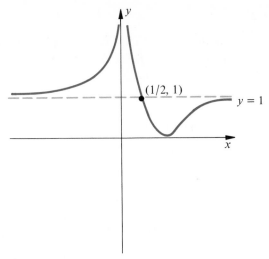

Figure 29.8

**Table 29.6**

| $x$ | $\frac{1}{4}$ | $\frac{1}{2}$ | 2 | 3 | 4 | 5 | $-\frac{1}{2}$ | $-1$ | $-2$ | $-3$ | $-4$ |
|---|---|---|---|---|---|---|---|---|---|---|---|
| $y$ | 9 | 1 | $\frac{1}{4}$ | $\frac{4}{9}$ | $\frac{16}{16}$ | $\frac{16}{25}$ | 9 | 4 | $\frac{9}{4}$ | $\frac{16}{9}$ | $\frac{55}{16}$ |

**Intercepts:** When $x = 0$, $y$ is not defined. Therefore, we have no $y$ intercept. When $y = 0$, then $x = 1$. Thus we have an $x$ intercept located at $(1, 0)$.

**Asymptotes:** We see that the denominator $x^2$ will be equal to 0 when $x = 0$. Thus the line $x = 0$ or the $y$-axis is a vertical asymptote. For $x = \dfrac{-1 \pm \sqrt{y}}{y - 1}$ we cannot conclude that $y = 1$ is a horizontal asymptote, since the numerator $-1 + \sqrt{y} = 0$. However, for $x = \dfrac{-1 - \sqrt{y}}{y - 1}$ we can conclude that $y = 1$ is a horizontal asymptote.

**Symmetry:** We have no symmetry.

**Graphing Comments:** We first place our asymptotes in Figure 29.8 and then plot our $x$-intercept at $(1, 0)$. The domain tells us that the graph exists for all real values except $x = 0$. Since we have no symmetry, we must rely a little more heavily on our values in Table 29.6 to sketch the curve. Note that the curve intersects its horizontal asymptote at the point $(\frac{1}{2}, 1)$.  ●

**Example 29.13**

Discuss and sketch the graph of $y = \dfrac{x^2 - 1}{x - 1}$.

Solution

**Domain:** $\{x \mid x \neq 1\}$.

**Range:** $\{y \mid y \neq 2\}$.

**Intercepts:** When $x = 0$, then $y = 1$. Thus we have a $y$-intercept located at $(0, 1)$. When $y = 0$, then $x = \pm 1$. Since $x \neq 1$ we have an $x$-intercept located at $(-1, 0)$.

**Asymptotes:** None.

**Symmetry:** We have no symmetry.

**Graphing Comments:** We note that

$$y = \frac{x^2 - 1}{x - 1} = \frac{(x - 1)(x + 1)}{x - 1}$$

or that $y = x + 1$ if $x \neq 1$. Thus the graph of $y = \frac{x^2 - 1}{x - 1}$ is simply the graph of the straight line $y = x + 1$ without the point $(1, 2)$. This graph is shown in Figure 29.9. ●

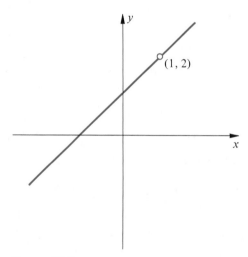

Figure 29.9

If the numerator of a rational function is one degree higher than the degree of the denominator, then the rational function usually has an *oblique* asymptote. To find such an asymptote if it exists, we divide the numerator by the denominator and proceed as outlined in the following example.

### Example 29.14

Determine the equation of the oblique asymptote for $y = \dfrac{2x^2 - x + 3}{x - 1}$.

Solution

We divide $2x^2 - x + 3$ by $x - 1$ to obtain a quotient of $2x + 1$ and a remainder of 4. Thus

$$y = 2x + 1 + \frac{4}{x - 1}$$

In this latter form we see that as $|x|$ grows very large the value of $\dfrac{4}{x-1}$ approaches 0 and the values of $y$ approach $2x+1$. This means that the graph of $y = \dfrac{2x^2 - x + 3}{x - 1}$ comes closer and closer to the line $y = 2x + 1$. Thus $y = 2x + 1$ is an oblique asymptote.   ●

**Remark 29.1**
Oblique lines (such as example 29.14) may be asymptotes for a given curve. In all of our future discussions, we will restrict the possibilities to horizontal and vertical asymptotes unless otherwise stipulated.

## EXERCISES FOR SECTION 29

Discuss the domain, range, intercepts, asymptotes, and symmetries of each of the following. Use these results to sketch the graph of each.

*Odd Numbers*

**1.** $y = \dfrac{3}{x - 2}$

**2.** $y = \dfrac{4}{x + 1}$

**3.** $y = \dfrac{3x}{x - 2}$

**4.** $y = \dfrac{1}{x^2 - 1}$

**5.** $y = \dfrac{1}{x^2 + 1}$

**6.** $y = \dfrac{x^2}{x^2 + 1}$

**7.** $y = \dfrac{x^2 - 2x}{x - 2}$

**8.** $y = \dfrac{x}{x^2 - 1}$

**9.** $y = \dfrac{x}{(x - 1)(x + 3)}$

**10.** $y = \dfrac{x^2 - 1}{9 - x^2}$

**11.** $y = \dfrac{x}{x^2 + 2}$

**12.** $y = \dfrac{x^2 + 1}{x}$

**13.** $y = \dfrac{x^4}{4}$

**14.** $y = \dfrac{x^2 - 4}{x - 1}$

**15.** $y = \dfrac{1}{x^2 + x - 2}$

**16.** $y = \dfrac{x^2 - 2x}{(x + 1)(x - 3)}$

**17.** $y = \dfrac{6}{x^2 + 2}$

**18.** $y = \dfrac{(x - 1)(x + 2)}{(x - 1)(x + 2)}$

**19.** $y = \dfrac{x^2}{x^2 - 1}$

**20.** $y = \dfrac{x^2}{(x - 1)(x - 3)}$

**21.** $y = \dfrac{2x^2}{(x - 2)(x - 6)}$

**22.** $y = \dfrac{1}{(x + 1)^3}$

**23.** $y = \dfrac{-1}{(x + 1)^3}$

**24.** $y = \dfrac{x^2 - 4}{x^2 - 1}$

**25.** Construct an example that verifies the following statement: If a curve is symmetric with respect to the origin, it is not necessarily symmetric with respect to both the $x$- and $y$-axes.

**26.** Any function satisfying the property $f(-x) = f(x)$ for all $x$ in the domain of $f$ is called an **even function.** Any function satisfying the property $f(-x) = -f(x)$ for all $x$ in the domain of $f$ is called an **odd function.** Determine which of the following functions are even and which are odd.

(a) $f(x) = x^4 + x^2$      (e) $f(x) = 4$

(b) $f(x) = x^3$           (f) $f(x) = \sqrt{x^2 - 9}$

(c) $f(x) = x^5 - x^3$     (g) $f(x) = 2x$

(d) $f(x) = x^2 - x$

**27.** Are all *even functions* symmetric with respect to the $y$-axis?

**28.** Are all *odd functions* symmetric with respect to the origin?

**29.** Use the method of Examples 29.7 and 29.8 to find all possible horizontal asymptotes for exercises 1–24.

In exercises 30 through 37, find all possible vertical, horizontal, and oblique asymptotes.

**30.** $y = \dfrac{x^2 + 1}{x}$        **31.** $y = \dfrac{3x^2 + 2x - 1}{x - 2}$

**32.** $y = \dfrac{x^3}{x^2 - 1}$        **33.** $y = \dfrac{x^3 - 8}{x^2 - 8}$

**34.** $y = \dfrac{x^3 + 8}{x^2}$        **35.** $y = \dfrac{4x^3 + x + 1}{x^2 - 9}$

**36.** $y = \dfrac{x^2 - 4}{x + 1}$        **37.** $y = \dfrac{x^2 - 4}{x - 1}$

**38.** Suppose $y = \dfrac{a_0 x^n + a_1 x^{n-1} + \cdots + a_n}{b_0 x^m + b_1 x^{m-1} + \cdots + a_m}$    where $a_0 \neq 0$ and $b_0 \neq 0$. Describe all possible horizontal asymptotes when

(a) $n = m$      (b) $n < m$      (c) $n > m$

**39.** If a rational function has a numerator of degree $n$ and a denominator of degree $m$ and if $n - m \geq 2$ (that is the numerator is of degree 2 or more greater than that of the denominator), then the rational function has **curvilinear** asymptotes. To find such asymptotes we divide the numerator by the denominator as illustrated in the following example.

### Example

Find the curvilinear asymptotes for $y = \dfrac{x^3 + 2x + 1}{x - 2}$.

Solution.    We divide $x^3 + 2x + 1$ by $x - 2$ to obtain a quotient $x^2 + 2x + 6$ and a remainder of 13. Thus

$$y = x^2 + 2x + 6 + \frac{13}{x - 2}$$

In this latter form we see that as $|x|$ grows very large the value of $\dfrac{13}{x - 2}$ approaches $0$ and the values of $y$ approach $x^2 + 2x + 6$. This means that the graph of $y = \dfrac{x^3 + 2x + 1}{x - 2}$ comes closer and closer to the parabola $y = x^2 + 2x + 6$. Thus $y = x^2 + 2x + 6$ is a curvilinear asymptote.    ●

Use the above method to determine the curvilinear asymptotes for each of the following:

(a) $y = \dfrac{x^3 + 1}{x}$  (b) $y = \dfrac{x^4 + x^2 + 1}{x^2 - 1}$

(c) $y = \dfrac{x^5 + x^3 - 1}{x^3 - 1}$  (d) $y = \dfrac{x^6}{x^3 + 1}$

# SECTION 30
## GRAPHS OF ALGEBRAIC FUNCTIONS

The graphic techniques established in Section 29 can be used to examine the behavior of the graphs of all algebraic functions. Since algebraic functions such as $y = \sqrt{x(x - 1)(x + 2)}$ and $y = x/\sqrt{x - 1}$ involve the extraction of even roots, it is necessary to examine the domains of such functions even more carefully than before for excluded regions. A close examination of the following examples will reveal some of the extreme behavior of the graphs of algebraic functions.

**Example 30.1**
Discuss and then sketch the graph of $y = \sqrt{x + 2}$.

Solution

**Domain:** $\{x \,|\, x + 2 \geq 0\} = \{x \,|\, x \geq -2\}$.

**Range:** $\{y \,|\, y \geq 0\}$.

**Intercepts:** When $x = 0$, then $y = \sqrt{2}$. Thus we have a $y$-intercept located at $(0, \sqrt{2})$. When $y = 0$, then $x = -2$. This gives us an $x$-intercept located at $(-2, 0)$.

**Symmetry:** None.

**Asymptotes:** None.

**Graphing Comments:** The domain restriction $x \geq -2$ tells us that the graph of $y = \sqrt{x + 2}$ in Figure 30.1 does not exist for $x < -2$. Since there are no asymptotes or symmetries, we simply plot our $x$- and $y$-intercepts along with the values in Table 30.1 to complete our sketch.  ●

**Example 30.2**
Discuss and sketch the graph of $y = \sqrt{x^2 - 4}$.

Solution

**Domain:** $\{x \,|\, x^2 - 4 \geq 0\} = \{x \,|\, x \geq 2 \text{ or } x \leq -2\}$.

**Range:** $\{y \,|\, y \geq 0\}$.

**Intercepts:** When $x = 0$, $y$ is nonreal. When $y = 0$, then $x = \pm 2$. Thus we have $x$-intercepts located at $(2, 0)$ and $(-2, 0)$.

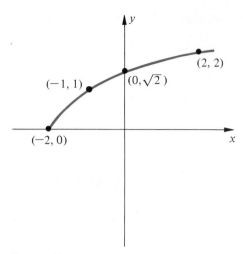

**Table 30.1**

| x | −1 | 2 |
|---|----|---|
| y | 1 | 2 |

Figure 30.1

**Asymptotes:** None.

**Symmetry:** We have symmetry with respect to the $y$-axis.

**Graphing Comments:** The domain restriction tells us that the graph does not exist for $-2 < x < 2$. We plot the $x$-intercepts and the values from Table 30.2 and then sketch the right portion or branch of this curve. Since the curve is symmetric with respect to the $y$-axis, we can reflect the right branch to obtain the left branch and complete the sketch of the curve, which is shown in Figure 30.2.   ●

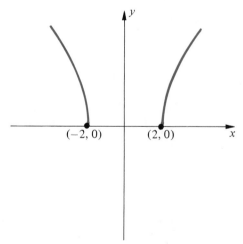

**Table 30.2**

| x | 3 | 4 |
|---|---|---|
| y | $\sqrt{5}$ | $\sqrt{12}$ |

Figure 30.2

**Example 30.3**

Discuss and sketch the graph of $y = \sqrt{x(x^2 - 1)}$.

Solution

**Domain:** $\{x \mid x(x^2 - 1) \geq 0\} = \{x \mid -1 \leq x \leq 0 \text{ or } x \geq 1\}$.

**Range:** $\{y \mid y \geq 0\}$.

**Intercepts:** When $y = 0$, then $\sqrt{x(x^2 - 1)} = 0$ or $x = 0, 1$, and $-1$. Thus we have $x$-intercepts located at $(0, 0)$, $(1, 0)$, and $(-1, 0)$. When $x = 0$, then $y = 0$. Thus $(0, 0)$ is also the location of the $y$-intercept.

**Asymptotes:** None.

**Symmetry:** None.

**Graphing Comments:** The domain restrictions tell us that the graph does not exist for $x < -1$ or $0 < x < 1$. We plot the $x$-intercepts and the values from Table 30.3 to obtain the sketch in Figure 30.3. ●

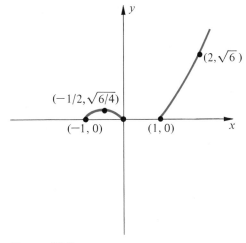

**Table 30.3**

| $x$ | $-\dfrac{1}{2}$ | $2$ | $3$ |
|---|---|---|---|
| $y$ | $\dfrac{\sqrt{6}}{4}$ | $\sqrt{6}$ | $\sqrt{24}$ |

Figure 30.3

**Example 30.4**

Discuss and sketch the graph of $y = \dfrac{1}{\sqrt{x + 1}}$.

**Solution**

**Domain:** $\{x \mid x + 1 > 0\} = \{x \mid x > -1\}$.

**Range:** We first solve for $x$ in terms of $y$ as follows:

$$y = \frac{1}{\sqrt{x + 1}}$$

$$y^2 = \frac{1}{x + 1}$$

$$xy^2 + y^2 = 1$$

$$xy^2 = 1 - y^2$$

$$x = \frac{1 - y^2}{y^2}$$

Since $x$ will be real for all $y$ except $y = 0$ and since $y \geq 0$, the range is $\{y \mid y > 0\}$.

**Intercepts:** When $y = 0$ there is no $x$ value, thus no $x$-intercept. When $x = 0$, then $y = 1$. Thus we have a $y$-intercept located at $(0, 1)$.

**Asymptotes:** We see that the denominator $\sqrt{x + 1}$ will be 0 when $x = -1$. Thus the line $x = -1$ is a vertical asymptote.

Since $x = \dfrac{1 - y^2}{y^2}$ we see that the denominator $y^2$ will be 0 when $y = 0$. Thus the line $y = 0$ is a horizontal asymptote.

**Symmetry:** None.

**Graphing Comments:** The domain restriction tells us that the graph does not exist for $x \leq -1$. We first place our asymptotes in Figure 30.4 and then we plot the $y$-intercept and the values from Table 30.4. The complete sketch is shown in Figure 30.4.  ●

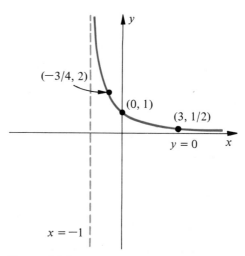

Figure 30.4

**Table 30.4**

| $x$ | $-\frac{3}{4}$ | 3 | 8 |
|-----|------|------|------|
| $y$ | 2 | $\frac{1}{2}$ | $\frac{1}{3}$ |

**Example 30.5**

Discuss and sketch the graph of $y = \dfrac{x}{\sqrt{9 - x^2}}$.

Solution

**Domain:** $\{x \mid 9 - x^2 > 0\} = \{x \mid -3 < x < 3\}$.

**Range:** We solve for $x$ in terms of $y$ as follows:

$$y = \frac{x}{\sqrt{9 - x^2}}$$

$$y^2 = \frac{x^2}{9 - x^2}$$

$$9y^2 - x^2 y^2 = x^2$$

$$9y^2 = x^2 + x^2 y^2$$

$$x^2(1 + y^2) = 9y^2$$

$$x^2 = \frac{9y^2}{1 + y^2}$$

$$x = \pm\sqrt{\frac{9y^2}{1 + y^2}}$$

Since $x$ will be real when $\dfrac{9y^2}{1 + y^2} \geq 0$, the range is $\left\{y \,\middle|\, \dfrac{9y^2}{1 + y^2} \geq 0\right\}$ or all real numbers.

**Intercepts:** When $x = 0$, then $y = 0$. Thus we have both our $x$- and $y$-intercept located at $(0, 0)$.

**Asymptotes:** We see that the denominator $\sqrt{9 - x^2}$ will be 0 when $x = \pm3$. Thus the lines $x = 3$ and $x = -3$ are vertical asymptotes. Since $x^2 = \dfrac{9y^2}{1 + y^2}$ and the denominator $1 + y^2$ cannot equal 0, we have no horizontal asymptotes.

**Symmetry:** If we replace $x$ by $-x$ and $y$ by $-y$ in the expression $y = \dfrac{x}{\sqrt{9 - x^2}}$, we obtain $-y = \dfrac{-x}{\sqrt{9 - (-x)^2}}$ or $y = \dfrac{x}{\sqrt{9 - x^2}}$. Thus we have symmetry with respect to the origin.

**Graphing Comments:** The domain restriction tells us that the graph does not exist for $x \leq -3$ or $x \geq 3$. We place our vertical asymptotes in Figure 30.5 and then plot our intercept and the values from Table 30.5 to sketch the portion of the graph appearing in the first quadrant. We then use the symmetry with respect to the origin to obtain the portion of the graph shown in the third quadrant. The complete sketch is shown in Figure 30.5. ●

**Table 30.5**

| $x$ | 1 | 2 |
|---|---|---|
| $y$ | $\dfrac{1}{\sqrt{8}}$ | $\dfrac{2}{\sqrt{5}}$ |

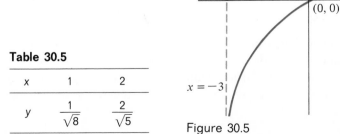

Figure 30.5

**Example 30.6**

Discuss and sketch the graph of $y = \dfrac{1}{\sqrt{x^2 - 4}}$.

## Solution

**Domain:** We see that $y$ will be a real number if $x^2 - 4 > 0$ or when $|x| > 2$. Thus the domain is $\{x \mid x < -2 \text{ or } x > 2\}$.

**Range:** We note that $\dfrac{1}{\sqrt{x^2 - 4}}$ will *always* be positive. Thus the range is $\{y \mid y > 0\}$.

**Intercepts:** If $x = 0$ then $y = \dfrac{1}{\sqrt{-4}}$ is nonreal. If $y = 0$ then we have $0 = \dfrac{1}{\sqrt{x^2 - 4}}$ which has no solution. Thus we have no x- or y-intercepts.

**Asymptotes:** We see that the denominator $\sqrt{x^2 - 4}$ will equal zero when $x^2 - 4 = 0$ or when $x = \pm 2$. Thus the lines $x = 2$ and $x = -2$ are vertical asymptotes. To find any possible horizontal asymptotes we solve for $x$ in terms of $y$ as follows

$$y = \frac{1}{\sqrt{x^2 - 4}}$$

$$y^2 = \frac{1}{x^2 - 4}$$

$$x^2 y^2 - 4y^2 = 1$$

$$x^2 y^2 = 1 + 4y^2$$

$$x^2 = \frac{1 + 4y^2}{y^2} \quad \text{or} \quad x = \pm \sqrt{\frac{1 + y^2}{y^2}}$$

Since the denominator of this latter expression will equal 0 when $y = 0$ we have a horizontal asymptote at $y = 0$ (the $x$-axis).

**Symmetry:** If we replace $x$ by $-x$ in the given equation we have

$$y = \frac{1}{\sqrt{(-x)^2 - 4}} \quad \text{or} \quad y = \frac{1}{\sqrt{x^2 - 4}}.$$

Since the resulting expression is the same as the original, we have symmetry with respect to the $y$-axis.

**Graphing Comments:** We first place our asymptotes in Figure 30.6. Since the domain is $x > 2$ or $x < -2$ we calculate the coordinates shown in Table 30.6 and then plot the corresponding points in Figure 30.6. Finally we use the symmetry with respect to the $y$-axis to determine the left portion of this curve.  ●

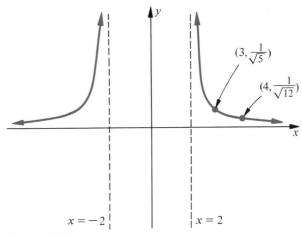

**Table 30.6**

| $x$ | 3 | 4 |
|---|---|---|
| $y$ | $\dfrac{1}{\sqrt{5}}$ | $\dfrac{1}{\sqrt{12}}$ |

Figure 30.6

**Example 30.7**

Discuss and sketch the graph of $y = \dfrac{1}{(x-1)^2}$.

**Solution**

**Domain:** $\{x \mid x \in \mathbb{R}, \ x \neq 1\}$.

**Range:** $\{y \mid y > 0\}$.

**Intercepts:** When $x = 0$, then $y = 1$. So we have a $y$-intercept located at $(0, 1)$. There is no $x$-intercept.

**Asymptotes:** We see that the denominator $(x - 1)^2$ will be equal to 0 when $x = 1$. Thus the line $x = 1$ is a vertical asymptote. If we solve for $x$ in terms of $y$ we obtain $x = 1 \pm \dfrac{1}{\sqrt{y}}$. The denominator $y$ will be equal to 0 when $y = 0$. Thus the line $y = 0$ is a horizontal asymptote.

**Symmetry:** None.

**Graphing Comments:** We place the asymptotes $x = 1$ and $y = 0$ in Figure 30.7, and then plot the $y$-intercept and the table of values (Table 30.7). We then sketch the curve, which is shown in Figure 30.7. ●

## EXERCISES FOR SECTION 30

Discuss and sketch the graphs of each of the following.

**1.** $y = \sqrt{x + 4}$     **3.** $y = \sqrt{x^2 + 4}$

**2.** $y = \sqrt{x^2 - 9}$     **4.** $y = \dfrac{1}{\sqrt{x^2 - 9}}$

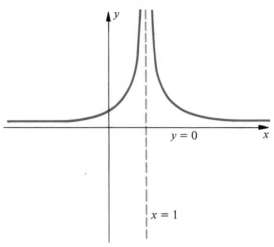

**Table 30.7**

| x | −1 | 2 | 3 | 4 |
|---|----|---|---|---|
| y | $\frac{1}{4}$ | 1 | $\frac{1}{4}$ | $\frac{1}{9}$ |

Figure 30.7

**5.** $y = \dfrac{1}{\sqrt{x + 4}}$      **11.** $y = \dfrac{1}{(x + 2)^2}$

**6.** $y = \dfrac{x}{\sqrt{4 - x^2}}$      **12.** $y = \dfrac{-1}{(x - 2)^2}$

**7.** $y = \dfrac{x}{\sqrt{x - 1}}$      **13.** $y = \sqrt[3]{(x - 2)^2}$

**8.** $y = \sqrt{x} - x$      **14.** $y = \dfrac{-1}{\sqrt{x^2 - 25}}$

**9.** $y = \sqrt{9 - x^2}$      **15.** $y = \dfrac{x}{\sqrt{x^2 - 9}}$

**10.** $y = \sqrt[3]{x^2}$      **16.** $y = \dfrac{-x}{\sqrt{9 - x^2}}$

# SECTION 31
## GRAPHS OF ALGEBRAIC RELATIONS

In the two previous sections of this chapter we have introduced some geometric properties which enable us to sketch the graphs of algebraic functions. These same geometric properties will also apply to algebraic relations such as $y^2 = \dfrac{x^2 - 4}{1 - x^2}$, $y^2 = x^3$, and $y^2 = \dfrac{x - 1}{(x - 3)(x + 1)}$. We will see that the geometric property of symmetry will play a more prominent role in the graphing of algebraic relations.

**Example 31.1**

Discuss and then sketch the graph of $y^2 = x^3$.

Solution

**Domain:** The domain of $y^2 = x^3$ or $y = \pm\sqrt{x^3}$ is $\{x \mid x^3 \geq 0\} = \{x \mid x \geq 0\}$.

**Range:** $\{y \mid y \in \mathbb{R}\}$.

**Intercepts:** We have both the $x$- and $y$-intercepts located at $(0, 0)$.

**Symmetry:** We have symmetry with respect to the $x$-axis.

**Asymptotes:** None.

**Graphing Comments:** The domain tells us that the graph does not exist for $x < 0$. Since we have no asymptotes we next plot the intercepts and the values from Table 31.1. Now we sketch that portion of the graph in the first quadrant and then use the symmetry with respect to the $x$-axis to obtain that portion of the graph in the fourth quadrant. The complete sketch is shown in Figure 31.1. Note that the algebraic relation can be expressed as the union of the two algebraic functions $y = \sqrt{x^3}$ and $y = -\sqrt{x^3}$. ●

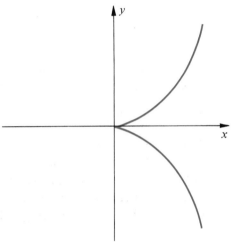

**Table 31.1**

| $x$ | 1 | 2 | 3 |
|---|---|---|---|
| $y$ | $\pm 1$ | $\pm\sqrt{8}$ | $\pm 27$ |

Figure 31.1

**Example 31.2**

Discuss and then sketch the graph of $y^2 = \dfrac{x^2 - 4}{1 - x^2}$.

Solution

**Domain:** $\left\{x \left| \dfrac{x^2 - 4}{1 - x^2} \geq 0 \quad \text{and} \quad 1 - x^2 \neq 0\right.\right\} = \{x \mid -2 \leq x < -1 \quad \text{or} \quad 1 < x \leq 2\}$.

**Range:** We first solve for $x$ in terms of $y$ as follows:

$$y^2 = \frac{x^2 - 4}{1 - x^2}$$

$$y^2 - x^2 y^2 = x^2 - 4$$

$$x^2 + x^2 y^2 = y^2 + 4$$

$$x^2(1 + y^2) = y^2 + 4$$

$$x^2 = \frac{y^2 + 4}{1 + y^2} \quad \text{or} \quad x = \pm\sqrt{\frac{y^2 + 4}{y^2 + 1}}$$

$x$ will be real when $\dfrac{y^2 + 4}{y^2 + 1} \geq 0$. Thus the range is

$$\left\{ y \left| \frac{y^2 + 4}{y^2 + 1} \geq 0 \right. \right\} = \{y \mid y \in \mathbb{R}\}$$

**Intercepts:** When $x = 0$, then $y^2 = -4$ and $y$ is nonreal. When $y = 0$, then $x^2 - 4 = 0$ or $x = \pm 2$. Thus we have no $y$-intercepts and two $x$-intercepts located at $(2, 0)$ and $(-2, 0)$.

**Asymptotes:** The denominator $1 - x^2$ will be equal to 0 when $x = 1$ or $-1$. Thus the lines $x = 1$ and $x = -1$ are vertical asymptotes. Since the denominator of $x^2 = \dfrac{y^2 + 4}{1 + y^2}$ will never be equal to 0, we have no horizontal asymptote.

**Symmetry:** We have symmetry with respect to both the $x$- and the $y$-axis as well as the origin.

**Graphing Comments:** The domain tells us that the graph does not exist for $x < -2$ or $-1 \leq x \leq 1$ or $x > 2$. We place the vertical asymptotes in Figure 31.2. We then plot the $x$-intercept and the values from Table 31.2. Now we can sketch the portion of the graph located in the first quadrant. The symmetry about the $x$-axis enables us to sketch the portion of the curve in the fourth quadrant, and finally we use the symmetry about the $y$-axis to obtain the entire left branch of this graph. The complete sketch is shown in Figure 31.2.   ●

**Table 31.2**

| $x$ | $\dfrac{5}{4}$ | $\dfrac{3}{2}$ |
|-----|-----|-----|
| $y$ | $\dfrac{\sqrt{39}}{3}$ | $\sqrt{\dfrac{7}{5}}$ |

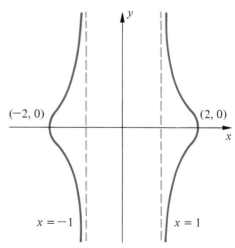

Figure 31.2

**Example 31.3**

Discuss and sketch the graph of $9y^2 + 4x^2 = 36$.

## Solution

**Domain:** We solve for $y$ in terms of $x$ to obtain $y = \pm\frac{1}{3}\sqrt{36 - 4x^2}$. Thus the domain is $\{x \mid 36 - 4x^2 \geq 0\} = \{x \mid -3 \leq x \leq 3\}$.

**Range:** We now solve for $x$ in terms of $y$ to obtain $x = \pm\frac{1}{2}\sqrt{36 - 9y^2}$. Thus the range is $\{y \mid 36 - 9y^2 \geq 0\} = \{y \mid -2 \leq y \leq 2\}$.

**Intercepts:** When $x = 0$, then $9y^2 = 36$ or $y = \pm 2$. When $y = 0$, then $4x^2 = 36$ or $x = \pm 3$. Thus we have $x$-intercepts at $(3, 0)$ and $(-3, 0)$ and $y$-intercepts located at $(2, 0)$ and $(-2, 0)$.

**Asymptotes:** None.

**Symmetry:** We have symmetry with respect to both the $x$- and the $y$-axis as well as with respect to the origin.

**Graphing Comments:** The domain tells us that the graph does not exist for $x < -3$ or $x > 3$. We plot the $x$- and $y$-intercepts and the values from Table 31.3 in Figure 31.3. We use these points to sketch that portion of the graph in the first quadrant. We then use the symmetry with respect to the $x$-axis to obtain that portion of the graph in the fourth quadrant. Finally, we use the symmetry with respect to the $y$-axis to sketch in the remaining portions in the second and third quadrants. The complete sketch is shown in Figure 31.3. This curve is called an **ellipse.** A detailed study of this curve is undertaken in Chapter 10.

It should be noted that with the symmetry information available we could have used other combinations of the symmetries to achieve the same sketch. For example, we could have reflected the portion of the graph in the first quadrant about the $y$-axis to obtain the portion in the second quadrant and then reflected both of these parts about the $x$-axis to complete the sketch. ●

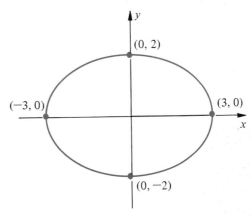

**Table 31.3**

| $x$ | 1 | 2 |
|---|---|---|
| $y$ | $\pm\frac{1}{3}\sqrt{32}$ | $\pm\frac{1}{3}\sqrt{20}$ |

Figure 31.3

## Example 31.4

Discuss and sketch the graph of $x^{2/3} + y^{2/3} = 1$.

Solution

**Domain:** We first solve for $y$ in terms of $x$ as follows:

$$x^{2/3} + y^{2/3} = 1$$
$$y^{2/3} = 1 - x^{2/3}$$
$$y^2 = (1 - x^{2/3})^3$$
$$y = \pm\sqrt{(1 - x^{2/3})^3}$$

$y$ will be real if $(1 - x^{2/3})^3 \geq 0$, so the domain is $\{x \mid (1 - x^{2/3})^3 \geq 0\}$ or $\{x \mid 1 - x^{2/3} \geq 0\} = \{x \mid -1 \leq x \leq 1\}$.

**Range:** If we solve for $x$ in terms of $y$ we obtain $x = \pm\sqrt{(1 - y^{2/3})^3}$. Thus the range is $\{y \mid (1 - y^{2/3})^3 \geq 0\} = \{y \mid -1 \leq y \leq 1\}$.

**Intercepts:** When $x = 0$, then $y^{2/3} = 1$ or $y = \pm 1$. When $y = 0$, then $x^{2/3} = 1$ or $x = \pm 1$. Thus we have intercepts located at $(1, 0), (-1, 0), (0, 1)$, and $(0, -1)$.

**Symmetry:** We first note that $x^{2/3} + y^{2/3} = 1$ is equivalent to $\sqrt[3]{x^2} + \sqrt[3]{y^2} = 1$. Now it is easy to see that we have symmetry with respect to both axes and with respect to the origin.

**Asymptotes:** None.

**Graphing Comments:** The domain tells us that the graph does not exist for $x < -1$ or $x > 1$. We plot the $x$- and $y$-intercepts and the values in Table 31.4 in Figure 31.4. We then sketch the portion of the graph in the first quadrant. Next we use the symmetry with respect to the $y$-axis to obtain that portion of the graph in the second quadrant. Finally, we use the symmetry with respect to the $x$-axis to reflect both the portions of the graph in the $x$-axis to obtain the "bottom half." The complete sketch is shown in Figure 31.4. The curve is called a **hypocycloid.** ●

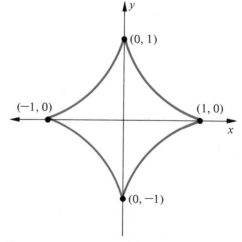

Figure 31.4

**Table 31.4**

| $x$ | 1 | 0 | $\dfrac{1}{8}$ |
|---|---|---|---|
| $y$ | 0 | $\pm 1$ | $\pm\dfrac{\sqrt{27}}{8}$ |

## EXERCISES FOR SECTION 31

Discuss and then sketch the graphs of each of the following algebraic relations.

1. $y^2 = \dfrac{x^2 - 1}{9 - x^2}$                11. $y^2 = (x - 2)^3$

2. $y^2 = \dfrac{6 - x^2}{x^4}$                12. $(y^2 - x^3)(y^2 + x) = 0$

3. $x^2 - y^2 = 1$                13. $(y^2 - x^2)(y - 2) = 0$

4. $9x^2 + 16y^2 = 144$                14. $y^2 = \dfrac{x + 1}{x - 1}$

5. $9x^2 - 4y^2 = 1$                15. $y^2 = \dfrac{x}{x - 2}$

6. $y^2 = \dfrac{x(x - 2)}{x + 3}$                16. $y^2 = \dfrac{x^2 + 1}{x^2 - 1}$

7. $y^2 = x^2(4 - x^2)$                17. $y^2 = \dfrac{x - 1}{(x - 3)(x + 1)}$

8. $x^{2/3} + y^{2/3} = 9$                18. $y^2 - x^2 = 9$

9. $x^{2/3} + y^{2/3} = 4$                19. $4x^2 + 9y^2 = 36$

10. $y^2 = x^3 + 4$

In exercises 20 through 25 sketch the region bounded by the graphs of the given equations. In each case state the domain and range of the determined region.

20. $y = \dfrac{1}{x^2},\ y = -x^2,\ x = 1,\ x = 2$        23. $y^2 = x^3,\ x = y^2$

21. $y = \dfrac{1}{x},\ y = x^2,\ x = 2$        24. $y^2 = x^3,\ y = 2x$

22. $y = \dfrac{1}{x^2 - 4},\ y = -4$        25. $y^2 = x^3,\ y = 3x - 4$

## SECTION 32
# ALGEBRA OF FUNCTIONS

We use the set of real numbers and the arithmetic operations of addition, subtraction, multiplication, and division to form other real numbers. Similarly, we use the set of all functions and the arithmetic operations to form other functions. In so doing, we shall focus our attention on functions whose domain and range are subsets of the real numbers. If $f$ and $g$ are functions with ordered pairs $(x, f(x))$ and $(x, g(x))$ respectively, then we can combine $f$ and $g$ to form a third function in the following ways.

**Sum:** $f + g = \{(x, f(x) + g(x) \,|\, (f + g)(x) = f(x) + g(x)\}$

**Difference:** $f - g = \{(x, f(x) - g(x) \,|\, (f - g)(x) = f(x) - g(x)\}$

**Product:** $fg = \{(x, f(x)g(x)) \mid (fg)(x) = f(x)g(x)\}$

**Quotient:** $\dfrac{f}{g} = \left\{\left(x, \dfrac{f(x)}{g(x)}\right) \Big| \left(\dfrac{f}{g}\right)(x) = \dfrac{f(x)}{g(x)}\right\}$

The domain of the sum, difference, and quotient functions is the **intersection** of the domains of $f$ and $g$. To find the domain of the quotient function $f/g$ we find the intersection of the domains of $f$ and $g$ and then we *exclude* all real zeros of $g$.

### Example 32.1

Let $f$ be defined by $f(x) = 4x$ and let $g$ be defined by $g(x) = \sqrt{x}$. Then

$$f + g = \{(x, f(x) + g(x)) \mid f(x) + g(x) = 4x + \sqrt{x}\}$$
$$f - g = \{(x, f(x) - g(x)) \mid f(x) - g(x) = 4x - \sqrt{x}\}$$
$$fg = \{(x, f(x)g(x)) \mid f(x)g(x) = 4x\sqrt{x} = 4x^{3/2}\}$$
$$\frac{f}{g} = \left\{\left(x, \frac{f(x)}{g(x)}\right) \Big| \frac{f(x)}{g(x)} = \frac{4x}{\sqrt{x}} = 4x^{1/2}\right\}$$

The respective domains for the above functions are:

$$D_{f+g} = [0, \infty) \qquad D_{fg} = [0, \infty)$$
$$D_{f-g} = [0, \infty) \qquad D_{f/g} = (0, \infty)$$

Note that since $g(x) = \sqrt{x}$ equals 0, when $x = 0$, this value had to be discarded from $D_{f/g}$.

### Example 32.2

Let $f$ be a function defined $f(x) = 3x$ and let $g$ be a function defined by $g(x) = x^2 - 4$. Sketch the graph of $f + g$.

Solution. We perform the addition of $f(x)$ and $g(x)$ as shown in Table 32.1. We then plot the ordered pairs to determine the sketch of the graph of $f + g$ shown in Figure 32.1.

**Table 32.1**

| $x$ | $f(x)$ | $g(x)$ | $f(x) + g(x)$ | Ordered Pairs $(x, f(x) + g(x))$ |
|---|---|---|---|---|
| $-4$ | $-12$ | $12$ | $-12 + 12 = 0$ | $(-4, 0)$ |
| $-3$ | $-9$ | $5$ | $-9 + 5 = -4$ | $(-3, -4)$ |
| $-2$ | $-6$ | $0$ | $-6 + 0 = -6$ | $(-2, -6)$ |
| $0$ | $0$ | $-4$ | $0 - 4 = -4$ | $(0, -4)$ |
| $1$ | $3$ | $-3$ | $3 - 3 = 0$ | $(1, 0)$ |
| $2$ | $6$ | $0$ | $6 + 0 = 6$ | $(2, 6)$ |

Another operation that combines the functions $f$ and $g$ to form a third function is the method of composition. Suppose $f$ and $g$ are functions with ordered pairs $(x, f(x))$ and $(x, g(x))$ respectively, then we can combine $f$ and $g$ to form a third function in the following ways:

**Composition of g by f:**

$$h = f \circ g = \{(x, h(x) \,|\, h(x) = f(g(x))\}$$

**Composition of f by g:**

$$k = g \circ f = \{(x, k(x) \,|\, k(x) = g(f(x))\} \quad \bullet$$

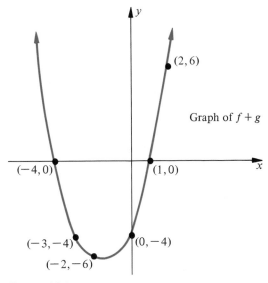

Graph of $f + g$

Figure 32.1

### Example 32.3

Let $f$ be defined by $f(x) = 4 - x^2$ and $g$ be defined by $g(x) = \sqrt{x}$. Then $h = f \circ g$ is defined by $h(x) = f(g(x))$. By direct substitution we have

$$h(x) = f(g(x)) = 4 - (\sqrt{x})^2 = 4 - x$$

Since $k = g \circ f$ is defined by $k(x) = g(f(x))$ we have

$$k(x) = g(f(x)) = \sqrt{4 - x^2}$$

To determine the domain for the composite function $f \circ g$ defined by $f(g(x))$ we first compute $f(g(x))$ for all $x$ in the domain of $g$ and then discard those $x$ in the domain of $g$ for which $f(g(x))$ has no meaning.

To determine the domain for the composite function $g \circ f$ defined by $g(f(x))$ we first compute $g(f(x))$ for all $x$ in the domain of $f$ and discard those $x$ in the domain of $f$ for which $g(f(x))$ has no meaning. $\quad \bullet$

### Example 32.4

Determine the domain of $f \circ g$ and $g \circ f$ in Example 32.3.

Solution. From Example 32.3 we have $h(x) = f(g(x)) = 4 - x$. Since the domain of $g$ is $x \geq 0$ and $h(x)$ is defined for $x \geq 0$, we say that the domain of $f \circ g$ is $x \geq 0$. Note that the function $h(x) = 4 - x$ would have $\mathbb{R}$ for its domain had it not been composed by $f$ and $g$. From Example 32.3 we also have $k(x) = g(f(x)) = \sqrt{4 - x^2}$. The domain of $f$ is $\mathbb{R}$ but the domain of $k(x) = g(f(x)) = \sqrt{4 - x^2}$ cannot be $\mathbb{R}$. To see this we note that $k(5) = \sqrt{4 - 25} = \sqrt{-21}$ will be nonreal. Thus we must discard all values of $x$ such that $4 - x^2 < 0$ or equivalently $x < -2$ or $x > 2$. Therefore, the domain of $k(x) = g(f(x)) = \sqrt{4 - x^2}$ is only $\{x \mid -2 \leq x \leq 2\}$. ●

**Example 32.5**

Suppose that $f(x) = x^3$ and $g(x) = 3x - 1$. Find $f(g(x))$ and $g(f(x))$.

Solution. $f \circ g = f[3x - 1] = (3x - 1)^3$ and $g \circ f = g[x^3] = 3x^3 - 1$. We note that $f[g(x)] \neq g[f(x)]$. In general, $f \circ g \neq g \circ f$.

**Example 32.6**

Suppose that $f[g(x)] = x^2 + 1$ and $g(x) = x - 3$. Find $f(x)$.

Solution. Let $x - 3 = r$ so that $x = r + 3$. Then $f[x - 3] = f(r) = (r + 3)^2 + 1$ or $f(r) = r^2 + 6r + 10$. Thus $f(x) = x^2 + 6x + 10$. Verify this by direct computation of $f[g(x)]$.

**Example 32.7**

Suppose that $g(f(x)) = x$ and $f(x) = 2x + 1$. Find $g(x)$.

Solution. Let $r = f(x) = 2x + 1$. Thus $2x = r - 1$ or $x = \frac{1}{2}(r - 1)$. Now since $g(f(x)) = x$ we have

$$g(2x + 1) = x$$

or

$$g(r) = \frac{1}{2}(r - 1)$$

Thus $g(x) = \frac{1}{2}(x - 1)$. ●

## EXERCISES FOR SECTION 32

In exercises 1 through 9 assume that $f$ and $g$ are defined by $f(x)$ and $g(x)$ respectively. For each of the following pairs of $f(x)$ and $g(x)$, find the defining functions for $f + g, f - g, fg, f \circ g$, and $g \circ f$ and then state their respective domains.

1. $f(x) = x$, $g(x) = x^2$        6. $f(x) = 3$, $g(x) = x^3$

2. $f(x) = x^2$, $g(x) = x$        7. $f(x) = x^2$, $g(x) = x^3$

3. $f(x) = |x|$, $g(x) = x^2 + 1$        8. $f(x) = 3x - 1$, $g(x) = x^2 - 3x + 4$

4. $f(x) = \sqrt{x}$, $g(x) = x^2 - 4$        9. $f(x) = 1 - x^2$, $g(x) = \sqrt{x}$

5. $f(x) = x^3$, $g(x) = 4$

10. If $f[g(x)] = x^2 - 1$ and $g(x) = x + 3$, find $f(x)$.

**11.** If $f[g(x)] = 1 + x^2$ and $g(x) = 1 + x^2$, find $f(x)$.

**12.** If $f[g(x)] = 1 + x^2$ and $g(x) = x$, find $f(x)$.

**13.** If $f(x) = 3x + 4$, find $g(x)$ such that $g(f(x)) = x$.

**14.** If $f(x) = x^2 - 4$, $x \geq 0$, find $g(x)$ such that $g(f(x)) = x$.

**15.** Construct an example which shows that $f[f(x)] \neq f(x) \cdot f(x)$.

**16.** Consider the function $i = \{[x, i(x)] | i(x) = x\}$. This function is often referred to as the "identity function" since all ordered pairs belonging to $i$ are of the form $(x, x)$. For each of the following pairs of functions, determine whether $f[g(x)] = g[f(x)] = i(x)$.
(a) $f(x) = 3x + 2$, $g(x) = \frac{1}{3}(x - 2)$
(b) $f(x) = x^2$, $x \geq 0$, and $g(x) = \sqrt{x}$
(c) $f(x) = x^2$, $x \leq 0$, and $g(x) = -\sqrt{x}$

**17.** Do the expressions $y = x - 1$ and $y = \dfrac{x^2 - 1}{x + 1}$ describe the same functions? Explain.

In exercises 18 through 22, use the tabular method of Example 32.2 to sketch the graph of the indicated function when $f$ is defined by $f(x)$ and $g$ is defined by $g(x)$.

**18.** $f - g$ when $f(x) = 2x$ and $g(x) = 3$.

**19.** $f + g$ when $f(x) = 5x$ and $g(x) = x^2 + 6$.

**20.** $fg$ when $f(x) = x$ and $g(x) = \dfrac{1}{x - 1}$.

**21.** $\dfrac{f}{g}$ when $f(x) = x^2$ and $g(x) = x^3$.

**22.** $f \circ g$ when $f(x) = x^2 - 2$ and $g(x) = \sqrt{x}$.

# SECTION **33**
## INVERSE FUNCTIONS

In Chapter 2 we defined a function as a rule that assigns or maps to each domain value a unique range value. This means that if we pick a number from the domain of the function, a corresponding number in the range is uniquely determined. We also noted that in a geometric sense, any vertical line would intersect the graph of our function $f$ in at most one point.

In many cases our functions will be one to one. This simply means that if we pick a number in the range of $f$, a corresponding number in the domain is uniquely determined. Geometrically, this means that any horizontal line will intersect the graph of $f$ in at most one point. If $f$ is a one-to-one function, it has an inverse function (denoted $f^{-1}$). If $f$ is not one to one, an inverse function does not exist (Figure 33.1 on page 164).

**Definition 33.1** Suppose a function $f$ is a set of ordered pairs of the form $(x, y)$ and $f$ is one to one. The **inverse function** $f^{-1}$ is the set of ordered pairs of the form $(y, x)$ obtained from $f$ by interchanging $x$ and $y$.

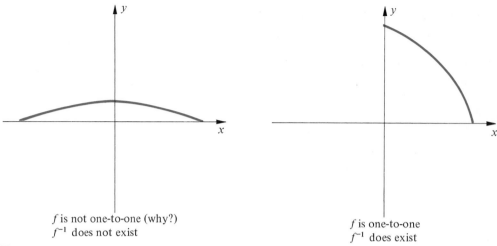

$f$ is not one-to-one (why?)
$f^{-1}$ does not exist

$f$ is one-to-one
$f^{-1}$ does exist

Figure 33.1

**Example 33.1**

Find the inverse function of $y = f(x) = 3x + 2$, if it exists.

Solution. Since $y = 3x + 2$ is one to one, the inverse does exist. To find the inverse function we interchange $x$ and $y$ and obtain $x = 3y + 2$. While this is the inverse function, we would like to follow the usual procedure of expressing $y$ in terms of $x$ if possible. Thus we go one step further and solve $x = 3y + 2$ for $y$.

$$y = \frac{x - 2}{3} \quad \text{or} \quad y = f^{-1}(x) = \frac{x - 2}{3} \quad \bullet$$

**Procedure for Finding the Equation of the Inverse Function**

If $y = f(x)$ is the equation for any function that has an inverse function the inverse function may be found as follows:

*Step 1.* Interchange the roles of $x$ and $y$.

*Step 2.* Solve for $y$ in terms of $x$ *if possible.* The resulting equation is the rule of the inverse function of $f$.

Every linear function except the constant function is one to one and thus must have an inverse function. But not every function has an inverse function. Consider $f(x) = x^2 + 2$. When we select the range value 3 we obtain the domain values $x = \pm 1$. Thus $f(x)$ is not a one-to-one function and does not have an inverse function. It can also be seen from Figure 33.2 that the line $y = 3$ intersects the graph of $f(x) = x^2 + 2$ at more than one point.

If a function is not one to one it may be possible to restrict the domain in such a way that the new function becomes one to one and thus has an inverse. If we restrict the domain of $f(x) = x^2 + 2$ to be $\{x \mid x \geq 0\}$, the function is one to one (Figure 33.3). By using the procedure outlined previously, we find the inverse function to be $f^{-1}(x) = \sqrt{x - 2}$.

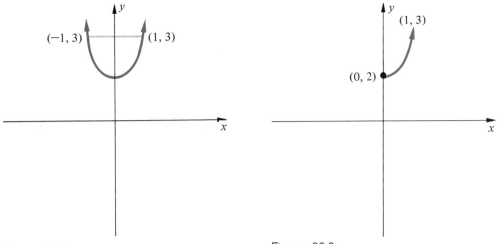

Figure 33.2                         Figure 33.3

We restricted the domain of $f(x)$ by considering only those values of $x \geq 0$. Since $f^{-1}(x) = \sqrt{x-2}$, the domain of $f^{-1}(x)$ must be given by $x - 2 \geq 0$ or $x \geq 2$. A little investigation will shed some light on this. The domain of $f(x)$ is $\{x \mid x \geq 0\}$ and the range is $\{y \mid y \geq 2\}$. For $f^{-1}(x)$, the domain is $\{x \mid x \geq 2\}$ and the range is $\{y \mid y \geq 0\}$. This is to be expected. If we remember that the inverse function is obtained by interchanging the roles of $x$ and $y$, the range of $f$ is the domain of $f^{-1}$ and the domain of $f$ is the range of $f^{-1}$. This is true in general.

### Example 33.2
Find the inverse of $y = \sqrt{x^2 - 1}$, $x \geq 1$. Find the domain and the range of $f^{-1}(x)$ and graph both $f(x)$ and $f^{-1}(x)$.

Solution.   We interchange $x$ and $y$ to obtain $x = \sqrt{y^2 - 1}$. To solve for $y$, we square both sides of the equation and obtain $x^2 = y^2 - 1$. Thus $y^2 = x^2 + 1$ or $y = \pm \sqrt{x^2 + 1}$. We can only use one sign. Since the domain of $f(x)$ is $x \geq 1$, we use the plus sign. The inverse function is given by $f^{-1}(x) = \sqrt{x^2 + 1}$.

The domain and range of $f^{-1}(x)$ can be determined most easily from the domain and range of $f(x)$. The domain of $f(x)$ is $\{x \mid x \geq 1\}$. The range is $\{y \mid y \geq 0\}$. Therefore, the domain of $f^{-1}(x)$ is $\{x \mid x \geq 0\}$ and the range is $\{y \mid y \geq 1\}$. The graphs of $f(x)$ and $f^{-1}(x)$ are shown in Figure 33.4.   •

There is an important relationship between the graph of a function and the graph of its inverse. Suppose that $f(a) = b$; that is, $f: a \to b$. If $f$ has an inverse function, $f^{-1}(b) = a$; that is, $f^{-1}: b \to a$. This simply means that, if the ordered pair $(a, b)$ belongs to $f$, then the ordered pair $(b, a)$ belongs to $f^{-1}$. But the points $(a, b)$ and $(b, a)$ are symmetric with respect to the line $y = x$ (Figure 33.5). This observation enables us to graph $f^{-1}$ by merely reflecting the graph of $f$ in the line $y = x$. This supposes of course that we know or can construct the graph of $f$.

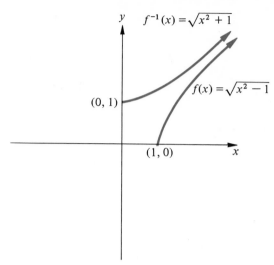

Figure 33.4

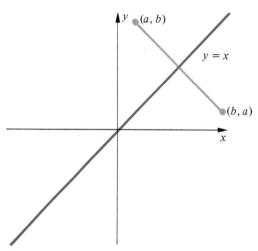

Figure 33.5

### Example 33.3

Find the inverse of $f = \left\{(x, y) \mid y = \dfrac{1}{x}\right\}$ and sketch its graph.

**Solution.** To find $f^{-1}$ we interchange $x$ and $y$ to obtain $x = \dfrac{1}{y}$. Solving for $y$, we have that $y = f^{-1}(x) = \dfrac{1}{x}$. Thus $f(x) = \dfrac{1}{x}$ is its own inverse function and the graph of $f^{-1}(x)$ is simply a reflection of $f(x)$ about the line $y = x$ or a reflection into itself (Figure 33.6). ●

### Example 33.4

Find the inverse function of $y = f(x) = x^3$ and graph both functions.

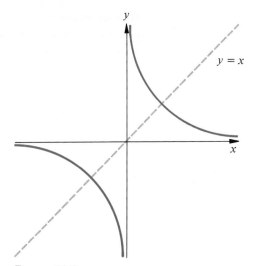

Figure 33.6

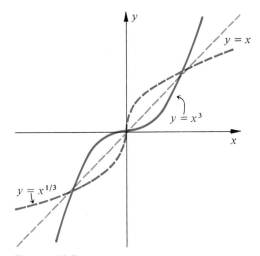

Figure 33.7

**Solution.** Interchanging $x$ and $y$, we have $x = y^3$ or $y = f^{-1}(x) = x^{1/3}$. The graph of $y = x^{1/3}$ is a reflection in the line $y = x$ (Figure 33.7).  ●

**Example 33.5**

Use $f$ and $f^{-1}$ from Example 33.4 to find $(f \circ f^{-1})(x)$ and $(f^{-1} \circ f)(x)$.

**Solution.**   $f(x) = x^3$   and   $f^{-1}(x) = x^{1/3}$.   Therefore,   $(f \circ f^{-1})(x) = f(x^{1/3}) = (x^{1/3})^3 = x$. $(f^{-1} \circ f)(x) = f^{-1}(x^3) = (x^3)^{1/3} = x$.  ●

**Remark 33.1**

Not every function will have an inverse function. But if the function does have an inverse function and we can graph the function $f$, we will always know what the graph of the inverse looks like. The only restriction on the method of finding an inverse function is that of algebra. It may be difficult or impossible to solve for $y$ in terms of $x$. In such situations we simply introduce a new name for the inverse function. We shall encounter this situation in Chapter 6 when we deal with the logarithmic function.

## EXERCISES FOR SECTION 33

Find the inverse function of each of the following functions if it exists. In each case state the domain and the range of the function and its inverse and show a sketch of both.

**1.** $y = f(x) = 5x + 1$      **14.** $y = f(x) = x^2, x \geq 1$

**2.** $y = f(x) = 2x - 3$      **15.** $y = f(x) = x^2, 0 \leq x \leq 1$

**3.** $y = f(x) = 3$      **16.** $y = f(x) = x^2, -1 \leq x \leq 1$

**4.** $y = f(x) = \frac{1}{2}x + 2$      **17.** $y = f(x) = \dfrac{3}{x}$

**5.** $y = f(x) = x^2 - 1, x \leq 0$      **18.** $y = f(x) = \dfrac{3}{x - 2}$

**6.** $y = f(x) = x^2 - 1, x \geq 0$      **19.** $y = f(x) = \dfrac{1}{x^2 - 1}$

**7.** $y = f(x) = (x - 2)^2, x \geq 2$      **20.** $y = f(x) = \dfrac{x(x - 2)}{x - 2}$

**8.** $y = f(x) = 5x$      **21.** $y = f(x) = \dfrac{x^2 - 1}{x^2}$

**9.** $y = f(x) = 4 - 3x$      **22.** $y = f(x) = \dfrac{x}{x^2 - 4}$

**10.** $y = f(x) = \sqrt{x + 1}$      **23.** $y = f(x) = \sqrt{4 - x^2}$

**11.** $y = f(x) = -\sqrt{x + 1}$      **24.** $y = f(x) = \sqrt{4 - x}$

**12.** $y = f(x) = \sqrt{x - 4}$      **25.** $y = f(x) = \dfrac{x}{\sqrt{4 - x^2}}$

**13.** $y = f(x) = -\sqrt{x - 4}$      **26.** $y = f(x) = \sqrt{16 - x^2}, x \geq 0$

**27.** Determine which of the following graphs have inverses. Sketch the inverse where it exists.

(a)                          (b)

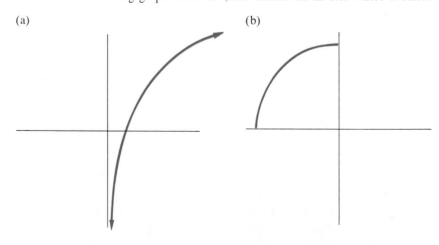

(c)

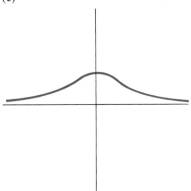

(d)

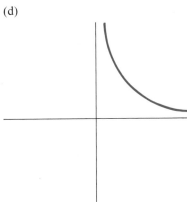

(e)

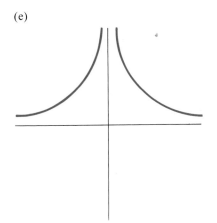

(f)

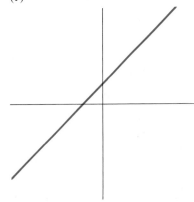

(g)

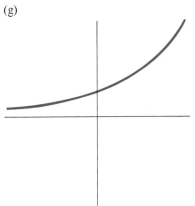

(h)

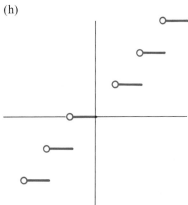

(i)                                      (j)

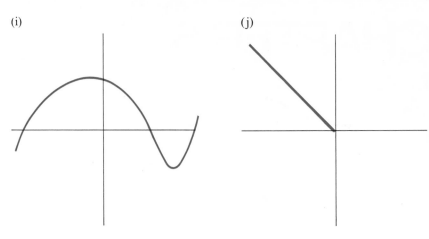

28. Draw the graph of a relation that is not a function but whose reflection in the line $y = x$ is a function.

29. If $g(x + 3) = x^2 + 2$, find $g(x)$.

30. If $f(x - 1) = x^2 + 1$, find $f(x)$.

31. Suppose that a function $f$ has the property that $f(xy) = f(x) + f(y)$ and that $f^{-1}$ exists.
    (a) Show that $f^{-1}(x + y) = f^{-1}(x) \cdot f^{-1}(y)$.
    (b) Show that $f(1) = 0$.
    (c) Show that if $f(2) = 1$, then $f(4) = 2$.

32. Suppose that a function $f$ has the property that $f(x/y) = f(x) - f(y)$ and that $f^{-1}$ exists.
    (a) Show that $f^{-1}(x - y) = \dfrac{f^{-1}(x)}{f^{-1}(y)}$.
    (b) Show that $f(1) = 0$.

33. Suppose that a function $f$ has the property that $f(x^p) = pf(x)$ and that $f^{-1}$ exists. Show that $[f^{-1}(x)]^p = f^{-1}[px]$.

34. Show that all linear functions of the form $y = f(x) = k - x, k \in \mathbb{R}$, are their own inverses.

35. Show that all functions of the form $y = f(x) = \dfrac{k}{x}, k \in \mathbb{R}$ are their own inverses.

# CHAPTER 6
# EXPONENTIAL AND LOGARITHMIC FUNCTIONS

## EXPONENTIAL FUNCTIONS

Previously, much of our functional analysis has dealt with algebraic functions. There are many functions which are nonalgebraic. We would now like to consider two such functions, the exponential and the logarithmic functions. We have seen functions of the type $y = x^2$, where the base is a variable and the exponent is constant. Now we consider the properties of functions of the type $y = 2^x$ or $y = b^x$, where $b$ is any positive constant. These functions are called exponential functions.

**Definition 34.1**  Let $b$ be any positive real number. $f(x) = b^x$ is called an **exponential function** with base $b$. Its domain is the set of all real numbers and its range is the set of all positive real numbers provided that $b \neq 1$.

**Example 34.1**
Sketch the graph of $f(x) = 2^x$.

Solution.  We use the values in Table 34.1 and the fact that all exponential functions are continuous to sketch the graph shown in Figure 34.1 on page 172. Note that $y = 0$ is a horizontal asymptote.  ●

**Table 34.1**

| $x$ | $-3$ | $-2$ | $-1$ | 0 | 1 | 2 | 3 |
|-----|------|------|------|---|---|---|---|
| $f(x)$ | $\frac{1}{8}$ | $\frac{1}{4}$ | $\frac{1}{2}$ | 1 | 2 | 4 | 8 |

**171**

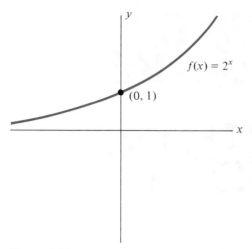

Figure 34.1

### Example 34.2
Graph $f(x) = (\frac{1}{2})^x$.

Solution. We use the values in Table 34.2 to sketch the graph shown in Figure 34.2. Here, again, $y = 0$ is a horizontal asymptote. ●

**Table 34.2**

| $x$ | $-3$ | $-2$ | $-1$ | 0 | 2 | 3 |
|---|---|---|---|---|---|---|
| $f(x) = (\frac{1}{2})^x$ | 8 | 4 | 2 | 1 | $\frac{1}{4}$ | $\frac{1}{8}$ |

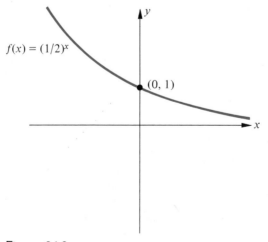

Figure 34.2

Figure 34.3 shows the typical shape of the graph of the exponential function $f(x) = b^x$ when $b > 1$. We say that this function is **increasing** because its graph rises from left to right.

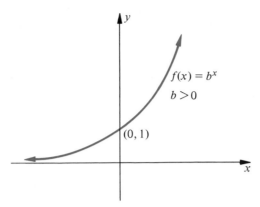

Figure 34.3

Figure 34.4 shows the typical shape of the graph of the exponential function $f(x) = b^x$ when $0 < b < 1$. We say that this function is **decreasing** because its graph drops from left to right.

The graphs shown in Figures 34.3 and 34.4 indicate that the graph *always* passes through the point (0, 1) and *always* has $y = 0$ ($x$-axis) for a horizontal asymptote.

There is a particular value of $b$ that is very important in mathematics. It is an irrational number, denoted by $e$. The *approximate value* of $e$ to five decimal places is 2.71828. A discussion of $e$ is undertaken in the branch of mathematics called calculus.

We also observe that $f(x) = b^x$ is a one-to-one function since any horizontal line will intersect the graphs shown in Figures 34.3 and 34.4 in at most one point. Thus $f(x) = b^x$ has an inverse function.

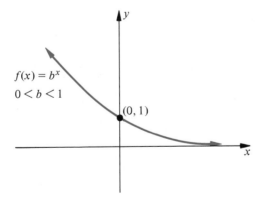

Figure 34.4

**Example 34.3**
Graph $f(x) = e^x$.

Solution.   We use Table III in the appendix to determine the approximate values of $e^x$ shown in Table 34.3 on page 174. We use the values in Table 34.3 to sketch the graph shown in Figure 34.5. We note that $f(x) = e^x$ is an increasing function and the $x$-axis is its horizontal asymptote.

**Table 34.3**

| $x$ | $-2$ | $-1$ | $0$ | $1$ | $2$ |
|---|---|---|---|---|---|
| $f(x) = e^x$ | 0.1 | 0.4 | 1 | 2.7 | 7.4 |

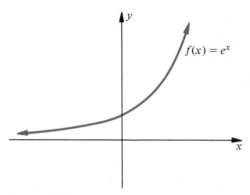

Figure 34.5

**Example 34.4**

Suppose that in a culture a certain type of bacteria grow in such a way that $N = N_0 e^{0.5t}$ where $N$ is the number of bacteria present at the end of $t$ hours. If there are 10,000 bacteria present initially how many bacteria will be present after 5 hours?

**Solution.** Since there are 10,000 bacteria present initially, we know that $N = 10{,}000$ when $t = 0$. We substitute these values into the equation $N = N_0 e^{0.5t}$ to obtain

$10{,}000 = N_0 e^{(0.5)(0)}$
$10{,}000 = N_0 e^0$
$10{,}000 = N_0(1) = N_0$

Since $N_0 = 10{,}000$ we now have

$N = 10{,}000 e^{0.5t}$

We now find $N$ when $t = 5$ as follows:

$N = 10{,}000 e^{(0.5)5}$
$N = 10{,}000 e^{2.5}$

From Table III we see that the approximate value of $e^{2.5}$ to three decimal places is 12.182. Thus $N = 10{,}000(12.182)$. Therefore, there are approximately 121,820 bacteria present in the culture after 5 hours. ●

**Remark 34.1**

The exponential function with base $e$ is often denoted by

$\exp = \{(x, f(x)) \mid f(x) = e^x\}$

and

$$\exp(x) = e^x$$

### Remark 34.2

In the discussions concerning the exponential function we described the function as increasing when $b > 1$ and decreasing when $0 < b < 1$ by noting whether the graph of the function was rising from left to right or dropping from left to right. A more formal definition is given in exercise 23.

The exponential function has a rather unusual property, given in the following theorem.

**Theorem 34.1** If $f(x)$ is an exponential function and $x_1$ and $x_2$ are any two real numbers, then $f(x_1 + x_2) = f(x_1) \cdot f(x_2)$.

Proof. Since $f(x)$ is an exponential function, it is of the form $f(x) = b^x$. Then $f(x_1 + x_2) = b^{x_1 + x_2}$ and $f(x_1) \cdot f(x_2) = b^{x_1} \cdot b^{x_2}$. Since $b^{x_1 + x_2} = b^{x_1} \cdot b^{x_2}$, we conclude that $f(x_1 + x_2) = f(x_1) \cdot f(x_2)$. ●

## EXERCISES FOR SECTION 34

1. Each of the given points lie on the graph of an exponential function. Find the bases of the exponential functions.
   (a) $(2, 25)$  (c) $(-2, \frac{1}{4})$  (e) $(0, 1)$
   (b) $(2, 8)$  (d) $(2, \frac{1}{4})$  (f) $(2, 100)$

2. If the graph of $f(x) = b^x$ passes through the point $(2, 9)$ find $f(4)$.

3. If the graph of $f(x) = b^x$ passes through the point $(-1, 4)$ find $f(-2)$.

In Exercises 4 through 13, (a) sketch the graph of the exponential function defined by $f(x)$; (b) determine whether the function is increasing or decreasing; (c) state the domain and the range.

4. $f(x) = 3^x$  9. $f(x) = (\frac{2}{3})^{-x}$

5. $f(x) = (\frac{1}{3})^x$  10. $f(x) = 10^x$

6. $f(x) = 4^{-x}$  11. $f(x) = 10^{-x}$

7. $f(x) = 5^x$  12. $f(x) = e^{-x}$

8. $f(x) = (\frac{1}{3})^{-x}$  13. $f(x) = e^{2x}$

14. Can an exponential function be symmetric with respect to the $x$-axis? The $y$-axis? Explain.

15. Describe the graph of $y = b^x$ if $b = 1$.

16. Why are exponential functions not defined for $b < 0$?

17. Suppose that a certain type of bacteria grows in such a way that $N = N_0 2^t$ represents the number of bacteria present at the end of $t$ hours. $N_0$ represents the initial number of bacteria.

(a) If there are 200,000 bacteria at the end of 2 hours, find the number of bacteria present at the end of 4 hours.

(b) How long does it take for the initial population to double? To quadruple?

18. The half-life of radium is approximately 1600 years; that is, half of the original amount remains after 1600 years. The amount $A$ of radium that remains after $t$ years is given by the equation

$$A = A_0(\tfrac{1}{2})^{kt}$$

where $A_0$ is the initial amount of radium. Suppose that we begin with 100 mg of radium. (a) Find the value of k. (b) How long will it take for 100 mg to disintegrate to 25 mg?

19. An object of temperature 150°C is placed in air of temperature 50°C and cools to 75°C in 5 minutes. If the temperature $T$ of the body after $t$ minutes is given by the equation $T = 100e^{kt} + 50$, find the temperature of the body after 15 minutes.

20. Show that the given functions are not exponential functions.
    (a) $f(x) = 2^{x-3}$  (c) $f(x) = 2^x \cdot 2^{x+1}$
    (b) $f(x) = 2^{3x+1}$  (d) $f(x) = 2^x + 2^{-x}$

21. Sketch the graphs of
    (a) $y = 2^{|x|}$  (b) $|y| = 2^{|x|}$  (c) $y = |2^x|$

22. Sketch the graphs of
    (a) $f = \{(x, y) | x = 4^y\}$
    (b) $f = \{(x, y) | x = 10^y\}$
    (c) $f = \{(x, y) | x = 2^y\}$

23. A function $f$ is said to be an *increasing* function on an interval $I$ if $f(x_1) < f(x_2)$ whenever $x_1 < x_2$ in $I$. A function $f$ is said to be a *decreasing* function on an interval $I$ if $f(x_1) > f(x_2)$ whenever $x_1 < x_2$ in $I$. Use the above definitions to show that
    (a) $f(x) = 3x - 1$ is an increasing function on $(-\infty, \infty)$
    (b) $f(x) = 4 - 2x$ is a decreasing function on $(-\infty, \infty)$
    (c) $f(x) = x^2$ is an increasing function on $[0, \infty)$
    (d) $f(x) = x^2$ is a decreasing function on $(-\infty, 0]$

24. Sketch the region bounded by the curves $y = 2^x$, $y = 1$, and $x = 2$. State the domain and range of this region.

25. Repeat Exercises 24 for the curves $y = e^x$, $y = e^{-x}$, $x = 0$, and $x = 1$.

SECTION **35A**
## LOGARITHMIC FUNCTIONS

In Section 34 we examined the typical shapes of the graphs of the exponential function $y = f(x) = b^x$ when $b > 1$ (Figure 34.3) and when $0 < b < 1$ (Figure 34.4). Since any horizontal line will intersect the graphs shown in Figures 34.3 and 34.4 in at most one point we know that $f(x) = b^x$ is a one-to-one function and thus it has an inverse function. This inverse is called the **logarithmic function to the base $b$** and is defined as follows:

**Definition 35A.1**  The inverse function of the exponential function $y = b^x$, $b > 0$ and $b \neq 1$, is given by $y = \log_b x$. This is read "the logarithm of $x$ to the base $b$." Notationally, we have $f = \{(x, y) \mid y = b^x\}$. Then $f^{-1} = \{(x, y) \mid x = b^y\} = \{(x, y) \mid y = \log_b x\}$. Thus $y = \log_b x$ is equivalent to $x = b^y$.

In Definition 35A.1 we see that the equation

$$y = \log_b x \tag{1}$$

is equivalent to the equation

$$b^y = x \tag{2}$$

We can substitute the value of $y$ in equation (1) into equation (2) to obtain

$$b^{\log x} = x \tag{35A.1}$$

where $b > 0, b \neq 1$, and $x > 0$.

Equation (35A.1) shows us that a logarithm is an exponent. That is, $\log_b x$ is the power to which we must raise $b$ to obtain $x$.

### Example 35A.1
Evaluate $\log_2 8$.

**Solution.** Let $y = \log_2 8$. Now $y = \log_2 8$ is equivalent to $2^y = 8$. Since $8 = 2^3$, we have $2^y = 2^3$. We now equate the exponents to obtain $y = 3$. Thus $\log_2 8 = 3$. It should be noted that to equate the exponents we made use of the fact that if $a > 0$ and $a \neq 1$, then $a^x = a^y$ implies that $x = y$. ●

### Example 35A.2
Evaluate $\log_8(\frac{1}{32})$.

**Solution.** Let $y = \log_8(\frac{1}{32})$. Now $y = \log_8(\frac{1}{32})$ is equivalent to $8^y = \frac{1}{32}$. Since $8 = 2^3$ and $\frac{1}{32} = 1/2^5 = 2^{-5}$, we have

$$8^y = \frac{1}{32}$$
$$(2^3)^y = 2^{-5}$$
$$2^{3y} = 2^{-5}$$

From the latter equation we see that $3y = -5$ or $y = -\frac{5}{3}$. Thus $\log_8(\frac{1}{32}) = -\frac{5}{3}$. ●

### Example 35A.3
Solve for $x$ when (a) $\log_2 x = 4$, (b) $\log_5 x = \frac{1}{3}$.

**Solution.** (a) The equation $\log_2 x = 4$ is equivalent to $4^2 = x$. Thus $x = 16$.
(b) The equation $\log_5 x = \frac{1}{3}$ is equivalent to $5^{1/3} = x$. Thus $x = 5^{1/3}$ or $\sqrt[3]{5}$.

### Example 35A.4
Solve for $b$ when $\log_b 4 = \frac{2}{5}$.

**Solution.** The equation $\log_b 4 = \frac{2}{5}$ is equivalent to $b^{2/5} = 4$. Thus

$$[(b^{2/5})]^{5/2} = 4^{5/2}$$
$$b = (4^{1/2})^5$$
$$b = 2^5 = 32. \quad \bullet$$

### Example 35A.5
Solve for $x$ when $\log_2(x^2 + 5) = 2$.

Solution. The equation $\log_2(x^2 + 5) = 2$ is equivalent to $2^2 = x^2 + 5$. Thus

$$x^2 + 5 = 4$$
$$x^2 = -1$$

The equation $x^2 = -1$ has no real solution. Therefore, the equation $\log_2(x^2 + 5) = 2$ has no real solution. $\quad \bullet$

### Remark 35A.1
It is important to be able to convert the **logarithmic equation** $y = \log_b x$ to the equivalent **exponential equation** $b^y = x$ and conversely.

In Section 33 we observed that we can obtain the graph of $f^{-1}$ by merely reflecting the graph of $f$ in the line $y = x$. Since $f(x) = \log_b x$ is the inverse of $f(x) = b^x$ where $b > 0$ and $b \neq 1$ we can sketch the graph of $f(x) = \log_b x$ by first sketching the graph of the $f(x) = b^x$ and reflecting this graph in the line $y = x$ to obtain the graph $f(x) = \log_b x$.

### Example 35A.6
Sketch the graph of $y = \log_2 x$.

Solution. From definition 35A.1 we know that $y = \log_2 x$ is equivalent to $2^y = x$. We also know that $2^y = x$ is the inverse of the exponential function $y = 2^x$. Thus the graph of $y = \log_2 x$ or $2^y = x$ can be obtained by reflecting the graph of $y = 2^x$ (See example 34.1 on page 171) in the line $y = x$. Figure 35A.1 shows the sketch of the graph of both $y = 2^x$ and $y = \log_2 x$.

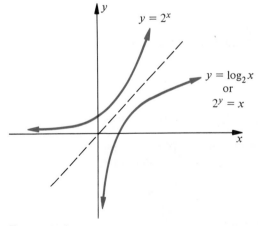

Figure 35A.1

From Figure 35A.1 we see that $y = \log_2 x$ is an increasing function whose domain is $\{x \mid x > 0\}$ and whose range is $\mathbb{R}$. •

**Example 35A.7**
Sketch the graph of $y = \log_{1/2} x$.

Solution. From Definition 35A.1 we know that $y = \log_{1/2} x$ is equivalent to $(\frac{1}{2})^y = x$. We also know that $(\frac{1}{2})^y = x$ is the inverse of the exponential function $y = (\frac{1}{2})^x$. Thus the graph of $y = \log_{1/2} x$ can be obtained by reflecting the graph of $y = (\frac{1}{2})^x$. (See Example 34.2 on page 172) in the line $y = x$. Figure 35A.2 shows in color this sketch of the graph of $y = \log_{1/2} x$. From Figure 35A.2 we see that $y = \log_{1/2} x$ is a decreasing function whose domain is $\{x \mid x > 0\}$ and whose range is $\mathbb{R}$. •

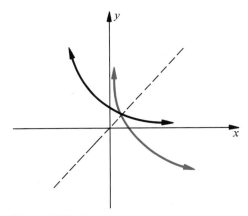

Figure 35A.2

We can also sketch the graph of $y = \log_b x$ directly by choosing convenient domain values to first form a table.

**Example 35A.8**
Sketch the graph of $y = \log_{10} x$.

Solution. We use the value in Table 35A.1 to sketch the graph shown in Figure 35A.3. We also note from Figure 35A.3 that $x = 0$ (the $y$-axis) is a vertical asymptote for $y = \log_{10} x$. •

**Table 35A.1**

| $x$ | 0.01 | 0.1 | 1 | 10 |
|---|---|---|---|---|
| $y = \log_{10} x$ | $-2$ | $-1$ | 0 | 1 |

From the graph of $y = \log_b x$, $b > 1$ shown in Figure 35A.4 we note the following properties of $\log_b x$.

1. If $0 < x < 1$, then $\log_b x < 0$.
2. If $x = 1$, then $\log_b x = 0$.

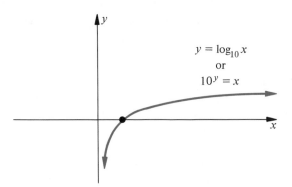

Figure 35A.3

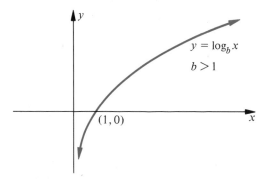

Figure 35A.4

3. If $x > 1$, then $\log_b x > 0$.
4. If $0 < x_1 < x_2$, then $\log_b x_1 < \log_b x_2$.
5. If $x_1 = x_2$ and if $x_1 > 0$ and $x_2 > 0$, then $\log_b x_1 = \log_b x_2$.

From the graph of $y = \log_b x$, $0 < b < 1$ shown in Figure 35A.5 we note the following properties of $\log_b x$.

1. If $0 < x < 1$, then $\log_b x > 0$.
2. If $x = 1$, then $\log_b x = 0$.
3. If $x > 1$, then $\log_b x < 0$.
4. If $0 < x_1 < x_2$, then $\log_b x_1 > \log_b x_2$.
5. If $x_1 = x_2$ and if $x_1 > 0$ and $x_2 > 0$, then $\log_b x_1 = \log_b x_2$.

Properties 1, 2, and 3 for $y = \log_b x$ when $b > 1$ or $0 < b < 1$ tell us that the domain of the logarithmic function is $\{x \mid x > 0\}$ and the range is $\mathbb{R}$.

Property 5 for $y = \log_b x$ tells us that the function is increasing when $b > 1$ and decreasing when $0 < b < 1$.

The most frequently encountered values for the base $b$ of logarithmic functions are 10 and $e$. If $b = 10$, then the logarithmic function is called the **common logarithmic function.** For common logarithmic functions we will agree to omit the base number 10 and write $\log x$ for $\log_{10} x$. When $b = e$ whose approximate value is 2.71828 then the logarithmic function is called the **natural logarithmic function.** For natural logarithmic function we will agree to use the notation $\ln x$ instead of $\log_e x$.

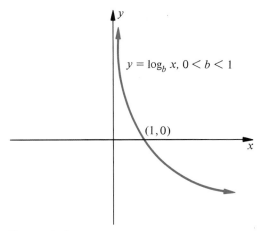

Figure 35A.5

## EXERCISES FOR SECTION 35A

For exercises 1 through 10, convert the given logarithmic equation to their equivalent exponential form.

**1.** $y = \log_3 x$     *1 to 6*     **6.** $\log \frac{1}{10} = -1$

**2.** $y = \log_4 6$     **7.** $\ln x = 1$

**3.** $3 = \log_2 8$     **8.** $\ln x = -2$

**4.** $-2 = \log_b 8$     **9.** $\log_a N = y$

**5.** $\log 100 = 2$     **10.** $\log_x N = \frac{1}{3}$

For exercises 11 through 20, convert the given exponential equations to their equivalent logarithmic form.

**11.** $2^5 = 32$     *odd #*     **16.** $2^{-3} = \frac{1}{8}$

**12.** $8^0 = 1$     **17.** $e^x = 3$

**13.** $b^y = 6$     **18.** $e^2 = y$

**14.** $16^{3/2} = 64$     **19.** $2^x = y$

**15.** $3^{-2} = \frac{1}{9}$     **20.** $4^y = x$

In exercises 21 through 34, find the value for the given logarithm.

**21.** $\log_2 16$     **28.** $\ln e^{-3}$

**22.** $\log_2(\frac{1}{8})$     **29.** $\log_6(\frac{1}{6})$

**23.** $\log 1000$     **30.** $\log_5(\frac{1}{125})$

**24.** $\log 0.001$     **31.** $\log_7(\frac{1}{7})$

**25.** $\log_2 \sqrt{2}$     **32.** $\log_3(\frac{1}{27})$

**26.** $\log_5 \sqrt[3]{5}$     **33.** $\log 1$

**27.** $\ln e^2$     **34.** $\ln 1$

In exercises 35 through 52, solve for $x$ in the given equation.

**35.** $\log_3 x = 2$      **44.** $\log_{9x} 3 = \dfrac{1}{\log_3 5}$

**36.** $\log_9 x = \frac{3}{2}$      **45.** $\ln x = \frac{1}{2}$

**37.** $\log_8 x = \frac{2}{3}$      **46.** $\ln 1 = x$

**38.** $\log_5 125 = x$      **47.** $3^{\log_3(x-2)} = 4$

**39.** $\log_9(\frac{1}{3}) = x$      **48.** $5^{\log_5 3} = x$

**40.** $\log_2(\frac{1}{8}) = x$      **49.** $x^{\log_2 9} = 9$

**41.** $\log_3 x = -2$      **50.** $x^{\log 100} = 100$

**42.** $\log_x 8 = \frac{3}{4}$      **51.** $7^{\log_x 4} = 4$

**43.** $\log_x 9 = -\frac{2}{3}$      **52.** $5^{2\log_5 x} = 16$

In Exercises 53 through 66, sketch the graph of the given function.

**53.** $y = \log_3 x$      **61.** $y = \ln(x+1)$

**54.** $y = \log_4 x$      **62.** $y = \log(-x)$

**55.** $y = \log_{1/3} x$      **63.** $y = \log|x|$

**56.** $y = \log_{1/4} x$      **64.** $y = \log x^2$

**57.** $y = \log_5 x$      **65.** $y = \ln(x^2+1)$

**58.** $y = \log_{1/5} x$      **66.** $y = \log(x^2-1)$

**59.** $y = \log(x+1)$      **67.** Prove that $\log_b\left(\dfrac{1}{x}\right) = \log_{1/b} x$

**60.** $y = \log(x-1)$

# SECTION **35B**
## FURTHER PROPERTIES OF LOGARITHMS

In section 35A we observed that a logarithm is an exponent. In this section we wish to develop some theorems and some more properties of logarithms. Then we will use these results to solve equations involving logarithms.

**Theorem 35B.1**   If $x$, $y$ and $b$ are positive real numbers and if $b \neq 1$, then

$$\log_b xy = \log_b x + \log_b y \qquad (1)$$

Equation (1) tells us that *the logarithm of the product of two numbers is equal to the sum of the logarithm of each.*

         Proof of Theorem 35B.1   Let $u = \log_b x$ and $v = \log_b y$. Then their respective exponential forms are $b^u = x$ and $b^v = v$.

$$xy = b^u b^v$$
$$xy = b^{(u+v)}$$

The logarithmic form of this latter equation is

$$\log_b xy = u + v$$

Substitution of $\log_b x$ for $u$ and $\log_b y$ for $v$ into this latter equation yields

$$\log_b xy = \log_b x + \log_b y. \quad \bullet$$

### Example 35B.1

Express $\log_2(8 \cdot 16)$ as a sum.

**Solution.** In equation (1), we can let $x = 8$ and $y = 16$, to obtain

$$\log_2 8 \cdot 16 = \log_2 8 + \log_2 16$$

or

$$\log_2 128 = \log_2 2^3 + \log_2 2^4 = 3 + 4 = 7. \quad \bullet$$

### Example 35B.2

Express $\log_3 4 + \log_3 5$ as a single logarithm.

**Solution.** We can use equation (1) letting $x = 4$ and $y = 5$ to obtain

$$\log_3 4 + \log_3 5 = \log_3 4 \cdot 5 = \log_3 20. \quad \bullet$$

**Theorem 35B.2**   If $x$, $y$ and $b$ are positive real numbers and if $b \neq 1$, then

$$\log_b \frac{x}{y} = \log_b x - \log_b y \qquad (2)$$

Equation (2) tells us that *the logarithm of a quotient equals the logarithm of the numerator minus the logarithm of denominator.*

**Proof of Theorem 35B.2.**   Let $u = \log_b x$ and $v = \log_b y$. Then their respective exponential forms are $b^u = x$ and $b^v = y$. Thus

$$\frac{x}{y} = \frac{b^u}{b^v}$$

$$\frac{x}{y} = b^{(u-v)}.$$

The logarithmic form of this latter equation is

$$\log_b \frac{x}{y} = u - v$$

We now substitute $\log_b x$ for $u$ and $\log_b y$ for $v$ into this equation to obtain

$$\log_b \frac{x}{y} = \log_b x - \log_b y. \quad \bullet$$

**Example 35B.3**

Express $\log \dfrac{8}{3}$ as the difference of two logarithms.

**Solution.** In equation (2) we can let $x = 8$ and $y = 3$ to obtain

$$\log \frac{8}{3} = \log 8 - \log 3. \quad \bullet$$

**Example 35B.4**

Express $\log(x^2 - 1) - \log(x + 1)$ as a single logarithm.

**Solution.** We use equation 2 as follows.

$$\log(x^2 - 1) - \log(x + 1) = \log \frac{x^2 - 1}{x + 1}.$$

Since $x^2 - 1 = (x + 1)(x - 1)$ we have $\log \dfrac{x^2 - 1}{x + 1} = \log (x - 1)$.

Thus, $\log(x^2 - 1) - \log(x + 1) = \log(x - 1). \quad \bullet$

**Theorem 35B.3** If $x$ and $b$ are positive real numbers, $b \neq 1$ and if $n$ is any real number, then

$$\log_b x^n = n \log_b x. \tag{3}$$

Equations (3) tells us that *the logarithm of the nth power of x is equal to n multiplied by the logarithm of x.*

**Proof of Theorem 35B.3.** Let $u = \log_b x$. Then its exponential form is $b^u = x$. Thus

$$x^n = (b^u)^n$$
$$x^n = b^{nu}$$

The logarithmic form of this latter equation is

$$\log_b x^n = nu$$

We now substitute $\log_b x$ for $u$ into this latter equation to obtain

$$\log_b x^n = n \log_b x \quad \bullet$$

**Example 35B.5**
(a) $\log_4 9 = \log_4 3^2 = 2 \log_4 3$
(b) $\ln x^3 = 3 \ln x$
(c) $\log \sqrt{7} = \log(7)^{1/2} = \frac{1}{2} \log 7$

**Example 35B.6**

Express $3 \log_b x + 2 \log_b y - \log_b z$ as a single logarithm.

**Solution.** We first apply Theorem 35B.3 to obtain

$\log_b x^3 + \log_b y^2 - \log_b z$

Now we can apply Theorem 35B.1 to obtain

$\log_b x^3 y^2 - \log_b z$

Finally, we apply Theorem 35B.2 to obtain

$\log_b \dfrac{x^3 y^2}{z}.$

### Example 35B.7

Express $\log_b \sqrt[4]{xy^2/z^3}$ in terms of the logarithms of $x$, $y$, and $z$.

Solution. Using Theorem 35B.3 we obtain

$$\log_b \sqrt[4]{\frac{xy^2}{z^3}} = \frac{1}{4} \log_b \frac{xy^2}{z^3}.$$

Now we apply Theorems 35B.1, 35B.2, and 35B.3 to the right member of this equation as follows.

$$\log_b \sqrt[4]{\frac{xy^2}{z^3}} = \frac{1}{4}[\log_b xy^2 - \log_b z^3]$$

$$= \frac{1}{4}[\log_b x + 2\log_b y - 3\log_b z]$$

$$= \frac{1}{4}\log_b x + \frac{1}{2}\log_b y - \frac{3}{4}\log_b z \quad \bullet$$

Equations that contain logarithmic expressions are called **logarithmic equations.** To solve such equations we try to obtain a single logarithmic expression on one side of the equation and express this latter equation in its equivalent exponential form.

### Example 35B.8

Solve for $x$ when $\log_3(2x + 7) - \log_3 x = 2$.

Solution. We apply Theorem 35B.2 to the left member of the equation to obtain

$$\log_3 \frac{2x + 7}{x} = 2.$$

We now write this equation in its equivalent exponential form and we have

$$\frac{2x + 7}{x} = 3^2$$
$$2x + 7 = 9x$$
$$7 = 7x \quad \text{or} \quad x = 1$$

### Example 35B.8

Solve for $y$ when $\log_5(y - 5) + \log_5(y - 1) = 1$.

Solution. We apply Theorem 35B.1 to the left member of the equation to obtain

$$\log_5(y - 5)(y - 1) = 1.$$

We now write the equation in its equivalent exponential form and we have

$$\begin{aligned}
(y - 5)(y - 1) &= 5^1 \\
y^2 - 6y + 5 &= 5 \\
y^2 - 6y &= 0 \\
y(y - 6) &= 0
\end{aligned}$$

The solutions to this latter equation are $y = 0$ and $y = 6$. The solution to the original equation is only $y = 6$. We rejected $y = 0$ as a solution because neither $\log_5(y - 5)$ nor $\log_5(y - 1)$ exist when $y = 0$.

## EXERCISES FOR SECTION 35B

In exercises 1 through 6, construct an example that will show that each statement is false.

**1.** $\log \dfrac{1}{x} = \dfrac{\log 1}{\log x}$

**2.** $\log_4(xy) = (\log_4 x)(\log_4 y)$

**3.** $n \ln x = (\ln x)^n$

**4.** $(\log x)(\log y) = \log x + \log y$

**5.** $\dfrac{\log x}{\log y} = \log x - \log y$

**6.** $\log_b(x^n) = (\log_b x)^n$

In exercises 7 through 14, given that $\log_b u = 1$, $\log_b v = 2$, and $\log_b w = 3$, evaluate each of the following by using the appropriate theorems of this section.

**7.** $\log_b uv$

**8.** $\log_b \dfrac{uv}{w}$

**9.** $(\log_b v)^2$

**10.** $\log_b \sqrt[6]{uvw}$

**11.** $\log_b \sqrt{vw}$

**12.** $\log_b \sqrt{\dfrac{v}{w}}$

**13.** $\dfrac{\log_b w}{\log_b u}$

**14.** $\log_b \dfrac{w}{v}$

In exercises 15 through 21, express the given expression as a single logarithm.

**15.** $2 \log x + \dfrac{1}{2} \log y$

**16.** $2 \log_4 x - \dfrac{1}{3} \log_4 z$

**17.** $2 \ln x + 3 \ln y - 4 \ln z$

**18.** $\log 2 - 2 \log x + \log z$

**19.** $\log_b 5 - \dfrac{2}{3} \log_b y + 3 \log_b z$

**20.** $\log x - 3 \log y + 3 \log 2$

**21.** $2 \ln x + 3 \ln y - 4 \ln 3 + 2 \ln z$

In exercises 22 through 30, express the given logarithm in terms of the logarithms of $x$, $y$ and $z$. Assume that $x$, $y$ and $z$ are positive real numbers.

**22.** $\log_b(4xy)$          **27.** $\log_b(x^{\frac{2}{3}})$

**23.** $\log_b(2xyz)$          **28.** $\log_b \sqrt[4]{y^5}$

**24.** $\log_b\left(\dfrac{4}{z}\right)$          **29.** $\log_b\left(\dfrac{x^2 y^3}{z^2}\right)$

**25.** $\log_b(y^3 z)$          **30.** $\log_b \sqrt{\dfrac{x}{y}}$

**26.** $\log_b(x^3 y^2)$

In exercises 31 through 40, solve each equation for $x$.

**31.** $\log_3 x = \log_3 2 + \log_3 5$

**32.** $\log x = \log 6 - \log 4$

**33.** $\log_5 x = \log_5 56 - 3 \log_5 2$

**34.** $\log(2x - 3) + \log 3 = \log 15$

**35.** $2 \log x = 4 \log 3$

**36.** $\log x - \log 4 = 0$

**37.** $\log(x^2 - 3x) = \log 8 - \log 2$

**38.** $\log_2(x^2 - 4) - \log_2(x + 2) = \log_2 3$

**39.** $\log_2(x^2 - 4) - \log_2(x + 2) = 3$

**40.** $\log x + \log(x - 1) = \log 6$

In exercises 41 through 47, solve each equation for $y$.

**41.** $2 \log x - \log y = 1$

**42.** $\log_2 x + \log_2 y + \log_2 3 = 1$

**43.** $\log_2 x + \log_2 y - \log_2 3 = 1$

**44.** $\log_2 x + \log_2 y - \log_2 3 = 0$

**45.** $2 \ln x + \ln y + 3 \ln 2 = 1$

**46.** $\log_4 x - \log_4 y - \log_4 3 = 1$

**47.** $3 \log x + 2 \log y - \log 5 = 2$

In exercises 48 through 53, solve for $x$.

**48.** $(\log x)(\log(x - 1)) = 0.$

**49.** $(\log x)(\log(x - 1)) > 0$

**50.** $(\log x)(\log(x - 1)) < 0$

**51.** $(\ln x)(\ln(x - 1)) = 0$

**52.** $(\ln x)(\ln(x - 1)) > 0$

**53.** $(\ln x)(\ln(x - 1)) < 0$

## SECTION 36
## COMPUTATIONS WITH LOGARITHMS

Theorems 35B.1, 35B.2, and 35B.3 provide us with the basis for using logarithms as a computational aid. The following examples illustrate various uses of logarithms in problem solving.

**Example 36.1**

1. $\log_2(4 \cdot 8) = \log_2 4 + \log_2 8 = 2 + 3 = 5$
2. $\log_2(\frac{4}{8}) = \log_2 4 - \log_2 8 = 2 - 3 = -1$
3. $\log_3 9^5 = 5 \log_3 9 = 5 \cdot 2 = 10.$

**Example 36.2**

If $\log 2 = 0.3010$, then

1. $\log 20 = \log 10(2) = \log 10 + \log 2 = 1 + 0.3010$
2. $\log 0.002 = \log(\frac{1}{1,000} \cdot 2) = \log 10^{-3} + \log 2 = -3 + 0.3010$
3. $\log 0.002 = \log(\frac{2}{1,000}) = \log 2 - \log 1,000 = 0.3010 - 3$
4. $\log 2^8 = 8 \log 2 = 8 \cdot (0.3010) = 2.4080$

For purposes of computation, it is advantageous to use 10 for a base. The reason for this lies in the fact that every positive number $N$ can be written in form $N = (10)^n x$, where $n$ is an integer and $1 \le x < 10$. For example, $9860 = 10^3(9.86)$, $4.5 = 10^0(4.5)$, and $0.00034 = 10^{-4}(3.4)$.

Thus $\log N = \log(10^n x) = \log 10^n + \log x = n + \log x$. For any $N$, $n$ can be easily found. The problem of determining $\log N$ reduces to finding $\log x$. Since $1 \le x < 10$, common logarithm tables include the values in this interval and it becomes a simple matter to "look up" $\log x$. The values given in a log table are only approximations.

**Definition 36.1** If $N = 10^n x$, where $1 \le x < 10$, then $n$ is called the **characteristic** and $\log x$ is called the **mantissa** of $\log N$.

**Example 36.3**

Use logarithms to find $\dfrac{(23.0)(13.7)}{(19.5)}$.

Solution. If $N = \dfrac{(23.0)(13.7)}{19.5}$, then by taking the log of both sides we have

$$\log N = \log \frac{(23.0)(13.7)}{19.5} = \log 23.0 + \log 13.7 - \log 19.5$$

$$= \log(2.3 \cdot 10^1) + \log(1.37 \cdot 10^1) - \log(1.95 \cdot 10^1)$$

Using the appropriate characteristics and the table of common logarithms (Table I) in the Appendix, we have $\log N = 1.3617 + 1.1367 - 1.2900 = 1.2084$. Since $\log N = 1.2084$, $N = 10^{1.2084}$. $N$ is called the **antilogarithm** of

1.2084 and is found by working backward from the table. Thus $N = 16.2$.  •

**Example 36.4**
Find $\sqrt[3]{61.4}$.

**Solution.**    If    $N = \sqrt[3]{61.4} = (61.4)^{1/3}$,    then    $\log N = \log(61.4)^{1/3} = \frac{1}{3}\log 61.4 = \frac{1}{3}(1.7882) = 0.5961$. Since $\log N = 0.5961$, $N = 3.95$.  •

Equations in which a variable occurs in an exponent are called **exponential equations.** Some exponential equations can be solved by considering the equivalent equation obtained by equating the logarithms of the two members and then solving the resulting equation.

**Example 36.5**
Solve for $x$ when $4^x = 7$

**Solution.**    Since $4^x = 7$ we equate the common logarithms of the two members to obtain

$$\log 4^x = \log 7$$

or, equivalently

$$x \log 4 = \log 7$$
$$x = \frac{\log 7}{\log 4}$$

We now use Table I to find $\log 7$ and $\log 4$, and we have

$$x = \frac{0.8451}{0.6021} \approx 1.4 \quad •$$

The general properties of logarithms may also be used to compute natural logarithms. We recall that natural logarithms are logarithms to the base $e$, where $e$ is approximately ($\approx$) 2.7183. Such logarithms are usually denoted ln. Tables of ln $N$ are usually written for $1 \leq N < 10$. However, care must be exercised in finding the logarithms of numbers larger or smaller than $N$ because the natural logarithms of integral powers of 10 are not integers.

If $N = x \cdot 10^n$, where $1 \leq x < 10$, then $\ln N = \ln(x \cdot 10^n) = \ln x + n \ln 10$. Since $\ln 10 = 2.3026$, we have $\ln N = \ln x + n(2.3026)$.

**Example 36.6**
$\ln 4.13 = 1.4183$ (Table II, Appendix)
$\ln 41.3 = \ln 4.13 + \ln 10 = 1.4183 + 2.3026 = 3.7209$
$\ln 0.0413 = \ln 4.13 + \ln 10^{-2} = 1.4183 - 2(2.3026)$
$= 1.4183 - 4.6052 = -3.1869$  •

To find $N$ if $\ln N$ is given and $0 \leq \ln N < 2.3026$, we simply refer to Table II. For values of $\ln N$ outside this interval, we must first express our given value in the form

$a + n(2.3026)$, where $0 \leq a < 2.3026$ and $n$ is any integer. Then $N$ is $10^n$ times the antilog of $a$. For example, if $\ln N = 0.8961 + 3(2.3026)$, then $N = 2.45 \cdot 10^3$. Similarly, if $\ln N = 0.8961 - 3(2.3026)$, then $N = 2.45 \cdot 10^{-3}$.

**Example 36.7**
Find $N$ if $\ln N = 4.3570$.

Solution

$$\ln N = 4.3570 = 2.3026 + 2.0544$$
$$N = 10 \cdot 7.81 = 78.1 \quad \bullet$$

**Example 36.8**
Find $N$ if $\ln N = -2.8950$.

Solution.  We first express $-2.8950$ in the form $a + n(2.3026)$, where $0 < a < 2.3026$. Thus

$$-2.8950 = -2(2.3026) + 1.7102$$
$$\ln N = 1.7102 - 2(2.3026)$$
$$N = 5.53 \cdot 10^{-2} = 0.0553 \quad \bullet$$

**Example 36.9**
Find $t$ if $e^{-2t} = 0.52$.

Solution.  Taking the natural logarithm of both sides of the equation, we have $\ln e^{-2t} = \ln 0.52$.

$\ln e^{-2t} = -2t \ln e = -2t$.

$\ln 0.52 = \ln\left(\dfrac{5.2}{10}\right) = \ln 5.2 - \ln 10 = 1.6487 - 2.3026 = -0.6539$

Thus

$$-2t = -0.6539 \quad \text{or} \quad t = \frac{-0.6539}{-2} = 0.327 \quad \bullet$$

**Example 36.10**
Assume that $N$, the number of bacteria present at a certain time $t$, is given by $N = N_0 e^{3t}$, where $N_0$ is the number of bacteria present when $t = 0$. If $t$ is measured in hours, find how long it will take for the number of bacteria to double?

Solution.  We are looking for $t$ when $N = 2N_0$, so $2N_0 = N_0 e^{3t}$ or $e^{3t} = 2$. Thus $\ln e^{3t} = \ln 2$.

$3t \ln e = \ln 2 \quad \text{or} \quad 3t = 0.6932$

Then

$$t = \frac{0.6932}{3} = 0.231 \text{ hours} \quad \bullet$$

The solution for the exponential equation $4^x = 7$ in Example 36.4 could have been obtained by equating the natural logarithms of the two members to obtain

$$\ln 4^x = \ln 7$$
$$x \ln 4 = \ln 7$$
$$x = \frac{\ln 7}{\ln 4}$$

We now use Table II to find $\ln 7$ and $\ln 4$, and we have

$$x = \frac{1.9459}{1.3863} \approx 1.4$$

We also could have expressed the exponential equation $4^x = 7$ in its equivalent logarithmic form to obtain $x = \log_4 7$. The problem with this form is that we would need a table of logarithmic values for base 4 to find $x$ (in this case $x = 1.4$). However, logarithmic tables other than bases 10 or $e$ are not usually accessible so we use either common or natural logarithms to solve such equations.

The procedure used in solving the equation $4^x = 7$ can be used to establish a formula relating logarithms with different bases. Let

$$y = \log_a x \tag{1}$$

or equivalently

$$a^y = x \tag{2}$$

Now equate the logarithms to the base $b$ of (2) to obtain

$$\log_b a^y = \log_b x$$
$$y \log_b a = \log_b x \tag{3}$$
$$y = \frac{\log_b x}{\log_b a}$$

We use (1) to replace $y$ in (3) by $\log_a x$ and we have

$$\log_a x = \frac{\log_b x}{\log_b a} \tag{36.1}$$

If we let $a = e$ and $b = 10$ in (36.1) we have

$$\ln x = \frac{\log x}{\log e} \tag{36.2}$$

If we let $a = 10$ and $b = e$ in (36.1) we have

$$\log x = \frac{\ln x}{\ln 10} \tag{36.3}$$

**Example 36.11**
Evaluate $\log_5 9$.

**Solution.** We use (36.1) to convert $\log_5 9$ to an expression involving common logarithms and then use Table I as follows.

$$\log_5 9 = \frac{\log 9}{\log 5} = \frac{0.9542}{0.6990} \approx 1.4$$

We also can use (36.1) to convert $\log_5 9$ to an expression involving natural logarithms and then use Table II as follows.

$$\log_5 9 = \frac{\ln 9}{\ln 5} = \frac{2.1972}{1.6094} \approx 1.4$$

## EXERCISES FOR SECTION 36

**1.** Use the tables in the Appendix to find
(a) $\log 23$      (d) $\ln 12$
(b) $\log 1.32$      (e) $\ln 137$
(c) $\log 0.0146$      (f) $\ln 0.0076$

**2.** Find $N$ if
(a) $\log N = 2.4829$      (d) $\ln N = 2.2565$
(b) $\log N = 0.7042$      (e) $\ln N = 4.0037$
(c) $\log N = -1.2865$      (f) $\ln N = -2.6037$

**3.** Use logarithms to solve each of the following equations.
(a) $2^x = 5$      (e) $e^x = 10$
(b) $10^x = 8^4$      (f) $e^{2.7183} = x$
(c) $e^x = 2.7183$      (g) $x^5 = 16$
(d) $e^{1/x} = 0.2$      (h) $x^3 = 12$

In exercises 4 through 12, solve for $x$.

**4.** $(\frac{1}{3})^x \geq 3$      **9.** $\log(\log x) = 2$

**5.** $(\frac{1}{2})^x \geq 8$      **10.** $(\log x)(\log x) = 2$

**6.** $(\frac{1}{2})^{x^2} \geq \frac{1}{16}$      **11.** $\log(\log x) = 10$

**7.** $10^{x^2-3x+2} = 1$      **12.** $\log_b(\log_b x) = b, \, b > 0, \, b \neq 1$

**8.** $10^{x^2-3x+2} > 1$

**13.** Write $N$ in an exponent form for each of the following.
(a) $\ln N = 3$      (d) $\log N + 1 = 1$
(b) $2 \ln N + 1 = 0$      (e) $3 \log N - 2 = 1$
(c) $\log N = -2.3147$

**14.** If $\ln 2 = 0.6932$, $\ln 10 = 2.3026$, find
(a) $\ln 20$      (e) $\ln 0.02$
(b) $\ln 200$      (f) $\log 2$
(c) $\ln 8$      (g) $\ln 0.2$
(d) $\ln \sqrt{2}$      (h) $\ln \frac{1}{2}$

**15.** On the same set of axes, graph $y = e^x$ and $y = \ln x$. State the domain and the range for each.

**16.** Use logarithms to compute the following numbers.
(a) $1.2e^{1.314}$      (b) $2e^{-2.3146}$      (c) $0.15e^3$

**17.** Let $N = N_0e^{kt}$ represent a formula for a certain population $N$ at any time $t$ ($t$ measured in years). $N_0$ is the population when $t = 0$. The population doubles in 12 years.
(a) Find the value for $k$.
(b) How long it will take for the initial population to triple?

**18.** Suppose that a radioactive substance decays according to the formula $N = Pe^{-kt}$, where $N$ represents the amount present after $t$ years and $P$ represents the initial amount.
(a) Suppose that after 25 years one-half of the original amount will remain (this is known as the *half-life*). Find $k$.
(b) How long will it take for the original amount to decay to $\frac{1}{8}$ of its original amount?

**19.** On the Richter scale the magnitude $M$ of an earthquake of intensity $I$ is given as

$$M = \log \frac{I}{I_0}$$

where $I_0$ is a minimum intensity used for comparison. If an earthquake has an intensity of $10^{7.8}$ times $I_0$, find its magnitude on the Richter scale.

**20.** Sketch the graphs of
(a) $|y| = \ln x$      (c) $|y| = e^x$
(b) $y = \ln |x|$      (d) $y = e^{|x|}$

# CHAPTER 7 CIRCULAR FUNCTIONS

## SECTION 37
### THE COORDINATE FUNCTION

In this chapter we wish to study the behavior of a collection of functions known as the **circular functions.** We shall define the circular functions so that both the domain and the range are subsets of the real numbers. We begin the study of these functions by constructing a function called the **coordinate function.** We first consider a circle of radius 1 whose center is at the origin of the rectangular coordinate system shown in Figure 37.1. This circle is called the **unit circle** and its equation is $x^2 + y^2 = 1$ (see Example 10.4). We know that, in general, the circumference or arc length of a circle is $2\pi r$, where $r$ is the length of the radius and $\pi$ is the special irrational real number represented by the decimal 3.14159.... Since the radius of the unit circle is equal to 1, its circumference or length of arc is $2\pi$ units.

Until now, we have only discussed distances measured along a straight line, but it should seem intuitively clear that between any two points $P$ and $Q$ on a circle there exists a *real* number that corresponds to the length of the arc $\widehat{PQ}$. We will assume that there is a one-to-one correspondence between the lengths of all the arcs of the unit circle measured in both counterclockwise and clockwise directions from the point $(1, 0)$ to the points $(x, y)$ on the unit circle and the set of real numbers. A careful treatment of arc length is beyond the scope of this text and is left for the calculus.

Now, if the point $P$ starts at the point $A(1, 0)$ and moves $|S|$ units around the circumference of the unit circle (Figure 37.1) in a counterclockwise direction if $S > 0$, clockwise if $S < 0$, we can locate the exact position of $P$ for any given value of $S$. Thus for every value of $S$ there is associated a point on the unit circle called the **terminal point** whose distance along the arc from $A(1, 0)$ to $P(S)$ is $S$ units. If $|S| > 2\pi$, we shall travel around the circle past $A$ until we have traveled the entire distance.

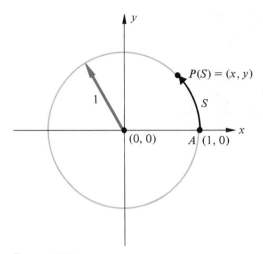

Figure 37.1

The above description defines a function that we will call the **coordinate function** denoted as follows.

$$P = \{[S, (x, y)] \mid P(S) = (x, y)\}$$

The domain of $P$ is the set of all real numbers. The range of $P$ is $\{(x, y) \mid x^2 + y^2 = 1\}$.

### Example 37.1
Draw the appropriate figure and then find the value of the coordinates of $P(\pi/2)$, $P(\pi)$, and $P(-\pi/2)$.

Solution.  We use Figure 37.2 to note that $P(\pi/2)$ is located at the point $B(0, 1)$, which is exactly one-fourth of the way around the unit circle, in a counterclockwise direction. Thus $P(\pi/2) = (0, 1)$. Similarly, $P(\pi)$ is located at the point $C(-1, 0)$, which is exactly one-half the way around the unit circle. Thus $P(\pi) = (-1, 0)$. Finally, $P(-\pi/2)$ is located at the point $D(0, -1)$, which is one-fourth of the way around the circle in a *clockwise* direction. We traveled clockwise since $S = -\pi/2 < 0$. Thus $P(-\pi/2) = (0, -1)$. We may also note from Figure 37.2 that $P(3\pi/2) = (0, -1)$. Thus $P(3\pi/2) = P(-\pi/2)$.  ●

### Example 37.2
Show that $P(\pi/2)$ and $P(5\pi/2)$ have identical coordinates.

Solution.  If $S = 5\pi/2$ or $2\pi + \pi/2$ we must travel from $A$ in a counterclockwise direction $2\pi$ units around the circle to the point $A$ and then $\pi/2$ units or one-fourth of the way farther. Thus $P(5\pi/2)$ is located at the point $B(0, 1)$ or $P(5\pi/2) = (0, 1)$. In Example 37.1 we established that $P(\pi/2) = (0, 1)$. Thus $P(5\pi/2) = P(\pi/2)$. A closer examination of the unit circle of Figure 37.2 shows us that not only are $P(\pi/2)$ and $P(5\pi/2)$ equal, but that $P(\pi/2) = P(\pi/2 \pm 2k\pi)$, $k = 0, 1, 2, \ldots$. In general, $P(S) = P(S \pm 2k\pi)$,

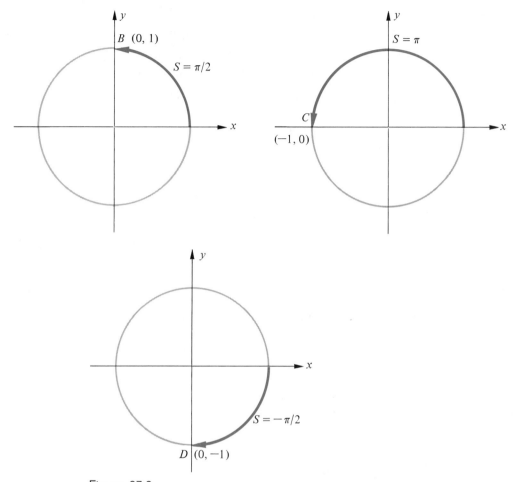

Figure 37.2

since $S \pm 2k\pi$ simply means traveling from the point $A(1, 0)$ $k$ times around the circle and then $|S|$ units in the appropriate direction, thus arriving at the point corresponding to $P(S)$. When $k = 1$, $P(S) = P(S + 2\pi)$. We say that $P$ is a **periodic function** with a period of $2\pi$. ●

**Definition 37.1** A function $f$ is said to be **periodic** if $f(x + p) = f(x)$ for some fixed $p > 0$ and for all $x$ in the domain of $f$. The smallest such $p$ is called the **period** of $f$.

In this section we have seen that it is easy to evaluate $P(S)$ for some special values of $S$ such as $0, \pi/2, \pi, 3\pi/2$, and $2\pi$. In general, the problem of evaluating $P(S)$ is not a simple one. We shall see later in the chapter how to handle this problem. We use the next two examples to illustrate how to find the coordinates of $P(S)$ for the special values $S = \pi/6$ and $S = \pi/3$.

**Example 37.3**
Find the coordinates for $P(\pi/6)$.

Solution. Let $P(\pi/6)$ be located at the point $D(h, k)$ and let the point $Q$ have the coordinates $(h, -k)$ in Figure 37.3. We see that the length of the arcs $\overset{\frown}{AD}$ and $\overset{\frown}{QA}$ are equal. (Why?) Thus the length of arc $\overset{\frown}{QD} = \overset{\frown}{QA} + \overset{\frown}{AD}$ or $\overset{\frown}{QD} = \pi/6 + \pi/6 = \pi/3$. Now the length of arc $\overset{\frown}{AB}$ is $\pi/2$. Thus the length of arc $\overset{\frown}{DB}$ is equal to the length of arc $\overset{\frown}{AB}$—length of arc $\overset{\frown}{AD}$ or $\pi/2 - \pi/6 = \pi/3$. We now recall from plane geometry that equal arcs of a circle subtend equal chords. Therefore, since

length $\overset{\frown}{QD}$ = length $\overset{\frown}{DB}$

the length of subtended chords are equal. Thus

length $\overline{QD}$ = length $\overline{DB}$

We now apply the distance formula to obtain the equation

$$2k = \sqrt{(h - 0)^2 + (k - 1)^2} \quad \text{or} \quad 4k^2 = h^2 + k^2 - 2k + 1 \tag{A}$$

Since $(h, k)$ lies on the unit circle, we have

$$h^2 + k^2 = 1 \tag{B}$$

Now we combine (A) and (B) to obtain the following:

$$4k^2 = 1 - 2k + 1$$
$$4k^2 = 2 - 2k$$
$$2k^2 = 1 - k$$
$$2k^2 + k - 1 = 0$$
$$(2k - 1)(k + 1) = 0$$
$$k = \tfrac{1}{2} \quad \text{or} \quad k = -1$$

We reject $k = -1$ since $P$ lies in the first quadrant. Finally, if $k = \tfrac{1}{2}$ and $h^2 + k^2 = 1$, we have

$$h^2 + (\tfrac{1}{2})^2 = 1$$
$$h^2 = 1 - \tfrac{1}{4} = \tfrac{3}{4}$$
$$h = \sqrt{3}/2 \quad \text{or} \quad -\sqrt{3}/2$$

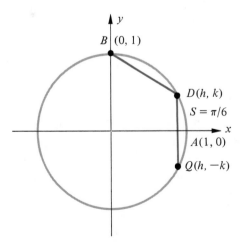

Figure 37.3

We reject $h = -\sqrt{3}/2$ since $P$ lies in the first quadrant. Thus $P(\pi/6) = (h, k) = (\sqrt{3}/2, 1/2)$. ●

## Example 37.4
Find the coordinates for $P(\pi/3)$.

Solution.   In Figure 37.4 we will let the coordinates for $P(\pi/3)$ be located at the point $C(h, k)$ on the unit circle. It should be noted that the point $C$ is located exactly one-third of the distance from point $B$ in a clockwise direction. Now the length of the arc $\overarc{BC}$ is equal to one-third of the length of arc $\overarc{BA}$ or arc length of $\overarc{BC} = 1/3(\pi/2)$ or $\pi/6$. We also note that $P(\pi/6)$ is located at point $D(\sqrt{3}/2, 1/2)$ and that the length of arc $\overarc{AD}$ is also $\pi/6$. Thus the length of both the arcs $\overarc{BC}$ and $\overarc{AD}$ are equal. Once again, we can conclude that if the lengths of the arcs are equal, so are the lengths of the respective subtended chords. Thus length of chord $\overline{BC}$ = length of chord $\overline{AD}$. Now we apply the distance formula to obtain

$$\sqrt{(h - 0)^2 + (k - 1)^2} = \sqrt{[(\sqrt{3}/2) - 1]^2 + [(1/2) - 0]^2}$$
$$\sqrt{h^2 + k^2 - 2k + 1} = \sqrt{(3/4) - \sqrt{3} + 1) + 1/4)} = \sqrt{2 - \sqrt{3}}$$
$$h^2 + k^2 - 2k + 1 = 2 - \sqrt{3} \tag{A}$$

Since the point $C$ lies on the unit circle we also know that

$$h^2 + k^2 = 1 \tag{B}$$

We now combine the results of (A) and (B) to obtain

$$1 - 2k + 1 = 2 - \sqrt{3}$$
$$-2k = -\sqrt{3}$$
$$k = \sqrt{3}/2$$

We can now substitute $k = \sqrt{3}/2$ into (B) to obtain

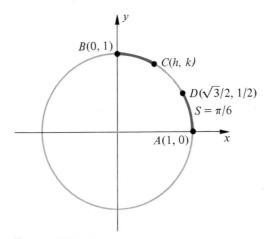

Figure 37.4

$$h^2 + (\sqrt{3}/2)^2 = 1$$
$$h^2 + 3/4 = 1$$
$$h^2 = 1/4$$
$$h = \pm 1/2$$

We reject $h = -1/2$ since the point $C$ is to be located in the first quadrant and we conclude that the coordinates of $C$ are $(1/2, \sqrt{3}/2)$; thus $P(\pi/3) = (1/2, \sqrt{3}/2)$. ●

### Example 37.5

Use the unit circle to determine the quadrant in which the coordinates of $P(S)$ are located if (a) $S = 2$, (b) $S = 4$, (c) $S = -5$, (d) $S = 12$.

Solution. We will assume that $\pi$ is approximated by 3.14. Since $S = 2 > 0$, we will move 2 units in a counterclockwise direction from the point $A(1, 0)$. We can see from Figure 37.5 that the coordinates for $P(2)$ must be located in the *second* quadrant, since $S = 2$ lies between $S = \pi/2$ and $S = \pi$. When $S = 4$ we must move 4 units in a counterclockwise direction from the point $A(1, 0)$. We see from Figure 37.6 that the coordinates of $P(4)$ are located in the *third* quadrant since $S = 4$ exceeds $\pi = 3.14$ and is less than $3\pi/2$. When $S = -5$ we must move 5 units in a clockwise direction (why?) from the point $A(1, 0)$. We see in Figure 37.7 that the coordinates of $P(-5)$ are located in the *first* quadrant since $|S| = 5$ exceeds $|-3\pi/2| = 4.71$. When $S = 12$ we must move 12 units in a counterclockwise direction from the point $A(1, 0)$. This is slightly less than $4\pi$ units (or two complete passes of the point $A$). Thus the coordinates of $P(12)$ are located in the *fourth* quadrant (Figure 37.8). ●

In Figure 37.9 we show the location on the unit circle of all the special values of the coordinate function.

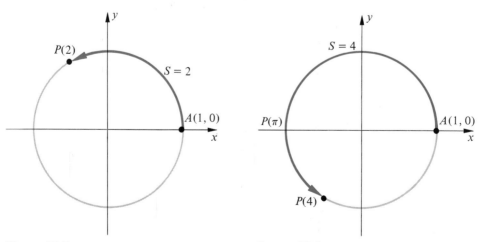

Figure 37.5                                    Figure 37.6

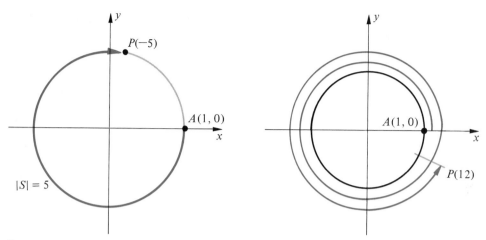

Figure 37.7                                    Figure 37.8

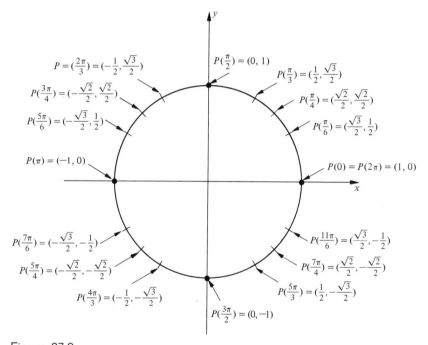

Figure 37.9

## EXERCISES FOR SECTION 37

**1.** Draw a unit circle and then find the coordinate of each of the following.
    (a) $P(-\pi)$       (c) $P(7\pi/2)$      (e) $P(-5\pi/2)$
    (b) $P(3\pi)$        (d) $P(6\pi)$       (f) $P(0)$

2. Use the unit circle to determine the quadrant in which the coordinates of $P(S)$ are located for each of the following.
   (a) $P(2.5)$    (c) $P(5)$    (e) $P(-10)$    (g) $P(-20)$
   (b) $P(-1)$    (d) $P(7)$    (f) $P(16)$

3. Use the facts that $P(\pi/6) = (\sqrt{3}/2, 1/2)$ and $P(\pi/3) = (1/2, \sqrt{3}/2)$ and that $P(S)$ is periodic to determine the coordinates of each of the following:
   (a) $P(13\pi/6)$    (b) $P(-5\pi/3)$    (c) $P(25\pi/6)$    (d) $P(-11\pi/6)$    (e) $P(13\pi/3)$

4. Use the techniques of Examples 37.3 and 37.4 to show that the coordinates for $P(\pi/4) = (\sqrt{2}/2, \sqrt{2}/2)$.

5. In which quadrant is $S$ located if
   (a) $P(S) = (-\sqrt{3}/2, 1/2)$?    (d) $P(S) = (-1/3, \sqrt{8}/3)$?
   (b) $P(S) = (-\sqrt{2}/2, \sqrt{2}/2)$?    (e) $P(S) = (\sqrt{3}/2, -1/2)$?
   (c) $P(S) = (-\sqrt{2}/2, -\sqrt{2}/2)$?

6. Use the results of this section and the symmetries of the unit circle to verify Table 37.1.

**Table 37.1**

| $S$ | $P(S)$ | $S$ | $P(S)$ |
|---|---|---|---|
| $0$ | $(1, 0)$ | $\dfrac{7\pi}{6}$ | $\left(-\dfrac{\sqrt{3}}{2}, -\dfrac{1}{2}\right)$ |
| $\dfrac{\pi}{6}$ | $\left(\dfrac{\sqrt{3}}{2}, \dfrac{1}{2}\right)$ | $\dfrac{5\pi}{4}$ | $\left(-\dfrac{\sqrt{2}}{2}, -\dfrac{\sqrt{2}}{2}\right)$ |
| $\dfrac{\pi}{4}$ | $\left(\dfrac{\sqrt{2}}{2}, \dfrac{\sqrt{2}}{2}\right)$ | $\dfrac{4\pi}{3}$ | $\left(-\dfrac{1}{2}, -\dfrac{\sqrt{3}}{2}\right)$ |
| $\dfrac{\pi}{3}$ | $\left(\dfrac{1}{2}, \dfrac{\sqrt{3}}{2}\right)$ | $\dfrac{3\pi}{2}$ | $(0, -1)$ |
| $\dfrac{\pi}{2}$ | $(0, 1)$ | $\dfrac{5\pi}{3}$ | $\left(\dfrac{1}{2}, -\dfrac{\sqrt{3}}{2}\right)$ |
| $\dfrac{2\pi}{3}$ | $\left(-\dfrac{1}{2}, \dfrac{\sqrt{3}}{2}\right)$ | $\dfrac{7\pi}{4}$ | $\left(\dfrac{\sqrt{2}}{2}, -\dfrac{\sqrt{2}}{2}\right)$ |
| $\dfrac{3\pi}{4}$ | $\left(-\dfrac{\sqrt{2}}{2}, \dfrac{\sqrt{2}}{2}\right)$ | $\dfrac{11\pi}{6}$ | $\left(\dfrac{\sqrt{3}}{2}, -\dfrac{1}{2}\right)$ |
| $\dfrac{5\pi}{6}$ | $\left(-\dfrac{\sqrt{3}}{2}, \dfrac{1}{2}\right)$ | $2\pi$ | $(1, 0)$ |
| $\pi$ | $(-1, 0)$ | | |

7. Show that if $P(S_1) = (x_1, y_1)$ and if $P(S_2) = (x_2, y_2)$, then $P(S_1 + S_2) = (h, k)$, where $h = x_1 x_2 - y_1 y_2$ and $k = y_1 x_2 + x_1 y_2$.

8. For each of the following, determine whether the function is periodic and if so, state its basic period.

(a)

(b)

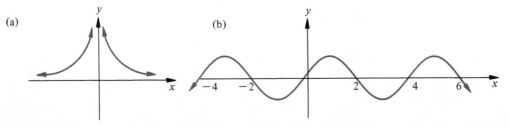

(c)

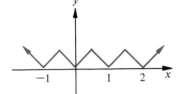

(d)

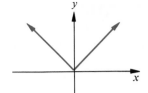

(e)

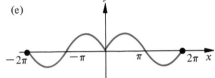

(f)

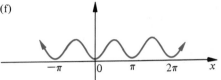

(g)

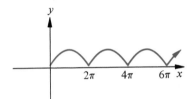

(h)

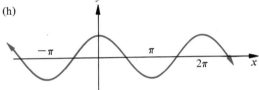

(i)

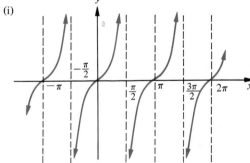

(j)

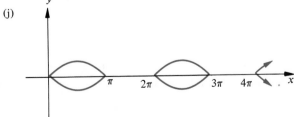

SECTION **38**

## THE CIRCULAR FUNCTIONS

In Section 37 we introduced the circular coordinate function

$$P = \{[S, (x, y)] \mid P(S) = (x, y)\}$$

which assigned to each real number $S$ a point $(x, y)$ on the unit circle $x^2 + y^2 = 1$. Thus the $P$ function associates real numbers with ordered pairs of real numbers. Since it is more convenient to work with functions that associate a set of real numbers to a set of real numbers, we will simplify matters here by defining two functions, one function that takes $S$ into $x$ and another that takes $S$ into the $y$ of the ordered pair $(x, y)$. In other words, we give special names to the $x$- and $y$-coordinates of the elements in the range of $P$.

**Definition 38.1**   If $P(S) = (x, y)$, then

$$\text{cosine} = \{(S, x) \mid x = \cos S\} \tag{38.1}$$
$$\text{sine} = \{(S, y) \mid y = \sin S\} \tag{38.2}$$

Cos $S$ is an abbreviation for cosine $S$ and sin $S$ is an abbreviation for sine $S$. Thus $x = \cos S$ is read "$x$ equals the cosine of $S$" and $y = \cos S$ is read "$y$ equals the cosine of $S$."

### Example 38.1
Evaluate $\cos \pi/6$ and $\sin \pi/6$.

**Solution.**   Since $P(\pi/6) = (\sqrt{3}/2, 1/2)$ we take the $x$ value $\sqrt{3}/2$ to be equal to the $\cos \pi/6$ and the $y$ value $(1/2)$ to be equal to $\sin \pi/6$. Thus

$$\cos \pi/6 = \sqrt{3}/2 \quad \text{and} \quad \sin \pi/6 = 1/2 \quad \bullet$$

### Example 38.2
Evaluate $\cos 3\pi/2$ and $\sin 3\pi/2$.

**Solution.**   We know that $P(3\pi/2) = (0, -1)$. Thus

$$\cos 3\pi/2 = 0 \quad \text{and} \quad \sin 3\pi/2 = -1 \quad \bullet$$

Since the sine and cosine functions were derived from the coordinate function, $P$, we see that their respective domains are all real numbers $S$. The range of the cosine function is the set of all $x$ values corresponding to points on the unit circle. The range of the sine function is the set of all $y$ values corresponding to the points on the unit circle. Thus the range of each is

$$\{r \mid -1 \leq r \leq 1\}$$

Since $\cos S$ and $\sin S$ are the $x$- and $y$-components of the points in the range of $P$, the functions sine and cosine are periodic with a period of $2\pi$. Since $P(S \pm 2k\pi) = P(S)$, $k = 0, 1, 2, \ldots$, we have

$$\cos (S \pm 2k\pi) = \cos S, \quad k = 0, 1, 2, \ldots \tag{38.3}$$
$$\sin (S \pm 2k\pi) = \sin S, \quad k = 0, 1, 2, \ldots \tag{38.4}$$

Since $P(S) = (x, y)$ and $(x, y) = (\cos S, \sin S)$ we can find the values for $\cos S$ and $\sin S$ for some special values of $S$ by referring to the table of Exercise 6 in Section 37. We list these values in Table 38.1.

**Table 38.1**

| S | P(S) | cos S | sin S | S | P(S) | cos S | sin S |
|---|------|-------|-------|---|------|-------|-------|
| 0 | $(1, 0)$ | 1 | 0 | $\dfrac{7\pi}{6}$ | $\left(-\dfrac{\sqrt{3}}{2}, -\dfrac{1}{2}\right)$ | $-\dfrac{\sqrt{3}}{2}$ | $-\dfrac{1}{2}$ |
| $\dfrac{\pi}{6}$ | $\left(\dfrac{\sqrt{3}}{2}, \dfrac{1}{2}\right)$ | $\dfrac{\sqrt{3}}{2}$ | $\dfrac{1}{2}$ | $\dfrac{5\pi}{4}$ | $\left(-\dfrac{\sqrt{2}}{2}, -\dfrac{\sqrt{2}}{2}\right)$ | $-\dfrac{\sqrt{2}}{2}$ | $-\dfrac{\sqrt{2}}{2}$ |
| $\dfrac{\pi}{4}$ | $\left(\dfrac{\sqrt{2}}{2}, \dfrac{\sqrt{2}}{2}\right)$ | $\dfrac{\sqrt{2}}{2}$ | $\dfrac{\sqrt{2}}{2}$ | $\dfrac{4\pi}{3}$ | $\left(-\dfrac{1}{2}, -\dfrac{\sqrt{3}}{2}\right)$ | $-\dfrac{1}{2}$ | $-\dfrac{\sqrt{3}}{2}$ |
| $\dfrac{\pi}{3}$ | $\left(\dfrac{1}{2}, \dfrac{\sqrt{3}}{2}\right)$ | $\dfrac{1}{2}$ | $\dfrac{\sqrt{3}}{2}$ | $\dfrac{3\pi}{2}$ | $(0, -1)$ | 0 | $-1$ |
| $\dfrac{\pi}{2}$ | $(0, 1)$ | 0 | 1 | $\dfrac{5\pi}{3}$ | $\left(\dfrac{1}{2}, -\dfrac{\sqrt{3}}{2}\right)$ | $\dfrac{1}{2}$ | $-\dfrac{\sqrt{3}}{2}$ |
| $\dfrac{2\pi}{3}$ | $\left(-\dfrac{1}{2}, \dfrac{\sqrt{3}}{2}\right)$ | $-\dfrac{1}{2}$ | $\dfrac{\sqrt{3}}{2}$ | $\dfrac{7\pi}{4}$ | $\left(\dfrac{\sqrt{2}}{2}, -\dfrac{\sqrt{2}}{2}\right)$ | $\dfrac{\sqrt{2}}{2}$ | $-\dfrac{\sqrt{2}}{2}$ |
| $\dfrac{3\pi}{4}$ | $\left(-\dfrac{2}{\sqrt{2}}, \dfrac{\sqrt{2}}{2}\right)$ | $-\dfrac{\sqrt{2}}{2}$ | $\dfrac{\sqrt{2}}{2}$ | $\dfrac{11\pi}{6}$ | $\left(\dfrac{\sqrt{3}}{2}, -\dfrac{1}{2}\right)$ | $\dfrac{\sqrt{3}}{2}$ | $-\dfrac{1}{2}$ |
| $\dfrac{5\pi}{6}$ | $\left(-\dfrac{\sqrt{3}}{2}, \dfrac{1}{2}\right)$ | $-\dfrac{\sqrt{3}}{2}$ | $\dfrac{1}{2}$ | $2\pi$ | $(1, 0)$ | 1 | 0 |
| $\pi$ | $(-1, 0)$ | $-1$ | 0 | | | | |

Another circular function of equal importance is the *tangent* function.

**Definition 38.2**  If $P(S) = (x, y)$, then tangent $= \{(S, h) \mid h = \tan S = \dfrac{y}{x}, x \neq 0\}$ (here $\tan S$ is an abbreviation for tangent $S$). Thus $\tan S$ is equal to the ratio of $y$ over $x$, where again $y$ is the second component and $x$ is the first component of the point $(x, y)$ on the unit circle. We may also state the defining equation for tangent as

$$\tan S = \frac{\sin S}{\cos S}, \quad \cos S \neq 0 \tag{38.5}$$

The domain of the tangent function is the set of real numbers $S$ excluding $S = \dfrac{\pi}{2} \pm k\pi$. (Why?) The range of the tangent function is all real numbers. The tangent function differs from the sine and cosine in its period. Examination of Figure 38.1 shows us that $P(S) = (x, y)$ and $P(S + \pi) = (-x, -y)$. Thus $\tan S = \dfrac{y}{x}$ and $\tan (S + \pi) = \dfrac{-y}{-x} = \dfrac{y}{x}$. Since $\tan (S + \pi) = \tan S$ or, more generally,

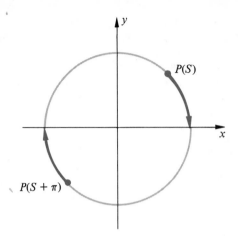

Figure 38.1

$$\tan (S \pm k\pi) = \tan S \tag{38.6}$$

we can conclude that the tangent function is periodic, with a period of $\pi$.

Table 38.1 and equations (38.3) and (38.4) can be used to compute the values of $\cos S$ and $\sin S$ for values differing by $2k\pi$.

**Example 38.3**

Evaluate $\cos 11\pi/2$ and $\sin 11\pi/2$.

**Solution.** We first note that $11\pi/2 = 4\pi + 3\pi/2$. We then use (38.3) and (38.4) to obtain

$$\cos \frac{11\pi}{2} = \cos \left( 4\pi + \frac{3\pi}{2} \right) = \cos \frac{3\pi}{2} = 0$$

$$\sin \frac{11\pi}{2} = \sin \left( 4\pi + \frac{3\pi}{2} \right) = \sin \frac{3\pi}{2} = -1 \quad \bullet$$

From Section 9 we recall the signs of the coordinates of the points in different quadrants. If we note in which quadrant $P(S)$ lies and the signs of $x$ or $y$, we can determine in which quadrants the circular functions are positive or negative. We show this in Table 38.2.

**Table 38.2**

| Quadrant for $P(S)$ | Values for $S$ | $\cos S = x$ | $\sin S = y$ | $\tan S = y/x$ |
|---|---|---|---|---|
| I | $0 \leq S \leq \frac{\pi}{2}$ | + | + | + |
| II | $\frac{\pi}{2} \leq S \leq \pi$ | − | + | − |
| III | $\pi \leq S \leq \frac{3\pi}{2}$ | − | − | + |
| IV | $\frac{3\pi}{2} \leq S \leq 2\pi$ | + | − | − |

Since the ordered pairs in the range of $P$ lie on the unit circle $x^2 + y^2 = 1$, and since $x = \cos S$ and $y = \sin S$, we have a basic identity relating $\sin S$ and $\cos S$.

**Theorem 38.1**  *For every real number $S$,*

$$\cos^2 S + \sin^2 S = 1 \tag{38.7}$$

It should be noted that for convenience of notation we write $\sin^2 S$ for $(\sin S)^2$. Thus if either $\sin S$ or $\cos S$ is known, and we can determine the quadrant in which $P(S) = (x, y)$ lies, we can find the value for the other by using (38.7) and Table 38.2.

### Example 38.4
If $\sin S = 3/4$ and $0 \leq S \leq \pi/2$, find $\cos S$.

**Solution.**  We observe that $0 \leq S \leq \pi/2$ means that $P(S)$ lies in the first quadrant and thus $\cos S \geq 0$. Next we use the result of (38.7) as follows:

$$\sin^2 S + \cos^2 S = 1$$
$$\cos^2 S = 1 - \sin^2 S$$
$$\cos S = \pm\sqrt{1 - \sin^2 S}$$
$$\cos S = \sqrt{1 - (3/4)^2}$$
$$\cos S = \sqrt{7/16} \text{ or } \sqrt{7}/4 \quad \bullet$$

### Example 38.5
If $\cos S = -4/5$ and $\pi \leq S \leq 3\pi/2$, find $\sin S$ and $\tan S$.

**Solution.**  We first observe that $\pi \leq S \leq 3\pi/2$ means that $P(S)$ lies in the third quadrant. Thus $\sin S \leq 0$ and $\tan S \geq 0$. We use (38.7) to find $\sin S$ as follows.

$$\sin^2 S + \cos^2 S = 1$$
$$\sin^2 S = 1 - \cos^2 S$$
$$\sin S = -\sqrt{1 - \cos^2 S}$$
$$\sin S = -\sqrt{1 - (-4/5)^2}$$
$$\sin S = -\sqrt{1 - 16/25} = -\sqrt{9/25} = -3/5$$

Now using (38.5), we have

$$\tan S = \frac{\sin S}{\cos S} = \frac{-3/5}{-4/5} = \frac{3}{4} \quad \bullet$$

There are three other circular functions. They are defined in terms of the coordinates of $P(S) = (x, y) = (\cos S, \sin S)$ as follows.

**Definition 38.3**  If $P(S) = (x, y) = (\cos S, \sin S)$, then

$$\text{cosecant} = \{(S, Z) \mid Z = \csc S = \frac{1}{y}, y \neq 0\} \tag{38.8}$$

$$\text{secant} = \{(S, V) \mid V = \sec S = \frac{1}{x}, x \neq 0\} \tag{38.9}$$

$$\text{cotangent} = \{(S, W) \mid W = \cot S = \frac{x}{y}, y \neq 0\} \qquad (38.10)$$

It should be noticed immediately from Definition 38.1 and 38.3 that we have the following equalities that are valid for all *permissible* values of $S$:

$$\sin S \csc S = 1 \qquad (38.11)$$

$$\cos S \sec S = 1 \qquad (38.12)$$

$$\tan S \cot S = 1 \qquad (38.13)$$

If the $\sin S$, $\cos S$, and $\tan S$ are known, any of the other three functions can be found. Thus there is no need for further discussion of the $\csc S$, $\sec S$, and $\cot S$.

We know that the unit circle, $x^2 + y^2 = 1$, is symmetric with respect to the $x$-axis. (Why?) Thus, if $P(S) = (x, y)$, then $P(-S) = (x, -y)$ and we have the following theorem, which is illustrated in Figure 38.2 for the case where $P(S)$ lies in the first quadrant.

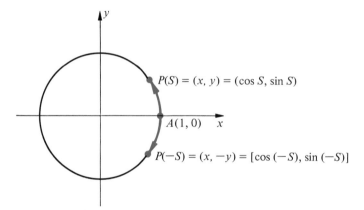

Figure 38.2

**Theorem 38.2**   For every real number $S$,

$$\cos(-S) = \cos S \qquad (38.14)$$
$$\sin(-S) = -\sin S \qquad (38.15)$$
$$\tan(-S) = -\tan S \qquad (38.16)$$

Any function $f$ that possesses the property that for every $x$ in its domain

$$f(-x) = f(x)$$

is called an **even function.** Any function $f$ that possesses the property that for every $x$ in its domain

$$f(-x) = -f(x)$$

is called an **odd function.** Thus cosine is an even function and sine and tangent are odd functions.

**Remark 38.1**

It should be noted that the circular functions have names that can be paired. In each pair we say that one function is the **cofunction** of the other. For example, the sine is cofunction of the cosine and the cosine is cofunction of the sine.

**Remark 38.2**

Because of equations (38.11), (38.12), and (38.13), the three circular functions csc $S$, sec $S$, and cot $S$ are called **reciprocal functions.**

## EXERCISES FOR SECTION 38

Exercises 1–22. Evaluate the following:

1. $\cos \dfrac{19\pi}{2}$

2. $\sin \dfrac{41\pi}{2}$

3. $\tan \dfrac{3\pi}{2}$

4. $\cos 23\pi$

5. $\sec 4\pi$

6. $\csc \dfrac{5\pi}{2}$

7. $\sin(-12\pi)$

8. $\cos(-12\pi)$

9. $\tan \dfrac{7\pi}{4}$

10. $\cot \dfrac{15\pi}{4}$

11. $\csc \dfrac{19\pi}{4}$

12. $\sec \dfrac{31\pi}{6}$

13. $\sec \dfrac{9\pi}{4}$

14. $\sec\left(-\dfrac{9\pi}{4}\right)$

15. $\csc \dfrac{32\pi}{3}$

16. $\tan\left(-\dfrac{30\pi}{3}\right)$

17. $\sin \dfrac{32\pi}{3}$

18. $\cos\left(-\dfrac{9\pi}{4}\right)$

19. $\sin\left(-\dfrac{19\pi}{2}\right)$

20. $\csc\left(-\dfrac{77\pi}{3}\right)$

21. $\sin\left(\dfrac{59\pi}{3}\right)$

22. $\tan\left(\dfrac{25\pi}{3}\right)$

23. If $P(S) = \left(\dfrac{3}{5}, -\dfrac{4}{5}\right)$, find the values for the six circular functions of $S$.

24. If $P(S) = \left(-\dfrac{5}{13}, -\dfrac{12}{13}\right)$, find the values for the six circular functions of $S$.

25. Determine which of the following are positive.
(a) $\cos 2$
(b) $\sin 7\pi/8$
(c) $\tan 3.6$
(d) $\sin 8$
(e) $\cos 9\pi/8$
(f) $\sin(-1)$
(g) $\cos(-2)$
(h) $\sec(-2)$
(i) $\cot 11\pi/4$
(j) $\csc(-2)$
(k) $\tan 7\pi/4$
(l) $\cos 5\pi/3$

**26.** Determine the value(s) of $S$ if $0 \leq S < 2\pi$, using Table 38.1.

(a) $\sin S = \sqrt{3}/2$     (f) $\cos S = -\dfrac{\sqrt{2}}{2}$     (k) $\cot S = 0$

(b) $\sin S = 1/2$     (g) $\sin S = -\sqrt{2}/2$     (l) $\sec S = \sqrt{2}$

(c) $\cos S = -1/2$     (h) $\sin S = -1/2$     (m) $\tan S = -\dfrac{\sqrt{3}}{3}$

(d) $\sin S = 0$     (i) $\cos S = 1/2$     (n) $\sec S = 2$

(e) $\cos S = -1$     (j) $\tan S = -1$     (o) $\tan S = \dfrac{\sqrt{3}}{3}$

**27.** Determine the quadrant(s) in which the corresponding $P(S)$ may lie under the following conditions.

(a) $\cos S > 0$             (d) $\sin S > 0$ and $\cos S < 0$
(b) $\cos S > 0$ and $\sin S < 0$     (e) $\sec S > 0$
(c) $\tan S < 0$ and $\sin S < 0$     (f) $\tan S > 0$

**28.** If $\sin S = \frac{1}{3}$ and $\cos S < 0$, find $\cos S$.

**29.** If $\sin S = \frac{1}{3}$ and $\cos S > 0$, find $\cos S$.

**30.** If $\cos S = \frac{3}{5}$ and $\tan S > 0$, find $\sin S$.

**31.** If $\cos S = \frac{3}{5}$ and $\tan S < 0$, find $\sin S$.

**32.** If $\sin S = \frac{12}{13}$ and $\cos S > 0$, find $\cos S$.

**33.** If $\sin S = \frac{12}{13}$ and $\sin S > \cos S$, find $\cos S$.

**34.** If $\sin S = -\frac{2}{3}$ and $\cos S > 0$, find $\cos S$.

**35.** If $\sin S = \frac{1}{3}$ and $\cos S > 0$, find $\tan S$.

**36.** If $\cos S = \frac{4}{5}$ and $\sin S < 0$, find $\tan S$.

**37.** Find the values of the $\sin S$, $\cos S$, and $\tan S$ for the following values of $S$.

(a) $7\pi/2$     (d) $5\pi$     (g) $80\pi$
(b) $9\pi/4$     (e) $-3\pi/2$     (h) $-21\pi$
(c) $\pi/6$     (f) $-\pi$     (i) $2\pi/3$

**38.** If $\sec S = -\frac{5}{4}$ find $\sin S$, $\cos S$, and $\tan S$.

**39.** If $\csc S = \frac{17}{15}$, find $\sin S$, $\cos S$, and $\tan S$.

**40.** If $\cot S = \frac{3}{2}$, find $\sin S$, $\cos S$, and $\tan S$.

**41.** Evaluate and simplify each of the following.

(a) $(\sin \pi/3)(\cos \pi/6)$

(b) $\dfrac{1 - \cos \pi/6}{2}$

(c) $\dfrac{1 + \cos \pi/6}{2}$

(d) $\sin^2 (2) + \cos^2 (2)$

(e) $(\sin \pi/6)(\cos \pi/3) + \cos(\pi/6)(\sin \pi/3)$

(f) $\dfrac{1 - \cos \pi/3}{1 + \cos \pi/3}$

(g) $\cos^2 \pi/6 - \sin^2 \pi/6$

**42.** (a) Evaluate $\sin (\pi/3 - \pi/6)$ by first noting that $\pi/3 - \pi/6 = \pi/6$.
   (b) Evaluate $\sin \pi/3 - \sin \pi/6$.
   (c) Does $\sin (\pi/3 - \pi/6) = \sin \pi/3 - \sin \pi/6$?
   (d) In general, does $\sin (S_1 - S_2) = \sin S_1 - \sin S_2$?

**43.** (a) Evaluate $\cos (2\pi/3 + \pi/3)$ by first noting that $2\pi/3 + \pi/3 = \pi$.
   (b) Evaluate $\cos 2\pi/3 + \cos \pi/3$.
   (c) Does $\cos (2\pi/3 + \pi/3) = \cos 2\pi/3 + \cos \pi/3$?
   (d) In general, does $\cos (S_1 + S_2) = \cos S_1 + \cos S_2$?

**44.** (a) Evaluate $\sin 2(\pi/6)$ by first noting that $2(\pi/6) = \pi/3$.
   (b) Evaluate $2 \sin \pi/6$.
   (c) Does $\sin 2(\pi/6) = 2 \sin \pi/6$?
   (d) In general, does $\sin 2S = 2 \sin S$?

**45.** If $\sin S = \frac{3}{5}$, find $\sin (-S)$.

**46.** If $\sin S = \frac{4}{5}$ and $\cos S < 0$, find $\sin (-S)$ and $\cos (-S)$.

**47.** Prove that the cosecant function is an *odd* function.

**48.** If $\tan S = \frac{7}{12}$, find $\tan (-S)$.

**49.** If $\cot S = -\frac{5}{12}$, find $\tan (-S)$.

**50.** (a) Find an expression in terms of $S$ for the length of any chord (in a unit circle) whose corresponding arc is the length $|S|$ units. [*Hint:* Use the distance formula to compute the length of chord $\overline{AB}$, where $A(1, 0)$ and $B(\cos S, \sin S)$ are as in Figure 38.3.]
   (b) Use the results of part (a) to find the length of the chord in a unit circle if the length of the corresponding arc is $S = \pi/4$, $S = 2\pi/3$.

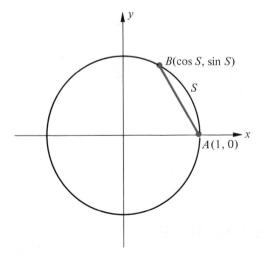

Figure 38.3

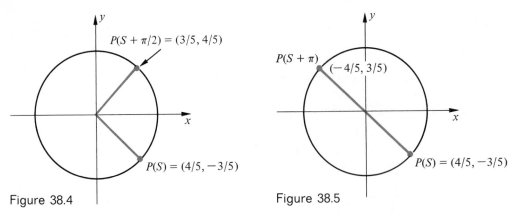

Figure 38.4  Figure 38.5

**51.** Use Figures 38.4 and 38.5 and congruent triangles to prove that if $P(S) = (4/5, -3/5)$, then

(a) $P(S + \pi/2) = (3/5, 4/5)$ (b) $P(S + \pi) = (-4/5, 3/5)$

(c) Use the results of parts (a) and (b) to show that

$$\sin (S + \pi/2) = \cos S \qquad \sin (S + \pi) = -\sin S$$
$$\cos (S + \pi/2) = -\sin S \qquad \cos (S + \pi) = -\cos S$$

**52.** Complete Table 38.3 by referring to Figure 37.9 on page 201. If $\tan S$ is not defined use a dash in the table to so indicate.

**Table 38.3**

| $S$ | $\cos S$ | $\sin S$ | $\tan S$ | $S$ | $\cos S$ | $\sin S$ | $\tan S$ |
|---|---|---|---|---|---|---|---|
| $0$ | $1$ | $0$ | | $\dfrac{7\pi}{6}$ | $-\dfrac{\sqrt{3}}{2}$ | $-\dfrac{1}{2}$ | |
| $\dfrac{\pi}{6}$ | $\dfrac{\sqrt{3}}{2}$ | $\dfrac{1}{2}$ | | $\dfrac{5\pi}{4}$ | $-\dfrac{\sqrt{2}}{2}$ | $-\dfrac{\sqrt{2}}{2}$ | |
| $\dfrac{\pi}{4}$ | $\dfrac{\sqrt{2}}{2}$ | $\dfrac{\sqrt{2}}{2}$ | | $\dfrac{4\pi}{3}$ | $-\dfrac{1}{2}$ | $-\dfrac{\sqrt{3}}{2}$ | |
| $\dfrac{\pi}{3}$ | $\dfrac{1}{2}$ | $\dfrac{\sqrt{3}}{2}$ | | $\dfrac{3\pi}{2}$ | $0$ | $-1$ | |
| $\dfrac{\pi}{2}$ | $0$ | $1$ | | $\dfrac{5\pi}{3}$ | $\dfrac{1}{2}$ | $-\dfrac{\sqrt{3}}{2}$ | |
| $\dfrac{2\pi}{3}$ | $-\dfrac{1}{2}$ | $\dfrac{\sqrt{3}}{2}$ | | $\dfrac{7\pi}{4}$ | $\dfrac{\sqrt{2}}{2}$ | $-\dfrac{\sqrt{2}}{2}$ | |
| $\dfrac{3\pi}{4}$ | $-\dfrac{\sqrt{2}}{2}$ | $\dfrac{\sqrt{2}}{2}$ | | $\dfrac{11\pi}{6}$ | $\dfrac{\sqrt{3}}{2}$ | $-\dfrac{1}{2}$ | |
| $\dfrac{5\pi}{6}$ | $-\dfrac{\sqrt{3}}{2}$ | $\dfrac{1}{2}$ | | $2\pi$ | $1$ | $0$ | |
| $\pi$ | $-1$ | $0$ | | | | | |

## SUM AND DIFFERENCE FORMULAS

We will often wish to consider circular functions of two numbers. In this section we shall see that such functions can be expressed in terms of the functions of each number separately.

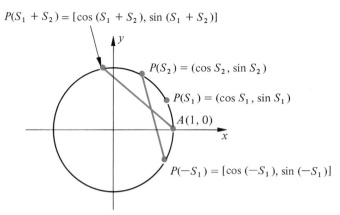

$P(S_1 + S_2) = [\cos (S_1 + S_2), \sin (S_1 + S_2)]$

$P(S_2) = (\cos S_2, \sin S_2)$

$P(S_1) = (\cos S_1, \sin S_1)$

$A(1, 0)$

$P(-S_1) = [\cos (-S_1), \sin (-S_1)]$

Figure 39.1

More specifically, we know that the function $\cos (2\pi/3 + \pi/3)$ *does not equal* $\cos 2\pi/3 + \cos \pi/3$ (see Exercise 19, Section 38).

We begin the development of these formulas by first developing a formula for $\cos (S_1 + S_2)$. Figure 39.1 shows that the points $P(S_1) = (\cos S_1, \sin S_1)$ and $P(S_2) = (\cos S_2, \sin S_2)$ lie on the unit circle along with the points $P(S_1 + S_2)$ and $P(-S_1)$. We see from the figure that the length of arc from the point $(1, 0)$ to the point $P(S_1 + S_2)$ is equal to the length of the arc from the point $P(-S_1)$ to the point $P(S_2)$. Thus the lengths of their respective chords are also equal. (Why?) We can apply the distance formula to obtain

$$\sqrt{[\cos (S_1 + S_2) - 1]^2 + [\sin (S_1 + S_2)]^2} =$$
$$\sqrt{[\cos S_2 - \cos (-S_1)]^2 + [\sin S_2 - \sin (-S_1)]^2} \quad \textbf{(A)}$$

We now use (38.14) and (38.15) to replace $\cos (-S_1)$ by $\cos S_1$ and $\sin (-S_1)$ by $-\sin S_1$ in (A) and then square both sides to obtain

$$[\cos (S_1 + S_2) - 1]^2 + [\sin (S_1 + S_2)]^2 = [\cos S_2 - \cos S_1]^2 + [\sin S_2 + \sin S_1]^2$$

or

$$\cos^2 (S_1 + S_2) - 2 \cos (S_1 + S_2) + 1 + \sin^2 (S_1 + S_2)$$
$$= \cos^2 S_2 - 2(\cos S_2)(\cos S_1) + \cos^2 S_1 + \sin^2 S_2 + 2 \sin S_2 \sin S_1 + \sin^2 S_1$$

We now regroup our terms and use the fact that $\sin^2 S + \cos^2 S = 1$ to obtain

$$[\cos^2 (S_1 + S_2) + \sin^2(S_1 + S_2)] - 2 \cos (S_1 + S_2) + 1$$
$$= [\cos^2 S_2 + \sin^2 S_2] + [\cos^2 S_1 + \sin^2 S_1] - 2 \cos S_1 \cos S_2 + 2 \sin S_1 \sin S_2$$

or

$$1 - 2 \cos (S_1 + S_2) + 1 = 1 + 1 - 2 \cos_1 \cos S_2 + 2 \sin S_1 \sin S_2$$
$$-2 \cos (S_1 + S_2) = -2 \cos S_1 \cos S_2 + 2 \sin S_1 \sin S_2$$
$$\cos (S_1 + S_2) = \cos S_1 \cos S_2 - \sin S_1 \sin S_2$$

Thus we have the following theorem.

**Theorem 39.1**   If $S_1$ and $S_2$ are real numbers, then

$$\cos (S_1 + S_2) = \cos S_1 \cos S_2 - \sin S_1 \sin S_2 \tag{39.1}$$

It should be carefully noted that (39.1) holds for *any real value* of $S_1$ or $S_2$. It is in this respect that (39.1) is a truly basic formula which will enable us to develop other formulas for sums and differences of the circular functions.

**Theorem 39.2**    If $S_1$ and $S_2$ are real numbers, then

$$\cos (S_1 - S_2) = \cos S_1 \cos S_2 + \sin S_1 \sin S_2 \tag{39.2}$$

**Proof.**   We first note that $\cos (S_1 - S_2) = \cos[S_1 + (-S_2)]$. This enables us to apply (39.1) to obtain

$$\cos (S_1 - S_2) = \cos [S_1 + (-S_2)] = \cos S_1 \cos (-S_2) + \sin S_1 \sin (-S_2)$$

We now use (38.14) and (38.15) to obtain

$$\cos (S_1 - S_2) = \cos S_1 \cos S_2 + \sin S_1 \sin S_2$$

which is the desired result.   ●

**Example 39.1**
Evaluate $\cos 7\pi/12$.

**Solution.**   We first observe that $7\pi/12$ can be expressed as the *sum* of $\pi/4$ and $\pi/3$. Thus

$$\cos 7\pi/12 = \cos (\pi/4 + \pi/3)$$

We now apply (39.1) to obtain

$$\cos 7\pi/12 = \cos (\pi/4 + \pi/3) = (\cos \pi/4)(\cos \pi/3) - (\sin \pi/4)(\sin \pi/3)$$

We now use Table 38.1 to obtain

$$\cos 7\pi/12 = (\sqrt{2}/2)(1/2) - (\sqrt{2}/2)(\sqrt{3}/2)$$
$$= \sqrt{2}/4 - \sqrt{6}/4 \quad \text{or} \quad (\sqrt{2} - \sqrt{6})/4 \quad ●$$

**Theorem 39.3**    If $S$ is a real number, then

$$\cos (\pi/2 - S) = \sin S \tag{39.3}$$
$$\cos (\pi - S) = -\cos S \tag{39.4}$$
$$\cos (\pi + S) = -\cos S \tag{39.5}$$
$$\cos (2\pi - S) = \cos S \tag{39.6}$$

**Proof of 1.**   First we substitute $\pi/2$ for $S_1$ and $S$ for $S_2$ in formula (39.2) to obtain

$$\cos (\pi/2 - S) = (\cos \pi/2)(\cos S) + (\sin \pi/2)(\sin S)$$

or

$$\cos (\pi/2 - S) = (0)(\cos) + (1)(\sin S) = \sin S \quad ●$$

We leave the proofs for the remaining conclusions as exercises.

**Example 39.2**

Show that $\sin (\pi/2 - S) = \cos S$ for all real number $S$.

**Solution.** We note that formula (39.3) states that $\sin S = \cos (\pi/2 - S)$, so we replace $S$ by $\pi/2 - S$ to obtain

$$\sin [\pi/2 - S] = \cos [\pi/2 - (\pi/2 - S)]$$

or

$$\sin [\pi/2 - S] = \cos [\pi/2 - \pi/2 + S] = \cos S \quad \bullet$$

**Theorem 39.4** If $S_1$ and $S_2$ are real numbers, then

$$\sin (S_1 + S_2) = \sin S_1 \cos S_2 + \cos S_1 \sin S_2 \qquad (39.7)$$

**Proof**

First we observe that $\cos [\pi/2 - (S_1 + S_2)] = \cos [(\pi/2 - S_1) - S_2]$. Next we substitute $\pi/2 - S_1$ for $S_1$ and $S_2$ for $S_2$ in formula (39.2) to obtain

$$\cos [\pi/2 - (S_1 + S_2)] = \cos [(\pi/2 - S_1) - S_2]$$
$$= \cos (\pi/2 - S_1) \cos S_2 + \sin (\pi/2 - S_1) \sin S_2$$

We now use formula (39.3) and the result of Example 39.2 to obtain

$$\cos [\pi/2 - (S_1 + S_2)] = \sin S_1 \cos S_2 + \cos S_1 \sin S_2$$

Finally, we use formula (39.3) again to rewrite the left members of the above equality as follows:

$$\sin (S_1 + S_2) = \sin S_1 \cos S_2 + \cos S_1 \sin S_2$$

the desired result. $\quad \bullet$

**Theorem 39.5** If $S_1$ and $S_2$ are real numbers, then

$$\sin (S_1 - S_2) = \sin S_1 \cos S_2 - \cos S_1 \sin S_2 \qquad (39.8)$$

The proof of this theorem is left as exercise 39.11.

**Example 39.3**

Evaluate $\sin \pi/12$.

**Solution.** We first observe that $\pi/12$ can be expressed as a difference between the numbers $\pi/4$ and $\pi/6$. That is, $\pi/12 = \pi/4 - \pi/6$. Thus

$$\sin \pi/12 = \sin [\pi/4 - \pi/6]$$

We now apply formula (39.8) to obtain

$$\sin \pi/12 = \sin [\pi/4 - \pi/6] = (\sin \pi/4)(\cos \pi/6) - (\cos \pi/4)(\sin \pi/6)$$

or

$$\sin \pi/12 = (\sqrt{2}/2)(\sqrt{3}/2) - (\sqrt{2}/2)(1/2)$$
$$= \sqrt{6}/4 - \sqrt{2}/4 = (\sqrt{6} - \sqrt{2})/4 \quad \bullet$$

**Theorem 39.6**   If $S$ is a real number, then

$$\sin (\pi/2 - S) = \cos S \tag{39.9}$$
$$\sin (\pi - S) = \sin S \tag{39.10}$$
$$\sin (\pi + S) = -\sin S \tag{39.11}$$
$$\sin (2\pi - S) = -\sin S \tag{39.12}$$

The proof of part (1) is done in Example 39.2. We leave the remaining proofs as exercises.

**Example 39.4**
Express $\cos (\pi/2 + S)$ in terms of a function of $S$ only.

**Solution.**   We substitute $\pi/2$ for $S_1$ and $S$ for $S_2$ in formula (39.1) to obtain

$$\cos (\pi/2 + S) = (\cos \pi/2)(\cos S) - (\sin \pi/2)(\sin S)$$

or

$$\cos (\pi/2 + S) = (0)(\cos S) - (1) \sin S$$
$$= -\sin S \quad \bullet$$

**Example 39.5**
Evaluate $\sin (S_1 + S_2)$ if $\sin S_1 = 3/5$, $\sin S_2 = 5/13$, $0 \leq S_1 \leq \pi/2$, and $0 \leq S_2 \leq \pi/2$.

**Solution.**   Since $\sin S_1 = 3/5$ and $0 \leq S_1 \leq \pi/2$, we use formula (38.7) to obtain

$$(3/5)^2 + \cos^2 S_1 = 1$$
$$\cos^2 S_1 = 1 - 9/25$$
$$\cos S_1 = \sqrt{16/25} = 4/5 \qquad \text{(Why?)}$$

Similarly, since $\sin S_2 = 5/13$ and $0 \leq S_2 \leq \pi/2$, we have

$$(5/13)^2 + \cos^2 S_2 = 1$$
$$\cos^2 S_1 = 1 - 25/169$$
$$\cos S_2 = \sqrt{144/169} = 12/13$$

Now we substitute these results into formula (39.7) to obtain

$$\sin (S_1 + S_2) = \sin S_1 \cos S_2 + \cos S_1 \sin S_2$$
$$= (3/5)(12/13) + (4/5)(5/13)$$
$$= 36/65 + 20/65 = 56/65 \quad \bullet$$

We have established the **sum formulas** (39.1) and (39.7) for the cosine and the sine functions. We can now develop a sum formula for the tangent function.

**Theorem 39.7**   If $S_1$ and $S_2$ are real numbers, then

$$\tan (S_1 + S_2) = \frac{\tan S_1 + \tan S_2}{1 - (\tan S_1)(\tan S_2)} \tag{39.13}$$

Proof. We use formulas (39.1) and (39.7) and the fact that

$$\tan (S_1 + S_2) = \frac{\sin (S_1 + S_2)}{\cos (S_1 + S_2)} \text{ to obtain}$$

$$\tan (S_1 + S_2) = \frac{\sin S_1 \cos S_2 + \cos S_1 \sin S_2}{\cos S_1 \cos S_2 - \sin S_1 \sin S_2}$$

We now divide each term in both the numerator and the denominator by $(\cos S_1)(\cos S_2)$ to obtain

$$\tan (S_1 + S_2) = \frac{\dfrac{\sin S_1 \cos S_2}{\cos S_1 \cos S_2} + \dfrac{\cos S_1 \sin S_2}{\cos S_1 \cos S_2}}{\dfrac{\cos S_1 \cos S_2}{\cos S_1 \cos S_2} - \dfrac{\sin S_1 \sin S_2}{\cos S_1 \cos S_2}}$$

$$= \frac{(\tan S_1)(1) + (1)(\tan S_2)}{1 - (\tan S_1)(\tan S_2)}$$

or

$$\tan (S_1 + S_2) = \frac{\tan S_1 + \tan S_2}{1 - (\tan S_1)(\tan S_2)} \quad \bullet$$

**Theorem 39.8**  If $S_1$ and $S_2$ are real numbers, then

$$\tan (S_1 - S_2) = \frac{\tan S_1 - \tan S_2}{1 + (\tan S_1)(\tan S_2)} \tag{39.14}$$

The proof of this theorem is left as exercise 39.15.

**Summary of General Sum Formulas**

$$\sin (S_1 + S_2) = \sin S_1 \cos S_2 + \cos S_1 \sin S_2$$
$$\cos (S_1 + S_2) = \cos S_1 \cos S_2 - \sin S_1 \sin S_2$$
$$\tan (S_1 + S_2) = \frac{\tan S_1 + \tan S_2}{1 - (\tan S_1)(\tan S_2)}$$

**Summary of General Difference Formulas**

$$\sin (S_1 - S_2) = \sin S_1 \cos S_2 - \cos S_1 \sin S_2$$
$$\cos (S_1 - S_2) = \cos S_1 \cos S_2 + \sin S_1 \sin S_2$$
$$\tan (S_1 - S_2) = \frac{\tan S_1 - \tan S_2}{1 + (\tan S_1)(\tan S_2)}$$

## EXERCISES FOR SECTION 39

1. Find the value of $\cos \pi/12$ by letting $\pi/12 = \pi/4 - \pi/6$.

2. Find the value of $\tan \pi/12$ by letting $\pi/12 = \pi/4 - \pi/6$.

3. Find the value of $\sin 5\pi/12$ by letting $5\pi/12 = \pi/4 + \pi/6$.

4. Find the value of $\cos 5\pi/12$ by letting $5\pi/12 = \pi/4 + \pi/6$.

5. Find the value of $\tan 5\pi/12$ by letting $5\pi/12 = \pi/4 + \pi/6$.

6. Find the value of $\sin 7\pi/12$ by letting $7\pi/12 = \pi/3 + \pi/4$.

7. Find the value of $\cos 7\pi/12$ by letting $7\pi/12 = \pi/3 + \pi/4$.

8. Express each of the following in terms of a function of $S$ only.
   (a) $\cos(\pi/4 - S)$   (d) $\sin(3\pi/2 + S)$   (g) $\sec(S - \pi/4)$
   (b) $\sin(\pi/2 + S)$   (e) $\cos(\pi/2 - S)$   (h) $\csc(S - \pi/6)$
   (c) $\tan(\pi/4 + S)$   (f) $\tan(\pi + S)$   (i) $\tan(\pi - S)$

9. (a) Evaluate $\sin(S_1 + S_2)$ and $\cos(S_1 + S_2)$ if $\sin S_1 = 8/17, 0 \leq S_1 \leq \pi/2, \sin S_2 = 4/5$, and $0 \leq S_2 \leq \pi/2$.
   (b) In which quadrant does $P(S_1 + S_2)$ lie?

10. (a) Evaluate $\sin(S_1 - S_2)$ and $\cos(S_1 - S_2)$ if $\sin S_1 = 15/17, \pi/2 \leq S_1 \leq \pi, \cos S_2 = -3/5$, and $\pi \leq S_2 \leq 3\pi/2$.
    (b) In which quadrant does $P(S_1 - S_2)$ lie?

11. Prove Theorem 39.5.

12. Prove Theorem 39.3, parts 2, 3, and 4.

13. Prove Theorem 39.6, part 2.

14. Prove Theorem 39.6, parts 3 and 4.

15. Prove Theorem 39.8. [*Hint:* Use $\tan(-S) = -\tan S$.]

16. Prove that $\sin S_1 \cos S_2 = \frac{1}{2}\sin(S_1 + S_2) + \frac{1}{2}\sin(S_1 - S_2)$.

17. Prove that $\cos S_1 \sin S_2 = \frac{1}{2}\sin(S_1 + S_2) - \frac{1}{2}\sin(S_1 - S_2)$.

18. Prove that $\cos S_1 \cos S_2 = \frac{1}{2}\cos(S_1 + S_2) + \frac{1}{2}\cos(S_1 - S_2)$.

19. Prove that $\sin S_1 \sin S_2 = \frac{1}{2}\cos(S_1 - S_2) - \frac{1}{2}\cos(S_1 + S_2)$.

20. Prove that $\sin S_1 + \sin S_2 = 2[\sin \frac{1}{2}(S_1 + S_2)][\cos \frac{1}{2}(S_1 - S_2)]$.

21. Prove that $\cos S_1 + \cos S_2 = 2[\cos \frac{1}{2}(S_1 + S_2)][\cos \frac{1}{2}(S_1 - S_2)]$.

22. Prove that $\sin S_1 - \sin S_2 = 2[\cos \frac{1}{2}(S_1 + S_2)][\sin \frac{1}{2}(S_1 - S_2)]$.

23. Prove that $\cos S_1 - \cos S_2 = -2[\sin \frac{1}{2}(S_1 + S_2)][\sin \frac{1}{2}(S_1 - S_2)]$.

In Exercises 24 through 29, use the results of Exercises 16 and 17 to express each product as a sum or difference.

24. $(\sin 3)(\cos 2)$   27. $(\sin 5x)(\cos 8x)$

25. $(\sin 4)(\cos 7)$   28. $(\cos 5)(\sin 4)$

26. $(\sin 3x)(\cos 2x)$   29. $(\cos 4x)(\sin 3x)$

In Exercises 30 through 37, use the results of Exercises 20 through 23 to express the sum or difference as a product.

30. $\sin 6 + \sin 7$   31. $\sin(-6) - \sin 5$

**32.** $\cos 4 + \cos 5$          **35.** $\cos 2x + \cos 3x$

**33.** $\cos(-3) - \cos 8$        **36.** $\sin 4x - \sin x$

**34.** $\sin 5x + \sin 2x$        **37.** $\cos 3x - \cos 5x$

**38.** Prove that $a \sin(ks) + b \cos(ks) = A \sin(ks + c)$, where

$$A = \sqrt{a^2 + b^2} \quad \text{and} \quad \sin c = \frac{b}{\sqrt{a^2 + b^2}}$$

**39.** Show that

$$\frac{\sin(x + h) - \sin x}{h} = \frac{\cos h - 1}{h}(\sin x) + \frac{\sin h}{h}(\cos x)$$

**40.** Show that if $(r, \theta)$ satisfies the system

$$x = r \cos \theta$$
$$y = r \sin \theta$$

then $(-r, \pi + \theta)$ does also.

SECTION **40**
# REDUCTION FORMULAS AND TABLES
# FOR CIRCULAR FUNCTIONS

In Section 38 we computed the exact values of the circular functions for certain real numbers (see Table 38.1). In general, the evaluation of the circular functions for any real number in their domain is not a simple matter. The reason for this is that the circumference or the arc length of the unit circle is the irrational number $2\pi$, which cannot be expressed as an even multiple of any integer. We have found that $\sin \pi/6 = 1/2 = 0.5$, an exact decimal value, while $\cos \pi/6 = \sqrt{3}/2$ can only be approximated as a decimal since $\sqrt{3}/2$ is an irrational number. The usual method for finding the exact values or the approximated values to any desired number of decimal places for the circular functions is developed in calculus and is beyond the scope of this text. These values are usually listed in tables. Tables list these values to a specified number of places, such as four, five, or six. Table IV in the Appendix is a five-place table.

Any table such as Table IV must, out of necessity, give values for the circular function in a restricted domain. We can express the value of a circular function of any real number $S$ in terms of a circular function of a real number $\bar{S}$, where $0 \leq \bar{S} \leq \pi/2$. We do this with the aid of some of the formulas developed in previous sections. These formulas are known as **reduction formulas.** We restate these formulas below.

**Reduction Formulas**

$$\sin[S + 2k\pi] = \sin S, \quad k \in J \qquad (38.3)$$
$$\cos[S + 2k\pi] = \cos S, \quad k \in J \qquad (38.4)$$
$$\sin(\pi - S) = \sin S \qquad (39.10)$$

$$\cos(\pi - S) = -\cos S \qquad (39.4)$$
$$\sin(\pi + S) = -\sin S \qquad (39.11)$$
$$\cos(\pi + S) = -\cos S \qquad (39.5)$$
$$\sin(2\pi - S) = -\sin S \qquad (39.12)$$
$$\cos(2\pi - S) = \cos S \qquad (39.6)$$
$$\sin(-S) = -\sin S \qquad (38.14)$$
$$\cos(-S) = \cos S \qquad (38.15)$$

If necessary, we use formulas (38.3) and (38.4) to reexpress the circular function in terms of a real number $S$, where $0 \le S < 2\pi$. Since $S$ is now a real number between 0 and $2\pi$ we can visualize in Figure 40.1 four possible positions for $P(S)$.

Associated with each arc length $S$ is its reference arc $\overline{S}$, where $\overline{S}$ is the shortest undirected arc length from the point $P(S)$ to the $x$-axis. In Figure 40.1 we see that when $P(S)$ lies in the first quadrant, the arc $S$ and its reference arc $\overline{S}$ are the same. When $P(S)$ lies in the second quadrant the arc length $S$ can be expressed in terms of its reference arc $\overline{S}$ by noting that $S + \overline{S} = \pi$. Hence, $\overline{S} = \pi - S$ where $0 \le \overline{S} < \dfrac{\pi}{2}$.

We can now use formulas (39.10) and (39.4) to express $\sin S$ and $\cos S$ in terms of the reference arc $\overline{S}$ as follows.

$$\sin S = \sin(\pi - \overline{S}) = \sin \overline{S}$$
$$\cos S = \cos(\pi - \overline{S}) = -\cos \overline{S}$$

Similarly, if $P(S)$ lies in the third quadrant, we can use formulas (39.11) and (39.5) to express $\sin S$ and $\cos S$ in terms of the reference arc $\overline{S}$ as follows:

$$\sin S = \sin(\pi + \overline{S}) = -\sin \overline{S}$$
$$\cos S = \cos(\pi + \overline{S}) = -\cos \overline{S}$$

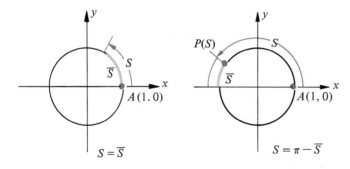

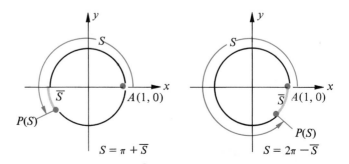

Figure 40.1

If $P(S)$ lies in the fourth quadrant we use formulas (39.12) and (39.6) to express sin $S$ and cos $S$ in terms of the reference arc $\overline{S}$ as follows:

$$\sin S = \sin (2\pi - \overline{S}) = -\sin \overline{S}$$
$$\cos S = \cos (2\pi - \overline{S}) = \cos \overline{S}$$

**Summary for Finding the Reference Arc $\overline{S}$**
To express any arc length $S$ in terms of its reference arc $\overline{S}, 0 \leq \overline{S} \leq \pi/2$, we first find $S$ such that $0 \leq S \leq 2\pi$ (if necessary). Then

1. If $0 \leq S \leq \pi/2$, $\overline{S} = S$.
2. If $\pi/2 \leq S \leq \pi$, $\overline{S} = \pi - S$.
3. If $\pi \leq S \leq 3\pi/2$, $\overline{S} = S - \pi$.
4. If $3\pi/2 \leq S \leq 2\pi$, $\overline{S} = 2\pi - S$.

**Example 40.1**
Express $\cos (31\pi/6)$ in terms of cos $S$, where $0 \leq S \leq \pi/2$.

**Solution.**  We first use formula (38.4) and the fact that $31\pi/6 = 4\pi + 7\pi/6$ to obtain

$$\cos (31\pi/6) = \cos [2(2\pi) + 7\pi/6] = \cos 7\pi/6$$

We next note that since $S = 7\pi/6$, $P(S)$ lies in the third quadrant and has a reference arc $\overline{S} = \pi/6$ (Figure 40.2). Thus $S = 7\pi/6 = \pi + \pi/6$ and

$$\cos 7\pi/6 = \cos (\pi + \pi/6) = -\cos \pi/6$$

Therefore, $\cos (31\pi/6) = \cos (7\pi/6) = -\cos \pi/6$  ●

**Example 40.2**
Express $\sin (-8\pi/3)$ in terms of sin $S$, where $0 \leq S \leq \pi/2$.

**Solution.**  We first use formula (38.14) to obtain $\sin (-8\pi/3) =$

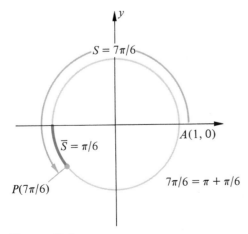

Figure 40.2

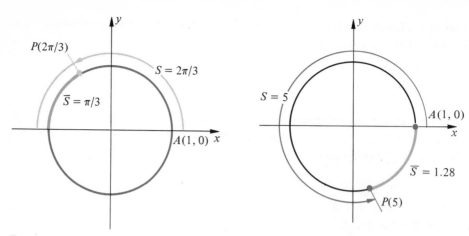

Figure 40.3                          Figure 40.4

$-\sin(8\pi/3)$. We next note that $8\pi/3 = 2\pi + 2\pi/3$. Thus $\sin(-8\pi/3) = -\sin[2\pi + 2\pi/3] = -\sin 2\pi/3$. (Why?) Since $S = 2\pi/3$, $P(S)$ lies in the second quadrant and has a reference arc $\overline{S} = \pi/3$. (See Figure 40.3). Thus $S = 2\pi/3 = \pi - \pi/3$ and by formula (39.10) we have

$$-\sin 2\pi/3 = -\sin(\pi - \pi/3) = -\sin \pi/3$$

Finally, we have

$$\sin(-8\pi/3) = -\sin \pi/3 \quad \bullet$$

### Remark 40.1

Whenever we wish to express $\sin S$ or $\cos S$ in terms of a real number $\overline{S}$ where $0 \le \overline{S} \le \pi/2$ and $S < 0$, we will *always first* apply the reduction formulas $\cos(-S) = \cos(S)$ and $\sin(-S) = -\sin S$ to obtain a positive $S$ (see Example 40.2).

### Example 40.3

Express $\cos 5$ in terms of $\cos S$, $0 \le S \le \pi/2$.

Solution.   For $S = 5$, $P(S)$ lies in the fourth quadrant, since $3\pi/2 \le 5 \le 2\pi$ or $4.71 \le 5 \le 6.28$. Since $5 = 2(3.14) - 1.28$, we see that the reference arc for 5 is $\overline{S} = 1.28$ (see Figure 40.4). We now apply formula (39.6) with $\pi = 3.14$ to obtain $\cos 5 = \cos[2(3.14) - 1.28] = \cos 1.28 \quad \bullet$

Table IV in the Appendix lists the actual values of the six circular functions for some of the special real numbers considered in the previous sections, as well as the approximate values for the six circular functions for any real number from 0 to 1.60 in increments of 0.01.

### Example 40.4

Use Table IV to evaluate $\tan 0.25$, $\sin 0.69$, $\sec 1.48$, and $\cot 1.20$.

Solution.   From Table IV we have

$$\tan 0.25 = 0.25534$$
$$\sin 0.69 = 0.63654$$
$$\sec 1.48 = 11.0288$$
$$\cot 1.40 = 0.17248 \quad \bullet$$

It should be noted that the column of values for the real numbers $S$ is entitled "radians." A radian will be defined in Chapter 8 when we discuss angular measure.

In reading values from a table, most numbers obtained are approximations. We say that $\tan 0.25 = 0.25534$, whereas $\tan 0.25$ is actually given by $0.25534 \pm \epsilon$, $0 < \epsilon < 0.000005$. If we desire greater accuracy, we either need more extensive tables or a method of computation that increases accuracy. One method of computation is called **linear interpolation** and is illustrated in the following example.

**Example 40.5**
Find $\sin 1.237$.

Solution

$$0.01 \left[ 0.007 \left[ \begin{array}{l} \sin 1.23 = 0.94249 \\ \sin 1.237 = \end{array} \right] d \middle| 0.00329 \right.$$
$$\sin 1.24 = 0.94578$$

As $S$ increases by 0.01, sine increases by 0.00329. In interpolating we assume that the rate of increase for $S$ and $\sin S$ is proportional and write

$$\frac{0.007}{0.01} = \frac{d}{0.00329}$$

Thus $d = 0.002303$ and $\sin 1.237 = 0.94479$.  $\bullet$

**Example 40.6**
Find $S$ if $\cos S = 0.31327$ and $0 \leq S \leq 1.57$.

Solution.   We proceed in essentially the same manner illustrated in Example 40.5. From the table we have

$$0.01 \left[ d \left[ \begin{array}{l} \cos 1.25 = 0.31532 \\ \cos S = 0.31327 \end{array} \right] -0.00205 \middle| -0.00950 \right.$$
$$\cos 1.26 = 0.30582$$

Thus

$$\frac{d}{0.01} = \frac{-0.00205}{-0.00950} = \frac{205}{950}$$

$$d = 0.002$$

Therefore, $S = 1.25 + 0.002 = 1.252$.  $\bullet$

**Example 40.7**

Find $S$ $(0 \leq S < 2\pi)$ if $\sin S = 0.91276$.

**Solution.** We use Table IV and see that $\sin S = 0.91276$ for $S = 1.15$. Since $\sin S$ is also positive in the second quadrant we have $S = \pi - \overline{S}$, where $\overline{S} = 1.15$. Thus $S = 3.14 - 1.15 = 1.99$. Hence the solutions are $S = 1.15$ and $1.99$. ●

## EXERCISES FOR SECTION 40

1. For each of the following, use the appropriate reduction formula to express the given circular function in terms of its reference arc $\overline{S}$, where $0 \leq \overline{S} \leq \pi/2$. Take $\pi \approx 3.14$ and $\pi/2 \approx 1.57$.
   (a) $\cos 3\pi/4$           (e) $\cos 2$              (i) $\tan(-15)$
   (b) $\sin 5\pi/6$           (f) $\sin 3$              (j) $\sin 31\pi/6$
   (c) $\cos(-12\pi/5)$        (g) $\cos(-15)$           (k) $\sin(-11\pi/6)$
   (d) $\sin 2$                (h) $\sin(-15)$           (l) $\cos 1$

2. Use Table IV to evaluate the following.
   (a) $\sin 0.25$      (e) $\sin 1$        (i) $\cos 1.75$
   (b) $\cos 0.66$      (f) $\cos 1$        (j) $\sin 4.75$
   (c) $\sec 1.31$      (g) $\tan 2$        (k) $\cos 6$
   (d) $\tan 1.02$      (h) $\sin(-3)$      (l) $\sin(-6)$

3. Express the given circular functions in terms of their reference arcs $\overline{S}$ and then evaluate.
   (a) $\cos 5\pi/6$         (e) $\cos 5\pi/3$          (i) $\sin(-11\pi/4)$
   (b) $\sin 3\pi/4$         (f) $\tan 11\pi/6$         (j) $\sin(7\pi/4)$
   (c) $\cos(-5\pi/6)$       (g) $\sec 5\pi/6$          (k) $\sin 3\pi/2$
   (d) $\sin(-3\pi/4)$       (h) $\csc 5\pi/3$          (l) $\csc 3\pi/2$

4. Use the appropriate reduction formulas and Table IV to evaluate the following.
   (a) $\sin 10$        (d) $\sec 22$        (g) $\cot 14$
   (b) $\cos 12$        (e) $\sin 100$       (h) $\sin(-31)$
   (c) $\tan(-13)$      (f) $\csc(-21)$      (i) $\cos(-12)$

5. Find the real number $S$ where $0 \leq S \leq \pi/2 \approx 1.57$, when
   (a) $\tan S = 1.2864$     (c) $\cos S = 0.26761$     (e) $\sec S = 9.0441$
   (b) $\sin S = 0.98370$    (d) $\cos S = 0.11195$     (f) $\csc S = 10.0167$

In exercises 6 through 19, find all real numbers $S$ where $0 \leq S \leq 2\pi$.

6. $\tan S = 1.2864$              13. $\sec S = -9.0441$

7. $\sin S = 0.98370$            14. $\cos S = -.50000$

8. $\cos S = 0.11195$            15. $\tan S = -1.0000$

9. $\sec S = 9.0441$             16. $\cos S = 0$

10. $\tan S = -1.2864$           17. $\sin S = .50000$

11. $\sin S = -0.98370$          18. $\sec S = -2.0000$

12. $\cos S = -0.11195$          19. $\sin S = -1.0000$

**20.** (a) We have already established the formulas

$$\sin\left(\frac{\pi}{2} - S\right) = \cos S$$

$$\cos\left(\frac{\pi}{2} - S\right) = \sin S$$

Use these results to verify that

$$\tan\left(\frac{\pi}{2} - S\right) = \cot S$$

$$\cot\left(\frac{\pi}{2} - S\right) = \tan S$$

Note that the four formulas above can be stated as: **The cofunction of any number equals the function of $\pi/2$ minus the number.**

(b) Use the formulas of part (a) to express each of the following in terms of $S$, where $0 \le S < \pi/4$. When necessary $\pi/4 \approx 0.79$.

*Example*

$$\sin \pi/3 = \sin (\pi/2 - \pi/6) = \cos \pi/6$$
$$\cos 1.50 = \cos (1.57 - 0.07) = \sin 0.07$$

(i)  $\cos 1$        (iii) $\tan 1$        (v)  $\cos 3$
(ii) $\sin 1.2$      (iv) $\sin 2$        (vi) $\tan 2$

# SECTION 41
# IDENTITIES FOR THE CIRCULAR FUNCTIONS

In Section 38 we defined the circular functions sine and cosine in terms of the unit circle and then we defined the other circular functions in terms of the sine and cosine functions. We restate these relationships now.

$$\sin^2 S + \cos^2 S = 1 \qquad\qquad\qquad (38.7)$$

$$\tan S = \frac{\sin S}{\cos S}, \qquad\qquad \cos S \ne 0 \quad (38.5)$$

$$\cot S = \frac{\cos S}{\sin S}, \qquad\qquad \sin S \ne 0 \quad (38.13) \qquad\qquad (41.1)$$

$$\sec S = \frac{1}{\cos S}, \qquad\qquad \cos S \ne 0 \quad (38.12)$$

$$\csc S = \frac{1}{\sin S}, \qquad\qquad \sin S \ne 0 \quad (38.11)$$

These relationships are called **identities** since they are true for all permissible values of $S$. From these *basic* identities we can develop other simple identities. Consider the identity

$$\sin^2 S + \cos^2 S = 1$$

Suppose that we divide each term of this equation by $\sin^2 S$ to obtain

$$\frac{\sin^2 S}{\sin^2 S} + \frac{\cos^2 S}{\sin^2 S} = \frac{1}{\sin^2 S}$$

or

$$1 + \left(\frac{\cos S}{\sin S}\right)^2 = \left(\frac{1}{\sin S}\right)^2$$

$$1 + \cot^2 S = \csc^2 S \tag{41.2}$$

This new identity is true for all values of $S$, where $\sin S \neq 0$. (Why?) Consider again the identity

$$\sin^2 S + \cos^2 S = 1$$

Now suppose we divide each term of this equation by $\cos^2 S$ to obtain

$$\frac{\sin^2 S}{\cos^2 S} + \frac{\cos^2 S}{\cos^2 S} = \frac{1}{\cos^2 S}$$

or

$$\left(\frac{\sin S}{\cos S}\right)^2 + 1 = \left(\frac{1}{\cos S}\right)^2$$

We now use substitution and the basic identities $\tan S = \dfrac{\sin S}{\cos S}$ and $\sec S = \dfrac{1}{\cos S}$ to obtain

$$\tan^2 S + 1 = \sec^2 S \tag{41.3}$$

We see that $\tan S = \dfrac{\sin S}{\cos S}$ and $\cot S = \dfrac{\cos S}{\sin S}$. Since $\dfrac{\sin S}{\cos S} = \dfrac{1}{\dfrac{\cos S}{\sin S}}$, we have

$$\tan S = \frac{1}{\cot S}, \quad \cot S \neq 0$$

or

$$\tan S \cot S = 1 \tag{41.4}$$

The restated identities of (41.1) and the identities (41.2), (41.3), and (41.4) are called the **fundamental identities** for the circular functions. These identities are so named since they can be used to establish (prove) other identities.

The proofs of other identities can be accomplished by using any of the following methods:

1. We can transform the right member of the equality into the exact form of the left member.
2. We can transform the left member of the equality into the exact form of the right member.
3. We can transform each side *separately* into the same form.

Unfortunately, there are no "standard" procedures for the actual transforming. When in doubt, it may be helpful to express the circular functions in terms of the sine and the cosine and then apply one of these three methods.

### Example 41.1

Prove that $\dfrac{1 + \tan^2 S}{\tan^2 S} = \csc^2 S$ and state the restrictions (if any) on the values of $S$.

Solution.  We will prove this identity by transforming the left member into the exact form of the right member, as follows. We first divide $\tan^2 S$ into each term in the numerator to obtain

$$\frac{1 + \tan^2 S}{\tan^2 S} = \frac{1}{\tan^2 S} + 1 \tag{A}$$

We now use the identity $\tan S = \dfrac{1}{\cot S}$ and substitution in (A) to obtain

$$\frac{1 + \tan^2 S}{\tan^2 S} = \frac{1}{\tan^2 S} + 1 = \cot^2 S + 1 \tag{B}$$

Next we use (41.2) and substitution into (B) to obtain

$$\frac{1 + \tan^2 S}{\tan^2 S} = \frac{1}{\tan^2 S} + 1 = \cot^2 S + 1 = \csc^2 S \tag{C}$$

In (C) we now have expressed the left member of the identity into the exact form of the right member. Thus the proof is complete. This identity is true for all values of $S$ where $\tan S \neq 0$.   ●

### Example 41.2

Prove that $\sec S - \cos S = \sin S \tan S$ and state the restriction, if any, on the values of $S$.

Solution.  We will prove this identity by transforming each side separately into the *same* form. We see that the left side can be expressed in terms of the $\cos S$ and then simplified to obtain

$$\sec S - \cos S = \frac{1}{\cos S} - \frac{\cos S}{1} = \frac{1 - \cos^2 S}{\cos S} = \frac{\sin^2 S}{\cos S} \tag{A}$$

The right side can be expressed in terms of $\sin S$ and $\cos S$ and simplified to obtain

$$\sin S \tan S = \sin S \left(\frac{\sin S}{\cos S}\right) = \frac{\sin^2 S}{\cos S} \tag{B}$$

Since (A) and (B) are identical, the proof is complete. We note that $\dfrac{\sin^2 S}{\cos S}$ will not be defined if $\cos S = 0$. Thus the identity is true for all values of $S$ where $\cos S \neq 0$.   ●

We can use the fundamental identities to express all the circular functions in terms of any one of them.

**Example 41.3**

Express csc $S$ in terms of the cos $S$.

**Solution.** We use formulas (38.11) and (38.7) to obtain

$$\csc S = \frac{1}{\sin S}$$

Since $\sin^2 S = 1 - \cos^2 S$

$$\sin S = \pm \sqrt{1 - \cos^2 S}$$

Thus $\csc S = \pm \dfrac{1}{\sqrt{1 - \cos^2 S}}$

where the sign depends on the quadrant in which $S$ is located. ●

## EXERCISES FOR SECTION 41

1. Given the following conditions, evaluate the other five circular functions.
   (a) $\tan S = \frac{1}{2}$, $\sin S > 0$       (c) $\sin S = -\frac{1}{5}$, $\sec S > 0$
   (b) $\cot S = -\frac{3}{5}$, $\csc S < 0$       (d) $\tan S = \frac{4}{5}$, $\sin S < 0$

2. Find the values of the six circular functions if
   (a) $S = -9\pi/2$       (c) $S = 5\pi/3$
   (b) $S = 11\pi/2$       (d) $S = 21\pi$

3. Express each of the following in terms of the sin $S$ only.
   (a) $\cos^2 S$       (c) $\sec^2 S$

   (b) $\dfrac{\sec S + 1}{\sin S + \tan S}$       (d) $\tan S$

4. Express the other five circular functions in terms of sin $S$.

5. Express the other five circular functions in terms of tan $S$.

6. Express $\sin S(\csc S - \sin S)$ in terms of cos $S$ only.

Prove the following identities.

7. $\tan S \cot S \sec S \cos S = 1$

8. $\tan S \sin S + \cos S = \sec S$

9. $\dfrac{1 + \cos S}{\sin S} = \dfrac{\sin S}{1 - \cos S}$

10. $\sin S \sec S = \tan S$

11. $\dfrac{\sin S \sec S}{\tan S} = 1$

12. $1 - \tan S = \dfrac{\cos S - \sin S}{\cos S}$

**13.** $(1 - \cos^2 S)(1 + \cot^2 S) = 1$

**14.** $2 \sec^2 S = \dfrac{1}{1 + \sin S} + \dfrac{1}{1 - \sin S}$

**15.** $\dfrac{1 + \tan^2 S}{\csc^2 S} = \tan^2 S$

**16.** $\sec S = \dfrac{\cos S}{2(1 + \sin S)} + \dfrac{\cos S}{2(1 - \sin S)}$

**17.** $\csc S = \dfrac{\sin S}{2(1 + \cos S)} + \dfrac{\sin S}{2(1 - \cos S)}$

# SECTION **42**
## MULTIPLE-VALUE IDENTITIES

We are frequently interested in expressing the circular function of multiples of $S$ in terms of the circular functions of $S$. If we wish to express the $\sin 2S$ in terms of $\sin S$ and $\cos S$, we can use sum formula (39.7) and let $S_1 = S_2 = S$ to obtain

$$\sin 2S = \sin (S + S) = \sin S \cos S + \cos S \sin S$$

or

$$\sin 2S = 2 \sin S \cos S \qquad\qquad (42.1)$$

This formula is called the **double-value formula for the sine function.** We can derive the **double-value formula for the cosine function** by using formula (39.1) and letting $S_1 = S_2 = S$ to obtain

$$\cos 2S = \cos (S + S) = \cos S \cos S - \sin S \sin S$$

or

$$\cos 2S = \cos^2 S - \sin^2 S \qquad\qquad (42.2)$$

If we wish to express $\cos 2S$ only in terms of $\cos S$, we can substitute $1 - \cos^2 S$ for $\sin^2 S$ in (42.2) to obtain

$$\cos 2S = \cos^2 S - (1 - \cos^2 S)$$

or

$$\cos 2S = 2 \cos^2 S - 1 \qquad\qquad (42.3)$$

If we wish to express $\cos 2S$ only in terms of $\sin S$ we can substitute $1 - \sin^2 S$ for $\cos^2 S$ in (42.2) to obtain

$$\cos 2S = 1 - \sin^2 S - \sin^2 S$$

or

$$\cos 2S = 1 - 2 \sin^2 S \qquad\qquad (42.4)$$

The identities (42.2), (42.3), and (42.4) are equivalent.

We can derive the **double-value formula for the tangent function** by either using formula (39.13) or the fact that $\tan 2S = \sin 2S / \cos 2S$ to obtain

$$\tan 2S = \frac{2 \tan S}{1 - \tan^2 S} \tag{42.5}$$

We can now use the identities (42.3) and (42.4) to derive the **half-value formulas** for sine and cosine by simply substituting $\frac{1}{2}S$ for $S$ in the respective identities and then solving for $\sin \frac{1}{2}S$ and $\cos \frac{1}{2}S$. We substitute $\frac{1}{2}S$ for $S$ in (42.3) and then solve for $\cos \frac{1}{2}S$ to obtain

$$\cos [2(\tfrac{1}{2}S)] = 2 \cos^2 (\tfrac{1}{2}S) - 1$$
$$\cos S = 2 \cos^2 (\tfrac{1}{2}S) - 1$$
$$\cos^2 \frac{S}{2} = \frac{1 + \cos S}{2}$$

$$\cos \frac{S}{2} = \pm \sqrt{\frac{1 + \cos S}{2}} \tag{42.6}$$

where the sign of the radical depends on the quadrant in which $S/2$ is located.

When we substitute $\frac{1}{2}S$ for $S$ in (42.4) we obtain the **half-value formula for sine.**

$$\sin \frac{S}{2} = \pm \sqrt{\frac{1 - \cos S}{2}} \tag{42.7}$$

Finally, we can derive the **half-value formula for tangent** by using the facts that

$$\tan \frac{S}{2} = \frac{\sin S/2}{\cos S/2}$$

and the formulas (42.6) and (42.7) to obtain

$$\tan \frac{S}{2} = \pm \sqrt{\frac{1 - \cos S}{1 + \cos S}} \tag{42.8}$$

**Remark 42.1**

We observe that the substitution of $S$ for $S/2$ in (42.6), (42.7) and (42.8) yields the following alternate forms for these formulas.

$$\cos S = \pm \sqrt{\frac{1 + \cos 2S}{2}}$$

$$\sin S = \pm \sqrt{\frac{1 - \cos 2S}{2}}$$

$$\tan S = \pm \sqrt{\frac{1 - \cos 2S}{1 + \cos 2S}}$$

**Example 42.1**

Express $\cos 3S$ in terms of $\cos S$.

Solution. We first note that $3S = 2S + S$ and then apply identity (39.1) to obtain

$$\cos 3S = \cos (2S + S) = (\cos 2S)(\cos S) - (\sin 2S)(\sin S) \qquad \textbf{(A)}$$

We can substitute the identities (42.3) and (42.1), respectively, for $\cos 2S$ and $\sin 2S$ in (A) to obtain

$$
\begin{aligned}
\cos 3S &= (2\cos^2 S - 1)(\cos S) - (2\sin S \cos S)(\sin S) \\
&= 2\cos^3 S - \cos S - 2\sin^2 S \cos S \\
&= 2\cos^3 S - \cos S - 2(1 - \cos^2 S)\cos S \\
&= 2\cos^3 S - \cos S - 2\cos S + 2\cos^3 S
\end{aligned}
$$

or

$$\cos 3S = 4\cos^3 S - 3\cos S \quad \bullet$$

We now give some additional uses of the special identities of this section in the following examples.

### Example 42.2

Express $\cos^4 S$ as an expression involving only circular functions of $S$ raised to the first power.

Solution. From identity (42.6) we have $\cos^2 \left(\dfrac{S}{2}\right) = \dfrac{1 + \cos S}{2}$. It should be carefully noted that the cosine function can be reduced from a power of 2 to 1 by *doubling* $S$. Thus an *equivalent* form of identity (42.6) is

$$\cos^2 S = \frac{1 + \cos 2S}{2}$$

Thus

$$\cos^4 S = (\cos^2 S)^2 = \left(\frac{1 + \cos 2S}{2}\right)^2$$

or

$$\cos^4 S = \frac{1 + 2\cos 2S + \cos^2 2S}{4} \qquad \textbf{(A)}$$

We again note that the term $\cos^2 2S$ can be reduced from a power of 2 to 1 by doubling $2S$ to obtain $\cos^2 2S = (1 + \cos 4S)/2$. We combine this result with (A) to obtain

$$
\begin{aligned}
\cos^4 S &= \frac{1}{4} + \frac{2}{4}\cos 2S + \frac{1}{4}\left(\frac{1 + \cos 4S}{2}\right) \\
&= \frac{1}{4} + \frac{1}{2}\cos 2S + \frac{1}{8} + \frac{1}{8}\cos 4S \\
&= \frac{3}{8} + \frac{1}{2}\cos 2S + \frac{1}{8}\cos 4S
\end{aligned}
$$

or

$$\cos^4 S = \frac{1}{8}(3 + 4\cos 2S + \cos 4S) \quad \bullet$$

### Example 42.3

Express $\sin 2S$, $\cos 2S$, and $\tan 2S$ in terms of $x$ if $\sin S = x$.

**Solution.** First we must find $\cos S$ in terms of $x$. We do this by noting that $\cos^2 S = 1 - \sin^2 S$, then substituting $x$ for $\sin S$ to obtain

$$\cos^2 S = 1 - x^2$$

or

$$\cos S = \pm\sqrt{1 - x^2}$$

We now can use this result and identity (42.1) to obtain

$$\sin 2S = 2\sin S \cos S$$
$$= 2x(\pm\sqrt{1 - x^2}) = \pm 2x\sqrt{1 - x^2} \qquad \textbf{(A)}$$

We also use identity (42.4) to obtain

$$\cos 2S = 1 - 2\sin^2 S$$
$$= 1 - 2x^2 \qquad \textbf{(B)}$$

Finally, we can express $\tan 2S$ in terms of $x$ by using the results (A) and (B) and the fact that $\tan 2S = \sin 2S/\cos 2S$ to obtain

$$\tan 2S = \frac{\sin 2S}{\cos 2S} = \pm\frac{2x\sqrt{1 - x^2}}{1 - 2x^2} \quad \bullet$$

### Example 42.4

Evaluate $\cos \pi/12$ by using the **half-value formula for cosine.**

**Solution.** Since $\dfrac{\pi}{12} = \dfrac{1}{2}\left(\dfrac{\pi}{6}\right)$ we can substitute $\dfrac{\pi}{6}$ for $S$ first in (42.6) to obtain

$$\cos\frac{\pi}{12} = \cos\left[\frac{1}{2}\left(\frac{\pi}{6}\right)\right] = \sqrt{\frac{1 + \cos\pi/6}{2}}$$

$$= \sqrt{\frac{1 + (\sqrt{3}/2)}{2}} = \sqrt{\frac{2 + \sqrt{3}}{4}}$$

or

$$\cos\frac{\pi}{12} = \frac{1}{2}\sqrt{2 + \sqrt{3}}$$

where it should be noted that we select the positive sign of the radical since $0 \le S/2 \le \pi/2$ (that is, $S/2 = \pi/12$ lies in the first quadrant).   $\bullet$

### Summary of Identities

For easy reference we list here the identities that have been discussed in Sections 38 through 42.

1. $\sin^2 S + \cos^2 S = 1$
2. $1 + \cot^2 S = \csc^2 S$
3. $\tan^2 S + 1 = \sec^2 S$
4. $\sin(-S) = -\sin S$
5. $\cos(-S) = \cos S$
6. $\tan(-S) = -\tan S$
7. $\sin(S_1 + S_2) = \sin S_1 \cos S_2 + \cos S_1 \sin S_2$
8. $\sin(S_1 - S_2) = \sin S_1 \cos S_2 - \cos S_1 \sin S_2$
9. $\cos(S_1 + S_2) = \cos S_1 \cos S_2 - \sin S_1 \sin S_2$
10. $\cos(S_1 - S_2) = \cos S_1 \cos S_2 + \sin S_1 \sin S_2$

11. $\tan(S_1 + S_2) = \dfrac{\tan S_1 + \tan S_2}{1 - \tan S_1 \tan S_2}$

12. $\tan(S_1 - S_2) = \dfrac{\tan S_1 - \tan S_2}{1 + \tan S_1 \tan S_2}$

13. $\sin 2S = 2 \sin S \cos S$
14. $\cos 2S = \cos^2 S - \sin^2 S$
    $\qquad = 2\cos^2 S - 1$
    $\qquad = 1 - 2\sin^2 S$

15. $\tan 2S = \dfrac{2 \tan S}{1 - \tan^2 S}$

16. $\sin\left(\dfrac{S}{2}\right) = \pm\sqrt{\dfrac{1 - \cos S}{2}}$

17. $\cos\left(\dfrac{S}{2}\right) = \pm\sqrt{\dfrac{1 + \cos S}{2}}$

18. $\tan\left(\dfrac{S}{2}\right) = \pm\sqrt{\dfrac{1 - \cos S}{1 + \cos S}}$

19. $\tan S = \dfrac{\sin S}{\cos S}$

20. $\cot S = \dfrac{\cos S}{\sin S}$

21. $\cot S = \dfrac{1}{\tan S}$

22. $\sec S = \dfrac{1}{\cos S}$

23. $\csc S = \dfrac{1}{\sin S}$

## EXERCISES FOR SECTION 42

**1.** If $\sin S = 3/5$ and $\cos S < 0$, find
   (a) $\sin 2S$     (b) $\cos 2S$     (c) $\tan 2S$

**2.** If $\tan S = 5/12$ and $\sin S > 0$, find
   (a) $\sin 2S$     (b) $\cos 2S$     (c) $\tan 2S$

3. If $\cos S = 15/17$ and $0 < S < \pi/2$, find
   (a) $\sin (S/2)$     (b) $\cos (S/2)$     (c) $\tan (S/2)$

4. If $\tan 2S = 12/5$ and $\cos 2S > 0$, find
   (a) $\sin 4S$     (b) $\cos 4S$     (c) $\tan 4S$

5. If $\sec S = 13/5$ and $\sin S > 0$, find
   (a) $\sin (S/2)$     (b) $\cos (S/2)$     (c) $\sec (S/2)$

6. Express $\sin 2S$, $\cos 2S$, and $\tan 2S$ in terms of $x$ if $\tan S = x$.

7. Express $\sin 2S$, $\cos 2S$, and $\tan 2S$ in terms of $y$ if $\cos S = y$.

8. Express $\sin 2S$, $\cos 2S$, and $\cot 2S$ in terms of $x$ if $\cot S = x$ and $x > 0$.

9. Evaluate $\sin 2S$ and $\sin 3S$ if $\sin S = 4/5$ and $0 \le S \le \pi/2$.

10. Evaluate $\cos 2S$ and $\cos 3S$ if $\cos S = -3/5$ and $\pi \le S \le 3\pi/2$.

11. Suppose that $\sin S_1 = 1/2, \cos S_2 = -\sqrt{3}/2, \pi/2 \le S_1 \le \pi$, and $\pi \le S_2 \le 3\pi/2$. Evaluate each of the following.
    (a) $\sin (S_2 - S_1)$     (c) $\sin 2S_1$     (e) $\cos S_1/2$
    (b) $\cos (\pi - S_1)$     (d) $\sin S_2/2$     (f) $\sin (4\pi - S_1)$

12. Use the appropriate half-value identities to evaluate each of the following.
    (a) $\sin \pi/8$     (c) $\sin 5\pi/12$     (e) $\tan \pi/12$
    (b) $\cos \pi/8$     (d) $\cos 5\pi/12$     (f) $\cos 7\pi/12$

13. Express $\sin^4 S$ in terms of circular functions involving circular functions of $S$ to the first power.

14. Express $\cos^2 S$ in terms of circular functions involving circular functions of $S$ to the first power.

Simplify exercises 15 through 24.

15. $2(\sin 3x)(\cos 3x)$

16. $(\cos 2x)^2 - (\sin 2x)^2$

17. $2(\sin x/2)(\cos x/2)$

18. $4(\sin 4x)(\cos 4x)$

19. $2 \cos^2(3x) - 1$

20. $4 \sin^2 (x/2) + 2 \cos x$

21. $\dfrac{\sin 4x}{\sin 2x}$

22. $\dfrac{\sin 4x}{2 \cos 2x}$

23. $\sin 2x - (\sin x + \cos x)^2$

24. $\sin 2x + (\sin x - \cos x)^2$

Prove the following identities.

25. $\sin 2S = \dfrac{2 \tan S}{1 + \tan^2 S}$

26. $\dfrac{2}{1 - \cos 2S} = \csc^2 S$

27. $\cot S = \cot 2S + \csc 2S$

28. $\dfrac{\sec S}{\tan S} = \csc S$

29. $\tan^2 S - \sec^2 S = \cot^2 S - \csc^2 S$

30. $\cos 2S = \dfrac{1 - \tan^2 S}{1 + \tan^2 S}$

31. $\sin 2S = \dfrac{1}{\tan S + \cot 2S}$

32. $\csc^2 S \cot^2 S = \dfrac{1}{\tan^2 S - \sin^2 S}$

33. $\tan \left(S + \dfrac{\pi}{4}\right) = \dfrac{1 + \tan S}{1 - \tan S}$

34. $(\sin S + \cos S)^2 = 1 + \sin 2S$

35. $\sin S = \dfrac{1 + \tan S}{\sec S + \csc S}$

36. $\sin^2 S_1 - \sin^2 S_2 = \sin (S_1 + S_2) \sin (S_1 - S_2)$

**37.** $\cos (S_1 + S_2) \cos (S_1 - S_2)$

$= \cos^2 S_1 - \sin^2 S_2$

**38.** $\sin 3S = 3 \sin S - 4 \sin^3 S$

**39.** $\sin (S/4) \cos (S/4) = \frac{1}{2} \sin (S/2)$

**40.** $\dfrac{1}{\cos S} - \dfrac{\cos S}{1 \, Q \sin S} = \tan S$

**41.** $\sin^3 S + \cos^3 S = \sin S(1 - \sin S \cos S) +$

$\cos S(1 - \sin S \cos S)$

**42.** $\cos^4 S - \sin^4 S = \cos 2S$

**43.** $\csc S = \cot S + \dfrac{\sin S}{1 + \cos S}$

**44.** (a) Verify identity (42.8).

(b) Show that $\tan S/2 = \dfrac{1 - \cos S}{\sin S}$.

(c) Show that $\tan S/2 = \dfrac{\sin S}{1 + \cos S}$.

**45.** If $y = \csc S$ and if $x = \csc S \cot S$, where $\dfrac{\pi}{2} \le S \le \pi$, show that $\dfrac{x}{y\sqrt{y^2 - 1}} = -1$.

**46.** Suppose $z = \tan \left( \dfrac{x}{2} \right)$ where $-\pi < x < \pi$.

(a) Show that $\sin x = \dfrac{2z}{1 + z^2}$

(b) Show that $\cos x = \dfrac{1 - z^2}{1 + z^2}$

(c) Show that $\tan x = \dfrac{2z}{1 - z^2}$

**47.** Use the results of Exercise **46**(a), (b) and (c) to express each of the following in terms of $z$.

(a) $\dfrac{1}{2 + \cos x}$

(b) $\dfrac{\cos x}{2 - \cos x}$

(c) $\dfrac{1}{3 + 5 \sin x}$

(d) $\dfrac{1}{\sin x + \tan x}$

# SECTION **43**
# GRAPHS OF THE CIRCULAR FUNCTIONS SINE AND COSINE

We now wish to examine the graph of the circular functions sine and cosine. We will use the graphing techniques that were established in Chapter 5 to graph first the sine function and then the cosine function. In Section 38 we defined the sine function and noted that its domain was all real numbers and its range was all real numbers between 1 and $-1$ inclusive. If we use the function notation from our earlier work, we can express the sine and cosine functions in terms of the conventional $(x, y)$ coordinate system rather than the $(s, y)$ and $(s, x)$ coordinate systems. We do so as follows:

$$\text{sine} = \{(x, y) \mid y = \sin x, -1 \le y \le 1\}$$

Similarly,

$$\text{cosine} = \{(x, y) \mid y = \cos x, -1 \le y \le 1\}$$

236

**Graph of** $y = \sin x$

**Domain:** $\{x \mid x \in \mathbb{R}\}$.

**Range:** $\{y \mid -1 \leq y \leq 1\}$.

**Period:** The period of $y = \sin x$ is $2\pi$. The graph of the sine function over an interval of one period is called a *cycle* of the curve. Thus we will simply sketch the graph $y = \sin x$ for those domain values $0 \leq x \leq 2\pi$ and then repeat this cycle over and over both to the left and to the right of the origin.

**Intercepts:** We see that if $y = 0$, then $0 = \sin x$. Thus $x = 0$, $\pi$, or $2\pi$ in the interval $0 \leq x \leq 2\pi$ (see Table 38.1). Therefore, we have $x$-intercepts at $(0, 0)$, $(\pi, 0)$, and $(2\pi, 0)$. If $x = 0$, then $y = \sin 0 = 0$. Thus we have a $y$-intercept at $(0, 0)$.

**Graphing Comments:** Table 43.1 can be easily extended to include the values of $y = \sin x$ for the interval $\pi < x \leq 2\pi$ (see Table 38.1). We now plot our intercepts and the points given Table 43.1 to obtain one cycle of the graph of $y = \sin x$. We then repeat this cycle over and over to obtain the graph shown in Figure 43.1. The complete graph is called the **standard sine curve**.

**Table 43.1**

| $x$ | $0$ | $\dfrac{\pi}{6}$ | $\dfrac{\pi}{4}$ | $\dfrac{\pi}{3}$ | $\dfrac{\pi}{2}$ | $\dfrac{2\pi}{3}$ | $\dfrac{3\pi}{4}$ | $\dfrac{5\pi}{6}$ | $\pi$ | $\dfrac{7\pi}{6}$ | $\dfrac{5\pi}{4}$ | $\dfrac{4\pi}{3}$ | $\dfrac{3\pi}{2}$ | $\dfrac{5\pi}{3}$ | $\dfrac{7\pi}{4}$ | $\dfrac{11\pi}{6}$ | $2\pi$ |
|---|---|---|---|---|---|---|---|---|---|---|---|---|---|---|---|---|---|
| $y = \sin x$ | $0$ | $\dfrac{1}{2}$ | $\dfrac{\sqrt{2}}{2}$ | $\dfrac{\sqrt{3}}{2}$ | $1$ | $\dfrac{\sqrt{3}}{2}$ | $\dfrac{\sqrt{2}}{2}$ | $\dfrac{1}{2}$ | $0$ | $-\dfrac{1}{2}$ | $-\dfrac{\sqrt{2}}{2}$ | $-\dfrac{\sqrt{3}}{2}$ | $-1$ | $-\dfrac{\sqrt{3}}{2}$ | $-\dfrac{\sqrt{2}}{2}$ | $-\dfrac{1}{2}$ | $0$ |

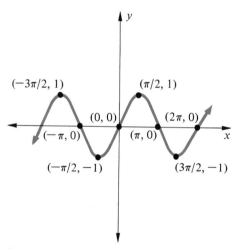

Figure 43.1

**Graph of** $y = \cos x$

**Domain:** $\{x \mid x \in \mathbb{R}\}$.

**Range:** $\{y \mid -1 \leq y \leq 1\}$.

**Period:** The period of $y = \cos x$ is $2\pi$. Thus we can sketch the graph of one cycle of

$y = \cos x$ by simply sketching its graph for the domain values $0 \le x \le 2\pi$ and repeating this cycle over and over again both to the left and right of the origin.

**Intercepts:** We see that if $y = 0$, then $0 = \cos x$. Thus $x = \pi/2$ or $3\pi/2$ in the interval $0 \le x \le 2\pi$. (Why?) Therefore, we have $x$-intercepts located at $(\pi/2, 0)$ and $(3\pi/2, 0)$. If $x = 0$, then $y = \cos 0 = 1$. Thus we have a $y$-intercept located at $(0, 1)$.

**Graphing Comments:** The table of values can be easily extended to include some special values for $y = \cos x$ in the interval $\pi < x \le 2\pi$ by using Table 38.1. We now plot our intercepts and the points in Table 43.2 to obtain one cycle of the graph of $y = \cos x$. We can then repeat this cycle over and over again to obtain the complete graph shown in Figure 43.2.

**Table 43.2**

| $x$ | 0 | $\frac{\pi}{6}$ | $\frac{\pi}{4}$ | $\frac{\pi}{3}$ | $\frac{\pi}{2}$ | $\frac{2\pi}{3}$ | $\frac{3\pi}{4}$ | $\frac{5\pi}{6}$ | $\pi$ | $\frac{7\pi}{6}$ | $\frac{5\pi}{4}$ | $\frac{4\pi}{3}$ | $\frac{3\pi}{2}$ | $\frac{5\pi}{3}$ | $\frac{7\pi}{4}$ | $\frac{11\pi}{6}$ | $2\pi$ |
|---|---|---|---|---|---|---|---|---|---|---|---|---|---|---|---|---|---|
| $y = \cos x$ | 1 | $\frac{\sqrt{3}}{2}$ | $\frac{\sqrt{2}}{2}$ | $\frac{1}{2}$ | 0 | $-\frac{1}{2}$ | $-\frac{\sqrt{2}}{2}$ | $-\frac{\sqrt{3}}{2}$ | $-1$ | $-\frac{\sqrt{3}}{2}$ | $-\frac{\sqrt{2}}{2}$ | $-\frac{1}{2}$ | 0 | $\frac{1}{2}$ | $\frac{\sqrt{2}}{2}$ | $\frac{\sqrt{3}}{2}$ | 1 |

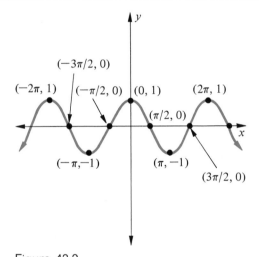

Figure 43.2

The **amplitude** of a periodic function is one-half of the absolute value of the difference between the maximum and minimum values. Thus for both the sine and cosine functions we have an amplitude equal to $\frac{1}{2}|1 - (-1)|$ or *one*.

The amplitude for $y = 3 \sin x$ is *three*, since each range value of $y = 3 \sin x$ is three times that of each range value of $y = \sin x$. In general, the amplitude of $y = A \sin x$ is $|A|$ units.

**Example 43.1**

Graph $f(x) = \sin 2x$ for $0 \le x \le \pi$.

Solution.   The domain of $f(x)$ is all real numbers and the range is

$\{y \mid -1 \leq y \leq 1\}$. We compute some of the values for $f(x) = \sin 2x$ by using special values of $x$ as follows:

$$f(0) = \sin 2(0) = \sin 0 = 0$$
$$f(\pi/6) = \sin 2(\pi/6) = \sin \pi/3 = \sqrt{3}/2$$
$$f(\pi/4) = \sin 2(\pi/4) = \sin \pi/2 = 1$$
$$f(\pi/3) = \sin 2(\pi/3) = \sin 2\pi/3 = \sqrt{3}/2$$
$$f(\pi/2) = \sin 2(\pi/2) = \sin \pi = 0$$

We list these results along with computations for additional special values of $x$ in Table 43.3.

It should be noted from this table that $f(x) = \sin 2x$ has x-intercepts at the points $(0, 0)$, $(\pi/2, 0)$, and $(\pi, 0)$. We now plot the points from the table to obtain the graph of $f(x) = \sin 2x$ shown in Figure 43.3.  ●

**Table 43.3**

| $x$ | 0 | $\dfrac{\pi}{6}$ | $\dfrac{\pi}{4}$ | $\dfrac{\pi}{3}$ | $\dfrac{\pi}{2}$ | $\dfrac{2\pi}{3}$ | $\dfrac{3\pi}{4}$ | $\dfrac{5\pi}{6}$ | $\pi$ |
|---|---|---|---|---|---|---|---|---|---|
| $y$ | 0 | $\dfrac{\sqrt{3}}{2}$ | 1 | $\dfrac{\sqrt{3}}{2}$ | 0 | $-\dfrac{\sqrt{3}}{2}$ | $-1$ | $-\dfrac{\sqrt{3}}{2}$ | 0 |

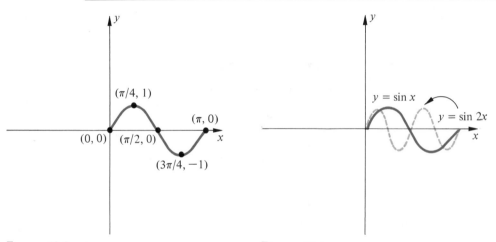

Figure 43.3                        Figure 43.4

It should be carefully noted that the amplitude of $f(x) = \sin 2x$ is equal to the amplitude of $f(x) = \sin x$, namely 1. But the period of $f(x) = \sin 2x$ is $\pi$ rather than $2\pi$. Thus the graph of $f(x) = \sin 2x$ completes its cycle every $\pi$ units. Thus the sine of a double value has a "compressing effect" on the basic sine curve. We show this comparison in Figure 43.4.

In general, $f(x) = \sin (bx)$, $b \neq 0$, has a period of $2\pi/|b|$. If $|b| > 1$, the graph of the basic sine curve is **compressed,** and if $|b| < 1$, the graph of the basic sine curve is **elongated.**

### Example 43.2
Determine the period of $f(x) = \sin (1/2) x$, and sketch its graph for one cycle.

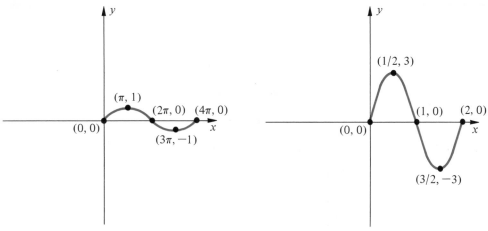

Figure 43.5                                    Figure 43.6

**Solution.** Since $b = \frac{1}{2}$, the period of $y = \sin \frac{1}{2}x$ is $2\pi/(\frac{1}{2}) = 4\pi$. The $x$-intercepts for $y = \sin \frac{1}{2}x$ for $0 \leq x \leq 4\pi$ are $x = 0, 2\pi$, and $4\pi$. (Why?) The amplitude is 1. We show the sketch in Figure 43.5. Note that $b = \frac{1}{2} < 1$ (that is, the sine of the half-value) has an elongating effect on the basic sine curve. ●

Comparison of the graph of the basic cosine curve $y = \cos x$ with $y = A \cos bx$, $A \neq 0$, $b \neq 0$, yields similar results. That is, the amplitude of $y = A \cos bx$ is $|A|$ units and its period is $2\pi/|b|$.

### Example 43.3
Find the amplitude and period of $y = 3 \sin \pi x$.

**Solution.** The amplitude is 3 units and the period is $2\pi/\pi = 2$. We use these results to sketch the graph of $y = 3 \sin \pi x$, $0 \leq x \leq 2$, shown in Figure 43.6. ●

### Example 43.4
Find the amplitude and the period of $y = 3 \cos 2x$, and sketch its graph $0 \leq x \leq 2\pi$.

**Solution.** The amplitude of $y = 3 \cos 2x$ is equal to 3 and its period is $2\pi/2 = \pi$. Since we are asked to sketch $y = 3 \cos 2x$ for $0 \leq x \leq 2\pi$, we must sketch two cycles of this curve. We do so in Figure 43.7. ●

Now, let us consider the graph of

$$y = f(x) = A \sin (bx + c), \quad A, b, c \in R$$

One complete cycle (or sine wave) of amplitude $|A|$ is obtained as $bx + c$ ranges from 0 to $2\pi$, that is,

$$0 \leq bx + c \leq 2\pi$$

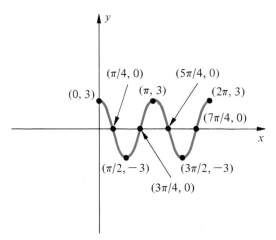

Figure 43.7

or

$$-c \leq bx \leq 2\pi - c \tag{A}$$

We can interpret (A) to mean that $bx$ must range from $-c$ to $2\pi - c$ to complete one cycle of $y = A \sin (bx + c)$. Finally, this variation is obtained by taking domain values from $-c/b$ to $(2\pi - c)/b$ since from (A) we have

$$-\frac{c}{b} \leq x \leq \frac{2\pi - c}{b} \quad \text{if } b > 0$$

or

$$-\frac{c}{b} \geq x \geq \frac{2\pi - c}{b} \quad \text{if } b < 0$$

Thus the graph of $y = A \sin (bx + c)$ can be achieved by *shifting* the graph of $y = A \sin bx$ to the right $-c/b$ units if $-c/b > 0$ and to the left $|-c/b|$ units if $-c/b < 0$. The number $-c/b$ is called the **phase shift.**

**Example 43.5**
Sketch the graph of $y = 4 \sin (2x + \pi/2)$.

Solution. We see that the amplitude of this sine curve is 4 units. We can find the period of this curve by letting $2x + \pi/2$ range from 0 to $2\pi$. If

$$2x + \pi/2 = 0 \quad \text{then } x = -\pi/4$$

and if

$$2x + \pi/2 = 2\pi \quad \text{then } 2x = 3\pi/2 \text{ or } x = 3\pi/4$$

Thus the period of this curve is the interval

$$-\frac{\pi}{4} \leq x \leq \frac{3\pi}{4}$$

which is $3\pi/4 - (-\pi/4) = \pi$ units in length. Since $b = 2$ and $c = \pi/2$, the phase shift is equal to $-c/b$ or $-(\pi/2)/2 = -\pi/4$. This shift is to the left

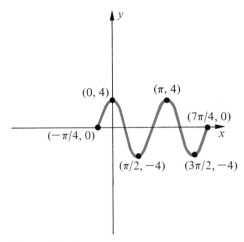

Figure 43.8

since $-c/b = -\pi/4 < 0$. The graph of two cycles of this curve is sketched in Figure 43.8 for the interval $-\pi/4 \leq x \leq 7\pi/4$. ●

## EXERCISES FOR SECTION 43

Sketch the graphs of each of the following and state the period and the amplitude for each.

**1.** $y = 3 \sin x$       **6.** $y = -\sin 2x$       **11.** $y = \cos \frac{1}{2} x$

**2.** $y = \sin \pi x$       **7.** $y = \sin \frac{1}{4} x$       **12.** $y = \cos 2x$

**3.** $y = -2 \sin x$       **8.** $y = 2 \sin \frac{1}{2} x$       **13.** $y = 2 \cos (-2x)$

**4.** $y = 3 \sin 2x$       **9.** $y = \frac{1}{2} \sin x$       **14.** $y = 2 \cos 3x$

**5.** $y = -\sin x$       **10.** $y = -\cos (-x)$

**15.** Compare the graphs of $y = \sin x$ and $y = \cos (\frac{\pi}{2} - x)$.

Sketch the graphs of each of the following, stating the period, amplitude, and phase shift.

**16.** $y = \sin \left( 2x - \frac{\pi}{2} \right)$       **19.** $y = 3 \cos \left( 2x - \frac{\pi}{2} \right)$

**17.** $y = \frac{3}{2} \sin \frac{1}{2}(x + 1)$       **20.** $y = 4 \sin \left( 2x + \frac{\pi}{4} \right)$

**18.** $y = 3 \cos \left( 2x + \frac{\pi}{2} \right)$       **21.** $y = -3 \sin \left( 2x - \frac{\pi}{2} \right)$

**22.** Use the fact that $\sin \left( \frac{\pi}{2} + x \right) = \cos x$ to show that the graph of $y = \cos x$ is simply a phase shift of the graph $y = \sin x$.

**23.** (a) Show that $y = \cos x$ is symmetric with respect to the $y$-axis.
    (b) Show that $y = \sin x$ is symmetric with respect to the origin.

**24.** Sketch the graphs of each of the following. State the period, amplitude, and the range.
    (a) $y = |\sin x|$       (b) $y = \sin |x|$       (c) $|y| = \sin x$

**25.** (a) Sketch the graph of $y = \sin^2 x$, $0 \le x \le 2\pi$. $\left[ Hint: y = \dfrac{1 - \cos 2x}{2}. \right]$

    (b) Sketch the graph of $y = \cos^2 x$, $0 \le x \le 2\pi$. $\left[ Hint: y = \dfrac{1 + \cos 2x}{2}. \right]$

**26.** Use the results of Exercise 38 in Section 39 to express $y = 3 \sin x + 4 \cos x$ in the form $y = A \sin (x + c)$ and then sketch its graph.

**27.** Repeat Exercise 26 when $y = \sin 2x + \cos 2x$.

**28.** Use a sketch of the graph of $y = \cos 2x$, $0 \le x \le 2\pi$ to solve the inequality $\cos 2x < 0$ for $0 \le x \le 2\pi$.

**29.** Use a sketch of the graph of $y = \sin 2x$, $-\pi \le x \le \pi$ to solve the inequality $\sin 2x \ge 0$ for $-\pi \le x \le \pi$.

**30.** (a) On the same set of axes, sketch the graphs of $y = e^x$ and $y = \sin x$.
    (b) Use the graph in part (a) to determine the number of solutions to the equation $\sin x - e^x = 0$.
    (c) Find the largest solution of the equation $\sin x - e^x = 0$ to the nearest tenth.

# SECTION 44
# GRAPHS OF THE OTHER CIRCULAR FUNCTIONS

We now wish to consider the graphs of the other four circular functions. We begin with the tangent curve.

**Graph of $y = \tan x$**

    **Domain:** $y = \tan x$ will be defined for all real $x$ except $\cos x = 0$. Since $\cos x = 0$ when $x = (\pi/2) + 2k\pi$ or $x = (3\pi/2) + 2k\pi$, $k \in J$, we have $\{x \,|\, x \in \mathbb{R}.\ x \ne \pm(\pi/2) + 2k\pi$ or equivalently $x \ne (2k + 1)(\pi/2)\}$ for the domain of $y = \tan x$.

    **Range:** $\{y \,|\, y \in \mathbb{R}\}$.

    **Period:** The period of $y = \tan x$ is $\pi$ units. Thus the graph of one cycle of $y = \tan x$ will occur over intervals of length $\pi$.

    **Intercepts:** We see that if $y = 0$, then $\tan x = 0$ or $\dfrac{\sin x}{\cos x} = 0$. Thus $\tan x = 0$ if $\sin x = 0$. We know that $\sin x = 0$ when $x = 0$ or $\pi$. In general, $\sin x = 0$ when $x = k\pi$. Therefore, we have $x$-intercepts located at $(0, 0)$, $(\pi, 0)$, $(-\pi, 0)$, etc. If $x = 0$, then $y = \tan 0 = 0$. Thus we have a $y$-intercept located at $(0, 0)$.

    **Asymptotes:** Since $y = \tan x = \dfrac{\sin x}{\cos x}$ is undefined when $\cos x = 0$ or when $x = \pm(\pi/2) + 2k\pi$. Thus we have *vertical asymptotes* at $x = \pm(\pi/2) + 2k\pi$ or at $x = -\pi/2$ and $x = \pi/2$ in the interval $-\pi/2 < x < \pi/2$.

    **Symmetry:** We have symmetry about the *origin* since we can replace $y$ by $-y$ and $x$ by $-x$ in the equation $y = \tan x$ to obtain

$$-y = \tan(-x)$$
$$-y = -\tan x \quad \text{(Why?)}$$

Thus

$$y = \tan x$$

We use Table 38.1 and the fact that $\tan x = \dfrac{\sin x}{\cos x}$ to obtain Table 44.1.

**Graphing Comments:** In the interval $-\pi/2 < x < \pi/2$ we have the lines $x = -\pi/2$, and $x = \pi/2$ as vertical asymptotes. Since we have symmetry about the origin we need only to sketch the graph in the region $0 \le x < \pi/2$ and then reflect this through the point $(0, 0)$ to obtain the remaining portion of the graph. These results will give us one period of $y = \tan x$. In Figure 44.1 we show the graph of $y = \tan x$ for $-\pi/2 < x < 3\pi/2$. Note that $y = \tan x$ is periodic, but *unbounded*. Thus $y = \tan x$ has *no amplitude*.

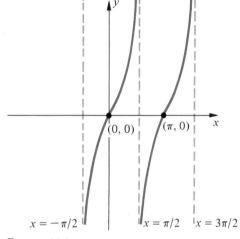

**Table 44.1**

| $x$ | $0$ | $\dfrac{\pi}{6}$ | $\dfrac{\pi}{4}$ | $\dfrac{\pi}{3}$ |
|---|---|---|---|---|
| $y = \tan x$ | $0$ | $\dfrac{1}{\sqrt{3}}$ | $1$ | $\sqrt{3}$ |

Figure 44.1

The graph of the three reciprocal circular functions, secant, cosecant, and cotangent, are of less importance than the graphs sine, cosine, and tangent. Thus we show the graphs of $y = \cot x$, $y = \sec x$, and $y = \csc x$, respectively, in Figures 44.2, 44.3, and 44.4. The details are left as an exercise.

**Graph of $y = \cot x$**

> **Domain:** $\{x \mid x \in \mathbb{R}, x \ne k\pi, k \in J\}$.
>
> **Range:** $\{y \mid y \in \mathbb{R}\}$.
>
> **Period:** $\pi$ units.
>
> **Asymptotes:** Vertical asymptotes at $x = k\pi$.

**Graph of $y = \sec x$**

> **Domain:** $\left\{x \mid x \in \mathbb{R}, x \ne (2k + 1)\left(\dfrac{\pi}{2}\right), k \in J\right\}$.

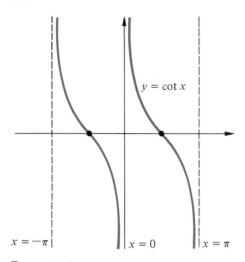

Figure 44.2                                       Figure 44.3

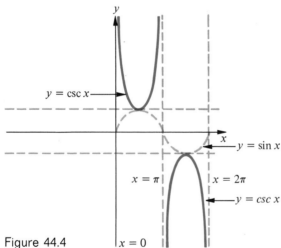

Figure 44.4

**Range:** $\{y \mid y \leq -1 \text{ or } y \geq 1\}$.

**Period:** $2\pi$ units.

**Asymptotes:** Vertical asymptotes at $x = (2k + 1)\left(\dfrac{\pi}{2}\right)$.

**Graph of $y = \csc x$**

**Domain:** $\{x \mid x \in \mathbb{R}, x \neq k\pi, k \in J\}$.

**Range:** $\{y \mid y \leq -1 \text{ or } y \geq 1\}$.

**Period:** $2\pi$ units.

**Asymptotes:** Vertical asymptotes at $x = k\pi$.

**Remark 44.1**

It should be observed that the graphs of these four circular functions are periodic, asymptotic, and unbounded.

The following is a summary of the domain and range of the circular functions.

**Domain and Range of the Circular Functions**

$$y = \sin x \begin{cases} \text{Domain} = \{x \mid x \in \mathbb{R}\} \\ \text{Range} = \{y \mid -1 \leq y \leq 1\} \end{cases}$$

$$y = \cos x \begin{cases} \text{Domain} = \{x \mid x \in \mathbb{R}\} \\ \text{Range} = \{y \mid -1 \leq y \leq 1\} \end{cases}$$

$$y = \tan x \begin{cases} \text{Domain} = \{x \mid x \neq (2k + 1)(\pi/2),\ k \in J\} \\ \text{Range} = \{y \mid y \in \mathbb{R}\} \end{cases}$$

$$y = \cot x \begin{cases} \text{Domain} = \{x \mid x \neq k\pi,\ k \in J\} \\ \text{Range} = \{y \mid y \in \mathbb{R}\} \end{cases}$$

$$y = \sec x \begin{cases} \text{Domain} = \{x \mid x \neq (2k + 1)(\pi/2),\ k \in J\} \\ \text{Range} = \{y \mid y \leq -1 \text{ or } y \geq 1\} \end{cases}$$

$$y = \csc x \begin{cases} \text{Domain} = \{x \mid x \neq k\pi,\ k \in J\} \\ \text{Range} = \{y \mid y \leq -1 \text{ or } y \geq 1\} \end{cases}$$

## EXERCISES FOR SECTION 44

1. Construct a graph of $y = \cot x$ by plotting points. Verify the properties stated with Figure 44.2.

2. Construct a graph of $y = \sec x$ by plotting points. Verify the properties stated with Figure 44.3.

3. Construct a graph of $y = \csc x$ by plotting points. Verify the properties stated with Figure 44.4.

Graph the following.

4. $y = \tan 2x$       8. $y = 2 \csc 2x$       12. $y = \tan (2x + \pi)$

5. $y = \cot 2x$       9. $y = -3 \tan \frac{1}{2}x$       13. $y = \sec (3x - \pi/2)$

6. $y = \sec 3x$       10. $y = \cot (2x - 1)$       14. $y = \cot (2x + \pi)$

7. $y = -\sec x$       11. $y = 3 \tan (x - \pi/2)$       15. $y = 2 \tan (x/2 - \pi/4)$

16. Find the zeros for $y = 4 \tan \frac{1}{2} (x - \pi/2)$, where $0 \leq x \leq 2\pi$.

17. Find the zeros for $y = 3 \tan \pi (x + 1)$ where $0 \leq x \leq 3$.

18. Show graphically that $\tan x = \cot (\pi/2 - x)$.

19. Show graphically that $\cot x = \tan (\pi/2 - x)$.

20. Show graphically that $\sec x = \csc (\pi/2 - x)$.

21. Show graphically that $\csc x = \sec (\pi/2 - x)$.

22. Use a sketch of the graph of $y = \tan x/2$, $0 \leq x \leq 2\pi$ to solve the inequality $\tan x/2 < 0$ for $0 \leq x \leq 2\pi$.

23. Use a sketch of the graph of $y = \cot 3x$, $0 \leq x \leq \pi$ to solve the inequality $\cot 3x \geq 0$ for $0 \leq x \leq \pi$.

SECTION **45**

# INVERSE CIRCULAR FUNCTIONS AND THEIR GRAPHS

In Chapter 5 (Section 33) we defined an inverse function. We recall that a function must be one to one if it is to have an inverse. If a function is not one to one, it may be possible to restrict its domain so that it is one to one and has an inverse function.

In this section we wish to consider the inverse circular functions. We begin by examining the possibility of an inverse function for $y = \sin x$. Using the graph of $y = \sin x$ and the geometric test established in Section 33, we see that $f(x) = \sin x$ is not a one-to-one function by noting that the horizontal line $y = 1$ intersects the graph of $f(x) = \sin x$ at more than one point (Figure 45.1).

In general, for a circular function to have an inverse function, the domain must be restricted. For the sine curve it is customary to choose the domain to be the interval $-\pi/2 \le x \le \pi/2$. The corresponding range is then $-1 \le y \le 1$.

We will call our inverse function the **inverse sine** or the **arcsine function** and denote it by $y = \sin^{-1} x$. The domain of $y = \sin^{-1} x$ is $-1 \le x \le 1$ and the range is $-\pi/2 \le y \le \pi/2$. To graph $y = \sin^{-1} x$, we reflect the graph of $y = \sin x$, $-\pi/2 \le x \le \pi/2$ about the line $y = x$. The graphs are shown in Figures 45.2 and 45.3.

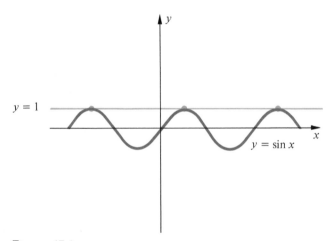

Figure 45.1

Starting with $y = \sin x$, $-\pi/2 \le x \le \pi/2$, we find the inverse sine function in the following manner. Interchanging $x$ and $y$ we have $x = \sin y$. Now we would like to solve for $y$ in terms of $x$, but this is algebraically impossible. Hence we "invent" a way of writing $y$ as a function of $x$. We say that $x = \sin y$ and $y = \sin^{-1} x$ (read "$y$ is equal to the inverse sine of $x$") are **equivalent**. More formally, we have the following definition.

**Definition 45.1  Inverse Sine Function.**  For each $x$, $-1 \le x \le 1$, $\sin^{-1} x$ is a real number $y$, $-\pi/2 \le y \le \pi/2$ such that $x = \sin y$. (We read $\sin^{-1} x$ to mean a real number $y$ whose sin is $x$.)

### Example 45.1

Evaluate $\sin^{-1}(\tfrac{1}{2})$, and $\sin^{-1}(1)$. and $\sin^{-1}(-1)$.

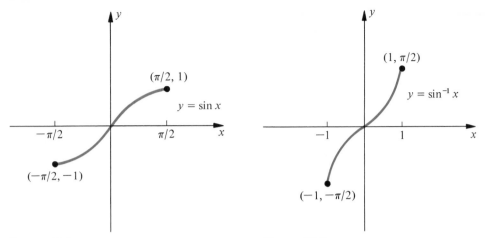

Figure 45.2                          Figure 45.3

Solution.   We read $\sin^{-1} 1/2$ as a real number $y$ whose sine is $1/2$ or $\sin y = 1/2$. Since $\sin \pi/6 = 1/2$ we have $y = \pi/6$. Thus $\sin^{-1}(1/2) = \pi/6$. To evaluate $\sin^{-1}(1)$ we look for a real number $y$ whose sine equals $1$ or $\sin y = 1$. Since $\sin \pi/2 = 1$ we have $y = \pi/2$. Thus $\sin^{-1}(1) = \pi/2$. To evaluate $\sin^{-1}(-1)$ we look for a real number $y$ whose sine equals $-1$ or $\sin y = -1$. We know that $\sin y = -1$ when $y = 3\pi/2$ or $-\pi/2$, but $-\pi/2 \le y \le \pi/2$, so we must use $y = -\pi/2$. Thus $\sin^{-1}(-1) = -\pi/2$. •

Our approach to the remaining inverses circular functions will parallel the discussion of the inverse sine function.

We note that $y = \tan x$ is a one-to-one function in the interval $-\pi/2 < x < \pi/2$. Thus we have an inverse tangent function denoted by $y = \tan^{-1} x$, whose graph is obtained by reflecting the graph of $y = \tan x$, $-\pi/2 < x < \pi/2$ about the line $y = x$. These graphs are shown in Figures 45.4 and 45.5. Note that $y = \tan^{-1} x$ has the lines $y = \pi/2$ and $y = -\pi/2$ as horizontal asymptotes.

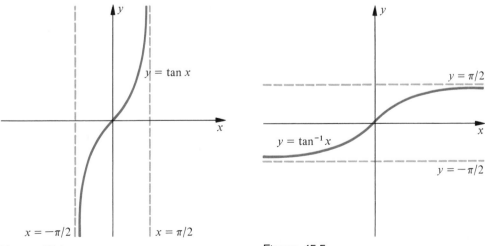

Figure 45.4                          Figure 45.5

**Definition 45.2   Inverse Tangent Function.**   For every real number $x$, $\tan^{-1} x$ is a real number $y$, $-\pi/2 < y < \pi/2$, such that $x = \tan y$. (We read $\tan^{-1} x$ to mean a real number $y$ whose tangent is $x$.)

The cosine function $y = \cos x$ can be made a one-to-one function by restricting its domain to the interval $0 \le x \le \pi$. Thus we have the following definition.

**Definition 45.3   Inverse Cosine Function.**   For each $x$, $-1 \le x \le 1$, $\cos^{-1} x$ is a real number $y$, $0 \le y \le \pi$, such that $x = \cos y$. (We read $\cos^{-1} x$ to mean a real number $y$ whose cosine is $x$.)

The graphs of $y = \cos x$, $0 \le x \le \pi$, and $y = \cos^{-1} x$, $-1 \le x \le 1$ are shown in Figures 45.6 and 45.7. The inverse circular functions for cotangent, secant, and cosecant can be constructed by appropriately restricting their domains. We leave these definitions and the corresponding graphs as exercises.

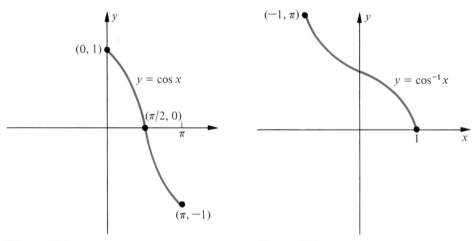

Figure 45.6                                      Figure 45.7

**Example 45.2**
Evaluate $\sin(\sin^{-1} \frac{3}{5})$.

Solution.   Let $y = \sin^{-1} \frac{3}{5}$, then $\sin(\sin^{-1} \frac{3}{5}) = \sin y$. If $y = \sin^{-1} \frac{3}{5}$, then $\sin y = \frac{3}{5}$. Thus

$$\sin(\sin^{-1} \tfrac{3}{5}) = \sin y = \tfrac{3}{5}. \quad \bullet$$

Since $f(x) = \sin x$, $-\pi/2 \le x \le \pi/2$ and $f^{-1}(x) = \sin^{-1} x$, $-1 \le x \le 1$ are inverses of each other we have

$$f[f^{-1}(x)] = \sin[\sin^{-1} x] = x, \quad -1 \le x \le 1 \tag{45.1}$$

and

$$f^{-1}[f(x)] = \sin^{-1}[\sin x] = x, \quad -\frac{\pi}{2} \le x \le \frac{\pi}{2} \tag{45.2}$$

We must be careful when we are evaluating composite functions such as Equations (45.1) and (45.2).

### Example 45.3
Evaluate $\sin^{-1}[\sin 5\pi/6]$.

Solution.   We cannot apply Equation (45.2) since $x = 5\pi/6$ is not in the interval $-\pi/2 \le x \le \pi/2$. However, we know that we can substitute $\sin \pi/6$ for $\sin 5\pi/6$ (since $\sin 5\pi/6 = \sin \pi/6$), in the given expression to obtain

$$\sin^{-1}\left[\sin \frac{5\pi}{6}\right] = \sin^{-1}\left[\sin \frac{\pi}{6}\right]$$

We can now apply (45.2) when $x = \pi/6$ to obtain

$$\sin^{-1}\left[\sin \frac{5\pi}{6}\right] = \frac{\pi}{6}. \quad \bullet$$

In Example 45.2 we could have directly applied Equation (45.1) to obtain $\sin(\sin^{-1}\frac{3}{5}) = \frac{3}{5}$ since $x = \frac{3}{5}$ is in the interval $-1 \le x \le 1$.

The composite functions involving $\cos x$, $\cos^{-1} x$, $\tan x$, and $\tan^{-1} x$, yield the following relationships

$$\tan\,[\tan^{-1} x] = x, \quad x \in \mathbb{R} \tag{45.3}$$

$$\tan^{-1}[\tan x] = x, \quad -\frac{\pi}{2} < x < \frac{\pi}{2} \tag{45.4}$$

$$\cos\,[\cos^{-1} x] = x, \quad -1 \le x \le 1 \tag{45.5}$$
$$\cos^{-1}[\cos x] = x, \quad 0 \le x \le \pi \tag{45.6}$$

### Example 45.4
Evaluate $\cos (\sin^{-1}\frac{3}{5})$.

Solution.    Let $y = \sin^{-1} 3/5$; then $\cos (\sin^{-1} 3/5) = \cos y$. Since $y = \sin^{-1} (3/5)$, we have $\sin y = 3/5$. We now use the identity $\sin^2 y + \cos^2 y = 1$ and the fact that $-\pi/2 \le y \le \pi/2$ to find $\cos y$ as follows:

$$\cos^2 y = 1 - \sin^2 y$$
$$\cos y = \pm\sqrt{1 - (3/5)^2}$$
$$= \pm\sqrt{1 - 9/25} = \pm\sqrt{16/25} = \pm 4/5$$

Since $-\pi/2 \le y \le \pi/2$, $\cos y$ must be in the first or fourth quadrant and, therefore, must be positive. Thus $\cos y = 4/5$. Finally,

$$\cos (\sin^{-1} 3/5) = \cos y = 4/5 \quad \bullet$$

### Example 45.5
Evaluate $\cos [\sin^{-1} (\frac{3}{5}) - \cos^{-1} (\frac{2}{3})]$.

Solution.   If we let $U = \sin^{-1} 3/5$ and $V = \cos^{-1} 2/3$, then $\sin U = 3/5$ and $\cos V = 2/3$. Once again, we use the identity $\sin^2 S + \cos^2 S = 1$ and the fact that $-\pi/2 \le U \le \pi/2$ and $0 \le V \le \pi$ to obtain

$$\cos U = \sqrt{1 - \sin^2 U} = \sqrt{1 - 9/25} = 4/5$$
$$\sin V = \sqrt{1 - \cos^2 V} = \sqrt{1 - (2/3)^2} = \sqrt{5/9} = \sqrt{5}/3$$

Thus

$$\cos[\sin^{-1}(3/5) - \cos^{-1}(2/3)] = \cos(U - V)$$

and

$$\cos(U - V) = \cos U \cos V + \sin U \sin V$$

or

$$\cos(U - V) = (4/5)(2/3) + (3/5)(\sqrt{5}/3) = \frac{8 + 3\sqrt{5}}{15}$$

or, finally,

$$\cos\left[\sin^{-1}\left(\frac{3}{5}\right) - \cos^{-1}\left(\frac{2}{3}\right)\right] = \frac{8 + 3\sqrt{5}}{15} \quad \bullet$$

**Example 45.6**
Express $\sin(\cos^{-1} x)$ in terms of $x$.

Solution.   If we let $y = \cos^{-1} x$, then $\cos y = x$, $0 \le y \le \pi$. Thus

$$\sin(\cos^{-1} x) = \sin y \tag{A}$$

Since

$$\sin y = \pm\sqrt{1 - \cos^2 y}$$
$$= \sqrt{1 - x^2} \quad \text{(Why?)} \tag{B}$$

Finally, we use the results of (A) and (B) to obtain

$$\sin(\cos^{-1} x) = \sqrt{1 - x^2} \quad \bullet$$

## EXERCISES FOR SECTION 45

Evaluate the following.

1. $\cos^{-1}\left(\frac{1}{2}\right)$

2. $\sin^{-1}\left(-\frac{1}{2}\right)$

3. $\tan^{-1}(0)$

4. $\tan^{-1}(1)$

5. $\sin^{-1}\left(\sqrt{3}/2\right)$

6. $\cos^{-1}\left(\sqrt{2}/2\right)$

7. $\sin\left(\cos^{-1}\frac{4}{5}\right)$

8. $\tan(\tan^{-1} 0.9)$

9. $\sin^{-1}(\sin 1)$

10. $\cos\left(\tan^{-1}\frac{2}{3}\right)$

11. $\cos[\cos^{-1}\left(\frac{1}{3}\right) + \tan^{-1}\left(\frac{1}{3}\right)]$

12. $\tan\left[\frac{1}{2}\tan^{-1}\frac{1}{3}\right]$

13. $\sin[2 \sin^{-1}\frac{4}{5}]$

14. $\tan[\sin^{-1}\left(\frac{15}{17}\right) + \sin^{-1}\left(\frac{8}{17}\right)]$

15. $\cos[\cos^{-1}(-0.2)]$

16. $\cos^{-1}[\tan \pi/4]$

17. $\sin^{-1}[\sin 4]$

18. $\cos^{-1}[\cos 5]$

19. $\tan^{-1}[\tan 2\pi/3]$

20. $\tan^{-1}[\tan 6]$

Rewrite each of the following in terms of $x$.

**21.** $\cos(\sin^{-1} x)$

**23.** $\cos(\tan^{-1} x)$

**22.** $\sin[2 \tan^{-1} x]$

**24.** $\cos[\frac{1}{2} \sin^{-1} x]$

**25.** (a) Consider $y = \cot x$ for $0 < x < \pi$. Write a definition for $\cot^{-1} x$.
   (b) Show that $\cot^{-1} x = \tan^{-1}(1/x)$ for $x > 0$.
   (c) Graph $y = \cot^{-1} x$.

**26.** Consider $y = \sec x$ for $0 \le x < \pi/2$ or $\pi \le x < 3\pi/2$. Write a definition for $\sec^{-1} x$.

**27.** Consider $y = \csc x$ for $\pi < x \le 3\pi/2$ or $0 < x \le \pi/2$. Write a definition for $\csc^{-1} x$.

Prove the following identities.

**28.** $\tan(\tan^{-1} 1 + \tan^{-1} x) = \dfrac{(1 + x)}{1 - x}$

**29.** $\cos^{-1} x = \dfrac{\pi}{2} - \sin^{-1} x$ for $0 \le x \le 1$

**30.** $\cot^{-1} x = \tan^{-1}\left(\dfrac{1}{x}\right)$ for $x > 0$

**31.** $\tan[\tan^{-1} x - \tan^{-1} y] = \dfrac{(x - y)}{1 + xy}$

**32.** $\sin^{-1} x + \cos^{-1} x = \pi/2$

**33.** $\sin^{-1}(-x) = -\sin^{-1}(x)$

**34.** $\tan^{-1}(1/7) + 2\tan^{-1}(1/3) = \pi/4$

**35.** $\tan^{-1}(1/3) + \tan^{-1} 3 = \pi/2$

**36.** Use the results of Exercise 27 to sketch the graph of $y = \csc^{-1} x$.

**37.** Use the results of Exercise 26 to sketch the graph of $y = \sec^{-1} x$.

**38.** Sketch the graphs of each of the following.
   (a) $y = \sin^{-1}(2x)$
   (c) $y = \cos^{-1}(x/2)$

   (b) $y = \tan^{-1}(x/2)$
   (d) $y = \tan^{-1}(2x)$

**39.** Sketch the graphs of each of the following.

   (a) $y = \sin(\sin^{-1} x)$
   (d) $y = \tan^{-1}(\sin x)$
   (b) $y = \sin^{-1}(\sin x)$
   (e) $y = \tan^{-1}(\cot x)$
   (c) $y = \sin(\tan^{-1} x)$
   (f) $y = \tan^{-1}(\tan x)$

**40.** Solve each of the following for $x$.
   (a) $y = \frac{2}{3}\sin^{-1}(2x - 2)$
   (d) $y = \sin^{-1} 4x$
   (b) $4y = (\pi/2) - 2\cos^{-1}(2x + 2)$
   (e) $y = \tan^{-1}(x - 4)$
   (c) $y = \pi + \sin^{-1} x$

**41.** If $x = 3\tan\theta$, express $\sin(2\theta) + \theta$ in terms of $x$.

**42.** If $x - 1 = 4\sin\theta$, express $2\tan\theta - \frac{1}{2}\theta$ in terms of $x$.

# SECTION 46
## EQUATIONS INVOLVING CIRCULAR FUNCTIONS

Earlier in this chapter we discussed equations involving circular functions that were true for all permissible values of any real number. These equations were called *identities*. In this section we will be concerned with equations that involve the circular and inverse circular functions that are not satisfied by all permissible values of any real number. These equations involving the circular functions can have an infinite number of solutions since the circular functions are periodic.

**Example 46.1**
Solve $2 \cos x - 1 = 0.$

Solution. We can solve the equation for $\cos x$ as follows:

$$2 \cos x - 1 = 0$$
$$2 \cos x = 1$$
$$\cos x = \tfrac{1}{2}$$

We know that $\cos = 1/2$ when $x = \pi/3$ or $5\pi/3$ in the interval $0 \le x \le 2\pi$. But since the cosine function is periodic with a period of $2\pi$ units, we can conclude that all solutions for the given equation are of the form

$$x = (\pi/3) + 2k\pi \quad \text{or} \quad x = (5\pi/3) + 2k\pi, \quad k \in J \quad \bullet$$

The solutions to a conditional equation involving circular functions in the interval $0 \le x \le 2\pi$ will be called the **primary** or **basic solution** of the equation. The complete set of solutions will be called the **general solution** of the given equation. In Example 46.1, $x = \pi/3$ and $x = 5\pi/3$ are the **primary solutions** for $2 \cos x - 1 = 0$ while $x = (\pi/3) + 2k\pi$ or $x = (5\pi/3) + 2k\pi$ are the general solutions of $2 \cos x - 1 = 0$.

**Example 46.2**
Find the basic solution for $\sin^2 x = \sin x$.

Solution. We solve the equation $\sin^2 x = \sin x$ as follows:

$$\sin^2 x = \sin x$$
$$\sin^2 x - \sin x = 0$$
$$\sin x(\sin x - 1) = 0$$

Now we set each factor equal to zero to obtain

$$\sin x = 0 \quad \text{or} \quad \sin x = 1$$

Since $\sin x = 0$ when $x = 0$, $x = \pi$ or $2\pi$ and $\sin x = 1$ when $x = \pi/2$ in the interval $0 \le x \le 2\pi$, we conclude that the basic solution is $\left\{0, \frac{\pi}{2}, \pi, 2\pi\right\}$.

Unless otherwise stipulated, we will assume that when solving conditional equations involving circular functions we seek the basic solution. We will find that it is impossible to classify the various types of equations involving circular functions as we do with algebraic equations. However, certain procedures are frequently used. We shall illustrate some of these procedures in the following examples.

The identities developed in the earlier sections of this chapter are useful in solving some equations. We can use these identities to rewrite equations involving more than one circular function as equivalent equations involving only one of the circular functions.

### Example 46.3
Find the general solution for $\cos 2x = \sin x$.

**Solution.** We use the identity $\cos 2x = 1 - 2\sin^2 x$ to rewrite $\cos 2x = \sin x$ as

$$1 - 2\sin^2 x = \sin x$$
or
$$2\sin^2 x + \sin x - 1 = 0 \tag{A}$$

We can now factor (A) to obtain

$$(2\sin x - 1)(\sin x + 1) = 0 \tag{B}$$

We set each of the factors of (B) to zero to obtain

$$\sin x = \tfrac{1}{2} \quad \text{or} \quad \sin x = -1 \tag{C}$$

Thus the general solution for equations (A) is the union of the general solution for each of the equations in (C). $\sin x = \tfrac{1}{2}$ when $x = \pi/6 + 2k\pi$ or $x = (5\pi/6) + 2k\pi$, $k \in J$. $\sin x = -1$ when $x = (3\pi/2) + 2k\pi$, $k \in J$. We describe the general solution as

$$x = \begin{cases} (\pi/6) + 2k\pi \\ (5\pi/6) + 2k\pi \quad k \in J \\ (3\pi/2) + 2k\pi \end{cases} \qquad \bullet$$

### Example 46.4
Solve $\sin x + \cos x = 1$.

**Solution.** We rewrite $\sin x + \cos x = 1$ as

$$\sin x = 1 - \cos x \tag{A}$$

We then square both sides of (A) to obtain

$$\sin^2 x = 1 - 2\cos x + \cos^2 x \tag{B}$$

Next we substitute $1 - \cos^2 x$ for $\sin^2 x$ in (B) to obtain

$$1 - \cos^2 x = 1 - 2\cos x + \cos^2 x$$
$$2\cos^2 x - 2\cos x = 0$$
$$2\cos x(\cos x - 1) = 0 \tag{C}$$

We now set each factor of (C) equal to zero and solve to obtain

$$2 \cos x = 0 \quad \text{or} \quad \cos x - 1 = 0$$
$$\cos x = 0 \qquad\qquad \cos x = 1$$
$$x = \pi/2 \text{ or } 3\pi/2, \quad x = 0,\ 2\pi$$

The basic solution "appears" to be $\{0, \pi/2, 3\pi/2, 2\pi\}$. However, if we substitute $x = 3\pi/2$ into the original equation we have

$$\sin 3\pi/2 + \cos 3\pi/2 = 1$$
$$-1 + 0 \neq 1$$

Thus $x = 3\pi/2$ is an *extraneous* root. The reader should verify that $x = 0$, $\pi/2$, and $2\pi$ satisfy the original equation. The basic solution is $\{0, \pi/2, 2\pi\}$. ●

We know from algebra that $x = 3$ and $x^2 = 9$ are not **equivalent equations** since they do not have the same solution sets. Thus whenever we square both sides of an equation, we may introduce *extraneous* roots or solutions. In Example 46.4 we generated the extraneous root $x = 3\pi/2$. Thus squaring both sides of equation *may* generate an extraneous solution. A check of the original equation will reveal them.

**Example 46.5**
Solve $\sec x + \tan x = 0$.

Solution.   We rewrite $\sec x + \tan x = 0$ and proceed as follows.

$$\sec x = -\tan x$$
$$\sec^2 x = \tan^2 x$$
$$1 + \tan^2 x = \tan^2 x \quad \text{(replace } \sec^2 x \text{ by } 1 + \tan^2 x\text{)}$$
$$1 = 0 \qquad\qquad \text{(false statement)}$$

Thus we conclude that the solution set is $\varnothing$. ●

**Example 46.6**
Find the general solution for $2 \sin 3x - 1 = 0$, and state the basic solution.

Solution.   We rewrite $2 \sin 3x - 1 = 0$ as

$$2 \sin 3x = 1$$
$$\sin 3x = \tfrac{1}{2}$$

Thus $\sin 3x = \tfrac{1}{2}$ when

$$3x = (\pi/6) + 2k\pi, \qquad k \in J$$
or
$$3x = (5\pi/6) + 2k\pi, \qquad k \in J$$

So that the general solution is given by

$$x = \frac{\pi}{18} + \frac{2k\pi}{3}, \qquad k \in J$$

or

$$x = \frac{5\pi}{18} + \frac{2k\pi}{3}, \qquad k \in J$$

The basic solution can be found by letting $k$ take on the values 0, 1, 2, 3, ... in the general solution until we have found all $x$ values in the interval $0 \le x \le 2\pi$. When

$$k = 0: \quad x = \frac{\pi}{18} \qquad\qquad \text{or} \quad x = \frac{5\pi}{18}$$

$$k = 1: \quad x = \frac{\pi}{18} + \frac{2\pi}{3} = \frac{13\pi}{18} \quad \text{or} \quad x = \frac{5\pi}{18} + \frac{2\pi}{3} = \frac{17\pi}{18}$$

$$k = 2: \quad x = \frac{\pi}{18} + \frac{4\pi}{3} = \frac{25\pi}{18} \quad \text{or} \quad x = \frac{5\pi}{18} + \frac{4\pi}{3} = \frac{29\pi}{18}$$

Finally, the basic solution for $2 \sin 3x - 1 = 0$ is $\{\pi/18, 5\pi/18, 13\pi/18, 17\pi/18, 25\pi/18, 29\pi/18\}$. •

## EXERCISES FOR SECTION 46

Find the *basic* solutions for each of the following.

1. $(2 \sin x - 1)(\cos x + 1) = 0$

2. $\sin x \,(\sin x + 1) = 0$

3. $2 \sin^2 x + 3 \sin x + 1 = 0$

4. $2 \cos^2 x + \cos x - 1 = 0$

5. $\cos^2 x - \cos x = 2$

6. $(2 \cos x + 1)(4 \sin^2 x - 1) = 0$

7. $\cos^2 x - 1 = 0$

8. $\sec^2 x - 4 = 0$

9. $1 + \tan x = \sec x$

10. $\sin x - \cos x = 1$

11. $\sin 4x = \sin 8x$

12. $2 \sin 4x = 1$

13. $2 \cos 2x = \sqrt{3}$

14. $\tan 3x = 1$

15. $\sin 2x + \sin x = 0$

16. $\tan x + 3 \cot x = 4$

17. $\sin x = 1 - 2 \cos x$

18. $\sin x + \cos x \tan x = 3$

19. $3 \sin x + \cos x = 0$

20. $\tan^2 x - 1 = 0$

21. $\sin 2x + 2 \sin x - \cos x - 1 = 0$

22. $\tan^2 x + \sec x - 3 = 0$

23. $\sin 4x = 1$

24. $\sin 2x + \sin x + 2 \cos x + 1 = 0$

25. $\tan 2x = \sqrt{3}$

26. $4 \sin x \cos x = 1$

27. $\cos 2x + \sin 2x = 0$

28. $2 \sin \frac{1}{2}x = 1$

29. $2 \cos^2 2x + 3 \cos 2x + 1 = 0$

30. $\cos 2x = \cos x - 1$

31. $\cos^{-1} x + \sin^{-1} 2x = \pi/6$ (*Hint:* Let $U = \cos^{-1} x$ and $V = \sin^{-1} 2x$.)

32. $\cos^{-1} x + \sin^{-1} 2x = \sin^{-1} \frac{1}{2}$

33. $4 \sin^{-1} x = \pi$

34. $\sin^{-1} x = \cos^{-1} \frac{x}{2}$

35. $\sin^{-1} (\cos^{-1} x) = 0$

**36.** $\cos^{-1}(\sin^{-1}x) = 0$

**37.** $\cos^{-1}(1) = \cos^{-1}(2x - x^2)$

**38.** $\sin^{-1}x = \sin^{-1}\left(\dfrac{1}{2x}\right)$

**39.** $\sin(\ln x) = 0$

**40.** $\cos(\ln x) = 0$

**41.** $\cos(\cos x) = 0$

**42.** $\sin(\sin x) = 0$

**43.** (a) Solve for $y$ when $\tan^{-1}x + \tan^{-1}y = -\pi/4$.
    (b) What conditions must be placed on $x$ in part (a)?
    (c) Show a sketch of $\tan^{-1}x + \tan^{-1}y = -\pi/4$. [*Hint:* Use the result of part (a).]

In Exercises 44 through 48, find the general solutions for each.

**44.** $\sin 2x = \cos x$

**45.** $2\sin 4x = 1$

**46.** $2\sin\dfrac{1}{2}x + 1 = 0$

**47.** $\cos^2 x + \cos x = 0$

**48.** $3\tan 2x = \sqrt{3}$

# CHAPTER 8
# THE TRIGONOMETRIC FUNCTIONS

## ANGLES AND THEIR MEASURE

In Chapter 7 we discussed the theory of the circular functions. In this chapter we wish to study an application of the circular functions where the real number $S$ is the measure of an angle. This application of the circular functions is a branch of mathematics that is known as **trigonometry.**

In geometry an angle is usually considered to be a geometric configuration formed by two half-lines (rays) emanating from the same initial point. In trigonometry we go a step further with this definition by adding that the angle defined by the two rays has a measure that corresponds to the amount of rotation about the initial point $O$ necessary to move the ray $l$ from its fixed position to the position of ray $m$. In Figure 47.1 we let $P$ and $Q$ be points that lie on the rays $l$ and $m$, respectively. We call $l$ the **initial side,** $m$ the **terminal side,** and $O$ the **vertex** of the angle $POQ$. The angle is said to be positive if the rotation is counterclockwise and negative if the rotation is clockwise. If we place ray $l$ so that it emanates from

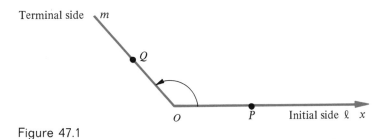

Figure 47.1

the origin and coincides with the positive $x$-axis, the angle $POQ$ is said to be in standard position. We generally denote angles by lowercase Greek letters such as $\alpha$, $\beta$, and $\theta$. If the terminal side of an angle is in a certain quadrant, we say the angle is in that quadrant. If the terminal side of the angle coincides with a coordinate axis, we call the angle a **quadrantal angle.** Two angles in standard position which have the same terminal side are called **coterminal.** The angles $\alpha$ and $\theta$ in Figure 47.2 are coterminal.

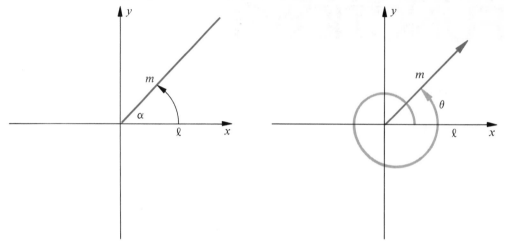

Figure 47.2

**Definition 47.1**  The radian **measure** of an angle such as $POQ$ in Figure 47.3 is determined by measuring the length of the intercepted arc $PQ$ of the unit circle.

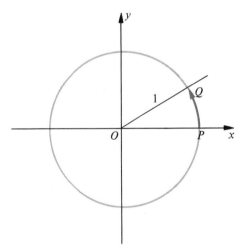

Figure 47.3

One method of angle measurement is to divide the circumference of the unit circle into $2\pi$ equal arcs. The measure of each of these arcs is defined to be 1 **radian.** We then say that the measure of an angle $POQ$ is the measure in radians of the arc $PQ$. For example, the measure of a right angle is $\frac{1}{2}\pi$ radians.

Another method of angle measurement is to divide the circumference of the unit circle into 360 equal arcs, the measure of each of these arcs is defined to be **1 degree,** denoted $1°$. Each degree is divided into 60 **minutes** ($60'$), and each minute is divided into 60 **seconds** ($60''$). We then say that the measure of an angle $POQ$ is the measure in degrees of the arc $PQ$. For example, the measure of a right angle is $90°$. In future discussions we will simply speak of the angle instead of "the measure of an angle" and we will use the same symbol for the angle and the measure of the angle.

The two scales of measurement for an angle $POQ$ are related by the following formulas. Since $2\pi$ radians $= 360$ degrees, we have

$$1 \text{ radian} = \frac{180}{\pi} \text{ degrees} \tag{47.1}$$

$$1 \text{ degree} = \frac{\pi}{180} \text{ radians} \tag{47.2}$$

**Example 47.1**
Convert $\theta = 3\pi/4$ and $\theta = 1$ to degrees.

Solution.   We use (47.1) to obtain

$$\frac{3\pi}{4} \text{ radians} = \frac{3\pi}{4}\left(\frac{180}{\pi}\right)° = 135°$$
and
$$1 \text{ radian} = \frac{180°}{\pi} \approx 57°18' \quad \bullet$$

**Example 47.2**
Convert $\theta = 450°$ and $\theta = -210°$ to radians.

Solution.   We use (47.2) to obtain

$$450° = 450(\pi/180) = 5\pi/2$$
and
$$-210° = -210(\pi/180) = -7\pi/6 \quad \bullet$$

**Remark 47.1**
It should be noted that cos 1 is read "cosine of 1 radian"; cos $1°$ is read "cosine of 1 degree."

In Figure 47.4 on page 260 we see that arcs $\overset{\frown}{PQ} = \theta$, $\overset{\frown}{P'Q'} = S$ determine the same central angle $\theta$. From plane geometry we have the ratio of the respective arcs equal to the ratio of the respective radii. Thus we have

$$\theta/S = 1/r$$
or

$$S = r\theta \tag{47.3}$$

The above discussion gives us the following theorem.

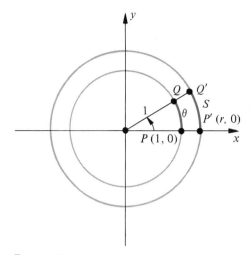

Figure 47.4

**Theorem 47.1** $S = r\theta$, where $S$ is the length of the arc of a circle whose radius is $r$ and whose central angle is equal to $\theta$ radians.

Another useful formula concerning the area of a sector of a circle can be developed if we recall the following result from plane geometry. If $\theta$ and $\theta_1$ are the radian measures of the central angles of the circle of radius $r$, and if $K$ and $K_1$ are the areas of the sectors determined by $\theta$ and $\theta_1$, then

$$\frac{K}{K_1} = \frac{\theta}{\theta_1} \tag{A}$$

Now, if we let $\theta_1 = 2\pi$, then $K_1 = \pi r^2$. We substitute these results into (A) to obtain

$$\frac{K}{\pi r^2} = \frac{\theta}{2\pi}$$

$$2\pi K = \pi r^2 \theta$$

$$\boxed{K = \frac{1}{2} r^2 \theta} \tag{47.4}$$

**Example 47.3**

Find the area of a circular sector that is generated by a central angle $\theta = 135°$ in a circle of radius $r = 2$.

Solution. We first must convert $\theta = 135°$ to radians.

$$\theta = 135° = (135)\left(\frac{\pi}{180}\right) = \frac{3\pi}{4}$$

Then we use formula (47.4) to obtain

$$K = \left(\frac{1}{2}\right) r^2 \theta = \left(\frac{1}{2}\right)(2)^2\left(\frac{3\pi}{4}\right) = \frac{3\pi}{2} \text{ square units} \quad \bullet$$

## EXERCISES FOR SECTION 47

Convert each of the following angles to degrees.

1. $5\pi/6$     4. $-2\pi$     7. $19\pi/6$

2. $\pi/4$     5. $0$     8. $-\pi/4$

3. $6\pi$     6. $11\pi/2$     9. $5\pi/3$

Convert each of the following angles to radians.

10. $30°$     13. $-80°$     16. $22.5°$

11. $120°$     14. $495°$     17. $15°$

12. $-315°$     15. $315°$     18. $75°$

19. Find the length of the circular arc that is generated by a central angle $\theta$ equal to $135°$ in a circle of radius 2.

20. Find the area of the circular sector generated by the central angle $\theta = 145°$ in a circle of radius $r = 1/\pi$.

21. Find all the angles $\theta_1$ that are coterminal with $\theta = 30°$; $\theta = 240°$.

22. Find all the angles $-720° \leq \theta_1 \leq 720°$ that are coterminal with $\theta_1 = 60°$; $\theta_1 = -75°$.

23. If the minute hand of a clock is 10 inches long, how far does the tip of the hand move in 30 minutes? in 40 minutes?

24. The length of a pendulum is 10 ft. If the pendulum bob makes an arc of 8 in, find the angle of the pendulum's motion in degrees; in radians.

25. A train is moving along a piece of circular track of radius 3000 ft at the rate of 10 mph. Through what angle does it turn in 1 minute?

## SECTION 48
# TRIGONOMETRIC FUNCTIONS

We will now consider the angle $\theta$ in the standard position shown in Figure 48.1 on page 262. Suppose that we choose any two distinct points $(x_1, y_1)$ and $(x_2, y_2)$ other than the origin that lie on the terminal side of $\theta$. With the use of similar triangles we can easily show that the following equations are true:

$$\frac{y_1}{x_1} = \frac{y_2}{x_2}; \quad \frac{x_1}{r_1} = \frac{x_2}{r_2}; \quad \frac{y_1}{r_1} = \frac{y_2}{r_2} \tag{48.1}$$

where

$$r_1 = \sqrt{x_1^2 + y_1^2} \quad \text{and} \quad r_2 = \sqrt{x_2^2 + y_2^2}$$

The significance of equation (48.1) is that the ratios $y/x$, $x/r$, and $y/r$ *do not depend* on the choice of the point on the terminal side of $\theta$. We use these ratios to define the trigonometric functions.

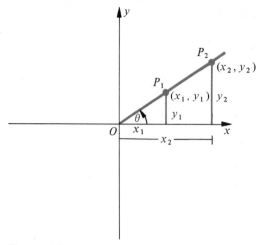

Figure 48.1

**Definition 48.1**   Let $\theta$ be angle in the standard position and let $(x, y) \neq (0, 0)$ be any point on the terminal side of $\theta$; then

$$\text{sine} = \left\{ (\theta, \sin \theta) \,|\, \sin \theta = \frac{y}{r} \right\}$$

$$\text{cosine} = \left\{ (\theta, \cos \theta) \,|\, \cos \theta = \frac{x}{r} \right\}$$

$$\text{tangent} = \left\{ (\theta, \tan \theta) \,|\, \tan \theta = \frac{y}{x}, x \neq 0 \right\}$$

$$\text{cotangent} = \left\{ (\theta, \cot \theta) \,|\, \cot \theta = \frac{x}{y}, y \neq 0 \right\}$$

$$\text{secant} = \left\{ (\theta, \sec \theta) \,|\, \sec \theta = \frac{r}{x}, x \neq 0 \right\}$$

$$\text{cosecant} = \left\{ (\theta, \csc \theta) \,|\, \csc \theta = \frac{r}{y}, y \neq 0 \right\}$$

where $r = \sqrt{x^2 + y^2}$.

The trigonometric functions have the *set of all angles in the plane for their domain* since every angle in the plane is congruent to an angle in standard position. These congruent angles have the same measure.

**Example 48.1**
Find the six trigonometric functions of the angle $\theta$ if its terminal side contains the point $(-4, 3)$.

Solution.   We see that $(x, y) = (-4, 3)$ and that $r = \sqrt{(-4)^2 + 3^2} = 5$. Thus we have

$$\sin \theta = \frac{y}{r} = 3/5 \qquad \csc \theta = \frac{r}{y} = 5/3$$

$$\cos \theta = \frac{x}{r} = -4/5 \quad \sec \theta = \frac{r}{x} = -5/4$$

$$\tan \theta = \frac{y}{x} = -3/4 \quad \cot \theta = \frac{x}{y} = -4/3 \quad \bullet$$

**Example 48.2**

Find the values of the six trigonometric functions of $\theta$ when $\theta = 45°$.

Solution.  In Figure 48.2 we show the angle $\theta = 45°$ in standard position along with a unit circle. The right triangle $POQ$ is formed by dropping a perpendicular line segment to the $x$-axis from the point $Q$ which is formed by the intersection of unit circle and the terminal side of $\theta$. Triangle $POQ$ is an isosceles triangle with $\overline{OP} = \overline{PQ}$. We use the Pythagorean Theorem and the fact that $\overline{OQ} = r = 1$ to obtain

$$(\overline{OP})^2 + (\overline{PQ})^2 = 1^2$$
$$2(\overline{OP})^2 = 1$$
$$(\overline{OP})^2 = 1/2$$
$$\overline{OP} = 1/\sqrt{2} = \sqrt{2}/2$$

Then

$$\overline{OP} = \overline{PQ} = \sqrt{2}/2 \quad \text{and} \quad x = \overline{OP} = \sqrt{2}/2, \quad y = \overline{PQ} = \sqrt{2}/2$$

We now use Definition 48.1 to obtain

$$\sin 45° = \frac{y}{r} = \sqrt{2}/2 \qquad\qquad \csc 45° = \frac{r}{y} = 2/\sqrt{2}$$

$$\cos 45° = \frac{x}{r} = \sqrt{2}/2 \qquad\qquad \sec 45° = \frac{r}{x} = 2/\sqrt{2}$$

$$\tan 45° = \frac{y}{x} = (\sqrt{2}/2)/(\sqrt{2}/2) = 1 \quad \cot 45° = \frac{x}{y} = (\sqrt{2}/2)/(\sqrt{2}/2) = 1$$

$$\bullet$$

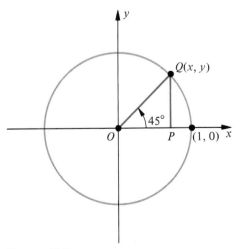

Figure 48.2

We have defined the trigonometric functions of an angle $\theta$ in terms of the coordinates of any point $(x, y) \neq (0, 0)$ on the terminal side of $\theta$. If we now choose $(x, y)$ to be a point on the unit circle $x^2 + y^2 = 1$, we see that the radius, $r$, will be equal to 1 and from definition 48.1 we have

$$\sin \theta = \frac{y}{1} \quad \text{or} \quad y = \sin \theta$$

and

$$\cos \theta = \frac{x}{1} \quad \text{or} \quad x = \cos \theta$$

We see that the range of each of the trigonometric functions is *equal* to the range of the corresponding circular function defined in Section 38. Since we measure the angle $\theta$ in terms of degrees and in terms of radians, we have

$$\begin{array}{ll} \sin \theta = \sin S & \cot \theta = \cot S \\ \cos \theta = \cos S & \sec \theta = \sec S \\ \tan \theta = \tan S & \csc \theta = \csc S \end{array} \qquad (48.2)$$

where $S$ is the length of the arc intercepted by $\theta$ on the unit circle.

We can then say, for example, that the sine of the angle whose measure is 60 degrees is equal to the circular function $\sin (\pi/3)$, since $S = \pi/3$ is the length of the arc intercepted by $\theta = 60°$ on the unit circle. Notationally, we have

$$\sin 60° = \sin \pi/3$$

In order to clarify the unit of angular measure for domain values, $\theta$, of the trigonometric functions, we will use the degree symbol and write $\sin 60°$, $\cos 45°$, and so on, when the angle $\theta$ is measured in degrees. If $\theta$ is measured in radians, we will use only a numeral such as $\sin 2$, $\cos 1$, and so on. This notation is consistent with that of the circular functions, since, for example, the sine of 2 radians is identical to the sine of the real number 2.

The correspondence between the trigonometric and the circular functions given in (48.2) allows us to apply all the identities involving the circular functions established in Chapter 7 to the trigonometric functions. (See summary in Section 42.)

### Example 48.3
Evaluate $\cos 135°$

Solution. We first rewrite the reduction formula

$$\cos (\pi - \bar{S}) = - \cos \bar{S}$$

as

$$\cos (180° - \bar{\theta}) = -\cos \bar{\theta}$$

Then we determine the reference angle $\bar{\theta} = 45°$ by noting that $135° = 180° - 45°$. We combine this with the fact that $\cos 45° = \sqrt{2}/2$ (see Example 48.2) to obtain

$$\cos 135° = \cos [180° - 45°] = -\cos 45° = -\sqrt{2}/2 \quad \bullet$$

**Example 48.4**

Prove that $\cos(90° - \theta) = \sin\theta$.

Solution.   We use the formula

$$\cos(S_1 - S_2) = \cos S_1 \cos S_2 + \sin S_1 \sin S_2$$

from Section 42 to obtain

$$\cos(90° - \theta) = (\cos 90°)(\cos\theta) + (\sin 90°)(\sin\theta)$$
$$\cos(90° - \theta) = 0 \cdot (\cos\theta) + (1)\sin\theta = \sin\theta \quad \bullet$$

**Example 48.5**

Without using tables, evaluate $\tan 15°$.

Solution.   Since $15° = \frac{1}{2}(30°)$ we can apply the half-value formula for tangent as follows:

$$\tan\frac{\theta}{2} = \sqrt{\frac{1 - \cos\theta}{1 + \cos\theta}}$$

$$\tan 15° = \tan\frac{1}{2}(30°) = \sqrt{\frac{1 - \cos 30°}{1 + \cos 30°}}$$

$$\tan 15° = \sqrt{\frac{1 - \dfrac{\sqrt{3}}{2}}{1 + \dfrac{\sqrt{3}}{2}}}$$

$$\tan 15° = \sqrt{\frac{1 - \sqrt{3}}{1 + \sqrt{3}}} = \sqrt{\sqrt{3} - 2}$$

The positive sign in the formula was selected since $\theta = 15°$ lies in the first quadrant.   $\bullet$

## EXERCISES FOR SECTION 48

1. Find the values of the six trigonometric functions of $\theta$ when $\theta = 30°$ by referring to Figure 48.3 on page 266. *Note:* Length $\overline{PQ} = \frac{1}{2}$ since the side opposite the $30°$ angle in the right triangle $POQ$ is equal to one-half the length of the hypotenuse $\overline{OP} = 1$.

2. Find the values of the six trigonometric functions of $\theta$ where $\theta = 60°$ by referring to Figure 48.4 on page 266. *Note:* Length $\overline{OP} = \frac{1}{2}$.

3. Find the values of the six trigonometric functions of $\theta$ when $\theta = 90°$ by referring to Figure 48.5 on page 266.

4. Find the values of the six trigonometric functions of $\theta$ when $\theta = 0°$ by referring to Figure 48.6 on page 266.

5. Complete Table 48.1 on page 267 by using the results of Exercises 1 to 4 and Example 48.2 of this section as well as the appropriate reduction formulas from Section 42.

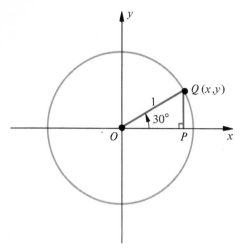

Figure 48.3                                      Figure 48.4

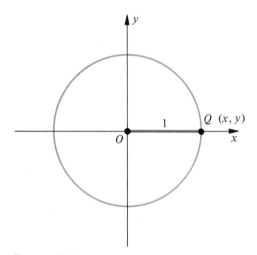

Figure 48.5                                      Figure 48.6

**6.** Determine which of the following are true. Explain.

   (a) $\sin 7\pi/6 = -\sin 30°$         (d) $\sin(-75°) = \cos 15°$

   (b) $\tan 240° = -\tan 60°$          (e) $\cos 7\pi/6 = \cos 30°$

   (c) $\cos(-75°) = -\cos 75°$

**7.** Find the six trigonometric functions of the angle $\theta$ if its terminal side contains the point

   (a) $(3, 4)$           (d) $(12, -5)$

   (b) $(-3, -4)$      (e) $(-5, 12)$

   (c) $(-3, 4)$

Evaluate the following without the use of tables.

**8.** $\sin 15°$

**9.** $\sin(30° + 45°)$

**10.** $\cos 22.5°$

**Table 48.1**

| $\theta$ | $\sin\theta$ | $\cos\theta$ | $\tan\theta$ | $\sec\theta$ | $\csc\theta$ | $\cot\theta$ |
|---|---|---|---|---|---|---|
| 0° | | | | | | |
| 30° | | | | | | |
| 45° | $\dfrac{\sqrt{2}}{2}$ | $\dfrac{\sqrt{2}}{2}$ | 1 | $\dfrac{2}{\sqrt{2}}$ | $\dfrac{2}{\sqrt{2}}$ | 1 |
| 60° | | | | | | |
| 90° | | | | | | |
| 120° | | | | | | |
| 135° | | | | | | |
| 150° | | | | | | |
| 180° | | | | | | |
| 210° | | | | | | |
| 225° | | | | | | |
| 240° | | | | | | |
| 270° | | | | | | |
| 300° | | | | | | |
| 315° | | | | | | |
| 330° | | | | | | |
| 360° | | | | | | |

**11.** $\tan 22.5°$

**12.** $\cos(60° + 45°)$

**13.** $\tan(45° - 30°)$

**14.** $(\sin 12°)(\cos 18°) + (\cos 12°)(\sin 18°)$

**15.** $(\cos 45°)(\cos 15°) + (\sin 45°)(\sin 15°)$

**16.** $\cos^2 15° - \sin^2 15°$

**17.** $2(\sin 22.5°)(\cos 22.5°)$

**18.** $\cos 75°$

Prove each of the following:

**19.** $\sin(90° - \theta) = \cos\theta$

**20.** $\tan(90° - \theta) = \cot\theta$

**21.** $\cos(360° - \theta) = \cos\theta$

**22.** $\sin(180° + \theta) = -\sin\theta$

**23.** $\tan(180° - \theta) = -\tan\theta$

## SECTION **49**
### RIGHT TRIANGLES

A right triangle is a triangle one of whose angles is a right angle. The side opposite the right angle is called the **hypotenuse** and the other two sides are termed the **legs** of the right triangle. For all discussions involving right triangles we will use the symbols shown in Figure 49.1, where the symbols $a$, $b$, and $c$ are the lengths of the sides opposite the respective angles $A$, $B$, and $C$.

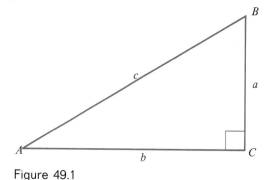

Figure 49.1

Definition 48.1 could be restated in terms of the side opposite an angle, the adjacent side, and the hypotenuse. For convenience we shall use *opp*, *adj*, and *hyp*, respectively.

**Definition 49.1**   Trigonometric functions of angle $A$ in the right triangle $ABC$ of Figure 49.1 are

$$\sin A = \text{opp/hyp} = \frac{a}{c} \quad \csc A = \text{hyp/opp} = \frac{c}{a}$$

$$\cos A = \text{adj/hyp} = \frac{b}{c} \quad \sec A = \text{hyp/adj} = \frac{c}{b}$$

$$\tan A = \text{opp/adj} = \frac{a}{b} \quad \cot A = \text{adj/opp} = \frac{b}{a}$$

The formulas in Definition 49.1 are extremely useful in working with right triangles and are worth memorizing.

In the example that follows, we will see that given the measures of three parts of the right triangle of Figure 49.1 we can calculate the remaining parts of the triangle.

### Example 49.1

If $a = 5$, $c = 13$, and $C = 90°$, find $b$, $A$, and $B$ in the right triangle $ABC$ shown in Figure 49.2.

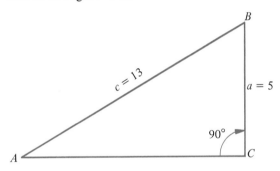

Figure 49.2

Solution. We first use the Pythagorean Theorem to compute $b$ as follows:

$$c^2 = a^2 + b^2$$
$$169 = 25 + b^2$$
$$b = \sqrt{169 - 25} = \sqrt{144} = 12$$

Next we observe that

$$\sin A = 5/13 \approx 0.3846$$

We use Table V in the Appendix and linear interpolation to obtain $A = 22°40'$. Finally, we observe from Figure 49.2 that

$$\tan B = \frac{b}{a} = \frac{12}{5} = 2.400$$

Again, we use Table V and linear interpolation to obtain $B = 67°23'$.  ●

Table V in the Appendix lists the approximate values (actual values when possible) for the six trigonometric functions for any angle measured to the nearest degree from 0 to 90. It should be noted that we will use the right-hand column when we wish to evaluate the value of a trigonometric function whose angle is greater than 45° but less than or equal to 90°.

**Example 49.2**

If $x = 4\tan\theta$, find $\sin 2\theta$.

**Solution.** Since $\tan\theta = \dfrac{x}{4}$, $\theta$ can be interpreted as an acute angle of a right triangle (see Figure 49.3) having an opposite side of length $x$ and an adjacent side of length 4. We use the Pythagorean Theorem to calculate the length of the hypotenuse to be $\sqrt{x^2 + 16}$. From Figure 49.3 we see that

$$\sin\theta = \frac{x}{\sqrt{x^2 + 16}} \quad\text{and}\quad \cos\theta = \frac{4}{\sqrt{x^2 + 16}}$$

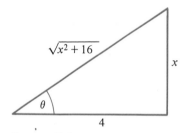

Figure 49.3

We now substitute these results into the double-value formula for sine to obtain

$$\sin 2\theta = 2\sin\theta\cos\theta$$

$$\sin 2\theta = 2\left(\frac{x}{\sqrt{x^2 + 16}}\right)\left(\frac{4}{\sqrt{x^2 + 16}}\right) = \frac{8x}{x^2 + 16}$$

The triangle shown in Figure 49.3 is sometimes called the **reference triangle** for the angle $\theta$.   ●

**Example 49.3**

In Figure 49.4 angle A of triangle $ABC$ has been placed in standard position. Find the coordinates of point $B$ in terms of angle $A$.

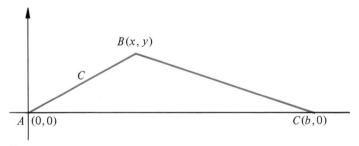

Figure 49.4

**Solution.** We drop an altitude from the vertex $B$ to the $x$-axis at point $D$ to form the right triangle $ABD$ (see Figure 49.5). From Figure 49.5 we see that

$$\sin A = \frac{y}{c} \quad\text{and}\quad \cos A = \frac{x}{c}$$

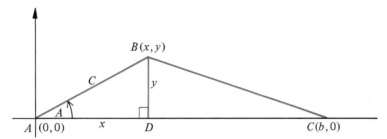

Figure 49.5

or equivalently $y = c \sin A$ and $x = c \cos A$. Thus point $B$ is located at $(c \cos A, c \sin A)$. •

Right triangles are useful in solving many types of applied problems. One such type of problem deals with the angles of **elevation** and **depression.** The angle of elevation is the angle formed by the horizontal and the line of sight when the object is above the horizontal. (See Figure 49.6.) The angle of depression is the angle formed by the horizontal and the line of sight when the object is below the horizontal (Figure 49.7).

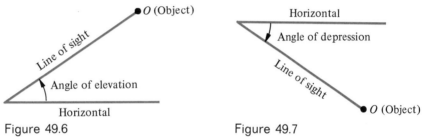

Figure 49.6                                        Figure 49.7

### Example 49.4

From a point at ground level 150 ft from the base of a tower, the angle of elevation of the top of the tower is $48°20'$. Find the height of the tower.

Solution.   In Figure 49.8 we show the right triangle $ABC$, where $A$ is the angle of elevation and $a$ is the height of the tower. We observe that

$$\tan 48°20' = \frac{a}{150}$$

or

$$a = 150(\tan 48°20') = 150(1.124) \approx 169$$

Thus, the height of the tower is approximately 169 ft. •

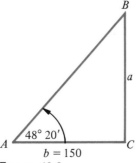

Figure 49.8

## EXERCISES FOR SECTION 49

In each of the following, solve for the remaining parts of the right triangle. Assume that the labeling of the parts of the right triangle is the same as discussed in this section (see Figure 49.1).

1. $A = 32°$, $c = 8$      5. $B = 63°$, $c = 16$

2. $A = 35°$, $a = 50$      6. $b = 12$, $c = 16$

3. $a = 15$, $b = 24$      7. $A = 46°$, $a = 54$

4. $A = 52°$, $b = 5.2$   , 8. $a = 32$, $b = 40$

9. What is the height of a telephone pole whose horizontal shadow is 65 feet long when the angle of elevation of the sun is 55°?

10. The string on a kite is taut and makes an angle of 25° with the horizontal. Find the height of the kite if 150 ft is let out and the end of the string is 4.5 ft above the ground.

11. A guy wire is fastened to a pole at a point 10 ft from the ground and to the ground 14 ft from the bottom of the pole. Find the length of the wire and the angle it makes with the pole.

12. A glider is approaching an airport from a distance of 3 miles at a height of 2600 ft. At what angle with the horizontal must the pilot maintain if he is to reach the airport?

13. Show that the area of a right triangle is $\frac{1}{4}c^2 \sin 2B$.

14. If $x = 3 \tan \theta$, express $3 \cos \theta - 3\theta$ in terms of $x$.

15. If $x = 2 \sin \theta$, express $\theta - \frac{1}{2} \sin 2\theta$ in terms of $x$.

16. If $x = 3 \sec \theta$, express $\cos 2\theta - \sin \theta$ in terms of $x$.

17. Find the area of the right triangle when $A = 40°$, $C = 90°$, and $a = 20$.

18. Superimpose triangle $ABC$ in Figure 49.1 on $\mathbb{R} \times \mathbb{R}$ with the point $A$ located at $(0, 0)$ and side $\overline{AC}$ on the positive $x$-axis. Now show that Definition 49.1 is equivalent to Definition 48.1.

## SECTION 50
## LAW OF SINES

In Section 49 we restricted our analysis to right triangles. We now wish to examine the *general* triangle shown in Figure 50.1. For all discussions of the general triangle, we will use the symbols $a$, $b$, and $c$ to represent the lengths of the sides opposite the angles $A$, $B$, and $C$, respectively.

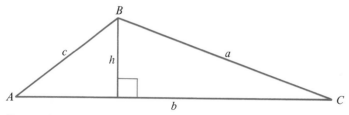

Figure 50.1

In Figure 50.1 we drop an altitude from the vertex $B$ to the opposite side $b$ and label the length of this altitude $h$. We see that $\sin A = \dfrac{h}{c}$ or $h = c \sin A$ and $\sin C = \dfrac{h}{a}$ or $h = a \sin C$. Thus $c \sin A = a \sin C$. If we drop altitudes $h$ from each of the other vertices ($A$ and $C$) to their respective opposite sides (extended if necessary), we obtain $b \sin A = a \sin B$ and $b \sin C = c \sin B$. We use these results to express the following law.

### Law of Sines

For *any* triangle the ratio of the length of any side to the sine of the angle opposite is the same as the corresponding ratio for any other side and corresponding angle. We express this as

$$\frac{a}{\sin A} = \frac{b}{\sin B} = \frac{c}{\sin C} \tag{50.1}$$

The law of sines enables us to solve a triangle if we are given two angles and one side or if we are given two sides and one opposite angle. The latter case is known as the **ambiguous case.**

Given $a$, $b$, and angle $A$ we use (50.1) to obtain

$$\frac{a}{\sin A} = \frac{b}{\sin B}$$

or

$$\sin B = \frac{b \sin A}{a}$$

We now consider the cases where $\sin B$ may be greater than 1, less than 1, or equal to 1.

*Case 1.* If $\sin B > 1$, we have *no* solution. (Why?)

*Case 2.* If $\sin B = 1$, $B = 90°$; we have a right triangle and one solution.

*Case 3.* If $\sin B < 1$, we obtain two values $B_1$ and $B_2$. These two values, $B_1$ and $B_2$, have a sum equal to 180°. $B_1$, $B_1 < B_2$, is always a solution. If $A + B_2 < 180°$, $B_2$ is a second solution. But if $A + B_2 \geq 180°$, $B_2$ is not a solution.

We use the following examples to illustrate these possibilities.

### Example 50.1

Suppose that $a = 5$, $b = 8$, and $A = 40°$. Find the remaining parts of the triangle $ABC$.

Solution.   From (50.1) we have

$$\sin B = \frac{b \sin A}{a} \approx \frac{8(0.6428)}{5} \approx 1.029$$

Thus $\sin B > 1$ and there is no solution.   •

### Example 50.2

Suppose that $a = 20$, $b = 27$, and $A = 30°$. Approximate the value of angle $B$.

**Solution.** From (50.1) we have

$$\sin B = \frac{b \sin A}{a} = \frac{27(0.5)}{20} = 0.6750$$

We note that $\sin B < 1$. Thus we have two possible solutions. We use Table V in the Appendix to obtain $B_1 \approx 43°$ and $B_2 \approx 137°$. $B_2$ is determined by using the identity

$$\sin(180° - B) = \sin B \quad \text{where} \quad B = 43°$$

to obtain

$$\sin(180° - 43°) = \sin 43°$$
or
$$\sin 137° = \sin 43°$$

Since $B_2 + A < 180°$, $B_2$ is also a solution. Thus we have *two* solutions, $B_1 \approx 43°$ and $B_2 \approx 137°$. ●

We have stated the ambiguous case for the law of sines occurs when we are given two sides of a triangle and the angle, say $A$, opposite one of these sides. The ambiguity arises depending on the length of $a$, in relation to the length of $b$, and the measure of $B$. In Figure 50.2 we show the various possibilities that may occur. Note that $B$ is an acute angle and $h$ is the length of a perpendicular drawn from the vertex $c$ to the opposite side.

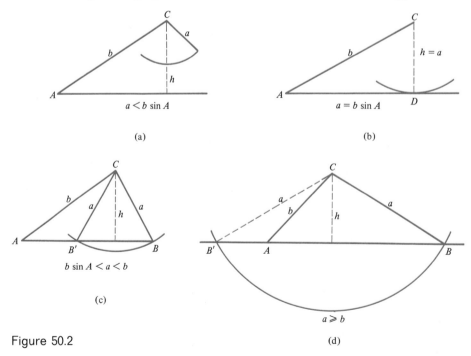

Figure 50.2

If $a < h$, the side a cannot intersect the side opposite $c$ and thus no triangle is possible. See Example 50.1. Since $h = b \sin A$, we have $a < b \sin A$. This is shown in Figure 50.2a.

If $a = h = b \sin A$, side a intersects the side opposite c at the point $D$. In this case, exactly one right triangle is possible as shown in Figure 50.2b. If $a > h$ (or $b \sin A$) and $a < b$, or equivalently if $b \sin A < a < b$, then the two triangles shown in Figure 50.2c are possible. Since both triangles contain the angle $A$, there are two values for angle $B$ and two values for side $c$. See Example 50.2.

If $a \geq b$, the triangle $ABC$ in Figure 50.2d does not contain the given angle $A$ and thus does not yield a solution. In this case only one triangle is possible. In the event $a = b$, this triangle is isosceles.

# SECTION 51
## LAW OF COSINES

If we again consider the general triangle of Figure 50.1, we can develop another important relationship between parts of this triangle and the cosine of an angle.

**Law of Cosines**
In any general triangle $ABC$ labeled in the usual manner we have

$$a^2 = b^2 + c^2 - 2bc \cos A \tag{51.1}$$

$$b^2 = a^2 + c^2 - 2ac \cos B \tag{51.2}$$

$$c^2 = a^2 + b^2 - 2ab \cos C \tag{51.3}$$

We can verbalize statements (51.1), (51.2), and (51.3) as follows: *The square of the length of any side of a triangle is equal to the sum of the squares of the lengths of the other two sides minus twice the product of the length of the other two sides and the cosine of the angle between them.*

Proof for Law of Cosines.   Given the general triangle $ABC$ in Figure 51.1, below, we place angle $A$ in the standard position. The point $C$ has

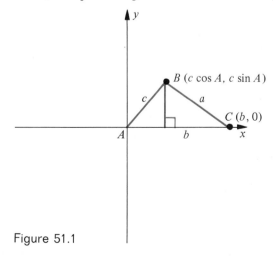

Figure 51.1

coordinates $(b, 0)$ and the point $B$ has coordinates $(c \cos A, c \sin A)$ see Example 49.3. The length of segment $\overline{BC}$ is equal to $a$ (i.e., $\overline{BC} = a$), but we can also express the length of segment $\overline{BC}$ by applying the distance formula from Section 10 to obtain

$$\text{length of } \overline{BC} = \sqrt{(c \cos A - b)^2 + (c \sin A - 0)^2} \tag{A}$$

We now equate the two expressions for the length of $\overline{BC}$ and square both sides of the equation to obtain

$$\begin{aligned}
a^2 &= (c \cos A - b)^2 + (c \sin A)^2 \\
&= c^2 \cos^2 A - 2\,bc \cos A + b^2 + c^2 \sin^2 A \\
&= b^2 + c^2\,[\sin^2 A + \cos^2 A] - 2\,bc \cos A
\end{aligned} \tag{B}$$

Finally, we use the identity $\sin^2 A + \cos^2 A = 1$ and (B) to obtain formula (51.1). That is,

$$a^2 = b^2 + c^2 - 2\,bc \cos A$$

Formulas (51.2) and (51.3) may be obtained in the same manner by placing the angles $B$ and $C$, respectively, in the standard position.

The law of cosines can be used to find the remaining parts of a given triangle if we are given the lengths of the three sides or the lengths of two sides and the included angle.

### Example 51.1

Suppose in triangle $ABC$ that $a = 8$, $b = 10$, and $c = 5$. Find angles $A$, $B$, and $C$.

Solution.   We can find angle $A$ by using (51.1) as follows:

$$\begin{aligned}
a^2 &= b^2 + c^2 - 2\,bc \cos A \\
64 &= 100 + 25 - 2(10)(5) \cos A
\end{aligned}$$

$$\cos A = \frac{125 - 64}{100} = 0.6100$$

We now use Table V to find $A \approx 53°$. We next use (51.2) to obtain

$$\begin{aligned}
b^2 &= a^2 + c^2 - 2\,ac \cos B \\
100 &= 64 + 25 - 2(8)(5) \cos B
\end{aligned}$$

$$\cos B = \frac{89 - 100}{80} = -0.1375$$

Again, we use Table V as well as the identity $\cos(180° - B) = -\cos B$ to obtain $B \approx 180° - 82° = 98°$. Finally, we use (51.3) to obtain

$$\begin{aligned}
c^2 &= a^2 + b^2 - 2\,ab \cos C \\
25 &= 64 + 100 - 2(8)(10) \cos C
\end{aligned}$$

$$\cos C = \frac{164 - 25}{160} = 0.8688$$

We now use Table V to obtain $C \approx 30°$. It should be noted that the sum of the approximates angles $A$, $B$, and $C$ is $181°$ instead of $180°$. This discrepancy is due to our approximating procedures.  ●

### Example 51.2

A tunnel is to be constructed through a mountain from point $A$ to point $B$. At a point from which both $A$ and $B$ are visible, the distance to $A$ is 321 ft, and to $B$, 339 ft. If the angle $ACB$ is $38°30'$, find the length of the tunnel.

Solution.  From Figure 51.2 we see that the length of the tunnel is $AB$. We can find $AB$ by applying the law of cosines as follows:

$$(AB)^2 = (AC)^2 + (CB)^2 - 2(AC)(CB) \cos C$$
$$= (321)^2 + (339)^2 - 2(321)(339)(0.7826)$$
$$= 103,041 + 114,921 - 170,323$$
$$= 47,639$$
$$AB = \sqrt{47,639} = 218 \text{ ft}$$

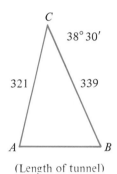

(Length of tunnel)

Figure 51.2

## Summary for Solving General Triangles

*Case 1.* If we are given one side and two angles, we can apply the **law of sines.**

*Case 2.* If we are given two sides and the angle opposite one of them, we can apply the **law of sines.**

*Case 3.* If we are given two sides and the included angle, we can apply the **law of cosines** to find the third side. We can then apply the law of sines to find the remaining angles.

*Case 4.* If we are given three sides, we can apply the **law of cosines** to find one of the angles. The remaining angles can be found by applying the law of sines.

## EXERCISES FOR SECTIONS 50 and 51

In each of the following problems solve for the indicated missing part. Assume that the labeling of the parts of the general (oblique) triangle is the same as discussed in Sections 50 and 51 (see Figure 50.1).

**1.** Solve for $b$ when $A = 62°$, $B = 75°$, and $a = 130$.

**2.** Solve for $a$ when $A = 40°$, $B = 53°$, and $b = 100$.

**3.** Solve for $B$ when $a = 62.5$, $b = 143.5$, and $A = 32°$.

**4.** Solve for $B$ when $a = 300$, $b = 320$, and $A = 35°$.

**5.** Solve for $A$ when $a = 3$, $b = 4$, and $c = 5$.

**6.** Solve for $B$ when $a = 3$, $b = 4$, and $c = 5$.

**7.** Solve for $C$ when $a = 14.1$, $b = 16.8$, and $c = 15$.

**8.** Solve for $B$ when $a = 20$, $b = 27$, and $A = 30°$.

**9.** Solve for $B$ when $a = 20$, $b = 25$, and $A = 45°$.

**10.** Solve for $A$ when $a = 24$, $b = 20$, and $B = 120°$.

**11.** Solve for $A$, $B$, and $C$ when $a = 4$, $b = 5$, and $c = 6$.

**12.** Solve for $A$, $B$, and $C$ when $a = 30$, $b = 50$, and $c = 65$.

**13.** Solve for $b$, when $a = 7.6$, $c = 9.2$, and $B = 46°$.

**14.** Find the lengths of the diagonals of a parallelogram whose sides are 18 and 24 and that has a vertex angle equal to 48°.

**15.** A ship leaves a point $A$ and travels 140 miles in the direction 115° and from there travels 200 miles in the direction 135°. How far is the ship from its original point?

**16.** Two cars leave together from point $A$ and travel along straight highways that are at an angle of 83° with one another. If their speeds are 50 and 60 mph, respectively, how far apart will they be after 30 minutes?

**17.** (a) Prove that the area of *any* triangle is equal to one-half the product of any two sides and the sine of their included angle.
(b) Find the area of the triangle $ABC$ when $c = 32$, $b = 42$, and $A = 150°$.

**18.** Show that if $A = 90°$, the law of cosines reduces to the Pythagorean Theorem (that is, $a^2 = b^2 + c^2$).

**19.** Two sides of a parallelogram are 68 and 83 inches and one of the diagonals is 42 inches. Find the angles of the parallelogram.

**20.** (a) Show that the area of any triangle $ABC$ is $A = \sqrt{S(S - a)(S - b)(S - c)}$, where the semiperimeter $S$ equals $\frac{1}{2}(a + b + c)$.
(b) Use the result of part (a) to find the area of an equilateral triangle.

**21.** Show that in any triangle $ABC$

(a) $\dfrac{a + b}{c} = \dfrac{\cos \frac{1}{2}(A - B)}{\sin \frac{1}{2} C}$

(b) $\dfrac{a - b}{c} = \dfrac{\sin \frac{1}{2}(A - B)}{\cos \frac{1}{2}C}$

[*Hint:* Rewrite $\dfrac{a + b}{c}$ as $\dfrac{a}{c} + \dfrac{b}{c}$ and use the law of sines.]

**22.** Prove that

$$\frac{\tan \left(\frac{1}{2}\right)(A - B)}{\tan \left(\frac{1}{2}\right)(A + B)} = \frac{a - b}{a + b}$$

This formula is known as the **law of tangents.**

**23.** (a) Show that the area $K$ of the segment of the circle shown in Figure 51.3 is

$$K = \frac{r^2(\theta - \sin \theta)}{2}$$

[*Hint:* Recall from Section 48 that the area of a sector of a circle is $A = \frac{1}{2}r^2\theta$.]

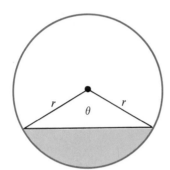

Figure 51.3

(b) How many gallons of water are in a horizontal cylindrical tank 20 feet long and 8 feet in diameter if the depth of water is 1 foot? (Recall that the volume is equal to the product of the length and the cross-sectional area.)

# SECTION 52
# VECTORS

In science and engineering many physical quantities can be described by both a *number* and a *direction*. For example, we refer to a wind velocity of 35 miles per hour westward or to a force of 50 pounds downward. Such quantities that have a magnitude and direction are called **vector quantities** and they may be represented by a straight-line segment of definite length called a **vector.** A vector can be represented graphically by a directed line segment. In Figure 52.1 we have indicated a directed straight-line segment with an initial point at $O$ and a terminal point at $P$. In this text such vectors will be indicated by either a letter with the symbol $\rightarrow$ over it or by using the initial and terminal points of the directed line segment with the symbol $\rightarrow$ over. Thus in Figure 52.1 we have $\vec{V}$ or $\overrightarrow{OP}$. The magnitude of $\overrightarrow{OP}$ or $\vec{V}$ will be denoted by $|\overrightarrow{OP}|$ or $|\vec{V}|$.

Many physical quantities can be described by a single number. These quantities are known as **scalars.** A scalar is a quantity having *magnitude only.* Examples of scalar quantities are length, temperature, time, and so on. A **zero vector,** denoted $\vec{0}$, is a vector whose initial and terminal points coincide. A zero vector has a magnitude of zero but an *arbitrary* direction.

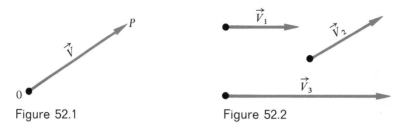

Figure 52.1                    Figure 52.2

We say that two vectors $\vec{A}$ and $\vec{B}$ are equal if they have the same magnitude and direction or if they are both zero vectors. (This concept of equal vectors is subject to certain limitations in some engineering applications.) In Figure 52.2 we see that $|\vec{V_1}| = |\vec{V_2}|$, but $\vec{V_1} \neq \vec{V_2}$ since $\vec{V_1}$ and $\vec{V_2}$ have different directions. In Figure 52.2 we also see that vectors $\vec{V_1}$ and $\vec{V_3}$ have the same direction, but $|\vec{V_1}| \neq |\vec{V_3}|$. Thus $\vec{V_1} \neq \vec{V_3}$.

# SECTION 53
## VECTOR OPERATIONS

Many times we will want to find the "sum" of two vectors or the "product of a vector and a scalar." To do this we formulate the following definitions.

**Definition 53.1 Addition of Vectors.** To add two vectors $\vec{A}$ and $\vec{B}$ we move (if necessary) one of the vectors parallel to itself until both vectors have the same initial point and then draw the parallelogram determined by these two vectors. The **sum** or **resultant** is formed by the diagonal of the parallelogram that has the same initial point as the vectors $\vec{A}$ and $\vec{B}$. This method is known as the **parallelogram method.**

**Example 53.1**
Find the sum of the vectors $\vec{A}$ and $\vec{B}$ shown in Figure 53.1.

Solution.   In Figure 53.2 we move $\vec{B}$ parallel to itself until the initial points of both $\vec{A}$ and $\vec{B}$ coincide and then draw the parallelogram they determine. Then we take the diagonal with the same initial point as $\vec{A}$ and $\vec{B}$ for the sum $\vec{A} + \vec{B} = \vec{C}$ shown in Figure 53.3.   ●

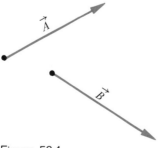

Figure 53.1

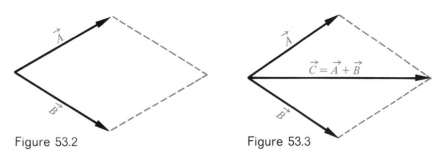

Figure 53.2             Figure 53.3

**Definition 53.2   Product of a Scalar and a Vector.**   The product of a scalar $c$ and a vector $\vec{v}$ denoted $c\vec{v}$ is a vector parallel to $\vec{v}$. If $c > 0$, then $c\vec{v}$ has the same **direction** as $\vec{v}$. If $c < 0$, then $c\vec{v}$ has the **opposite direction** of $\vec{v}$. And if $c = 0$, then $c\vec{v}$ is the zero vector.

We say that two vectors are parallel if they have the same or opposite directions. Thus Definition 53.2 tells us that *if two vectors $\vec{A}$ and $\vec{B}$ are parallel, then one is a scalar multiple of the other.* For convenience, we say that the zero vector $\vec{0}$ is parallel to every vector.

**Example 53.2**
Given $\vec{V}$ in Figure 53.4, find $3\vec{V}$ and $-\vec{V}$.

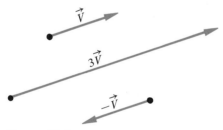

Figure 53.4

Solution.   In Figure 53.4 we show $3\vec{V}$, which is parallel to $\vec{V}$ and has a magnitude three times as great as $\vec{V}$. In Figure 53.4 we also show $-\vec{V}$, which has the opposite direction of $\vec{V}$ and the *same* magnitude.   ●

**Example 53.3**
Given vectors $\vec{A}$ and $\vec{B}$ in Figure 53.5, find the vector $\vec{A} - \vec{B}$.

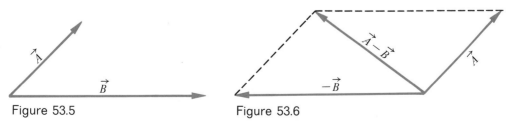

Figure 53.5             Figure 53.6

Solution.   Since $\vec{A} - \vec{B} = \vec{A} + (-\vec{B})$, we reverse the direction of $\vec{B}$ and then apply the parallelogram method. We show this result in Figure 53.6.   ●

**Remark 53.1**

When finding the sum or resultant of two vectors $\vec{A}$ and $\vec{B}$ such as in Example 53.1, we frequently refer to the vectors $\vec{A}$ and $\vec{B}$ as the *components of the resultant vector* $\vec{C}$.

Multiplication of a vector by a scalar adjusts the length or magnitude of the vector. Frequently we find it convenient to multiply a vector by the reciprocal of its magnitude. Such a vector is called a **unit vector.** For convenience, we will denote a unit vector by the lower case form of the capital letter that designates the vector. Thus a unit vector for $\vec{B}$ is given by

$$\vec{b} = \frac{1}{|\vec{B}|}(\vec{B}) = \frac{\vec{B}}{|\vec{B}|} \tag{53.1}$$

The operations of vector addition and multiplication of a vector and a scalar conform to certain laws of algebra. We list these as follows, without proof.

Given the scalars (real numbers) $a$ and $b$ and the vectors $\vec{A}$, $\vec{B}$, and $\vec{C}$, we have

1. $\vec{A} + \vec{B} = \vec{B} + \vec{A}$.
2. $(\vec{A} + \vec{B}) + \vec{C} = \vec{A} + (\vec{B} + \vec{C})$.
3. $a(b\vec{A}) = b(a\vec{A}) = (ab)\vec{A}$.
4. $a(\vec{A} + \vec{B}) = a\vec{A} + a\vec{B}$.
5. $(a + b)\vec{A} = a\vec{A} + b\vec{A}$.

Vector analysis can be used to ease the burden of proof in plane geometry. Consider the following example.

**Example 53.4**

Prove that the line joining the midpoints of the nonparallel sides of the trapezoid shown in Figure 53.7 is parallel to the parallel sides and one-half the sum of their lengths.

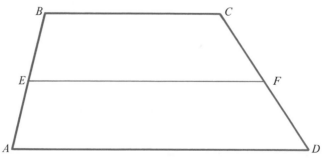

Figure 53.7

Solution. Consider the trapezoid $ABCD$ shown in Figure 53.7, where points $E$ and $F$ are the midpoints of the nonparallel sides. We must show that

1. $\overrightarrow{EF}\|\overrightarrow{BC}$ and $\overrightarrow{EF}\|\overrightarrow{AD}$.
2. $\overrightarrow{EF} = \frac{1}{2}(\overrightarrow{BC} + \overrightarrow{AD})$.

We see from Figure 53.7 that

$$\overrightarrow{EF} = \overrightarrow{EB} + \overrightarrow{BC} + \overrightarrow{CF} \tag{A}$$

and also that

$$\overrightarrow{EF} = \overrightarrow{EA} + \overrightarrow{AD} + \overrightarrow{DF}$$
or
$$\overrightarrow{EF} = -\overrightarrow{AE} + \overrightarrow{AD} - \overrightarrow{FD} \tag{B}$$

Since $E$ and $F$ are the midpoints of $\overrightarrow{AB}$ and $\overrightarrow{CD}$, respectively, we have

$$\overrightarrow{AE} = \overrightarrow{EB} \quad \text{and} \quad \overrightarrow{CF} = \overrightarrow{FD} \tag{C}$$

We now add results (A) and (B) and then substitute from result (C) to obtain

$$2\overrightarrow{EF} = \overrightarrow{EB} - \overrightarrow{AE} + \overrightarrow{BC} + \overrightarrow{AD} + \overrightarrow{CF} - \overrightarrow{FD}$$
$$2\overrightarrow{EF} = \vec{0} + \overrightarrow{BC} + \overrightarrow{AD} + \vec{0}$$
$$\overrightarrow{EF} = \frac{1}{2}(\overrightarrow{BC} + \overrightarrow{AD}) \tag{D}$$

which verifies part 2. Result (D) tells us that

$\overrightarrow{EF}$ is parallel to $\overrightarrow{BC} + \overrightarrow{AD}$   (Why?))

Thus $\overrightarrow{EF}$ must be parallel to $\overrightarrow{BC}$ and to $\overrightarrow{AD}$.   ●

## EXERCISES FOR SECTION 52 AND 53

1. Suppose that $\vec{V}$ has a magnitude of 3 and the same direction as the positive $x$-axis and that $\vec{W}$ has a magnitude of 2 and makes a counterclockwise angle of $45°$ with the positive $x$-axis (see Figure 53.8). Draw the vectors.
   (a) $2\vec{W},\ 2\vec{W} + \vec{V}$       (d) $\frac{1}{3}\vec{V},\ \vec{W} + \frac{1}{3}\vec{V}$
   (b) $-\frac{1}{2}\vec{W},\ \vec{V} - \frac{1}{2}\vec{W}$       (e) $\frac{1}{3}\vec{V},\ \frac{1}{3}\vec{V} - \vec{W}$
   (c) $\vec{V} - \vec{W},\ \vec{W} - \vec{V}$       (f) $3\vec{W},\ 3\vec{W} + 3\vec{V},\ \vec{W} + \vec{V}$

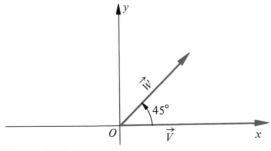

Figure 53.8

**2.** (a) Under what condition(s) will $c|\vec{A}| = |c\vec{A}|$?

   (b) Under what condition(s) will $|\vec{A}| - |\vec{B}| = |\vec{A} - \vec{B}|$?

   (c) Under what condition(s) will $|\vec{A} + \vec{B}| = |\vec{A} - \vec{B}|$, where $\vec{A} \neq \vec{0}$ and $\vec{B} \neq \vec{0}$?

Use vectors to prove the following theorems.

**3.** The line joining the midpoints of two sides of a triangle is parallel to the third side and half its length.

**4.** If the midpoints of the sides of a quadrilateral are joined in order, a parallelogram is formed.

**5.** The intersection point of two medians of any triangle trisects both of them.

**6.** The diagonals of a parallelogram bisect each other.

# SECTION 54

## THE BASIS VECTORS AND THE SCALAR PRODUCT

Any vector can be decomposed into any number of pairs of component vectors. One useful decomposition of a vector into component vectors is along the $x$- and the $y$-axes of the coordinate plane. To do this we introduce the unit vectors $\vec{i}$ and $\vec{j}$.

**Definition 54.1** The vectors $\vec{i}$ and $\vec{j}$ have a magnitude of one unit and the direction of the positive $x$-axis and the positive $y$-axis, respectively (Figure 54.1).

    All vectors in the plane can be expressed as sums of scalar multiples of $\vec{i}$ and $\vec{j}$. Such a set of vectors is called a basis for the plane.

**Theorem 54.1** If $P_1(x_1, y_1)$ and $P_2(x_2, y_2)$ are any two points in a plane, then

$$\vec{P_1P_2} = (x_2 - x_1)\vec{i} + (y_2 - y_1)\vec{j} \tag{54.1}$$

    **Proof of Theorem 54.1.** In Figure 54.2 we draw a line through $P_2$ parallel to the $y$-axis and a line through $P_1$ parallel to the $x$-axis to determine $P(x_2, y_1)$ their point of intersection. We see that

$$\vec{P_1P_2} = \vec{P_1P} + \vec{PP_2} \tag{A}$$

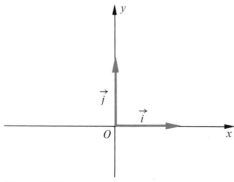

Figure 54.1

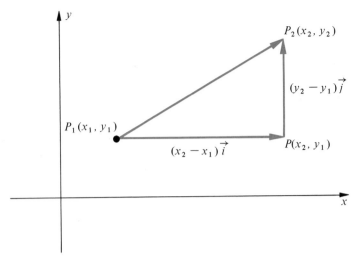

Figure 54.2

Since $P_1P$ is parallel to the $x$-axis (that is, same direction as $\vec{i}$ ) and has a length of $(x_2 - x_1)$ we have

$$\overrightarrow{P_1P} = (x_2 - x_1)\vec{i} \qquad \text{(B)}$$

Similarly,

$$\overrightarrow{PP_2} = (y_2 - y_1)\vec{j} \qquad \text{(C)}$$

We now substitute (B) and (C) into (A) to obtain

$$\overrightarrow{P_1P_2} = (x_2 - x_1)\vec{i} + (y_2 - y_1)\vec{j} \quad \bullet$$

**Example 54.1**

The vector with an initial point at $P(3, 2)$ and a terminal point at $Q(4, 7)$ is
$$\overrightarrow{PQ} = (4 - 3)\vec{i} + (7 - 2)\vec{j} = \vec{i} + 5\vec{j} \ \text{(Figure 54.3)}.$$

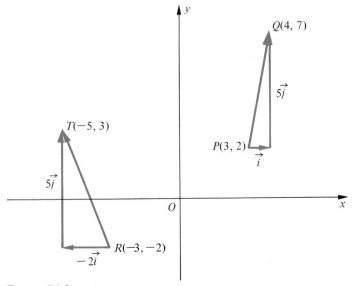

Figure 54.3

### Example 54.2

The vector with an initial point at $R(-3, -2)$ and a terminal point at $T(-5, 3)$ is

$$\vec{RT} = (-5 - (-3))\vec{i} + (3 - (-2))\vec{j} = -2\vec{i} + 5\vec{j}$$

(See Figure 54.3 on page 285.)

Another way of describing any vector $\vec{P_1 P_2}$ is to draw **position vectors** $\vec{OP_1}$ and $\vec{OP_2}$ to the points $P_1$ and $P_2$, respectively. (A position vector is a vector whose initial point is some reference point, usually the origin of the coordinate system, and whose terminal point is any point $P$ in the plane.) From Figure 54.4 we see that

$$\vec{OP_1} + \vec{P_1 P_2} = \vec{OP_2}$$

or

$$\vec{P_1 P_2} = \vec{OP_2} - \vec{OP_1} \tag{54.2}$$

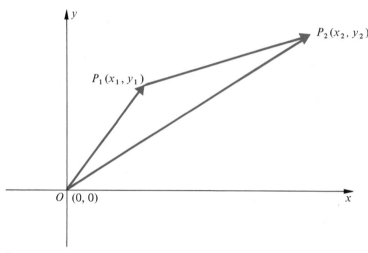

Figure 54.4

**Theorem 54.2**   If $\vec{V}_1 = a_1\vec{i} + b_1\vec{j}$ and $\vec{V}_2 = a_2\vec{i} + b_2\vec{j}$, then

$$\vec{V}_1 + \vec{V}_2 = (a_1 + a_2)\vec{i} + (b_1 + b_2)\vec{j}$$

### Example 54.3

If $\vec{V}_1 = 2\vec{i} - 3\vec{j}$ and $\vec{V}_2 = 4\vec{i} + 5\vec{j}$, then

1. $\vec{V}_1 + \vec{V}_2 = (2\vec{i} - 3\vec{j}) + (4\vec{i} + 5\vec{j}) = 6\vec{i} + 2\vec{j}$.
2. $\vec{V}_1 - \vec{V}_2 = (2\vec{i} - 3\vec{j}) - (4\vec{i} + 5\vec{j}) = -2\vec{i} - 8\vec{j}$.
3. $2\vec{V}_1 = 2(2\vec{i} - 3\vec{j}) = 4\vec{i} - 6\vec{j}$.
4. $2\vec{V}_1 - 3\vec{V}_2 = 2(2\vec{i} - 3\vec{j}) - 3(4\vec{i} + 5\vec{j}) = -8\vec{i} - 21\vec{j}$.

**Theorem 54.3   Magnitude of a Vector.**   If a vector has an initial point $P_1(x_1, y_1)$ and a terminal point $P_2(x_2, y_2)$, then its magnitude (length) is

$$|\vec{P_1 P_2}| = \sqrt{(x_2 - x_1)^2 + (y_2 - y_1)^2} \tag{54.3}$$

Proof. Since the magnitude of $\overrightarrow{P_1P_2}$ is simply the distance between the points $P_1$ and $P_2$, we use the distance formula established in Section 10. ●

If $\vec{V} = x\vec{i} + y\vec{j}$ (that is, as sum of multiples of $\vec{i}$ and $\vec{j}$), then its *length* or magnitude is given by

$$|\vec{V}| = \sqrt{x^2 + y^2} \tag{54.4}$$

## Example 54.4

Consider the vectors $\vec{V}_1 = 2\vec{i} - 3\vec{j}$ and $\vec{V}_2 = 4\vec{i} + 5\vec{j}$ of Example 54.3. Find $|\vec{V}_1|$, $|\vec{V}_2|$, $|\vec{V}_1 + \vec{V}_2|$, and $|2\vec{V}_1 - 3\vec{V}_2|$.

Solution. We use formula (54.4) and the results of Example 54.3 to obtain

$$|\vec{V}_1| = \sqrt{(2)^2 + (-3)^2} = \sqrt{13}$$
$$|\vec{V}_2| = \sqrt{(4)^2 + (5)^2} = \sqrt{41}$$
$$|\vec{V}_1 + \vec{V}_2| = \sqrt{(6)^2 + (2)^2} = \sqrt{40}$$
$$|2\vec{V}_1 - 3\vec{V}_2| = \sqrt{(-8)^2 + (-21)^2} = \sqrt{505}$$

Note that $|\vec{V}_1 + \vec{V}_2| \neq |\vec{V}_1| + |\vec{V}_2|$. ●

## Example 54.5

Find a unit vector $\vec{b}$ if $\vec{B} = 4\vec{i} + 3\vec{j}$.

Solution. We recall from Section 53 that we can obtain a unit vector $\vec{b}$ in the same direction as $\vec{B}$ by dividing $\vec{B}$ by its own length. We use formula (54.1) and the fact that $|\vec{B}| = \sqrt{4^2 + 3^2} = 5$ to obtain

$$\vec{b} = \frac{4\vec{i} + 5\vec{j}}{5} = \frac{4}{5}\vec{i} + \vec{j} \quad ●$$

Our definition of vector addition shows us how to associate with two given vectors a third vector we call their sum. We are now going to formulate a very useful operation which will associate a scalar (number) with a given pair of vectors. We will call this operation the scalar or dot product of the two vectors.

**Definition 54.2** The **scalar product** of two vectors $\vec{A}$ and $\vec{B}$, denoted $\vec{A} \cdot \vec{B}$, is equal to the product of their magnitudes and the cosine of the angle determined by $\vec{A}$ and $\vec{B}$. That is,

$$\vec{A} \cdot \vec{B} = |\vec{A}||\vec{B}|\cos\theta \tag{54.5}$$

## Example 54.6

Compute $\vec{i} \cdot \vec{i}$, $\vec{j} \cdot \vec{j}$, $\vec{i} \cdot \vec{j}$, and $\vec{j} \cdot \vec{i}$.

Solution. We note that the angle $\theta$ determined by $\vec{i}$ and itself, and $\vec{j}$ and itself, is 0 and that the angle determined by $\vec{i}$ and $\vec{j}$ is $\pi/2$. We use these facts and formula (54.5) to obtain

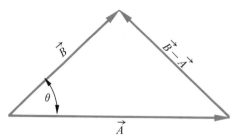

Figure 54.5

$$\vec{i} \cdot \vec{i} = |\vec{i}||\vec{i}| \cos 0 = (1)(1)(1) = 1$$
$$\vec{j} \cdot \vec{j} = |\vec{j}||\vec{j}| \cos 0 = (1)(1)(1) = 1$$
$$\vec{i} \cdot \vec{j} = |\vec{i}||\vec{j}| \cos \pi/2 = (1)(1)(0) = 0$$
$$\vec{j} \cdot \vec{i} = |\vec{j}||\vec{i}| \cos \pi/2 = (1)(1)(0) = 0 \quad \bullet$$

Definition 54.2 is a geometric one. We can also develop an algebraic expression for scalar product. Suppose that $\vec{A} = a_1\vec{i} + a_2\vec{j}$ and $\vec{B} = b_1\vec{i} + b_2\vec{j}$ in Figure 54.5, then $\vec{B} - \vec{A} = (b_1 - a_1)\vec{i} + (b_2 - a_2)\vec{j}$. We now apply the law of cosines (see Section 51 to the triangle in Figure 54.5 to obtain

$$|\vec{B} - \vec{A}|^2 = |\vec{A}|^2 + |\vec{B}|^2 - 2|\vec{A}||\vec{B}| \cos \theta \tag{A}$$

We now use formulas (54.3) and (54.4) to obtain

$$|\vec{B} - \vec{A}| = \sqrt{(b_1 - a_1)^2 + (b_2 - a_2)^2} \tag{B}$$
$$|\vec{A}| = \sqrt{a_1^2 + a_2^2} \qquad |\vec{B}| = \sqrt{b_1^2 + b_2^2} \tag{C}$$

Finally, we substitute results (B) and (C) along with (54.5) into (A) and simplify to obtain

$$\vec{A} \cdot \vec{B} = a_1 b_1 + a_2 b_2 \tag{54.6}$$

### Example 54.7
Find $\vec{V}_1 \cdot \vec{V}_2$ when $\vec{V}_1 = 2\vec{i} + \vec{j}$ and $\vec{V}_2 = 3\vec{i} + 4\vec{j}$.

Solution.   We use formula (54.6) to obtain

$$\vec{V}_1 \cdot \vec{V}_2 = 2(3) + (1)(4) = 10 \quad \bullet$$

### Example 54.8
Use (54.6) to verify that $\vec{i} \cdot \vec{j} = 0$.

Solution.   Since $\vec{i} = 1\vec{i} + 0\vec{j}$ and $\vec{j} = 0\vec{i} + 1\vec{j}$, we use (54.6) to obtain
$$\vec{i} \cdot \vec{j} = 1(0) + (0)(1) = 0 \quad \bullet$$

The scalar or dot product of two vectors has the following properties:

1. $\vec{A} \cdot \vec{B} = \vec{B} \cdot \vec{A}$ (54.7)
2. $(c\vec{A}) \cdot \vec{B} = \vec{A} \cdot (c\vec{B}) = c(\vec{A} \cdot \vec{B})$ (54.8)

$$3.\ \vec{A} \cdot (\vec{B} + \vec{C}) = (\vec{A} \cdot \vec{B}) + (\vec{A} \cdot \vec{C}) \tag{54.9}$$

We leave the proofs of these properties as exercises.

The scalar product is useful for finding the angle between two vectors. We see from (54.5) that

$$\vec{A} \cdot \vec{B} = |\vec{A}||\vec{B}| \cos \theta$$

or

$$\cos \theta = \frac{\vec{A} \cdot \vec{B}}{|\vec{A}||\vec{B}|} \tag{54.10}$$

**Theorem 54.4**   If $\theta$ is the angle between two vectors $\vec{A}$ and $\vec{B}$, then

$$\cos \theta = \frac{\vec{A} \cdot \vec{B}}{|\vec{A}||\vec{B}|} = \frac{\vec{A}}{|\vec{A}|} \cdot \frac{\vec{B}}{|\vec{B}|} = \vec{a} \cdot \vec{b} \tag{54.11}$$

Theorem 54.4 says that "the cosine of the angle $\theta$ can be computed by finding the scalar product of their respective *unit vectors*."

There are really two angles formed by the vectors $\vec{A}$ and $\vec{B}$ when their initial points coincide (see Figure 54.6), namely $\theta$ and $2\pi - \theta$, but $\cos(2\pi - \theta) = \cos \theta$; therefore, we can select either angle.

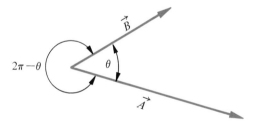

Figure 54.6

**Example 54.9**

Find the cosine of the angle $\theta$ between $\vec{A} = 2\vec{i} - 3\vec{j}$ and $\vec{B} = 3\vec{i} + \vec{j}$.

Solution.   We first unitize both vectors $\vec{A}$ and $\vec{B}$ to obtain

$$\vec{a} = \frac{2\vec{i} - 3\vec{j}}{\sqrt{13}} \quad \text{and} \quad \vec{b} = \frac{3\vec{i} + \vec{j}}{\sqrt{10}}$$

and then we use equation (54.11) to obtain

$$\cos \theta = \vec{a} \cdot \vec{b} = \frac{2(3) + (-3)(1)}{(\sqrt{13})(\sqrt{10})} = \frac{3}{\sqrt{130}} \quad \bullet$$

**Example 54.10**

Suppose that triangle $ABC$ is formed by the points $A(2, 2)$, $B(-2, -1)$, and $C(-1, 6)$. Find angle $B$.

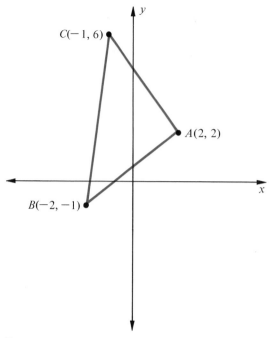

Figure 54.7

**Solution.** We know that angle $B$ is formed by the vectors $\overrightarrow{BA}$ and $\overrightarrow{BC}$ (see Figure 54.7). We use (54.1) to obtain

$$\overrightarrow{BA} = 4\vec{i} + 3\vec{j}$$
$$\overrightarrow{BC} = \vec{i} + 7\vec{j}$$

We now use (54.10) to obtain

$$\cos B = \frac{\overrightarrow{BA} \cdot \overrightarrow{BC}}{|\overrightarrow{BA}||\overrightarrow{BC}|} = \frac{4(1) + (3)(7)}{5\sqrt{50}} = \frac{25}{25\sqrt{2}} = \frac{1}{\sqrt{2}}$$

Therefore, $B = 45°$ or $\pi/4$.   ●

**Theorem 54.5**   If $\vec{A} \cdot \vec{B} = 0$ and if neither $\vec{A}$ nor $\vec{B}$ is the zero vector, then $\vec{A}$ is perpendicular to $\vec{B}$.

    **Proof.** We know that $\vec{A} \cdot \vec{B} = 0$, $|\vec{A}| \neq 0$, $|\vec{B}| \neq 0$, and $\vec{A} \cdot \vec{B} = |\vec{A}||\vec{B}| \cos \theta$. Thus we have

$$|\vec{A}||\vec{B}| \cos \theta = 0$$
$$\cos \theta = 0 \quad \text{(Why?)}$$
$$\theta = \pi/2 \text{ or } 90°$$

The converse of Theorem 54.5 also holds.   ●

**Example 54.11**

Find a vector $\vec{W}$ that is perpendicular to $\vec{V} = 2\vec{i} + \vec{j}$.

Solution.   Let $\vec{W} = x\vec{i} + y\vec{j}$. If $\vec{W}$ is perpendicular to $\vec{V}$, then

$$\vec{V} \cdot \vec{W} = 0 \quad \text{or} \quad 2x + y = 0 \tag{A}$$

Now if $x = -1$, then $y = 2$. Thus $\vec{W} = -\vec{i} + 2\vec{j}$. It should be noted that since there are an infinite number of ordered pairs $(x, y)$ that satisfy (A), there are an infinite number of possible answers to this problem.   ●

### Example 54.12

Find a vector $\vec{H}$ that is perpendicular to $\vec{V} = 2\vec{i} + \vec{j}$ and has a magnitude of 6.

Solution.   We use the result of Example 54.11 to obtain a vector $\vec{W} = -\vec{i} + 2\vec{j}$ that is perpendicular to $\vec{V}$, and then we simply unitize $\vec{W}$ and multiply by the scalar 6 to obtain the desired result. That is,

$$\vec{H} = 6\frac{\vec{W}}{|\vec{W}|} = 6\frac{-i + 2j}{\sqrt{5}} = \frac{-6}{\sqrt{5}}\vec{i} + \frac{12}{\sqrt{5}}\vec{j}$$

The reader should verify that $\vec{H}$ is perpendicular to $\vec{V}$.   ●

Another use of the scalar or dot product is to find the projection of one vector on the direction of the other.

If we draw perpendiculars from the endpoints of $\vec{A}$ to the line on which $\vec{B}$ lies, then the vector on the line of $\vec{B}$ whose initial point is the foot of the perpendicular drawn from the initial point of $\vec{A}$ and whose terminal point is the foot of the perpendicular drawn from the terminal point of $\vec{A}$ is called the projection of $\vec{A}$ on the direction of $\vec{B}$.

**Theorem 54.6**   The projection of $\vec{A}$ on the direction of $\vec{B}$, $\vec{B} \neq \vec{0}$, is given by

$$\vec{A} \cdot \frac{\vec{B}}{|\vec{B}|} \tag{54.12}$$

Proof.   In Figure 54.8 below the initial and terminal points of $\vec{A}$ are projected onto the points $P$ and $Q$ of $\vec{B}$, respectively. We use trigonometry of the right triangle to obtain the projection of $\vec{A}$ on $\vec{B}$ as

$$\overline{PQ} = \overline{RS} = |\vec{A}| \cos \theta \tag{A}$$

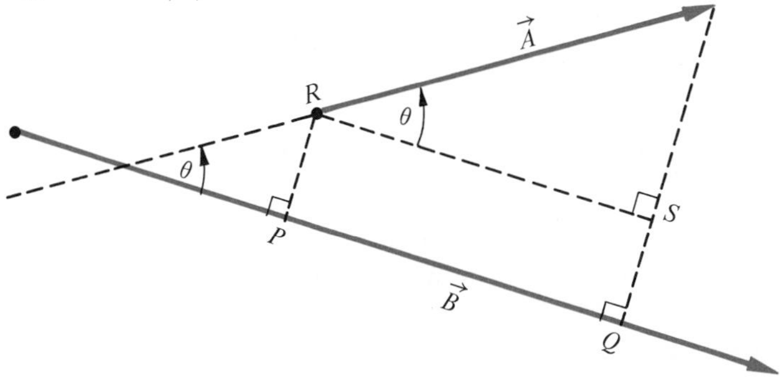

Figure 54.8

We now substitute $\cos\theta = \dfrac{\vec{A}\cdot\vec{B}}{|\vec{A}||\vec{B}|}$ into (A) to obtain

$$\overline{PQ} = |\vec{A}|\,\frac{\vec{A}\cdot\vec{B}}{|\vec{A}||\vec{B}|} = \vec{A}\cdot\frac{\vec{B}}{|\vec{B}|}$$

It should be noted that the projection of $\vec{A}$ onto $\vec{B}$ is simply the *dot product of $\vec{A}$ and $\vec{B}$'s unit vector $\vec{b}$.* In addition, if the angle $\theta$ between $\vec{A}$ and $\vec{B}$ is obtuse (that is, $90° < \theta < 180°$), then $\cos\theta$ is negative and the projection of $\vec{A}$ onto $\vec{B}$ will be negative.  ●

**Example 54.13**
Find the projection of $\vec{A} = 3\vec{i} + 2\vec{j}$ onto $\vec{B} = 4\vec{i} - 3\vec{j}$.

Solution.   We use (54.12) to obtain

$$\overline{PQ} = (3\vec{i} + 2\vec{j})\cdot\tfrac{1}{5}(4\vec{i} - 3\vec{j}) = \tfrac{6}{5} \quad ●$$

The **vector projection** of $\vec{A}$ onto $\vec{B}$ is the vector $\overrightarrow{PQ}$ shown in Figure 54.9. We can find this vector by simply multiplying the projection $\overline{PQ}$ times $\vec{B}$ unitized. That is,

$$\overrightarrow{PQ} = \left(\vec{A}\cdot\frac{\vec{B}}{|\vec{B}|}\right)\left(\frac{\vec{B}}{|\vec{B}|}\right) \tag{54.13}$$

Figure 54.9

**Example 54.14**
Use the result of Example 54.13 to find the vector projection of $\vec{A}$ onto $\vec{B}$.

Solution.   From Example 54.13 the projection of $\vec{A}$ onto $\vec{B}$ is $\overline{PQ} = \tfrac{6}{5}$. Thus the vector projection is.

$$\overrightarrow{PQ} = \frac{6}{5}\left(\frac{4\vec{i} - 3\vec{j}}{5}\right) = \left(\frac{24}{25}\right)\vec{i} - \left(\frac{18}{25}\right)\vec{j}. \quad ●$$

Another application of the scalar or dot product of two vectors deals with the concept of the work done in order to move an object against an opposing force.

**Definition 54.3**  **Work** is the product of a displacement by a component of the force in the direction of the displacement.

If $\vec{F}$ is the force and if $\vec{D}$ is a *linear* displacement with an angle $\theta$ between their directions, then

$$W = \vec{F} \cdot \vec{D} = |\vec{F}||\vec{D}| \cos \theta \qquad (54.14)$$

Note that work $W$ is a *scalar quantity.*

### Example 54.15

Find the work done if $|\vec{F}| = 3$ lb, $|\vec{D}| = 5$ ft, and the angle $\theta$ between the direction of $\vec{F}$ and $\vec{D}$ is $30°$.

Solution.   We use formula (54.14) to obtain

$$W = |\vec{F}||\vec{D}| \cos \theta$$
$$= 3 \cdot 5 \, (\cos 30°) = 15 \sqrt{3}/2 \text{ ft-lb.} \quad \bullet$$

## EXERCISES FOR SECTION 54

Express the vector $\vec{PQ}$ in terms of the basis vector $\vec{i}$ and $\vec{j}$ if

**1.** $P(2, 3)$ and $Q(1, 7)$      **4.** $P(-3, 2)$ and $Q(0, 0)$

**2.** $P(3, 5)$ and $Q(3, -4)$      **5.** $P(-3, -5)$ and $Q(-2, -1)$

**3.** $P(4, 2)$ and $Q(6, 2)$      **6.** $P(0, 1)$ and $Q(-3, -4)$

Find $\vec{A} + \vec{B}, \vec{A} - \vec{B}, |\vec{A}|, |\vec{B}|,$ and $|\vec{A} + \vec{B}|$ when

**7.** $\vec{A} = 2\vec{i} - 3\vec{j}; \vec{B} = 6\vec{i} + 3\vec{j}$

**8.** $\vec{A} = 2\vec{i} + 3\vec{j}; \vec{B} = 3\vec{i} - 2\vec{j}$

**9.** $\vec{A} = \vec{i} + \vec{j}; \vec{B} = -\vec{i} - \vec{j}$

**10.** Find a unit vector $\vec{b}$ for $\vec{B} = -3\vec{i} + 4\vec{j}$.

**11.** Find a unit vector $\vec{a}$ for $\vec{A} = 8\vec{i} + 6\vec{j}$.

**12.** Find the cosines of the angles of the triangles formed by the following points (use the methods of this section).
(a) $A(2, -1)$, $B(-2, 3)$, and $C(2, 6)$
(b) $A(-1, -1)$, $B(3, 3)$, and $C(0, 0)$
(c) $A(3, 0)$, $B(4, 6)$, and $C(-4, -1)$

**13.** Find a point $P_1$ that is three-fourths of the way from $P(3, 4)$ to $Q(5, -2)$.

**14.** Find a point $P_1$ that divides the line joining the points $P(-5, 9)$ and $Q(6, 6)$ in the ratio $3 : 2$.

**15.** Find (1) the projection of $\vec{A}$ on $\vec{B}$'s direction and (2) the vector projection of $\vec{A}$ on $\vec{B}$'s direction when
(a) $\vec{A} = -2\vec{i} + 6\vec{j}; \vec{B} = \vec{i} + \vec{j}$
(b) $\vec{A} = -2\vec{i} + \vec{j}; \vec{B} = -3\vec{i} - 2\vec{j}$
(c) $\vec{A} = \vec{i} + \vec{j}; \vec{B} = -\vec{i} - \vec{j}$

**16.** Determine whether $\vec{A}$ and $\vec{B}$ are perpendicular.
(a) $\vec{A} = 3\vec{i} - 2\vec{j}; \vec{B} = -3\vec{i} + 2\vec{j}$

(b) $\vec{A} = \vec{i} + \vec{j}$; $\vec{B} = \vec{i} - \vec{j}$

(c) $\vec{A} = \frac{2}{5}\vec{i} + \frac{3}{5}\vec{j}$; $\vec{B} = -\frac{3}{5}\vec{i} + \frac{2}{5}\vec{j}$

17. Find a vector that is perpendicular to $\vec{V} = 4\vec{i} + 2\vec{j}$ and has a magnitude of 10.

18. Show that if $\vec{V} = x\vec{i} + y\vec{j}$, then $x = \vec{V} \cdot \vec{i}$ and $y = \vec{V} \cdot \vec{j}$ (that is, $x$ and $y$ are simply the projections of $\vec{V}$ onto the direction of the $x$- and the $y$-axis, respectively).

19. Show that $|\vec{V}| = \sqrt{\vec{V} \cdot \vec{V}}$.

20. Show that if $\vec{A} = x\vec{i} + y\vec{j}$, then $\vec{a} = (\cos\theta)\vec{i} + (\sin\theta)\vec{j}$.

21. Show that $(\vec{A} \cdot \vec{B}) \cdot \vec{C} \neq \vec{A} \cdot (\vec{B} \cdot \vec{C})$.

22. Give an example which shows that if $\vec{A} \cdot \vec{B} = 0$, then $\vec{A} \neq \vec{0}$ and $\vec{B} \neq \vec{0}$.

23. Show that if $\vec{V_1} = a_1\vec{i} + b_1\vec{j}$, $\vec{V_2} = a_2\vec{i} + b_2\vec{j}$, $\vec{W} = -b_1\vec{i} + a_1\vec{j}$, and $\vec{W} \cdot \vec{V_2} = 0$, then $\vec{V_1}$ and $\vec{V_2}$ are parallel.

24. Show that $(\vec{A} \cdot \vec{B})^2 \leq |\vec{A}|^2|\vec{B}|^2$ (Cauchy's Inequality).

25. Show that two vectors $\vec{A}$ and $\vec{B}$ will be parallel if $\cos^2\theta = 1$.

26. Let $\vec{V}$ and $\vec{W}$ be two unit vectors that make angles of $\theta$ and $\alpha$, respectively, with $\vec{i}$. Express $\vec{V}$ and $\vec{W}$ in terms of $\vec{i}$ and $\vec{j}$ (see Exercise 20) and then compute $\vec{V} \cdot \vec{W}$ to derive the formula for $\cos(\theta - \alpha)$.

27. Find the value(s) for $x$ such that the following vectors are *perpendicular*.
   (a) $\vec{V} = x\vec{i} + 3\vec{j}$; $\vec{W} = -3\vec{i} + 6\vec{j}$
   (b) $\vec{V} = x\vec{i} + 2\vec{j}$; $\vec{W} = x\vec{i} - 8\vec{j}$
   (c) $\vec{V} = 3\vec{i} + x\vec{j}$; $\vec{W} = 4\vec{i} - x\vec{j}$

28. Find the work done on an object by the constant force $2\vec{i} + 3\vec{j}$ lb displaced in a straight line from (3, 4) ft to (5, 8) ft.

29. Find the work done on a particle acted upon by the constant forces $\vec{A} = 5\vec{i} + \vec{j}$ lb and $\vec{B} = 3\vec{i} + 2\vec{j}$ lb during a displacement of $7\vec{i} - 4\vec{j}$ ft.

30. An object is acted upon by the forces $\vec{A} = 3\vec{i} + 5\vec{j}$ tons and $\vec{B} = 2\vec{i} + 7\vec{j}$ tons and displaced from the point (0, 5) ft to (7, 8) ft. Determine the total work done by these forces.

31. Prove equations (54.7), (54.8), and (54.9).

32. Two vectors $\vec{V_1} = a_1\vec{i} + b_1\vec{j}$ and $\vec{V_2} = a_2\vec{i} + b_2\vec{j}$ are said to be **independent** if the only solution to $c_1\vec{V_1} + c_2\vec{V_2} = \vec{o}$ is $c_1 = c_2 = 0$. Determine which of the following pairs of vectors are independent.
   (a) $\vec{V_1} = 2\vec{i} + \vec{j}$; $\vec{V_2} = \vec{i} + 2\vec{j}$
   (b) $\vec{V_1} = 3\vec{i} - 2\vec{j}$, $\vec{V_2} = -2\vec{i} + 3\vec{j}$
   (c) $\vec{V_1} = \vec{i} + \vec{j}$, $\vec{V_2} = 2\vec{i} + 2\vec{j}$

33. A vector $\vec{V} = a\vec{i} + b\vec{j}$ is said to be a **linear combination** of the vectors $\vec{V_1} = a_1\vec{i} + b_1\vec{j}$ and $\vec{V_2} = a_2\vec{i} + b_2\vec{j}$ if there are scalars $c_1$ and $c_2$ such that
$$\vec{V} = c_1\vec{V_1} + c_2\vec{V_2}$$
   (a) Show that $\vec{V} = 2\vec{i} + 3\vec{j}$ is a linear combination of $\vec{V_1} = \vec{i} + \vec{j}$ and $\vec{V_2} = 3\vec{i} - 4\vec{j}$.
   (b) Show that $\vec{V} = 4\vec{i} - 5\vec{j}$ is not a linear combination of $\vec{V_1} = \vec{i} - 2\vec{j}$ and $\vec{V_2} = 2\vec{i} - 4\vec{j}$.

# CHAPTER 9 COMPLEX NUMBERS

## THE NEED FOR COMPLEX NUMBERS

Let us suppose that $a$ and $b$ are integers. Then any equation of the form $x + a = 0$ may be solved using only the set of all integers. If we attempt to solve an equation of the form $ax + b = 0$, where $a \neq 0$, we may not get an integer for a result. For example, $2x - 4 = 0$ has an integral solution, whereas $2x - 3 = 0$ does not.

In general, to solve an equation of the form $ax + b = 0$ we need the set of **rational numbers.** This introduction of the rational numbers represents an extension of the number system. But, even with this extension, an equation such as $x^2 - 2 = 0$ will have no solution. Hence we are led to a larger set of numbers called the **real numbers.**

Now we ask: Does the set of real numbers enable us to solve all algebraic equations? We can easily see that the answer is no by considering the equation $x^2 + 1 = 0$. For any real value for $x$, either $x < 0$, $x = 0$, or $x > 0$. Therefore, $x^2 \geq 0$ and $x^2 + 1 > 0$. We are forced to conclude that the equation has no real solutions.

At this point we have two possibilities. Either the equation is not solvable or our number system must be extended to include values that will make the equation solvable. We follow the latter course by introducing the symbol $i$, which is equal to $\sqrt{-1}$ and has the property that $i^2 = -1$. This is the basis for the set of complex numbers.

**Definition 55.1**  A **complex number** is any number of the form $a + bi$, where $a$ and $b$ are real numbers and $i^2 = -1$. If we denote this set by the letter $C$, we have $C = \{(a + bi) \mid a \in \mathbb{R}, b \in \mathbb{R}, i^2 = -1\}$.

In our work with complex numbers, we will refer to $a + bi$ as the "standard form" of the complex number. We agree to call $a$ the "real part" and $b$ the "imaginary part." If

**295**

$a \neq 0$ and $b = 0$, we have a real number. If $a = 0$ and $b \neq 0$, we have a **pure imaginary number.** For example, $2 + 3i$ is a complex number that has 2 for its real part and 3 for its imaginary part. A number such as 5 may be written as $5 + 0i$ and thus falls within the domain of definition of a complex number. We will, however, usually refer to 5 as a real number. Similarly, $3i$ may be written as $0 + 3i$. This in no way detracts from the fact that $3i$ is a pure imaginary number.

Figure 55.1 represents the relationships that exist between the various parts of our number system.

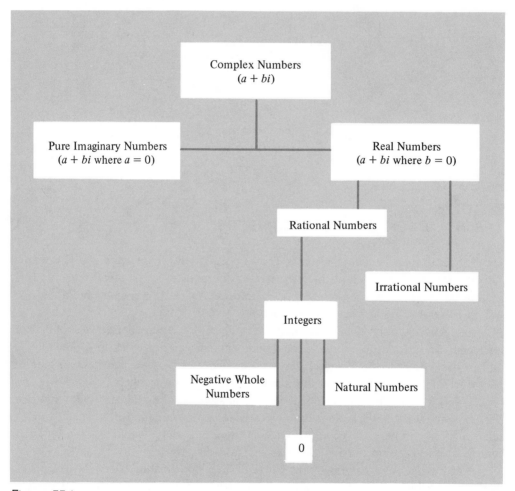

Figure 55.1

# SECTION 56
## THE ALGEBRA OF COMPLEX NUMBERS

The basis for the operations on complex numbers is given in the following definition.

### Definition 56.1

**Equality:** $a + bi = c + di$ if and only if $a = c$ and $b = d$.

**Addition:** $(a + bi) + (c + di) = (a + c) + (b + d)i.$

**Multiplication:** $(a + bi) \cdot (c + di) = (ac - bd) + (ad + bc)i.$

The definitions of equality and addition may be thought of as "natural" definitions. For equality, we are saying that two conditions must be satisfied. The real parts must be equal *and* the imaginary parts must be equal. This has considerable importance in future mathematics. For addition, we simply add the real parts and then add the imaginary parts.

The definition of multiplication is nothing more than the usual polynomial multiplication with $i^2$ replaced by $-1$, whenever it occurs.

**Example 56.1**
1. $(2 + 3i) + (1 - 5i) = 3 - 2i.$
2. $(4 - 2i) - (1 - 3i) = 3 + i.$
3. $(1 + 2i)(3 - 2i) = 3 + 4i - 4i^2 = (3 + 4) + 4i = 7 + 4i.$
4. $i^6 = (i^2)^3 = (-1)^3 = -1.$ ●

The following example illustrates the definition of equality for complex numbers. Keep in mind that a single equation involving complex numbers always generates two equations with real numbers.

**Example 56.2**
Find $x$ and $y$ such that $(1 + 2i) + (x + yi) = 3(2 - i).$

Solution.   Performing the indicated operations we have $(x + 1) + (y + 2)i = 6 - 3i$. Equating real and imaginary parts yields the equations $x + 1 = 6$ and $y + 2 = -3$. Thus $x = 5$ and $y = -5$. ●

The operation of division can be handled by making use of our initial definitions. If, for example, we wish to divide $3 + 2i$ by $2 - i$, we expect to obtain a result in the standard form $a + bi$. On this basis we have $\dfrac{3 + 2i}{(2 - i)} = a + bi$. Multiplying both sides by $2 - i$, we have $(3 + 2i) = (2 - i)(a + bi)$. Thus $3 + 2i = (2a + b) + (-a + 2b)i$. Equating real and imaginary parts yields the following system of equations:

$$2a + b = 3 \quad \text{and} \quad -2 + 2b = +2$$

Solving simultaneously we have $a = \frac{4}{5}$ and $b = \frac{7}{5}$.

The procedure just outlined will work each time that we have a problem in division. However, to simplify the division we define the conjugate of a complex number.

**Definition 56.2**   If $z = a + bi$, then $\bar{z} = a - bi$. $\bar{z}$ is called the **conjugate** of $z$.

The product of $z$ and $\bar{z}$ is always the real number $a^2 + b^2$. To effect division we multiply numerator and denominator by the conjugate of the denominator. This enables us to express the indicated quotient in "standard form," where the real and imaginary parts of the number can be read directly.

**Example 56.3**
Write each of the following in the form $a + bi$.

(a) $\dfrac{2}{3 + i}$      (b) $\dfrac{1 + i}{2 - 3i}$      (c) $\dfrac{2 - 3i}{i}$

Solution

(a) $\dfrac{2}{3 + i} = \dfrac{2(3 - i)}{(3 + i)(3 - i)} = \dfrac{6 - 2i}{10} = \dfrac{3}{5} - \dfrac{1}{5}i$

(b) $\dfrac{1 + i}{2 - 3i} = \dfrac{1 + i}{2 - 3i} \cdot \dfrac{2 + 3i}{2 + 3i} = \dfrac{-1 + 5i}{13} = -\dfrac{1}{13} + \dfrac{5}{13}i$

(c) $\dfrac{2 - 3i}{i} = \dfrac{2 - 3i}{i} \cdot \dfrac{-i}{-i} = \dfrac{-3 - 2i}{1} = -3 - 2i$ ●

With the introduction of the symbol $i$ and the operations thus defined, we have gained much. We can now easily show that every quadratic equation of the form $ax^2 + bx + c = 0$ has a solution. From section 15 we know that the general solution for the quadratic equation is given by

$$x = \frac{-b \pm \sqrt{b^2 - 4ac}}{2a}.$$

If the discriminant $b^2 - 4ac \geq 0$, $\sqrt{b^2 - 4ac}$ is a real number and the solutions to the quadratic equation are real. On the other hand, if $b^2 - 4ac < 0$, the solutions are complex numbers. This is demonstrated in the following manner. $b^2 - 4ac < 0$. Multiplying by $-1$ we have $4ac - b^2 > 0$. Now

$$\sqrt{b^2 - 4ac} = \sqrt{(-1)(4ac - b^2)} = i\sqrt{4ac - b^2}$$

**Example 56.4**
Find the solutions of the equation $x^2 - 2x + 2 = 0$.

Solution. The quadratic formula enables us to write

$$x = \frac{2 \pm \sqrt{4 - 8}}{2} = \frac{2 \pm \sqrt{-4}}{2} = \frac{2 \pm 2i}{2} = 1 + i \quad \text{and} \quad 1 - i \quad ●$$

In general, the set of complex numbers allows us to solve not only every equation but every algebraic equation of degree $n$ with real or complex coefficients.

## EXERCISES FOR SECTION 56

In Exercises 1–35 perform the indicated operations and write the answer in the form $a + bi$.

**1.** $(2 + 3i) + (4 - 5i)$          **4.** $(3 - i) \div (-2 + 5i)$

**2.** $(3 - 5i) + (6 + 7i)$          **5.** $-(3 - 2i) - (-4 + 5i)$

**3.** $(1 + 2i) - (3 - 2i)$          **6.** $6 + (3 - i)$

**7.** $(2 + i) - (0 + 2i)$

**8.** $(3 + 11i) - 13i$

**9.** $(1 + i) + 2(3 - i) - (6 + 4i)$

**10.** $(2 + 3i)(4 - 2i)$

**11.** $(3 - i)(5 - 7i)$

**12.** $(\sqrt{5} - i)(\sqrt{5} + i)$

**13.** $(\sqrt{2} + i)(\sqrt{2} - i)$

**14.** $3(2 - 7i)$

**15.** $3i(2 - 7i)$

**16.** $(2i)(3i)$

**17.** $(2 - i)^2$

**18.** $(2 - i)^3$

**19.** $i^3$

**20.** $i^5$

**21.** $(2 + 3i)(3 - 5i)(1 - i)$

**22.** $i^{10}$

**23.** $\dfrac{3 - 2i}{2 + i}$

**24.** $\dfrac{2 - i}{4 + 5i}$

**25.** $\dfrac{-4 - 2i}{3 - i}$

**26.** $\dfrac{3i}{6 + 9i}$

**27.** $\dfrac{5i}{2 - i}$

**28.** $\dfrac{3 + 5i}{6i}$

**29.** $\dfrac{2 - 3i}{4i}$

**30.** $\dfrac{6}{7i}$

**31.** $\dfrac{2}{3i}$

**32.** $(2 - 3i)^{-1}$

**33.** $(1 + i)^{-2}$

**34.** $\dfrac{-2}{i^4}$

**35.** $\dfrac{-i^3}{2i^7}$

In Exercises 36–42 find the real numbers $x$ and $y$ that satisfy the given equation.

**36.** $(x + iy)(2 - 3i) = 5 + 2i$

**37.** $(x + iy)(1 - 2i) = 6 - 3i$

**38.** $(x + iy)(3i) = 4$

**39.** $(x + iy)(-4i) = 13i$

**40.** $x - 2 + 3iy = 7i$

**41.** $2x - 3iy + 5 - 2i = 1 - i$

**42.** $3x - 2iy = 6 - 4i$

For Exercises 43–52, $f$ is a function whose domain is the set of complex numbers. Find the zeros of the given functions.

**43.** $f(x) = x^2 + x + 1$

**44.** $f(x) = x^2 - 4x + 5$

**45.** $f(x) = x^3 - 1$

**46.** $f(x) = x^3 - 4x^2 - 5x$

**47.** $f(x) = x^4 - 1$

**48.** $f(x) = 2ix + 3$

**49.** $f(x) = 3x - 2 + i$

**50.** $f(x) = (2 - 3i)x - 4i$

**51.** $f(x) = x^2 + 9$

**52.** $f(x) = x^4 + x^3 - x^2 + x - 2$

**53.** (a) Show that both $3 + 3i$ *and* $3 - 3i$ satisfy the equation $x^2 - 6x + 18 = 0$.

(b) Show that $(3 - 5i)$ and $(3 + 5i)$ do not both satisfy the equation $x^2 + (-8 + 3i)x + (25 - 19i) = 0$.

**54.** Show that $x^2 - x + 1 + i = 0$ has $1 - i$ for a root and then find the other root.

**55.** For any complex number $z$ show that the conjugate of the conjugate of $z$ is equal to $z$.

**56.** Prove that the conjugate of the sum of two complex numbers equals the sum of their conjugates.

**57.** Prove that the conjugate of the product of two complex numbers equals the product of their conjugates.

For exercises 58 through 60, use the fact that if an algebraic equation with real coefficients has a complex root $a + bi$, $b \neq 0$, then it also has its conjugate $a - bi$ for a root.

**58.** If $1 + i$ is a root of $x^4 - 4x^2 + 8x - 4 = 0$ find the remaining roots.

**59.** If $2i$ is a root of $x^3 + x^2 + ax + b = 0$ where $a$ and $b$ are real numbers, find the value of $a$ and $b$.

**60.** If $1 - i$ is a root of $x^4 - x^3 - x^2 + 4x - 2 = 0$, find the remaining roots.

**61.** Show that

(a) $i^n = \pm 1$, $n$ an even integer    (b) $i^n = \pm i$, $n$ an odd integer

# SECTION 57
## POLAR COORDINATES

We can locate a point in the plane by means of its rectangular coordinates. The general form of this type of coordinate is the ordered pair $(x, y)$, where $x$ is a measure of the perpendicular distance from the $y$-axis and $y$ is a measure of the perpendicular distance from the $x$-axis. For certain operations with complex numbers we will find a system involving polar coordinates to be much more useful.

In this system a point $P$ is located by its distance from a fixed point $O$ and the angle (measured in degrees or radians) that the line from $O$ to $P$ makes with a fixed line. The fixed point $O$ is called the **pole** and the fixed line is called the **polar axis.** The distance $OP = r$ is the **radius vector** and the angle $\theta$ is the polar angle, sometimes called the **amplitude.** The polar coordinates of the point $P$ are written $(r, \theta)$ (Figure 57.1).

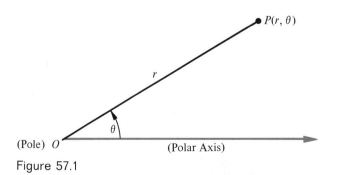

Figure 57.1

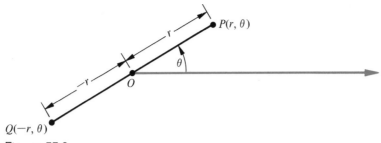

Figure 57.2

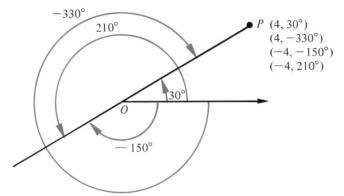

Figure 57.3

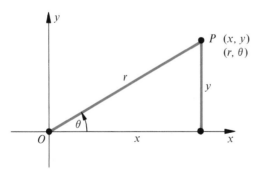

Figure 57.4

As in trigonometry, $\theta$ is positive when the measure is in a counterclockwise direction and negative when the measure is in a clockwise direction. $r$ may be positive or negative. When $r$ is positive, we measure from the pole to the point $P$ along the terminal side of the angle. If $r$ is negative, the distance is measured in the opposite direction, that is, along the terminal side extended through $O$ (Figure 57.2).

The polar form always yields a unique point, but the coordinates of a given point are not unique. We notice that $(2, 30°)$, $(2, 390°)$ and $(2, -330°)$ all represent the same point. In general, $(r, \theta)$ and $(r, \theta \pm 2k\pi)$, $k = 0, 1, 2, \ldots$, are just different ways of designating a single point. Figure 57.3 illustrates other possibilities.

The relationship between rectangular and polar coordinates is easily seen if we consider a point $P$ in terms of both sets of coordinates (Figure 57.4).

From Figure 57.4 we have the following sets of equations:

$$x = r \cos \theta$$
$$y = r \sin \theta \tag{57.1}$$

$$r = \sqrt{x^2 + y^2}$$
$$\tan \theta = \frac{y}{x} \quad \text{or} \quad \theta = \tan^{-1}\left(\frac{y}{x}\right) \tag{57.2}$$

The first set of equations enables us to convert polar coordinates to rectangular form; the second set yields a method of conversion from rectangular to polar coordinates. In finding $\theta$, we first determine the appropriate quadrant from the given point $(x, y)$.

**Example 57.1**
Convert $(2, 30°)$ and $(4, 225°)$ to rectangular form.

**Solution.** For $(2, 30°)$ we have $x = 2 \cos 30° = 2 \cdot \dfrac{\sqrt{3}}{2} = \sqrt{3}$ and $y = 2 \sin 30° = 2 \cdot \dfrac{1}{2} = 1$. Thus $(2, 30°) = (\sqrt{3}, 1)$. For $(4, 225°)$ we have $4 \cos 225° = 4 \cdot \left(-\dfrac{\sqrt{2}}{2}\right) = -2\sqrt{2}$ and $y = 4 \sin 225° = 4 \cdot \left(-\dfrac{\sqrt{2}}{2}\right) = -2\sqrt{2}$. Thus $(4, 225°) = (-2\sqrt{2}, -2\sqrt{2})$. ●

**Example 57.2**
Convert $(-1, \sqrt{3}$ to polar form.

**Solution.** $r = \sqrt{(-1)^2 + (\sqrt{3})^2} = \sqrt{1 + 3} = \sqrt{4} = 2$. To find $\theta$, we first note that $(-1, \sqrt{3})$ lies in the second quadrant. Since $\theta = \tan^{-1}\left(\dfrac{\sqrt{3}}{-1}\right)$ we have $\theta = 2\pi/3$. Thus $(-1, \sqrt{3}) = (2, 2\pi/3)$. Note that $(-1, \sqrt{3})$ is also equal to $\left(2, \dfrac{8\pi}{3}\right)$. That is, the polar form of the rectangular point $(-1, \sqrt{3})$ is *not* unique. ●

## EXERCISES FOR SECTION 57

**1.** Plot each of the following points, and describe each point by using a different pair of polar coordinates.

(a) $(1, 0)$          (d) $(-4, -30°)$

(b) $\left(0, \dfrac{\pi}{4}\right)$      (e) $(6, \pi)$

(c) $\left(-2, \dfrac{\pi}{3}\right)$     (f) $(2, 30°)$

**2.** Convert each of the following points to rectangular form.

(a) $\left(0, \dfrac{\pi}{4}\right)$        (d) $(-4, -30°)$

(b) $\left(0, \dfrac{\pi}{6}\right)$        (e) $(6, \pi)$

(c) $\left(-2, \dfrac{\pi}{3}\right)$        (f) $\left(2, \dfrac{4\pi}{3}\right)$

**3.** The distance between $P_1(x_1, y_1)$ and $P_2(x_2, y_2)$ is given by $d = \sqrt{(x_2 - x_1)^2 + (y_2 - y_1)^2}$. If the coordinates of $P_1$ and $P_2$ are $(r_1, \theta_1)$ and $(r_2, \theta_2)$, show that $d = \sqrt{r_1^2 + r_2^2 - 2r_1r_2 \cos(\theta_2 - \theta_1)}$.

**4.** Using the result in Exercise 3, find the distance between the following points.

(a) $P_1 = (1, 0)$    and   $P_2 = \left(1, \dfrac{\pi}{4}\right)$

(b) $P_1 = \left(2, \dfrac{\pi}{3}\right)$   and   $P_2 = \left(4, \dfrac{\pi}{3}\right)$

(c) $P_1 = \left(5, \dfrac{\pi}{4}\right)$   and   $P_2 = \left(5, \dfrac{3\pi}{4}\right)$

# SECTION 58
# THE ARGAND DIAGRAM: MULTIPLICATION AND DIVISION OF COMPLEX NUMBERS

In defining the complex numbers $x + yi$ we refer to $x$ as the real part and $y$ as the imaginary part. Since these real numbers are treated differently under algebraic operations, it is necessary to keep the order straight. In this context we may think of the complex numbers $x + yi$ as the ordered pair $(x, y)$. Thus the geometric representation of the complex number is either the point $P(x, y)$ in the $xy$-plane or the vector $\overrightarrow{OP}$ from the origin to $P$. The $x$-axis is called the **real axis** and the $y$-axis is the **imaginary axis.** The figures obtained are called **Argand diagrams** (Figure 58.1).

In terms of polar coordinates, we have seen the conversion equations $x = r \cos \theta$

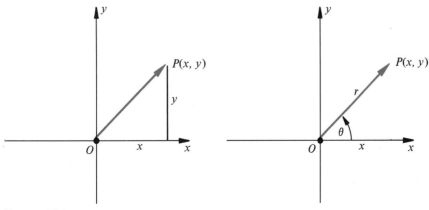

Figure 58.1

and $y = r \sin \theta$. Thus $x + yi = r(\cos \theta + i \sin \theta)$, which we express in the more convenient form $r(\operatorname{cis} \theta)$. $r(\operatorname{cis} \theta)$ is called the **trigonometric** or **polar form** of the complex number $x + yi$.

**Definition 58.1**   The **absolute value** or **modulus** of the complex number $x + yi$ is the length $r$ of the vector $\overrightarrow{OP}$ and is written $|x + yi|$. Thus $r = |x + yi| = \sqrt{x^2 + y^2}$, where $r$ is always real and nonnegative.

**Definition 58.2**   The angle $\theta$ between the positive $x$-axis and the vector $\overrightarrow{OP}$ is called the **argument** of $x + yi$. Since $\tan \theta = \dfrac{y}{x}$, $\theta = \tan^{-1}\left(\dfrac{y}{x}\right)$.

The modulus of a complex number is unique, but the argument is not. We recognize that any integral multiple of $2\pi$ may be added to $\theta$ to yield another appropriate angle.

**Example 58.1**
Convert $\sqrt{3} + i$ and $-1 + i$ to polar form.

Solution.   We represent $\sqrt{3} + i$ as the ordered pair $(\sqrt{3}, 1)$. The point lies in the first quadrant (Figure 58.2). For $\theta$ we have $\tan \theta = \dfrac{1}{\sqrt{3}}$ or $\theta = \pi/6$ and $r = \sqrt{3 + 1} = 2$. We now have $\sqrt{3} + i = 2 \operatorname{cis} \pi/6$. For $-1 + i$ we plot the point $(-1, 1)$ (Figure 58.3). $r = \sqrt{(-1)^2 + (1)^2} = \sqrt{2}$. To find $\theta$ we note that our angle lies in the second quadrant. If we can determine the reference angle ($\theta_1$), it becomes a simple matter to find $\theta$. But $\tan \theta_1 = 1/1$. (Why?) Thus $\theta_1 = \pi/4$ and $\theta = \pi - \pi/4 = 3\pi/4$. Therefore, $-1 + i = \sqrt{2} \operatorname{cis} 3\pi/4$.   ●

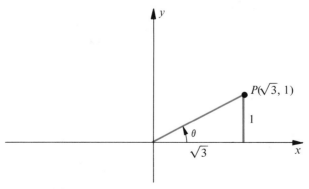

Figure 58.2

**Example 58.2**

Convert $4 \operatorname{cis} \dfrac{\pi}{6}$ and $2 \operatorname{cis} \dfrac{4\pi}{3}$ to rectangular form.

The geometric interpretation of the product and quotient of two complex numbers is shown in Figures 58.4 and 58.5.

Equation (58.1) tells us that to multiply two complex numbers in polar form, we multiply their absolute values and add their arguments.

Equation (58.3) tells us that to divide two complex numbers in polar form, we divide their absolute values and subtract their arguments in the appropriate order.

## EXERCISES FOR SECTION 58

Convert each of the following to polar form.

1. $4$
8. $\frac{1}{2} - \frac{1}{2}\sqrt{3}i$
15. $3\frac{\sqrt{3}}{2} + \frac{3}{2}i$

2. $-2$
9. $-8\sqrt{3} + 8i$
16. $-3 - 3i$

3. $3i$
10. $i$
17. $-3\frac{\sqrt{3}}{2} - \frac{3}{2}i$

4. $\sqrt{3} - i$
11. $-\frac{\sqrt{2}}{2} - \frac{\sqrt{2}}{2}i$
18. $\operatorname{cis} 0° + \operatorname{cis} 180°$

5. $1 + \sqrt{3}$
12. $-4i$
19. $2 \operatorname{cis} 90° + 5 \operatorname{cis} 180°$

6. $4 - 4\sqrt{3}i$
13. $8 - 8\sqrt{3}i$
20. $2 \operatorname{cis} 60° + 2 \operatorname{cis} 120°$

7. $-2 - 2i$
14. $3 + 4i$

Convert each of the following to rectangular form.

21. $\operatorname{cis} 0°$
31. $2 \operatorname{cis} 135°$

22. $\operatorname{cis} \frac{\pi}{3}$
32. $8 \operatorname{cis} 270°$

23. $4 \operatorname{cis} \frac{5\pi}{6}$
33. $-3 \operatorname{cis} 135°$

24. $2 \operatorname{cis} \frac{5\pi}{4}$
34. $4 \operatorname{cis} 130°$

25. $4 \operatorname{cis} \pi$
35. $4 \operatorname{cis} 120° + 2 \operatorname{cis} 30°$

26. $2 \operatorname{cis} \frac{\pi}{2}$
36. $6 \operatorname{cis} 135° - 4 \operatorname{cis} 45°$

27. $8 \operatorname{cis} \frac{\pi}{4}$
37. $\operatorname{cis} 180° + \operatorname{cis} 180°$

28. $\operatorname{cis} \frac{3\pi}{4}$
38. $4 \operatorname{cis} 270° + 2 \operatorname{cis} 90° - \operatorname{cis} 180°$

29. $\operatorname{cis}\left(-\frac{\pi}{3}\right)$
39. $4 \operatorname{cis} 270° + 2 \operatorname{cis} 90° + \operatorname{cis} 180°$

30. $\operatorname{cis} \frac{\pi}{6}$
40. $4 \operatorname{cis}(-60°) + 6 \operatorname{cis}(-90°)$

Find the indicated products and quotient.

**41.** $(\text{cis } 300°)(2 \text{ cis } 90°)$

**47.** $\dfrac{3 \text{ cis } 150°}{2 \text{ cis } 30°}$

**42.** $\left(2 \text{ cis } \dfrac{\pi}{3}\right)\left(4 \text{ cis } \dfrac{\pi}{4}\right)$

**48.** $\dfrac{2 \text{ cis } \dfrac{\pi}{6}}{5 \text{ cis } \dfrac{\pi}{4}}$

**43.** $\left(\text{cis}\left[-\dfrac{\pi}{6}\right]\right)(5 \text{ cis } 120°)$

**49.** $\dfrac{4 \text{ cis } \dfrac{\pi}{3}}{2 \text{ cis } \dfrac{4\pi}{3}}$

**44.** $\left(4 \text{ cis } \dfrac{\pi}{9}\right)\left(\dfrac{1}{2} \text{ cis } \dfrac{\pi}{6}\right)$

**50.** $\dfrac{\text{cis } 90°}{\text{cis } 120°}$

**45.** $\left(2 \text{ cis } \dfrac{4\pi}{3}\right)(\text{cis }[-2\pi])$

**51.** $\dfrac{(2 \text{ cis } 30°)(4 \text{ cis } 90°)}{8 \text{ cis } 60°}$

**46.** $(\text{cis } 180°)(\text{cis } 30°)(\text{cis } 45°)$

**52.** $\dfrac{(4 \text{ cis } 15°)(8 \text{ cis } 135°)}{6 \text{ cis } 90°}$

For exercises 53 through 56, convert each complex number to its polar form and perform the operations indicated.

**53.** $(2 + 2i)(1 - i)$

**55.** $\dfrac{1 + 2i}{2 - 3i}$

**54.** $(-8\sqrt{3} + 8i)(1 - \sqrt{3}i)$

**56.** $\dfrac{-3 + 3\sqrt{3}i}{\sqrt{3} + i}$

**57.** Show that for any complex number $z, |z| = \sqrt{z \cdot \bar{z}}$.

# SECTION **59**
## POWERS AND ROOTS OF COMPLEX NUMBERS: DE MOIVRE'S THEOREM

Suppose $z = r \text{ cis } \theta$. If $n$ is a positive integer, we may use our product formula to find $z^n$. In so doing, we obtain

$$z^n = (r \text{ cis } \theta)^n = \underbrace{\dfrac{(r \text{ cis } \theta \cdot r \text{ cis } \theta \cdot \cdots \cdot r \text{ cis } \theta)}{n \text{ times}}}$$

$$= \underbrace{\dfrac{r^n \text{ cis } (\theta + \theta + \cdots + \theta)}{n \text{ summands}}} = r^n \text{ cis } n\theta$$

This result is known as De Moivre's Theorem.

**Theorem 59.1  De Moivre's Theorem.**  $(r \text{ cis } \theta)^n = r^n \text{ cis } n\theta$, $n$ a positive integer.

We shall not prove the theorem here but illustrate its usefulness with the following examples.

**Example 59.1**
Evaluate $(1 + i)^6$.

Solution

Since $1 + i = \sqrt{2} \operatorname{cis} 45°$, we have

$$(1 + i)^6 = (\sqrt{2} \operatorname{cis} 45°)^6$$
$$= 2^3 \operatorname{cis} (6 \cdot 45°)$$
$$= 8 \operatorname{cis} 270°$$
$$= -8i \quad \bullet$$

**Example 59.2**
Evaluate $(-2\sqrt{3} + 2i)^3$.

Solution

Since $-2\sqrt{3} + 2i = 4 \operatorname{cis} 150°$, we have

$$(-2\sqrt{3} + 2i)^3 = (4 \operatorname{cis} 150°)^3 = 4^3 \operatorname{cis} (3 \cdot 150°)$$
$$= 64 \operatorname{cis} 450° = 64 \operatorname{cis} 90° = 64i \quad \bullet$$

We have noted that a single equation between two complex numbers always yields a pair of equations between real numbers. Using this fact and De Moivre's Theorem we may obtain formulas for $\cos n\theta$ and $\sin n\theta$ as polynomials of degree $n$ in $\cos \theta$ and $\sin \theta$.

**Example 59.3**
$(\cos \theta + i \sin \theta)^2 = \cos 2\theta + i \sin 2\theta$ according to De Moivre's Theorem. If we expand $(\cos \theta + i \sin \theta)^2$ by multiplying, we have $\cos^2 \theta - \sin^2 \theta + 2i \sin \theta \cos \theta$. Thus $\cos 2\theta + i \sin 2\theta = \cos^2 \theta - \sin^2 \theta + 2i \sin \theta \cos \theta$. Equating real and imaginary parts, we have $\cos 2\theta = \cos^2 \theta - \sin^2 \theta$ and $\sin 2\theta = 2 \sin \theta \cos \theta$. Compare these results with formulas 42.1 and 42.2.

We are now able to consider the problem of determining the roots of complex numbers. Recall that the $n$th root of a number $a$ is a number $b$ such that $b^n = a$. For example, the number 4 has two square roots, 2 and $-2$, since $2^2 = (-2)^2 = 4$. Similarly, $\operatorname{cis} 0°$, $\operatorname{cis} 120°$, and $\operatorname{cis} 240°$ are each cube roots of 1. This is easily verified by observing that $(\operatorname{cis} 0°)^3 = (\operatorname{cis} 120°)^3 = (\operatorname{cis} 240°)^3 = 1$.

In general, if $r \operatorname{cis} \theta$ is any complex number not equal to zero, and $n$ is any positive integer, there exists exactly $n$ different complex numbers each of which is an $n$th root of $r \operatorname{cis} \theta$.

To see this, let us suppose that $R \operatorname{cis} \alpha$ is an $n$th root of $r \operatorname{cis} \theta$. Then $(R \operatorname{cis} \alpha)^n = r \operatorname{cis} \theta$ and $R^n \operatorname{cis} (n \cdot \alpha) = r \operatorname{cis} \theta$.

Now $R^n = r$ or $R = r^{1/n}$, where $r^{1/n}$ is the unique real $n$th root of $r$.

In regard to the angles, we have noted that the argument of a complex number is not unique, for $r \operatorname{cis} \theta = r \operatorname{cis} (\theta + 2\pi k)$ for any integer $k$. Thus $n\alpha = \theta + 2\pi k$, $k$ an integer, and therefore $\alpha = \dfrac{\theta}{n} + \dfrac{2\pi k}{n}$.

We conclude, then, that all $n$th roots of $r \operatorname{cis} \theta$ are given by the equation

$$(r \operatorname{cis} \theta)^{1/n} = r^{1/n} \operatorname{cis}\left(\frac{\theta}{n} + \frac{2\pi k}{n}\right), \quad k \text{ an integer}$$

Further investigation will show that we need only take any $n$ consecutive values for $k$ to account for the $n$ distinct roots of $r \operatorname{cis} \theta$. For convenience we start with $k = 0$. Thus

$$(r \operatorname{cis} \theta)^{1/n} = r^{1/n} \operatorname{cis}\left(\frac{\theta}{n} + \frac{2\pi k}{n}\right), \quad k = 0, 1, 2, \cdots, n - 1 \qquad (59.1)$$

**Example 59.4**

Find the three cube roots of 1.

Solution.  $1 = 1 + 0i = 1 \operatorname{cis}(0 + 2\pi k)$. For the three cube roots of 1 we have $1^{1/3} \operatorname{cis}\left(\dfrac{0 + 2\pi k}{3}\right)$, $k = 0, 1, 2$. Thus

$$r_1 = 1 \operatorname{cis} 0 = 1$$

$$r_2 = 1 \operatorname{cis} \frac{2\pi}{3} = \cos \frac{2\pi}{3} + i \sin \frac{2\pi}{3} = -\frac{1}{2} + i\frac{\sqrt{3}}{2}$$

$$r_3 = 1 \operatorname{cis} \frac{4\pi}{3} = \cos \frac{4\pi}{3} + i \sin \frac{4\pi}{3} = -\frac{1}{2} - i\frac{\sqrt{3}}{2}$$

are the solutions we seek. The results may be checked by multiplication.  ●

The solutions found in Example 59.4 represent the three solutions of the equation $x^3 - 1 = 0$.

The cube roots of 1 lie on a circle with center at the origin and radius equal to 1 (Figure 59.1). In general, the $n$th roots of $r \operatorname{cis} \theta$ lie on a circle centered at the origin and having a radius equal to $r^{1/n}$. The roots are uniformly spaced about the circumference of the circle and are separated from each other by a central angle of $\dfrac{2\pi}{n}$.

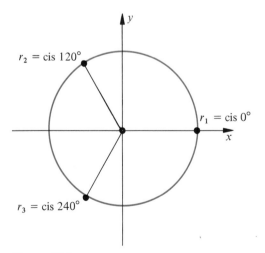

Figure 59.1

**Example 59.5**
Find the four fourth roots of $-8 + 8\sqrt{3}i$.

Solution. We first determine the polar form of our complex numbers. Here we have $r = \sqrt{256} = 16$ and $\theta = 120°$. Thus $-8 + 8\sqrt{3}i = 16$ cis $120°$. For our roots we have

$$(16 \text{ cis } 120°)^{1/4} = 16^{1/4} \text{ cis } \left(\frac{120°}{4} + \frac{k(360°)}{4}\right), \quad k = 0, 1, 2, 3 \qquad \textbf{(A)}$$

We note that $16^{1/4} = 2$ and we substitute the $k$ values into (A) to obtain

$r_1 = 2$ cis $30°$
$r_2 = 2$ cis $120°$
$r_3 = 2$ cis $210°$
$r_4 = 2$ cis $300°$

If we were to continue in this fashion, $r_5 = 2$ cis $390° = 2$ cis $30° = r_1$. We have accounted for all roots. The placement of the roots on the circumference of a circle of radius 2 is shown in Figure 59.2. ●

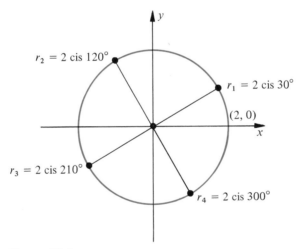

Figure 59.2

**Example 59.6**
Find all roots for $x^5 = 32$.

Solution. We are asked to find the five fifth roots of 32. This can be most easily done if we first convert $32 = 32 + 0i$ to polar form.

$32 + 0i = 32$ cis $0°$
$(32)^{1/5} = (32 \text{ cis } 0°)^{1/5}$

Using our formula we have

$$(32 \text{ cis } 0°)^{1/5} = 32^{1/5} \text{ cis } \left(\frac{0}{5} + \frac{2\pi k}{5}\right), \quad k = 0, 1, 2, 3, 4$$

Therefore,

$$r_1 = 2 \text{ cis } 0°$$
$$r_2 = 2 \text{ cis } 72°$$
$$r_3 = 2 \text{ cis } 144°$$
$$r_4 = 2 \text{ cis } 216°$$
$$r_5 = 2 \text{ cis } 288°$$

The roots are shown in Figure 59.3.  ●

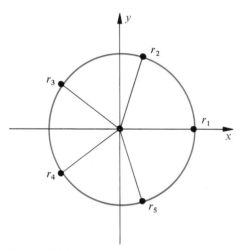

Figure 59.3

## EXERCISES FOR SECTION 59

Find each of the following powers and express the result in rectangular form.

**1.** $(2 \text{ cis } 60°)^5$

**2.** $(3 \text{ cis } 20°)^4$

**3.** $\left(\dfrac{1}{2} \text{ cis } \dfrac{\pi}{4}\right)^3$

**4.** $\left[2 \text{ cis } \left(-\dfrac{\pi}{5}\right)\right]^4$

**5.** $(1 + i)^3$

**6.** $(\sqrt{3} - i)^4$

**7.** $\left(-\dfrac{1}{2} - i\dfrac{\sqrt{3}}{2}\right)^3$

**8.** $\left(-\dfrac{1}{2} + i\dfrac{\sqrt{3}}{2}\right)^3$

**9.** $\left(\dfrac{1}{2} + i\dfrac{\sqrt{3}}{2}\right)^3$

**10.** $(\cos 10° + i \sin 10°)^6$

**11.** Use the procedure outlined in Example 59.3 to derive formulas for $\cos 3\theta$ and $\sin 3\theta$.

For each of the following, find all roots in polar form. If the angle is such that a table is not needed, express the result in rectangular form.

**12.** The cube roots of $-27i$.

**13.** The fourth roots of $-16$.

**14.** The square roots of $i$.

**15.** The square roots of $-i$.

**16.** The cube roots of $-8$.          **17.** The square roots of $i^5$.

**18.** The cube roots of $-4 + 4\sqrt{3}i$.

Find all roots of the following equations.

**19.** $x^3 - 8 = 0$          **23.** $x^2 - 1 - \sqrt{3}i = 0$

**20.** $x^4 - 16 = 0$          **24.** $x^5 + \dfrac{\sqrt{2}}{2} + \dfrac{\sqrt{2}}{2}i = 0$

**21.** $x^6 - 1 = 0$          **25.** $ix^3 - 1 = 0$

**22.** $4x^2 - 9 = 0$          **26.** $x^4 - 2x^2 + 4 = 0$

**27.** Show that

(a) $\cos\theta = \dfrac{1}{2}[\operatorname{cis}\theta + \operatorname{cis}(-\theta)]$

(b) $\sin\theta = \dfrac{1}{2i}[\operatorname{cis}\theta - \operatorname{cis}(-\theta)]$

**28.** Show that the reciprocals of the three cube roots of unity are also roots of unity.

**29.** Show that the reciprocal of any $n$th root of unity is also an $n$th root of unity.

# CHAPTER 10
# ELEMENTS OF ANALYTIC GEOMETRY

## SECTION 60
## THE STRAIGHT LINE

Our usual procedure in the discussion of functions and relations begins with a given equation. Using this equation as a base, we would then proceed to find the important geometric properties. We would now like to "reverse" this procedure and begin with a geometric description of a curve or its properties and attempt to determine the corresponding algebraic entity, the equation. We will, in effect, study these geometric quantities by using algebraic methods. In so doing we hope to discover some of the relationships between algebra and geometry and how one enriches the other.

Our study begins with the straight line. This would seem to be a logical choice, since, intuitively, the straight line is the simplest type of curve that we encounter. In Section 13 we discussed some of the properties of the straight line. We now begin a more detailed investigation of the straight line with the following definition.

**Definition 60.1**  Let $L$ be any line not parallel to the axes. The **inclination** of $L$ is the smallest angle ($\theta$) between the positive direction of the $x$-axis and $L$ measured in a counterclockwise direction.

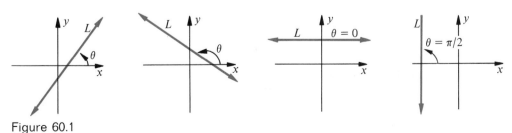

Figure 60.1

315

If $L$ is parallel to or coincides with the $x$-axis, its inclination is defined to be 0. If $L$ is parallel to or coincides with the $y$-axis, its inclination is 90°. In any event, $\theta$ always has the property that $0 \leq \theta < \pi$ (see Figure 60.1, page 315).

Another and sometimes more useful measure of the inclination of a line is its slope.

**Definition 60.2**    If $L$ is any line not parallel to the $y$-axis, then the **slope** ($m$) is given by $\tan \theta$, where $\theta$ is the angle of inclination of $L$.

**Remark 60.1**

The only lines for which the slope is not defined are those parallel to the $y$-axis. In this case $\theta = 90°$ and $\tan 90°$ is undefined.

If $L$ passes through the points $P_1(x_1, y_1)$ and $P_2(x_2, y_2)$, then

$$m = \tan \theta = \frac{y_2 - y_1}{x_2 - x_1} \quad \text{(See Figure 60.2)}$$

In many instances, $y_2 - y_1$ is denoted by $\Delta y$ (the change in $y$) and $x_2 - x_1$ is denoted by $\Delta x$ (the change in $x$). Thus we have

$$m = \frac{y_2 - y_1}{x_2 - x_1} = \frac{\Delta y}{\Delta x}$$

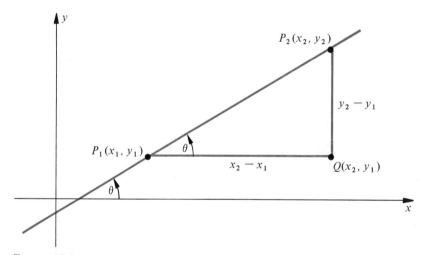

Figure 60.2

**Example 60.1**

Find the slope of the line passing through the points $(1, 3)$ and $(4, 5)$.

Solution

$$m = \frac{y_2 - y_1}{x_2 - x_1} = \frac{5 - 3}{4 - 1} = \frac{2}{3}$$

We obtain the same result if we change the order of subtraction. Thus

$$m = \frac{3 - 5}{1 - 4} = \frac{-2}{-3} = \frac{2}{3} \quad \bullet$$

The change in $y$ given by $(y_2 - y_1)$ is called the **rise** of the straight line, since it gives the vertical distance traveled in going from $P_1$ to $P_2$. Similarly, the change in $x$ given by $(x_2 - x_1)$ is called the **run** of the straight line, since it gives the horizontal distance traveled in going from $P_1$ to $P_2$. Thus

$$m = \frac{y_2 - y_1}{x_2 - x_1} = \frac{\Delta y}{\Delta x} = \frac{\text{rise}}{\text{run}}$$

We may interpret the rise over the run as the change in $y$ per unit change in $x$. For example, a slope of 2 means a rise of 2 units for each unit of run. A slope of $\frac{2}{3}$ means a rise of $\frac{2}{3}$ units for each unit of run. If the slope is negative, the line "falls" rather than rises as we move from left to right. A slope of $-3$ means a fall of 3 units for each unit of run (Figure 60.3).

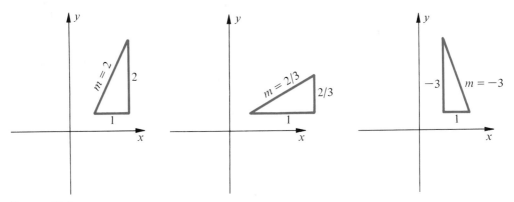

Figure 60.3

The value of the slope $m$ is independent of the choice of $P_1$ and $P_2$. To show this, suppose that we choose two different points on $L$. We denote these points by $P_3(x_3, y_3)$ and $P_4(x_4, y_4)$ (see Figure 60.4, page 318). From $P_1$ and $P_2$ we have $m = \dfrac{y_2 - y_1}{x_2 - x_1}$ and from $P_3$ and $P_4$ we have $m' = \dfrac{y_4 - y_3}{x_4 - x_3}$. In Figure 60.4 we see that triangle $P_1 P_2 Q$ is similar to triangle $P_3 P_4 Q'$. Since corresponding sides of similar triangles are proportional,

$$\frac{y_2 - y_1}{x_2 - x_1} = \frac{y_4 - y_3}{x_4 - x_3}$$

Thus $m = m'$.

### Example 60.2
Show that the points $(0, 1)$, $(2, 5)$, and $(-2, -3)$ lie on the same straight line.

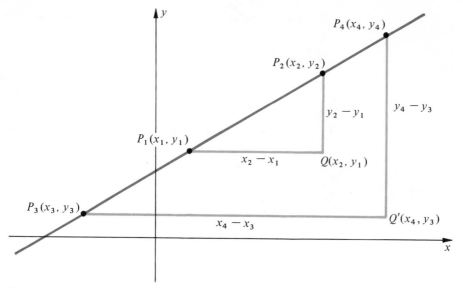

Figure 60.4

Solution.   The slope of the line from $(0, 1)$ to $(2, 5)$ is

$$\frac{5 - 1}{2 - 0} = \frac{4}{2} = 2$$

The slope of the line from $(2, 5)$ to $(-2, -3)$ is

$$\frac{-3 - 5}{-2 - 2} = \frac{-8}{-4} = 2$$

Since the slopes of the two lines are equal and they have a common point, the points lie on the same line.   ●

Any straight line is uniquely determined by two points. If the two points $P_1$ and $P_2$

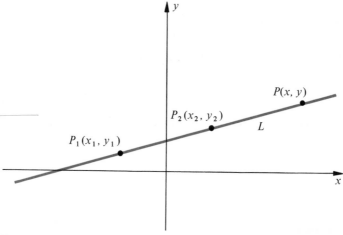

Figure 60.5

are such that the line segment $P_1P_2$ is not parallel to the $y$-axis, then we may determine the equation of the straight line by choosing a third point on the line and equating the slopes of line segments. This is possible since we have shown that the slope of a line is independent of the points selected. Since $P_1$, $P_2$, and $P$ lie on $L$, the slope of $\overline{P_1P_2}$ is equal to the slope of $\overline{P_1P}$ (see Figure 60.5). In terms of the coordinates of the points,

$$\frac{y - y_1}{x - x_1} = \frac{y_2 - y_1}{x_2 - x_1} \tag{60.1}$$

We call this the **two-point form** for the equation of a staight line.

### Example 60.3
Find the equation of the straight line passing through the points $(-1, 3)$ and $(2, 5)$.

Solution.   Using the two-point form (60.1) for the equation of a straight line, we have $\dfrac{y - 3}{x + 1} = \dfrac{2}{3}$. Although this is the equation of the line, we may write it a bit more concisely by taking the cross multiples and collecting all terms on the left of the equal sign. Thus $2x + 2 = 3y - 9$ or $2x - 3y + 11 = 0$. [*Note:* In using the two-point form we assumed that $P_1$ was $(-1, 3)$ and $P_2$ was $(2, 5)$. If we reverse the $P_1$ and $P_2$, we have $\dfrac{(y - 5)}{(x - 2)} = \dfrac{-2}{-3}$ or $-2x + 4 = -3y + 15$. We may write this as $-2x + 3y - 11 = 0$ or $2x - 3y + 11 = 0$, as before. The graph of this line is shown in Figure 60.6.] ●

A line is also uniquely determined if the slope $m$ and a point $P(x_1, y_1)$ are specified.

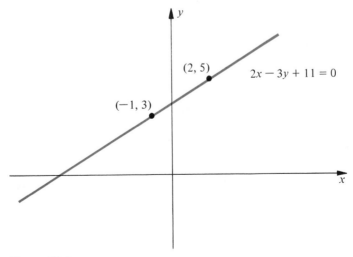

Figure 60.6

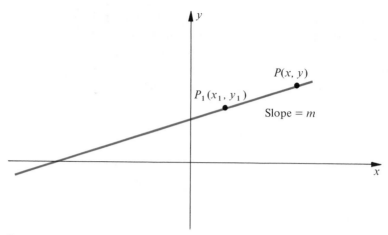

Figure 60.7

Again, we generate the equation of such a line by choosing an arbitrary point $P$ and equating the given slope $m$ with the slope of $\overline{P_1 P}$ (Figure 60.7). Thus

$$m = \frac{y - y_1}{x - x_1} \quad \text{or} \quad \boxed{(y - y_1) = m(x - x_1)} \tag{60.2}$$

This is called the **point-slope form** of the equation of a straight line.

### Example 60.4

Find the equation of the straight line passing through the point $(2, 3)$ and whose slope is $-\frac{1}{3}$.

Solution.  From our point-slope form (60.2) we have $y - 3 = -\frac{1}{3}(x - 2)$. Thus $3y - 9 = -x + 2$ or $x + 3y - 11 = 0$. The graph is shown in Figure 60.8.  ●

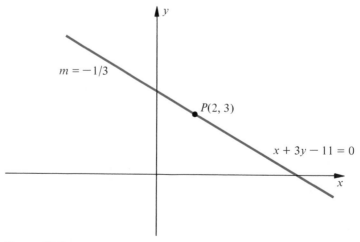

Figure 60.8

Now we consider a special case of the point-slope form. Suppose that $L$ passes through the $y$-axis. Then the $y$-intercept of $L$ will be of the form $(0, b)$. If $L$ has a slope of $m$, we have $y - b = m(x - 0)$, from Equation (60.2). Hence

$$y = mx + b \qquad (60.3)$$

This equation is called the **slope-intercept form** and has the following characteristics:

1. $y$ is given as a function of $x$.
2. The coefficient of $x$ is the slope of the line.
3. $b$ is the $y$ value of the intercept point on the $y$-axis.

### Example 60.5
Find the equation of a line with $m = 3$ and $y$-intercept at $(0, 4)$.

Solution. From (60.3) we have $y = mx + b$, where $m = 3$ and $b = 4$. Hence $y = 3x + 4$ is the equation we seek. ●

### Example 60.6
Find the slope and $y$-intercept of $2x + 3y - 5 = 0$.

Solution. We may write $2x + 3y - 5 = 0$ in slope-intercept form as $3y = -2x + 5$ or $y = -\frac{2}{3}x + \frac{5}{3}$. Hence $m = -\frac{2}{3}$ and the $y$-intercept is located at $(0, \frac{5}{3})$. ●

For any line parallel to the $y$-axis, $m$ will be undefined since $m = \dfrac{\Delta y}{\Delta x}$, and in this case $\Delta x = 0$.

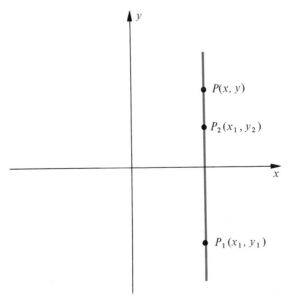

Figure 60.9

We can, however, find the equation of such a line. The equation we seek depends upon any two points that lie on the line. If $P_1(x_1, y_1)$ lies on $L$, then $P_2$ must have coordinates of the form $(x_1, y_2)$. Therefore, $P(x, y)$ is any point on the line if and only if $x = x_1$. Here we have an equation of a line parallel to the $y$-axis which is independent of $y$. Geometrically, this is accounted for by the fact that the $y$ value of $P$ has no restriction (see Figure 60.9, page 321).

Algebraically, the relation $x = x_1$ is the abbreviated form of the equation $0y + x = x_1$. Since the coefficient of $y$ is 0, $y$ may take on any real value.

### Example 60.7

Find the equation of the line parallel to the $y$-axis and passing through the point $(1, 6)$.

Solution. Since the line is parallel to the $y$-axis, its equation is of the form $x = x_1$. From the given point we have $x_1 = 1$, and thus $x = 1$ is the equation we seek. •

We have used the concept of slope to study various forms of the equation of a straight line. We now extend the use to cover parallel and perpendicular lines.

**Theorem 60.1**  If $L_1$ is parallel to $L_2$, then $m_1 = m_2$ or $m_1$ and $m_2$ are undefined.

The proof is immediate and is left to the exercises.

### Example 60.8

Find the equation of the line parallel to $2x - 3y + 4 = 0$ and passing through the point $(1, 5)$.

Solution. Since $2x - 3y + 4 = 0$ can be written as $y = \frac{2}{3}x + \frac{4}{3}$, the line we seek has a slope of $\frac{2}{3}$. From equation (60.2) we have $(y - 5) = \frac{2}{3}(x - 1)$ or, equivalently, $2x - 3y + 13 = 0$. •

### Example 60.9

The lines $x - 2y + 4 = 0$ and $3x - 6y + 1 = 0$ are parallel since they may be written as $y = \frac{1}{2}x + 2$ and $y = \frac{1}{2}x + 1/6$ and have equal slopes. •

**Theorem 60.2**  Suppose that $L_1$ and $L_2$ have slopes of $m_1$ and $m_2$, respectively. If $L_1$ is perpendicular to $L_2$ and the lines are not parallel to the axes, then $m_1 = -\dfrac{1}{m_2}$.

Proof. Let us suppose that the point of intersection of $L_1$ and $L_2$ lies on the $x$-axis as shown in Figure 60.10. From the figure we see that $B + \alpha_1 = 90°$; that is, $\alpha_1$ and $B$ are complementary angles. We have, therefore, $\tan \alpha_1 = \tan(90° - B) = \cot B = \dfrac{1}{\tan B}$. But $B = (180° - \alpha_2)$ (Figure 60.10). Hence

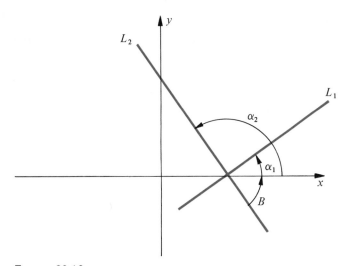

Figure 60.10

$$\frac{1}{\tan B} = \frac{1}{\tan (180° - \alpha_2)} = -\frac{1}{\tan \alpha_2}$$

Thus we have $\tan \alpha_1 = -\dfrac{1}{\tan \alpha_2}$. Since $\tan \alpha_1 = m_1$ and $\tan \alpha_2 = m_2$, we

have $m_1 = -\dfrac{1}{m_2}$ or $m_1 m_2 = -1$  ●

### Example 60.10
Find the equation of the line perpendicular to $x - 2y + 3 = 0$ and passing through the point $(2, -1)$.

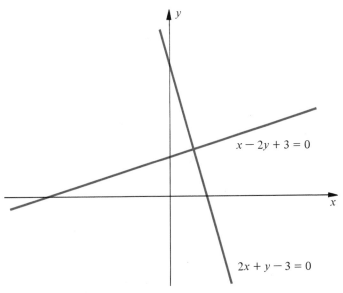

Figure 60.11

**Solution.** Since $x - 2y + 3 = 0$ is equivalent to $y = \frac{1}{2}x + \frac{3}{2}$, the line we seek has a slope of $-2$. Employing Equation (60.2) we have $(y + 1) = -2(x - 2)$ or $2x + y - 3 = 0$. The lines are shown in Figure 60.11, page 323. ●

The equation $Ax + By + C = 0$, where $A$, $B$, and $C$ are constants and both $A$ and $B$ are not equal to 0, is the **general form** of a first-degree equation in $x$ and $y$. It is called a **linear equation** because its graph is always a straight line. This can be shown as follows. Suppose that $B \neq 0$. We solve $Ax + By + C = 0$ for $y$ and obtain $y = -\dfrac{A}{B}x - \dfrac{C}{B}$. This is the equation of a straight line in slope-intercept form where $m = -\dfrac{A}{B}$ and $b = -\dfrac{C}{B}$. If $B = 0$ and $A \neq 0$, we have $Ax + C = 0$ or $x = -\dfrac{C}{A}$. We have seen that an equation of this form is the graph of a straight line parallel to the $y$-axis.

### Example 60.11

Find the distance from the point $P(-3, 5)$ to the line $L$, $x - y + 2 = 0$.

**Solution.** The distance we wish to find is the length of segment $\overline{PQ}$ shown in Figure 60.12. We use the following steps to find this length.

*Step 1:* We find the equation of line, $L_1$, that is perpendicular to $L$ and passes through the point $P(-3, 5)$ by noting that the slope of $L_1$ is the negative reciprocal or $L$. The slope of $L$ is 1. Thus the slope of $L_1$ is $-1$. We can now apply equation (60.2) to obtain the equation of $L_1$ as follows

$$(y - 5) = -1(x + 3) \text{ or } x + y - 2 = 0$$

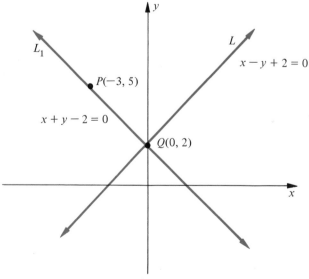

Figure 60.12

*Step 2:* The point of intersection of the lines $L$ and $L_1$ (denoted as $Q$ in Figure 60.12) can be found by solving $x - y + 2 = 0$ and $x + y - 2 = 0$ simultaneously. In so doing we have $x = 0$ and $y = 2$.

*Step 3:* The desired distance is the distance between $P(-3, 5)$ and $Q(0, 2)$. We find this distance by applying the distance formula as follows

$$\overline{PQ} = \sqrt{(0 + 3)^2 + (2 - 5)^2} = \sqrt{18} = 3\sqrt{2}. \quad \bullet$$

The distance from a point to a line is given generally by the following theorem.

**Theorem 60.3** The distance from the point $P_1(x_1, y_1)$ to the line $L$ whose equation is $Ax + By + C = 0$ is given by

$$\left| \frac{Ax_1 + By_1 + C}{\sqrt{A^2 + B^2}} \right| \tag{60.4}$$

The proof of Theorem 60.3 is outlined in the exercises.

The distance found in Example 60.10 can be checked by using formula (60.4). From the point $P$ we have $x_1 = -3$ and $y_1 = 5$ and from the equation for $L$ we have $A = 1$, $B = -1$, and $C = 2$. Thus

$$d = \left| \frac{1 \cdot (-3) + (-1)(5) + 2}{\sqrt{1^2 + (-1)^2}} \right| = \left| \frac{-6}{\sqrt{2}} \right| = \frac{6}{\sqrt{2}} = 3\sqrt{2}$$

## EXERCISES FOR SECTION 60

In Exercises 1–8, find the slope (if it exists) of the line through the given points. Draw the line.

1. $(2, 3)$, $(1, 5)$      5. $(4, 2)$, $(-5, -1)$

2. $(5, 2)$, $(7, 2)$      6. $(0, 0)$, $(1, 1)$

3. $(0, 0)$, $(4, 1)$      7. $(1.3, 1.2)$, $(2.5, 4.6)$

4. $(-2, 2)$, $(3, 3)$      8. $(3, 5)$, $(3, 7)$

In Exercises 9–18, find an equation of the line that satisfies the given conditions.

9. Through the points given in Exercises 1–8.

10. Through the point $(1, 2)$ and parallel to the $x$-axis.

11. Through the point $(1, 2)$ and perpendicular to the $x$-axis.

12. Through the point $(2, 3)$ and having a slope of $-1$.

13. Through the point $(1, 4)$ and parallel to the line $x + 2y - 4 = 0$.

14. Through the point $(0, 0)$ and parallel to the line $y = 2x - 3$.

15. The $x$-intercept is $-2$ and the $y$-intercept is 5.

16. The $x$-intercept is $-2$ and the slope is 4.

17. Through the point $(2, 5)$ and perpendicular to $x + y - 4 = 0$.

18. Through the point $(-2, -1)$ and perpendicular to $2x - 3y + 6 = 0$.

19. Find the equation of the three sides of the triangle with vertices $(1, 4)$, $(3, 0)$, and $(-2, 3)$.

20. Use slopes to show that the four points $(3, 1)$, $(4, 3)$, $(2, 4)$, and $(1, 2)$ are vertices of a rectangle.

21. Find the equation of the perpendicular bisector of the line segment joining $(4, 2)$ and $(-2, 6)$.

22. Find the area of the triangle formed by the lines $4x - 7y - 17 = 0$, $3x - 2y - 16 = 0$, and $2x + 3y + 11 = 0$.

23. Find the perimeter of the triangle formed in Exercise 22.

24. Graph $x^2 - y^2 = 0$.        25. Graph $xy = 0$.

26. Show that the points $(0, 2)$, $(2, 6)$, and $(-3, -4)$ are collinear by showing that the third point satisfies an equation for the line through the other two.

27. Use formula (60.4) to find the distance from $P$ to $L$ for each of the following.
    (a) $P(1, 3)$ and $L : 2x + y - 4 = 0$       (d) $P(1, 3)$ and $L : x = 5$
    (b) $P(2, 5)$ and $L : x - y + 3 = 0$       (e) $P(2, 5)$ and $L : y = 7$
    (c) $P(0, 0)$ and $L : x + y - 4 = 0$

28. For which values of $a$ are $ax - 2y + 8 = 0$ and $3ax + 6y + 4 = 0$ the equations of perpendicular lines?

29. Suppose we are given the lines $L_1$ and $L_2$ as follows:
    $L_1 : a_1 x + b_1 y + c_1 = 0$     $L_2 : a_2 x + b_2 y + c_2 = 0$
    (a) Show that $L_1$ and $L_2$ are parallel or coincidental if and only if there exists a number $k \neq 0$ such that $a_2 = ka_1$ and $b_2 = kb_1$.
    (b) Show that $L_1$ and $L_2$ are perpendicular if and only if $a_1 a_2 + b_1 b_2 = 0$.

30. Prove Theorem 60.1.

31. An outline of the proof of Theorem 60.3 proceeds as follows. Suppose that $B \neq 0$ in

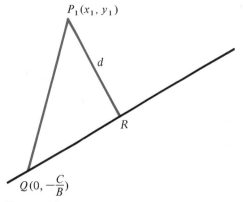

Figure 60.13

$Ax + By + C = 0$. Then a point on $L$ is $Q(0, -C/B)$ (Figure 60.13). The distance is found most easily in terms of vectors.

$$d = \text{projection of } \overrightarrow{QP_1} \text{ on } \overrightarrow{RP_1}$$

$$\vec{QP} = (x_1 - 0)\vec{i} + \left(y_1 + \frac{C}{B}\right)\vec{j}$$

Since $\vec{RP_1}$ is perpendicular to $L$, its slope is $\dfrac{B}{A}$. Thus $\vec{RP_1}$ may be written as $A\vec{i} + B\vec{j}$. (Why?)

Therefore

$$d = \text{projection of } \vec{QP_1} \text{ on } \vec{RP_1} = \left| \frac{\vec{QP_1} \cdot \vec{RP_1}}{|\vec{RP_1}|} \right| = \left| \frac{Ax_1 + By_1 + C}{\sqrt{A^2 + B^2}} \right|$$

The case where $B = 0$ is left to the reader.

32. Suppose that line $L$ has slope $m$. Show that $\vec{v} = c(\vec{i} + m\vec{j})$, $c$ any constant $\neq 0$, is a vector with the same direction as $L$.

33. Find the cosine of the angle between the given lines.
    (a) $L_1 : x + 2y - 3 = 0$; $L_2 : 2x - 3y + 4 = 0$
    (b) $L_1 : 2x + y - 4 = 0$; $L_2 : x - 2y + 6 = 0$
    (c) $L_1 : x = 0$; $L_2 : x - y = 0$

34. Use vector analysis to show that the midpoint of the line segment with endpoints $P_1(x_1, y_1)$ and $P_2(x_2, y_2)$ is given by $P\left(\dfrac{x_1 + x_2}{2}, \dfrac{y_1 + y_2}{2}\right)$. Compare this to Exercise 1 in section 10.

35. Use vector analysis to find the length of the medians of the triangle formed by $A(2, 2)$, $B(4, 6)$, and $C(5, 0)$.

# SECTION 61
## CONIC SECTIONS

We will continue our discussion of the relationship between algebra and geometry by studying types of curves called conic sections. **Conic sections** are the curves obtained by intersecting a plane with a right circular cone. The cone is formed by rotating a given line about a line that intersects it (Figure 61.1).

Figure 61.1

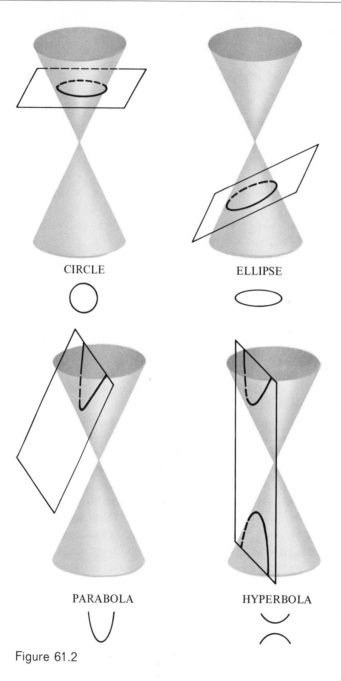

CIRCLE

ELLIPSE

PARABOLA

HYPERBOLA

Figure 61.2

There are exactly four ways in which a plane can intersect the cone if the plane does not contain the point of intersection. The curves thus obtained are the circle, ellipse, parabola, and hyperbola. These four cases are shown in Figure 61.2.

It is our purpose to study these curves by means of their definitions and equations. As we shall see, every second-degree equation that does not factor and whose graph consists of more than one point must represent a conic section. With a bit of investigation we shall soon learn how to recognize the curve from the given equation.

# SECTION 62
## THE CIRCLE

**Definition 62.1** A **circle** is the set of all points in a plane at a given distance from a fixed point. The fixed point is called the **center** of the circle and the measure of the given distance is called the **radius** of the circle.

If the center $C$ is at the point $(h, k)$ and the radius is $r$ (Figure 62.1), the equation is easily determined by means of the distance formula. The point $P(x, y)$ lies on the circle if and only if $\overline{CP} = r$ or, equivalently, if and only if $\sqrt{(x - h)^2 + (y - k)^2} = r$. (See Figure 62.2.) Squaring both sides yields the more convenient form of the equation of the circle.

$$(x - h)^2 + (y - k)^2 = r^2 \tag{62.1}$$

If the center of the circle is at the origin, then $h = k = 0$ and (62.1) reduces to

$$x^2 + y^2 = r^2 \tag{62.2}$$

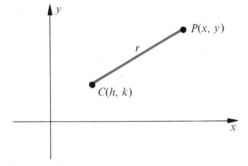

Figure 62.1                              Figure 62.2

### Example 62.1
Find the equation of the circle with center at $(2, -1)$ and radius equal to 5.

Solution. Here $h = 2, k = -1$, and $r = 5$. Substituting into (62.1) we have

$$(x - 2)^2 + [y - (-1)]^2 = 5^2 \quad \text{or} \quad (x - 2)^2 + (y + 1)^2 = 25$$

Squaring yields

$$x^2 - 4x + 4 + y^2 + 2y + 1 = 25 \quad \text{or} \quad x^2 + y^2 - 4x + 2y - 20 = 0$$

The circle is shown in Figure 62.3 on page 330. •

### Example 62.2
Find the center and radius of the circle whose equation is $x^2 + y^2 + 4x - 6y - 3 = 0$.

Solution. If we are able to rewrite the equation in the form of equation

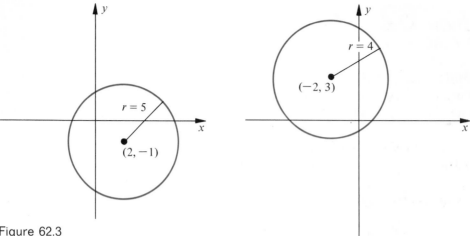

Figure 62.3

Figure 62.4

(62.1) we will be able to read the coordinates of the center and the radius directly. To accomplish this, we must complete the square. We first rewrite

$$x^2 + y^2 + 4x - 6y - 3 = 0 \text{ as } (x^2 + 4x) + (y^2 - 6y) = 3 \qquad \textbf{(A)}$$

To complete the square for the $x$ terms, we take half the $x$ coefficient and square, getting $(\frac{4}{2})^2 = 4$. Similarly, for the $y$ terms we have $(-\frac{6}{2})^2 = 9$. Adding these numbers to both sides of (A) yields $(x^2 + 4x + 4) + (y^2 - 6x + 9) = 3 + 4 + 9$. Writing the left-hand side in factored form, we have $(x + 2)^2 + (y - 3)^2 = 16$. Comparing this with equation (62.1) we see that the center is at $(-2, 3)$ and the radius is 4. This information enables us to sketch the circle in Figure 62.4. ●

**Example 62.3**
Find the equation of the circle if the point $P(1, 2)$ and $Q(3, 8)$ are the endpoints of one of its diameters.

Solution.  The center of the circle is the midpoint of $\overline{PQ}$. Hence $C$ is at $(2, 5)$. The radius is one-half the diameter. Thus $r = \frac{1}{2}\overline{PQ} = \frac{1}{2}\sqrt{4 + 36} = \frac{1}{2}\sqrt{40} = \sqrt{10}$. Therefore, the equation of the circle is

$$(x - 2)^2 + (y - 5)^2 = 10. \quad ●$$

**Example 62.4**
Find the equation of the circle passing through the points $P(5, 3)$, $Q(6, 2)$, and $R(3, -1)$.

Solution.  The equation of the circle is given by $(x - h)^2 + (y - k)^2 = r^2$, where we must determine the three constants $h$, $k$, and $r$. Three conditions are necessary to determine these values. Since the circle must pass through the given points, $h, k,$ and $r$ may be determined by substituting the coordinates of the points for $x$ and $y$ and solving simultaneously the three resulting equations. Since $P$ lies on the circle we have

$$(5 - h)^2 + (3 - k)^2 = r^2 \quad \text{or} \quad 34 - 10h + h^2 - 6k + k^2 = r^2$$

Similarly, for $Q$ we have

$$(6 - h)^2 + (2 - k)^2 = r^2 \quad \text{or} \quad 40 - 12h + h^2 - 4k + k^2 = r^2$$

and for $R$,

$$(3 - h)^2 + (-1 - k)^2 = r^2 \quad \text{or} \quad 10 - 6h + h^2 + 2k + k^2 = r^2$$

We now have three equations in three unknowns which we solve using the techniques of section 14.

$$34 - 10h + h^2 - 6k + k^2 = r^2 \tag{a}$$
$$40 - 12h + h^2 - 4k + k^2 = r^2 \tag{b}$$
$$10 - 6h + h^2 + 2k + k^2 = r^2 \tag{c}$$

Subtracting (c) from (b) we obtain $30 - 6h - 6k = 0$, and subtracting (b) from (a) we obtain $-6 + 2h - 2k = 0$. If we solve these simultaneously, we find that $h = 4$ and $k = 1$. $r = \sqrt{5}$ and the equation we seek is

$$(x - 4)^2 + (y - 1)^2 = 5 \quad \bullet$$

It is important for us to recognize that not all equations of the form we have shown represent a circle. Such equations may represent a point or the null set. For example, completing the square for $x^2 + y^2 + 6x - 2y + 10 = 0$ yields

$$(x^2 + 6x + 9) + (y^2 - 2y + 1) = -10 + 9 + 1 = 0$$

or

$$(x + 3)^2 + (y - 1)^2 = 0$$

This equation determines the single point $(-3, 1)$. (Why?)

Had we began with $x^2 + y^2 + 6x - 2y + 11 = 0$, we would have found that

$$(x^2 + 6x + 9) + (y^2 - 2y + 1) = -11 + 9 + 1 \quad \text{or} \quad (x + 3)^2 + (y - 1)^2 = -1$$

Since the left-hand side must be nonnegative for all $(x, y)$ and the right-hand side is negative, the solution set for the equation is empty; that is, there are no points that satisfy the equation.

The last two examples are sometimes referred to as **degenerate cases of the circle.**

## EXERCISES FOR SECTION 62

In Exercises 1–13, find the equation of the circle given that

**1.** $r = 2$, $C(1, 3)$      **5.** $r = 5$, $C(\sqrt{2}, \sqrt{3})$

**2.** $r = 5$, $C(-2, 4)$      **6.** $r = 3$, $C(0, 2)$

**3.** $r = 1$, $C(0, 0)$      **7.** $r = 3$, $C(2, 0)$

**4.** $r = 10$, $C(-1, -1)$      **8.** $r = a$, $C(h, k)$

**9.** $C(1, 2)$, $P(3, 5)$, where $P$ lies on the circle.

**10.** $C(-2, 5)$, $P(-1, 3)$, where $P$ lies on the circle.

**11.** $C(2, 7)$, $P(2, 1)$, where $P$ lies on the circle.

**12.** $P(1, 3)$ and $Q(4, 7)$ are the endpoints of a diameter.

**13.** $P(-2, -1)$ and $Q(-2, 6)$ are the endpoints of a diameter.

Consider the equations in Exercises 14–25. If the given equation is a circle, find the radius and the coordinates of the center, and sketch. If the equation is satisfied by a single point or no points, so state.

**14.** $x^2 + y^2 - 4x + 2y - 20 = 0$      **20.** $x^2 + y^2 = 0$

**15.** $x^2 + y^2 - 6x - 8y - 10 = 0$      **21.** $x^2 + y^2 + 4x - 12 = 0$

**16.** $x^2 + y^2 - 6x - 8y = 0$      **22.** $x^2 + y^2 + 4y - 12 = 0$

**17.** $2x^2 + 2y^2 + 4x - 8y - 6 = 0$      **23.** $x^2 + y^2 + 4y + 12 = 0$

**18.** $x^2 + y^2 + 8x - 4y + 4 = 0$      **24.** $x^2 + y^2 + 4x + 12 = 0$

**19.** $3x^2 + 3y^2 - 4x + 2y + 6 = 0$      **25.** $x^2 + y^2 + 4x + 4 = 0$

**26.** Find the equation of the circle passing through the points

    (a) $P(1, 3)$, $Q(-8, 0)$, and $R(0, 6)$
    (b) $P(1, -1)$, $Q(1, 8)$, and $R(-3, 3)$
    (c) $P(7, 1)$, $Q(6, 2)$, and $R(-1, -5)$
    (d) $P(1, 2)$, $Q(3, 1)$, and $R(-3, -1)$
    (e) $P(0, 0)$, $Q(4, 0)$, and $R(0, 4)$

**27.** Find the equations of the circle passing through $P(0, 0)$ and $Q(16, 0)$ with $r = 10$.

**28.** Find the equations of the circle passing through $P(1, 8)$ and $Q(-6, 7)$ with $r = 5$.

**29.** Find the equation of the circle concentric with $x^2 + y^2 - 2x - 4y - 4 = 0$ and passing through $P(4, 7)$.

**30.** Find the equation of the circle concentric with $x^2 + y^2 + 4x + 6y - 21 = 0$ and passing through $(-2, 4)$.

**31.** Find the equation of a circle whose center is at $(7, -2)$ and is tangent to the line $5x + 12y + 15 = 0$.

**32.** Find the equation of a circle whose center is at $(3, 2)$ and is tangent to the line $3x + 4y = 12$.

**33.** Find the equation of a circle that passes through the intersection of the lines $3x + y - 7 = 0$ and $x - 2y = 0$ and has the same center as $x^2 - 10x + y^2 - 4y - 33 = 0$.

**34.** Find the solution set for the following system of inequalities.

$$x^2 + y^2 - 4x - 2y - 11 \geq 0$$
$$2x^2 + 2y^2 - 8x - 4y - 40 \leq 0$$

**35.** Find the equation of all circles with

    (a) Center on the $x$-axis and $r = 5$.
    (b) Center on the $y$-axis and $r = 2$.
    (c) Center at the origin and $r = a$.

**36.** Show that if $b^2 + c^2 > 4ad$ and $a \neq 0$, the equation $ax^2 + ay^2 + bx + cy + d = 0$ defines a circle with center at

$$\left( \frac{-b}{2a}, \frac{-c}{2a} \right) \quad \text{and} \quad r = \sqrt{\frac{b^2 + c^2 - 4ad}{4a^2}}$$

# SECTION 63
## TRANSLATION OF AXES

The shape of any curve is not affected by the position of the coordinate axes. In many cases the choice of a new coordinate axes may simplify the solution of a problem. Since the placement of the axes is arbitrary, we might prefer to move them. This type of movement is called a **transformation of axes.** As we shall see, a transformation enables us to change an equation or algebraic expression by substituting for the given variables their values in terms of another set of variables.

One of the most useful transformations is a translation. A **translation of axes** is simply a new set of coordinate axes $(x', y')$ chosen parallel to the $x$- and the $y$-axis. The geometric effect is the movement of the origin $O$ to the point $O'(h, k)$ (Figure 63.1).

With the translation of axes, each point in the plane has two sets of coordinates: $(x, y)$ and $(x', y')$. This relationship is shown in Figure 63.2 on page 334.

Suppose that $P$ is any point in the plane. Then its coordinates are $(x, y)$ in terms of the given axes with origin at $O$ and $(x', y')$ in terms of the new axes drawn with origin at $(h, k)$. To determine $x$ and $y$ in terms of $x', y', h$, and $k$ we observe from Figure 63.2 that

$$x = \overline{QP} = \overline{QQ'} + \overline{Q'P} = h + x'$$

Similarly,

$$y = \overline{RP} = \overline{RR'} + \overline{R'P} = k + y'$$

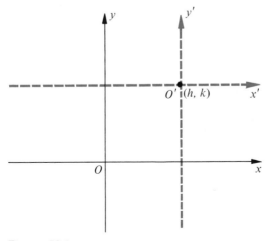

Figure 63.1

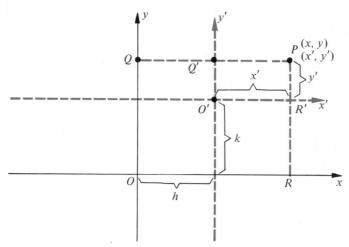

Figure 63.2

Thus

$$x = x' + h \quad \text{or} \quad x' = x - h$$

$$y = y' + k \quad \text{or} \quad y' = y - k$$

(63.1)

These are called the **equations of translation.**

We can translate any equation in $x$ and $y$ to an equation in $x'$ and $y'$ by replacing $x$ by $x' + h$ and $y$ by $y' + h$. We must be careful to note that we have changed only the equation. The graph of the equation in $x$ and $y$ is exactly the same as the graph of the corresponding equation in $x'$ and $y'$.

### Example 63.1

The equation of a circle takes on its simplest form when the center is at the origin [see (62.2)]. Let us consider the circle of a radius $r$ with center at $(h, k)$. The equation of this circle is given by $(x - h)^2 + (y - k)^2 = r^2$. A translation of axes with $O'$ at $(h, k)$ will locate the center of the circle at $O'$ (Figure 63.3). The equation of the circle in terms of $x'$ and $y'$ is obtained by replacing $x$ by $x' + h$ and $y$ by $y' + k$ or, equivalently, by replacing $x - h$ by $x'$ and $y - k$ by $y'$. Thus our new equation is $x'^2 + y'^2 = r^2$.

### Example 63.2

Determine a translation of axes that will transform the equation $x^2 + y^2 + 6x - 2y - 6 = 0$ into an equation without first-degree terms.

Solution.    Since we do not know which translation will work, we substitute for $x$ and $y$ the values $x' + h$ and $y' + k$.

$$(x' + h)^2 + (y' + k)^2 + 6(x' + h) - 2(y' + k) - 6 = 0 \qquad \textbf{(A)}$$

Expanding,

$$x'^2 + 2hx' + h^2 + y'^2 + 2ky' + k^2 + 6x' + 6h - 2y' - 2k - 6 = 0$$

Collecting the coefficients of the powers of $x'$ and $y'$ yields

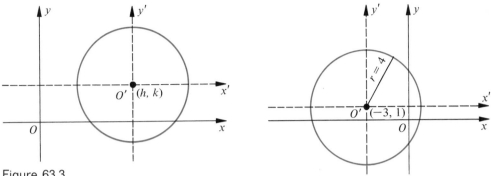

Figure 63.3

Figure 63.4

$$x'^2 + y'^2 + (2h + 6)x' + (2k - 2)y' + h^2 + k^2 + 6h - 2k - 6 = 0$$

In order to eliminate the 1st degree terms involving $x'$ and $y'$ we must have $2h + 6 = 0$ and $2k - 2 = 0$. From these we obtain $h = -3$ and $k = 1$. Substituting into (A), our equation becomes $x'^2 + y'^2 - 16 = 0$. Our translation is seen to be a new coordinate axes with center at $(-3, 1)$ (Figure 63.4).  •

The procedure outlined in Example 63.2 will always work. However, the method of completing the square is usually more efficient. To see this, let us again examine the equation $x^2 + y^2 + 6x - 2y - 6 = 0$. Completing the square yields $(x^2 + 6x + 9) + (y^2 - 2y + 1) = 6 + 9 + 1$  or  $(x + 3)^2 + (y - 1)^2 = 16$. Letting $x' = x + 3$ and $y' = y - 1$ we have $x'^2 + y'^2 = 16$ as before. The preferred method is obvious.

The only drawback to completing the square is that not every equation is such that we can complete the square. We illustrate this in the following example.

### Example 63.3
Find a translation of axes that will remove the 1st degree term from $xy - 4y = 3$.

Solution.   Setting $x = x' + h$ and $y = y' + k$ we have

$$(x' + h)(y' + k) - 4(y' + k) = 3$$
$$x'y' + kx' + hy' + hk - 4y' - 4k = 3$$

Collecting terms, $x'y' + kx' + (h - 4)y' + hk - 4k = 3$. In order to remove the 1st degree terms we now set $k = 0$ and $h = 4$ and our equation takes the form $x'y' = 3$. This equation is easily sketched if we draw an $x'y'$-coordinate axes whose origin is at the point $(4, 0)$ (why?) in the $xy$-plane. This graph is shown in Figure 63.5, page 336. Compare this example with Example 30.8.  •

### Example 63.4
Given the equation $4x^2 + y^2 + 8x + 10y + 13 = 0$, find an equation of the graph with respect to the $x$ and $y$ axes after a translation of axes to the new origin $(-1, -5)$.

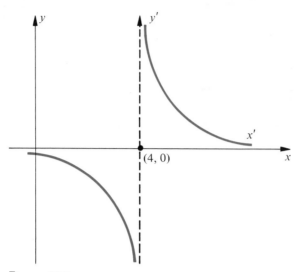

Figure 63.5

Solution. A point $P$ represented by $(x, y)$ with respect to the old axis is represented by $(x', y')$ with respect to the new axes. We apply the equations in (63.1) when $h = -1$ and $k = -5$ to obtain

$x = x' - 1$ and $y = y' - 5$

We now substitute these values into the given equation to obtain

$4(x' - 1)^2 + (y' - 5)^2 + 8(x' - 1) + 10(y' - 5) + 13 = 0$

or

$4x'^2 - 8x' + 4 + y'^2 - 10y' + 25 + 8x' - 8 + 10y' - 50 + 13 = 0$

or

$4x'^2 + y'^2 = 16$

or

$\dfrac{x'^2}{4} + \dfrac{y'^2}{16} = 1$  •

## EXERCISES FOR SECTION 63

In Exercises 1 through 10, determine the translation equations and the new origin that enables us to transform the first equation into the second.

1. $y = (x - 1)^2$; $y' = x'^2$

2. $(x - 3) = (y + 1)^2$; $x' = y'^2$

3. $(y - 1)^2 = (x + 3)^3$; $y'^2 = x'^3$

4. $(x - 1)^2 - (y + 2)^2 = 4$; $x'^2 - y'^2 = 4$

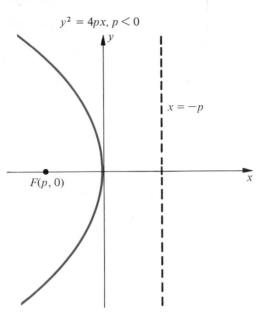

$y^2 = 4px, p < 0$

$x = -p$

$F(p, 0)$

Figure 64.3

$$\sqrt{(x - p)^2 + y^2} = |x + p|$$

Squaring both sides yields

$$(x - p)^2 + y^2 = (x + p)^2$$

Expanding and simplifying we have

$$x^2 - 2px + p^2 + y^2 = x^2 + 2px + p^2$$

or

$$y^2 = 4px \tag{64.1}$$

The line through the focal point drawn perpendicular to the directrix is called the **axis of symmetry** or the **axis of the parabola.** The point of intersection of this line and the curve is called the **vertex.** In this case the vertex is at $(0, 0)$. Note that the graph is symmetric with respect to the $x$-axis and opens to the right if $p > 0$. For $p < 0$ the graph is shown in Figure 64.3. The vertex is at $(0, 0)$, the axis of symmetry is the $x$-axis, and the curve opens to the left.

The analysis for the cases where the focal point lies on the $y$-axis and the directrix line is perpendicular to the $y$-axis does not differ significantly from what we have done and is left to the reader. These cases are shown in Figures 64.4 and 64.5 on page 340.

### Example 64.1

Discuss and sketch $y^2 = 8x$.

Solution.   The equation is of the form $y^2 = 4px$. If we take $4p = 8$, we have $p = 2$. The vertex is at $(0, 0)$ and the focal point is on the axis of symmetry (the $x$-axis), $p$ units from the vertex, that is at $(2, 0)$. The directrix line is given

$x^2 = 4py, p > 0$

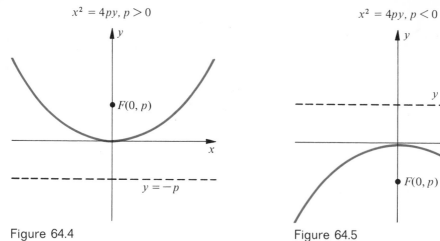

Figure 64.4

$x^2 = 4py, p < 0$

Figure 64.5

$y^2 = 8x$

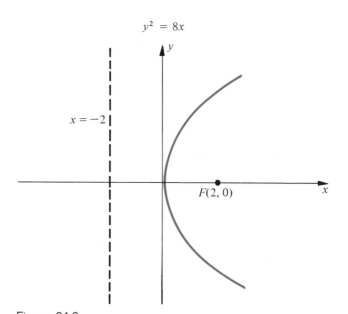

Figure 64.6

by $x = -2$ and the curve opens to the right. (Why?) The graph is shown in Figure 64.6.　●

**Example 64.2**
Discuss and sketch $x^2 = -16y$.

Solution.　The equation is of the form $x^2 = 4py$. Taking $4p = -16$ we have $p = -4$. Thus the vertex is at $(0, 0)$, and the focal point lies on the $y$-axis (the axis of symmetry) 4 units below the vertex at $(0, -4)$. The equation of the directrix line is given by $y = 4$ and the curve opens downward. The graph is shown in Figure 64.7.　●

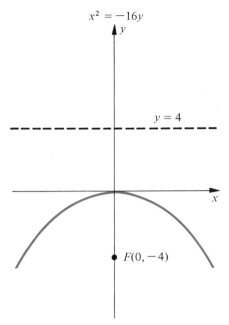

Figure 64.7

If the vertex of the parabola is at the point $(h, k)$, the equations previously obtained no longer apply. However, we can make use of our translation equations to generate various forms of the equation of such a parabola.

Suppose that the vertex of our parabola opening to the right is at $(h, k)$ and its axis is the line $y = k$ (see Figure 64.8). Let us introduce the new coordinate axes $x'$ and $y'$ parallel to the original axes and having their origin at the point $(h, k)$. Since the vertex of the parabola is now at the origin of the $x'y'$ system, its equation is $y'^2 = 4px'$, where $p$ is the distance from the vertex to the focus. The relationship between the old and new coordinate systems is

$$x = x' + h, \quad y = y' + k$$

or

$$x' = x - h, \quad y' = y - k$$

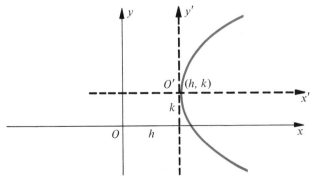

Figure 64.8

Substituting yields the desired equation in the original coordinate system:

$$(y - k)^2 = 4p(x - h), \, p > 0 \tag{64.2}$$

This equation has the following characteristics:

1. Vertex at $(h, k)$.
2. Axis parallel to the $x$-axis.
3. Focus $p$ units to the right of vertex at $(h + p, k)$.
4. Directrix $p$ units to the left of the vertex and perpendicular to the axis of symmetry. Its equation is $x = h - p$.
5. Parabola opens to the right.

    Other forms of the equation of the parabola are

$$(y - k)^2 = 4p(x - h), \, p < 0 \tag{64.3}$$

$$(x - h)^2 = 4p(y - k), \, p > 0 \tag{64.4}$$

$$(x - h)^2 = 4p(y - k), \, p < 0 \tag{64.5}$$

These forms are shown in Figure 64.9.

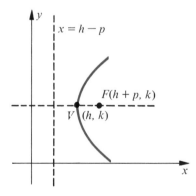
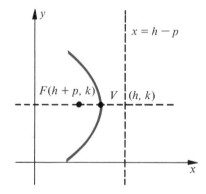
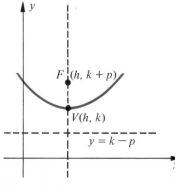
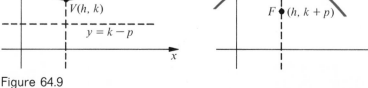

Figure 64.9

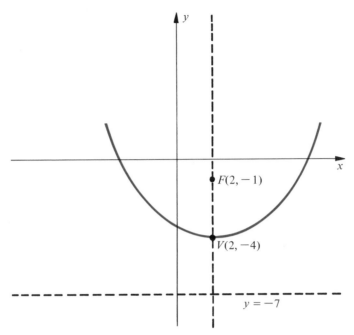

Figure 64.10

**Example 64.3**
Discuss and sketch $(x - 2)^2 = 12(y + 4)$.

Solution.   This equation is of the form $(x - h)^2 = 4p(y - k)$. Thus the parabola has vertex at $(2, -4)$ and opens upward. Setting $4p = 12$, we have $p = 3$. The focus is at the point $(2, -4 + 3)$ or $(2, -1)$. The equation of the directrix line is $y = k - p = -4 - 3 = -7$. The sketch is shown in Figure 64.10.   ●

**Example 64.4**
Discuss and sketch $(y - 2)^2 = -8(x + 1)$.

Solution.   This equation is of the form $(y - k)^2 = 4p(x - h)$. Therefore the vertex is at $(-1, 2)$ and the parabola opens to the left. (Why?) Setting $4p = -8, p = -2$, the focus is at $(-3, 2)$ and the equation of the directrix line is $x = 1$. The sketch is shown in Figure 64.11 on page 344.   ●

**Example 64.5**
Find an equation of the parabola whose directrix is the line $y = -2$ and whose focus is the point $(3, 6)$.

Solution.   Since the directrix is parallel to the $x$-axis our equation will have the form $(x - h)^2 = 4p(y - k)$. The vertex is halfway between the directrix and the focus. Thus the coordinates of the vertex are $(3, 2)$ (Figure 64.12). The directed distance from $V$ to $F$ is $p$, and thus $p = 6 - 2 = 4$. Therefore, the equation is $(x - 3)^2 = 16(y - 2)$.   ●

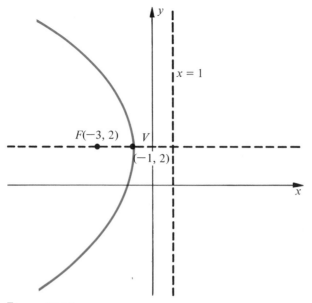

Figure 64.11

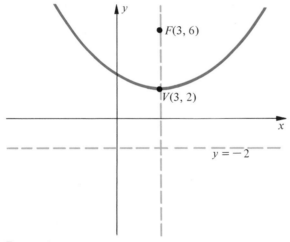

Figure 64.12

If the equation of a parabola is not given in one of the standard forms, we may recognize the equation by noting that it is quadratic in one variable and linear in the other. Whenever we have such a situation, we complete the square in the variable that appears quadratically and then put the linear terms in the form $4p(x - h)$ or $4p(y - k)$. The standard form of the equation then yields the information regarding vertex, focus, axis of symmetry, directrix, and direction of opening.

**Example 64.6**
Discuss and sketch $x^2 + 2x - 4y - 3 = 0$.

Solution. We proceed as we did for the equation of a circle, by completing the square.

$$x^2 + 2x = 4y + 3$$
$$x^2 + 2x + 1 = 4y + 3 + 1$$
$$(x + 1)^2 = 4y + 4$$
$$(x + 1)^2 = 4(y + 1)$$

Our equation is now of the form

$$(x - h)^2 = 4p(y - k)$$

The vertex is at $(-1, -1)$ and the parabola opens upward. Since $4p = 4$, $p = 1$ and the focus is at $(-1, 0)$. The equation of the directrix line is $y = -2$. The sketch is shown in Figure 64.13.  ●

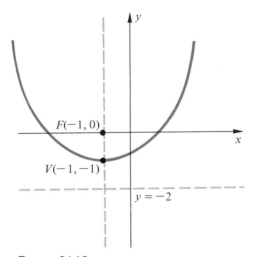

Figure 64.13

If the linear term of the nonquadratic variable is missing, the equation is no longer a parabola. In this case we have either a pair of parallel lines, a single line, or no graph. These are sometimes referred to as the **degenerate cases of a parabola.**

### Example 64.7
Discuss and sketch $x^2 + 2x - 3 = 0$.

Solution.  Since the equation does not involve both $x$ and $y$, it cannot represent a parabola. Factoring yields

$(x + 3)(x - 1) = 0$. Thus $x = -3, x = 1$ and

we have a pair of parallel lines (Figure 64.14 on page 346.)  ●

The parabola has many applications. The path of a projectile (neglecting air resistance) is a parabola. A hanging cable with uniform load will hang in the shape of a parabola. Many reflectors are parabolic in shape. Any ray or wave coming into a parabolic reflector parallel to the axis of symmetry will be reflected so that it passes through the focal point. Conversely, if a source of light rays is placed at the focus, those rays striking the reflector will be reflected into paths parallel to the axis of the parabola. This is the basic

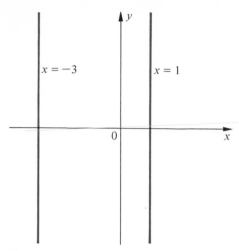

Figure 64.14

principle used for many types of lights, sound and radar detectors, and mirrors for telescopes and microscopes.

## EXERCISES FOR SECTION 64

In Exercises 1–20, discuss and sketch the given equations.

**1.** $y^2 = -4x$

**2.** $x^2 = 12y$

**3.** $x^2 = 4y$

**4.** $y^2 = 4x$

**5.** $(y - 2)^2 = 4(x - 3)$

**6.** $(x + 3)^2 = 6(y + 2)$

**7.** $(y + 1)^2 = -4(x + 1)$

**8.** $(y + 1)^2 = 8x$

**9.** $x^2 = 4y - 8$

**10.** $(y - 5)^2 = 2x + 3$

**11.** $x^2 + 6x - 4y - 3 = 0$

**12.** $y^2 + 4y - 8x + 36 = 0$

**13.** $y^2 - 2y + 8x - 39 = 0$

**14.** $y^2 + 8y - 6x + 4 = 0$

**15.** $4y^2 - 4y - 4x - 3 = 0$

**16.** $y^2 + 2y + 2x - 1 = 0$

**17.** $y^2 + 4y - 4 = 0$

**18.** $x^2 - x - 6 = 0$

**19.** $(x + y)^2 = 0$

**20.** $x^2 + y^2 = 4$

In Exercises 21–30, find an equation(s) of the parabola(s) described.

**21.** Focus $(0, 2)$, vertex $(0, 0)$.

**22.** Focus $(-2, 4)$, vertex $(-2, 3)$.

**23.** Focus $(0, 1)$, vertex $(-5, 1)$.

**24.** Focus $(4, 4)$, vertex $(4, 5)$.

**25.** Focus $(3, 5)$, vertex $(3, 9)$.

**26.** Vertex $(3, 5)$, directrix $x = -1$.

**27.** Vertex $(4, 0)$, directrix $x = 0$.

**28.** Vertex $(0, 1)$, directrix $x = -1$.

**29.** Vertex $(-2, 4)$, directrix $y = 12$.

**30.** Vertex $(1, -2)$, and passing through $(5, 2)$.

31. A chord of a parabola is the line segment joining any two points on the parabola. The *latus rectum* of a parabola is the chord through the focus drawn parallel to the directrix.
    (a) Show that the length of the latus rectum of a parabola is $4|p|$.
    (b) Find the length of the latus rectum for Exercises 1–10.
    (c) Find an equation of the parabola with endpoints of its latus rectum at $(2, 4)$ and $(8, 4)$; at $(3, 1)$ and $(3, 7)$.

32. Consider $y^2 = 4px$. Let $P_1(x_1, y_1)$ be any point on the parabola except the vertex. The tangent line to the parabola $P_1$ is the line through $P_1$ that has no other point in common with the parabola. The slope of this tangent line is given by $m = 2p/y_1$.
    (a) Show that the equation of the tangent line at $P_1(x_1, y_1)$ is $y_1 y = 2p(x + x_1)$.
    (b) Show that the intercepts of the tangent line are $(-x_1, 0)$ and $(0, y_1/2)$.
    (c) Find the equation of a tangent line to $y^2 = 8x$ at the point $(2, 4)$.

33. How high is a parabolic arch of span 12 feet and height 36 feet at a distance of 4 feet from the center of the span? At a distance of 3 feet from the center of the span?

34. When the load along a hanging cable is uniformly distributed, the cable hangs in the shape of an arc of a parabola. Suppose that we have such a cable with endpoints 100 feet apart horizontally and 40 feet above the lowest point.
    (a) Select a coordinate axes and then find the equation of the parabola.
    (b) Find the height of the cable above the low point at a distance 20 feet from the end measured horizontally.

35. Show that every equation of the following form represents a parabola.
    (a) $Ax^2 + Dx + Ey + F = 0$, $A \neq 0$, $E \neq 0$.
    (b) $Cy^2 + Dx + Ey + F = 0$, $C \neq 0$, $D \neq 0$.

36. (a) Show that the standard form of the parabola $(x - h)^2 = 4p(y - k)$ can be expressed as $x^2 + Ax + By + C = 0$.
    (b) Show that the standard form of the parabola $(y - k)^2 = 4p(x - h)$ can be expressed as $y^2 + Ax + By + C = 0$.

In Exercises 37 through 40, use the results of Exercises 36 to find the equations of the following parabolas.

37. Contains the points $(0, -1)$, $(0, 3)$ and $(6, 5)$ and has its axis of symmetry parallel to the $x$-axis.

38. Contains the points $(18, 1)$, $(3, -2)$ and $(11, 0)$ and has its axis of symmetry parallel to the $x$-axis.

39. Contains the points $(1, 0)$, $(2, -3)$ and $(0, 5)$ and has its axis of symmetry parallel to the $y$-axis.

40. Contains the points $(-1, 1)$, $(2, -14)$ and $(0, -2)$ and has its axis of symmetry parallel to the $y$-axis.

41. Discuss and sketch.
    (a) $y|y| = 4x$
    (b) $y^2 = 4|x|$

42. Give a geometric description of the set of all points $P(x, y)$ when the position vectors $\vec{A} = x\vec{i} + y\vec{j}$ and $\vec{B} = 2\vec{i} + \vec{j}$ satisfy each of the following.
    (a) $|\vec{A}| + |\vec{B}| = 6$      (c) $|\vec{A} + \vec{B}| = 6$
    (b) $\vec{A} \cdot \vec{B} = 6$      (d) $\vec{A} \cdot \vec{B} = \frac{1}{2}|\vec{A}|^2|\vec{B}|^2$

## SECTION 65
### THE ELLIPSE

**Definition 65.1**   An **ellipse** is the set of points $P(x, y)$ the sum of whose distance from two fixed points (called the **foci**) is constant.

The constant must be greater than the distance between the two given points. Let us choose the points $(-c, 0)$ and $(c, 0)$ to be the foci and $2a$ for the constant sum of the distance (Figure 65.1). The distance from $P$ to $(-c, 0)$ is $\sqrt{(x + c)^2 + (y - 0)^2}$, and the distance from $P$ to $(c, 0)$ is $\sqrt{(x - c)^2 + (y - 0)^2}$. If $P(x, y)$ represents a point on the ellipse, the sum of these two distances is equal to $2a$. Thus, we have

$$\sqrt{(x - c)^2 + y^2} + \sqrt{(x + c)^2 + y^2} = 2a$$

or

$$\sqrt{(x - c)^2 + y^2} = 2a - \sqrt{(x + c)^2 + y^2}$$

Simpler equivalent equations are found by squaring both sides of the equation and simplifying:

$$x^2 - 2cx + c^2 + y^2 = 4a^2 - 4a\sqrt{(x + c)^2 + y^2} + x^2 + 2cx + c^2 + y^2$$

or

$$4a\sqrt{(x + c)^2 + y^2} = 4a^2 + 4cx$$

Dividing by 4 and squaring again yields

$$a^2(x^2 + 2cx + c^2 + y^2) = a^4 + 2a^2cx + c^2x^2$$

Simplifying again, we have

$$x^2(a^2 - c^2) + a^2y^2 = a^2(a^2 - c^2)$$

Dividing by $a^2(a^2 - c^2)$, we have

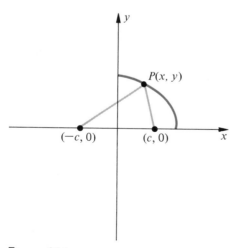

Figure 65.1

$$\frac{x^2}{a^2} + \frac{y^2}{a^2 - c^2} = 1$$

The triangle in Figure 65.1 has one side of length $2c$; the sum of the lengths of the other two sides is $2a$. Thus $2a > 2c$, $a > c$, $a^2 > c^2$, and $a^2 - c^2 > 0$. Since $a^2 - c^2$ is positive, we replace it by another positive number, $b^2$. Thus

$$\frac{x^2}{a^2} + \frac{y^2}{b^2} = 1 \quad \text{where } b_2 = a^2 - c^2 \tag{65.1}$$

We note that in squaring both sides of the equation twice we introduced no extraneous roots, since in both steps both sides of the equation were nonnegative.

By setting $y$ equal to 0 in Equation (65.1), we obtain $x = \pm a$. Therefore, the graph has $x$-intercepts at $(-a, 0)$ and $(a, 0)$. The line segment between $(-a, 0)$ and $(a, 0)$ is called the **major axis** of the ellipse, since it is the longer axis. Similarly, the graph has $y$-intercepts at $(0, -b)$ and $(0, b)$. The line segment between $(0, -b)$ and $(0, b)$ is called the **minor axis** of the ellipse. The endpoints of the major axis $(\pm a, 0)$ are called the **vertices** of the ellipse, and the endpoints of the minor axis $(0, \pm b)$ are called the **covertices** of the ellipse. The point of intersection of the major and minor axis is called the center of the ellipse and in this case is at $(0, 0)$ (see Figure 65.2). Note that the graph is symmetric with respect to the $x$-axis, $y$-axis, and origin.

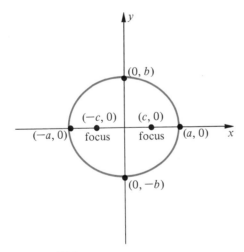

Figure 65.2

### Example 65.1
Discuss and sketch $9x^2 + 25y^2 = 225$.

Solution. Dividing by 225, we have $\frac{x^2}{25} + \frac{y^2}{9} = 1$, which is of the form of equation (65.1). It is apparent that the equation represents an ellipse with $a^2 = 25$, $b^2 = 9$, and $c^2 = a^2 - b^2 = 16$. Thus the ellipse has vertices at $(\pm 5, 0)$, covertices at $(0, \pm 3)$, and foci at $(\pm 4, 0)$. The major axis is of length

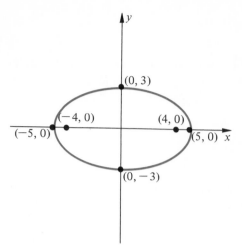

Figure 65.3

10, minor axis of length 6, and the center is at $(0, 0)$. The sketch is shown in Figure 65.3. ●

In addition to the properties mentioned, each ellipse has associated with it a number called the **eccentricity.** For every ellipse the eccentricity is given by $e = \dfrac{c}{a}$. We note that $0 < e < 1$. The eccentricity measures the shape of an ellipse. The closer $e$ is to 0, the "rounder" the ellipse becomes. As $e$ tends to 1, the ellipse becomes more oblong. The circle is sometimes called a special case of the ellipse with $e = 0$.

If the foci are located on the $y$-axis at $(0, c)$ and $(0, -c)$, an analogous derivation would yield the equation

$$\frac{x^2}{a_2 - c^2} + \frac{y^2}{a^2} = 1$$

or

$$\frac{x^2}{b^2} + \frac{y^2}{a^2} = 1 \quad \text{where } b^2 = a^2 - c^2 \tag{65.2}$$

One question is immediately apparent. How can we tell if we have

$$\frac{x^2}{a^2} + \frac{y^2}{b^2} = 1 \quad \text{or} \quad \frac{x^2}{b^2} + \frac{y^2}{a^2} = 1?$$

The answer is *size.* In both cases $a > b$. Thus the larger denominator is always $a^2$ and the smaller is always $b^2$.

The equation $\dfrac{x^2}{b^2} + \dfrac{y^2}{a^2} = 1$, $a^2 > b^2$, has the following properties: vertices at $(0, \pm a)$, covertices at $(\pm b, 0)$, foci at $(0, \pm c)$; the major axis is of length $2a$, minor axis of length $2b$, and center at $(0, 0)$. The sketch is shown in Figure 65.4.

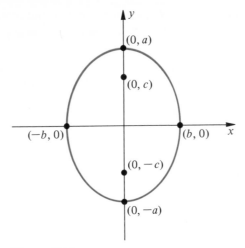

Figure 65.4

**Example 65.2**
Discuss and sketch $4x^2 + y^2 = 36$.

Solution.   We may write the equation equivalently as $\dfrac{x^2}{9} + \dfrac{y^2}{36} = 1$. Now
our equation is of form (65.2), with $a^2 = 36$, $b^2 = 9$, and $c^2 = a^2 - b^2 = 27$.
Thus the ellipse has vertices at $(0, \pm 6)$, covertices at $(\pm 3, 0)$, and foci at
$(0, \pm 3\sqrt{3})$. The major axis is of length 12, minor axis of length 6, and the
center is at $(0, 0)$. The sketch is shown in Figure 65.5.   •

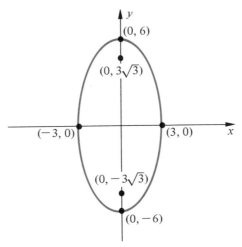

Figure 65.5

**Example 65.3**
Find an equation of the ellipse with vertices $(0, \pm 5)$ and foci $(0, \pm 3)$.

Solution.   Since the vertices are on the $y$-axis, the equation must be of the
form $x^2/b^2 + y^2/a^2 = 1$. In addition, $a = 5$, $c = 3$, and $b^2 = a^2 - c^2 =
25 - 9 = 16$. Our result is $x^2/16 + y^2/25 = 1$.   •

If the center of the ellipse is not at the origin and the major and minor axes are parallel to the $x$- and $y$-axes, we may use the method of Section 64 to write an equation for the ellipse.

Suppose that we have an ellipse with center at $(h, k)$ and foci at $(h - c, k)$ and $(h + c, k)$. The translation of axis $x' = x - h$ and $y' = y - k$ will readily show that an equation for this ellipse in terms of these new coordinates is $\dfrac{x'^2}{a^2} + \dfrac{y'^2}{b^2} = 1$. In terms of the original coordinates we have

$$\frac{(x - h)^2}{a^2} + \frac{(y - k)^2}{b^2} = 1 \quad \text{where } b^2 = a^2 - c^2 \tag{65.3}$$

Similarly, if the center is at $(h, k)$ and the foci at $(h, k - c)$ and $(h, k + c)$, an equation for the ellipse is

$$\frac{(x - h)^2}{b^2} + \frac{(y - k)^2}{a^2} = 1 \quad \text{where } b^2 = a^2 - c^2 \tag{65.4}$$

**Example 65.4**

Discuss and sketch the graph of $\dfrac{(x - 2)^2}{9} + \dfrac{(y - 1)^2}{25} = 1$

Solution. The equation represents an ellipse with center at $(2, 1)$ [compare with equation (65.4)]. $a^2 = 25$, $b^2 = 9$, and $c^2 = a^2 - b^2 = 16$. The vertices are $(2, 6)$ and $(2, -4)$ and the foci are $(2, 5)$ and $(2, -3)$. The covertices are to the right and left of the center at a distance of 3 units and are at $(-1, 1)$ and $(5, 1)$. The major axis is of length 10 and the minor axis of length 6. The graph is shown in Figure 65.6. ●

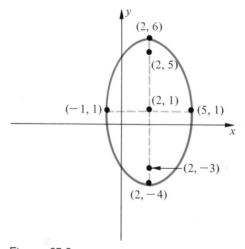

Figure 65.6

**Example 65.5**

Discuss and sketch $4x^2 + 9y^2 - 48x + 72y + 144 = 0$.

Solution. The equation can be put in a form that we recognize by completing the square

$$4(x^2 - 12x) + 9(y^2 + 8y) = -144$$
$$4(x^2 - 12x + 36) + 9(y^2 + 8y + 16) = -144 + 4(36) + 9(16)$$
$$4(x - 6)^2 + 9(y + 4)^2 = 144$$
$$\frac{(x - 6)^2}{36} + \frac{(y + 4)^2}{16} = 1 \qquad \textbf{(A)}$$

We now recognize (A) as the equation of an ellipse with center at $(6, -4)$, $a = 6$, $b = 4$, and $c = 2\sqrt{5}$. The vertices are $(0, -4)$, $(12, -4)$ and the foci are $(6 + 2\sqrt{5}, -4)$, $(6 - 2\sqrt{5}, -4)$. The covertices are $(6, 0)$ and $(6, -8)$. The sketch is shown in Figure 65.7. ●

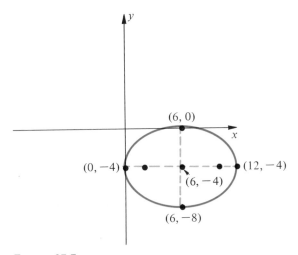

Figure 65.7

When completing the square the student should be aware of the possibility of degenerate cases. These include a single point or the empty set.

The ellipse has some interesting physical applications. Many arches have elliptical shapes. The planets travel in elliptic orbits with the sun as one focal point. The ellipse also has the focal property that a ray emanating from one focus is reflected to the other. Unfortunately, the treatment of these properties in detail involves many of the methods of calculus.

## EXERCISES FOR SECTION 65

In Exercises 1–20, discuss and sketch the given ellipse.

**1.** $\dfrac{x^2}{144} + \dfrac{y^2}{81} = 1$      **2.** $\dfrac{x^2}{25} + \dfrac{y^2}{4} = 1$

**3.** $\dfrac{x^2}{4} + \dfrac{y^2}{25} = 1$

**4.** $\dfrac{x^2}{16} + \dfrac{y^2}{49} = 1$

**5.** $9x^2 + 4y^2 = 36$

**6.** $4x^2 + 9y^2 = 36$

**7.** $x^2 + y^2 = 1$

**8.** $4x^2 + y^2 = 1$

**9.** $9x^2 + y^2 = 9$

**10.** $4x^2 + 16y^2 = 64$

**11.** $\dfrac{(x-1)^2}{16} + \dfrac{(y+2)^2}{4} = 1$

**12.** $x^2 + 4y^2 - 2x - 3 = 0$

**13.** $4x^2 + y^2 + 8x + 10y + 13 = 0$

**14.** $\dfrac{(x-2)^2}{25} + \dfrac{(y-4)^2}{9} = 1$

**15.** $\dfrac{(x+5)^2}{169} + \dfrac{(y+2)^2}{144} = 1$

**16.** $25x^2 + 9(y-2)^2 = 225$

**17.** $9x^2 + 4y^2 + 36x - 24y + 36 = 0$

**18.** $x^2 + 4y^2 - 4x - 8y - 92 = 0$

**19.** $x^2 + 5y^2 - 2x + 20y + 16 = 0$

**20.** $x^2 + y^2 - 6x - 27 = 0$

In Exercises 21–38, find an equation(s) of the ellipse(s) described. Sketch the graph.

**21.** Center $(0, 0)$, focus $(0, 2)$, $a = 4$.

**22.** Center $(0, 0)$, vertex $(0, 13)$, focus $(0, -12)$.

**23.** Center $(0, 0)$, vertex $(0, 13)$, focus $(0, -5)$.

**24.** Center $(0, 0)$, focus $(-3, 0)$, $a = 5$.

**25.** Center $(0, 2)$, focus $(0, 0)$, $a = 6$.

**26.** Center $(0, 0)$, covertex $(0, 5)$, focus $(12, 0)$.

**27.** Center $(0, 0)$, vertex $(5, 0)$, passes through $(\sqrt{15}, 2)$.

**28.** Vertices $(1, 0)$ and $(1, 8)$, covertices $(2, 4)$ and $(0, 4)$.

**29.** Vertices $(3, -4)$ and $(9, -4)$, covertices $(6, -6)$ and $(6, -2)$.

**30.** Foci $(0. \pm 4)$ and passes through $(\frac{12}{5}, 3)$.

**31.** Passes through $(1, 1)$, $(3, 4)$, $(1, 7)$, and $(-1, 4)$.

**32.** Passes through $(6, -1)$, $(-4, -5)$, $(6, -5)$, and $(-12, -3)$.

**33.** Focus $(4, 0)$, eccentricity $\frac{2}{5}$.

**34.** Vertex $(6, 0)$, eccentricity $\frac{2}{3}$.

**35.** Center $(-1, -1)$, vertex $(5, -1)$, and $e = \frac{1}{2}$.

**36.** Vertex at $(4, -3)$ and a co-vertex at $(-1, 0)$.

**37.** Ends of minor axis at $(-3, 5)$ and $(-3, -1)$ and a vertex at $(1, 2)$.

**38.** Ends of major axis at $(-6, -3)$ and $(4, -3)$ and a co-vertex at $(-1, -6)$.

**39.** Lines drawn from the foci to any point on an ellipse are called the **focal radii** of the ellipse. Find the equations of the focal radii drawn from the focal point $(6, 2)$ to the covertices on the ellipse $\dfrac{(x-1)^2}{45} + \dfrac{(y-2)^2}{20} = 1$.

**40.** Find the length of the chord perpendicular to the major axis of the ellipse $\dfrac{x^2}{a^2} + \dfrac{y^2}{b^2} = 1$ and passing through a focus. This chord is the latus rectum of the ellipse. How many exist? Find the equations of these chords.

**41.** Find the equations of the parabolas that have the same vertex and focus as that of the ellipse $4x^2 + 9y^2 = 36$.

**42.** An ornamental arch in the form of a semiellipse has a span of 30 ft and a greatest height of 9 ft. A walkway is to be built between two vertical supports that are equidistant from the ends of the arch and from each other. Find the height of the supports.

# SECTION 66
## THE HYPERBOLA

**Definition 66.1**  A **hyperbola** is the set of all points $P(x, y)$ in a plane such that the absolute value of the difference between the distance from $P$ to two fixed points (called the **foci**) is constant.

Let us again choose the foci to be $(c, 0)$ and $(-c, 0)$ and the constant to be $2a$ (Figure 66.1). If $P(x, y)$ is a point on the hyperbola, we have, according to our definition, $\overline{PF_1} - \overline{PF_2} = \pm 2a$. Algebraically, $\sqrt{(x+c)^2 + y^2} - \sqrt{(x-c)^2 + y^2} = \pm 2a$.

We now follow the pattern used in the derivation of the equation of an ellipse; we isolate radicals and square as follows:

$$\sqrt{(x+c)^2 + y^2} = \pm 2a + \sqrt{(x-c)^2 + y^2}$$
$$x^2 + 2cx + c^2 + y^2 = 4a^2 \pm 4a\sqrt{(x-c)^2 + y^2} + x^2 - 2cx + c^2 + y^2$$

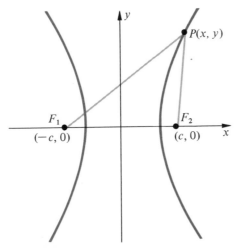

Figure 66.1

$$4cx - 4a^2 = \pm 4a\sqrt{(x-c)^2 + y^2}$$
$$cx - a^2 = \pm a\sqrt{(x-c)^2 + y^2}$$
$$c^2x^2 - 2a^2cx + a^4 = a^2(x^2 - 2cx + c^2 + y^2)$$
$$(c^2 - a^2)x^2 - a^2y^2 = a^2(c^2 - a^2)$$
$$\frac{x^2}{a^2} - \frac{y^2}{c^2 - a^2} = 1$$

From Figure 66.1 we note that in triangle $PF_1F_2$, $\overline{PF_1} < \overline{PF_2} + \overline{F_1F_2}$. Thus $\overline{PF_1} - \overline{PF_2} < \overline{F_1F_2}$ or

$$2a < 2c$$

and

$$a < c$$
$$c^2 - a^2 > 0$$

Since $c^2 - a^2$ is positive, we replace it by the positive number $b^2$ and in turn simplify our equation. Thus the equation of the hyperbola is

$$\frac{x^2}{a^2} - \frac{y^2}{b^2} = 1, \quad b^2 = c^2 - a^2 \tag{66.1}$$

Now let us use Equation (66.1) to discuss some of the properties of the hyperbola. First, we note that the curve is symmetric with respect to the $x$-axis, the $y$-axis, and the origin. The hyperbola crosses the $x$-axis, for if we substitute $y = 0$, we have $x^2/a^2 = 1$ or $x = \pm a$. The points $(a, 0)$ and $(-a, 0)$ are called the vertices and the line segment joining these points is called the **transverse axis** of the hyperbola and has a length of $2a$. The center of the hyperbola is at the midpoint of the transverse axis. The hyperbola does not cross the $y$-axis, for if we substitute $x = 0$ we have $-y^2/b^2 = 1$, which has no real solutions. Further, if we write Equation (66.1) in the form $y^2 = \dfrac{b^2}{a^2}(x^2 - a^2)$, $y$ will not be real when $x^2 - a^2 < 0$ or, equivalently, when $-a < x < a$.

The line segment joining the points $(0, b)$ and $(0, -b)$ is called the **conjugate axis.**

Every hyperbola also has a pair of asymptotes. For $\dfrac{x^2}{a^2} - \dfrac{y^2}{b^2} = 1$, the asymptotes are given by $y = \pm\dfrac{b}{a}x$.

A formal proof of this statement is beyond the scope of this text, but let us give some indication of its validity. We begin with $\dfrac{x^2}{a^2} - \dfrac{y^2}{b^2} = 1$. Dividing by $x^2$,

$$\frac{1}{a^2} - \frac{y^2}{x^2b^2} = \frac{1}{x^2} \quad \text{or} \quad \frac{y^2}{x^2b^2} = \frac{1}{a^2} - \frac{1}{x^2}$$

Multiplying by $b^2$,

$$\frac{y^2}{x^2} = \frac{b^2}{a^2} - \frac{b^2}{x^2}$$

Taking the square root of both sides,

$$\left|\frac{y}{x}\right| = \sqrt{\frac{b^2}{a^2} - \frac{b^2}{x^2}}$$

For large values of $x$, $|y/x|$ is very close to $b/a$, since $b^2/x^2$ tends to zero as $x$ grows "very large." Thus we have, with a wave of our hand, the asymptotes for the hyperbola given by $y = \pm\frac{b}{a}x$. [*Note:* The asymptotes are not a part of the curve; they are only an aid in sketching. To sketch the asymptotes quickly, we plot $(\pm a, 0)$ and $(0, \pm b)$ and then construct the rectangle determined by them (Figure 66.2). The diagonals of the rectangle are the asymptotes.]

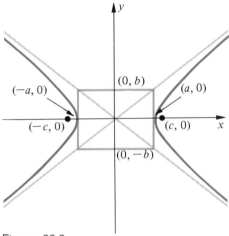

Figure 66.2

If the foci of the hyperbola are on the $y$-axis at $(0, -c)$ and $(0, c)$, the equation will be

$$\frac{y^2}{a^2} - \frac{x^2}{b^2} = 1, \quad b^2 = c^2 - a^2 \tag{66.2}$$

Our graph will then have branches opening upward and downward rather than right and left. In this case the asymptotes are given by $y = \pm\frac{a}{b}x$.

Like the ellipse, each hyperbola has associated with it a number called the eccentricity, given by $e = c/a$. For the hyperbola, $e > 1$.

Unlike the ellipse, the size of the numbers in the denominator have nothing to do with a determination of $a$ and $b$. $a^2$ is *always* the denominator of the positive term and $b^2$ is *always* the denominator of the negative term.

### Remark 66.1

The asymptotes for $\dfrac{x^2}{a^2} - \dfrac{y^2}{b^2} = 1$ are $y = \pm\dfrac{b}{a}x$. The asymptotes for $\dfrac{y^2}{a^2} - \dfrac{x^2}{b^2} = 1$ are $y = \pm\dfrac{a}{b}x$. To avoid confusion, use the following trick. Replace 1 by 0 in the standard form of the equation and solve for $y$.

**Example 66.1**

Discuss and sketch $\dfrac{x^2}{16} - \dfrac{y^2}{9} = 1$.

Solution. Comparing the given equation to (66.1), we note that $a^2 = 16$, $b^2 = 9$, and $c^2 = a^2 + b^2 = 25$. Thus the hyperbola has vertices at $(\pm 4, 0)$ foci at $(\pm 5, 0)$ and center at $(0, 0)$. Its asymptotes are $y = \pm \dfrac{3}{4}x$. The graph is shown in Figure 66.3. ●

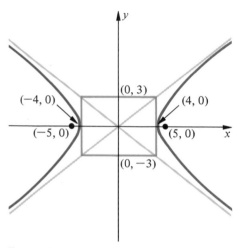

Figure 66.3

**Example 66.2**

Discuss and sketch $\dfrac{y^2}{9} - \dfrac{x^2}{16} = 1$.

Solution. This equation fits form (66.2). $a^2 = 9$, $b^2 = 16$, and $c^2 = 25$. The vertices are on the $y$-axis at $(0, \pm 3)$ and the foci are at $(0, \pm 5)$. The center is at $(0, 0)$ and its asymptotes are $y = \pm \dfrac{3}{4}x$ (Figure 66.4). ●

**Remark 66.2**

Note that the left-hand sides of the equations given in Examples 66.1 and 66.2 differ only by a minus sign. They are called **conjugate hyperbolas.** This may be due to the fact that the transverse axis of one is the conjugate axis of the other.

**Example 66.3**

Find the equation of the hyperbola with foci at $(\pm 3, 0)$ and vertex at $(2, 0)$.

Solution. Since the foci are on the $x$-axis, the equation must be of the form

$$\frac{x^2}{a^2} - \frac{y^2}{b^2} = 1$$

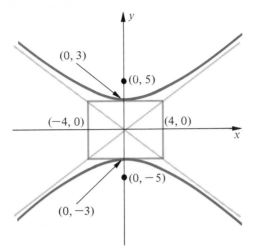

Figure 66.4

We are given that $a = 2$ and $c = 3$. Therefore, $b^2 = c^2 - a^2 = 9 - 4 = 5$. The equation is $\dfrac{x^2}{4} - \dfrac{y^2}{5} = 1$.  ●

The foci of a hyperbola may be anywhere in the plane. If they are on a line parallel to one of the coordinate axis, an equation similar to that of the ellipse may be written. If the foci are at $(h - c, k)$ and $(h + c, k)$, the center must be at $(h, k)$. We may introduce the translation $x' = x - h$ and $y' = y - k$ with origin $O'$ at $(h, k)$. In terms of these new coordinates, the equation of the hyperbola is

$$\frac{x'^2}{a^2} - \frac{y'^2}{b^2} = 1$$

In terms of the original coordinates we have

$$\frac{(x - h)^2}{a^2} - \frac{(y - k)^2}{b^2} = 1 \quad \text{where } b^2 = c^2 - a^2 \tag{66.3}$$

When the focal points are on a line parallel to the $y$-axis and the center is at $(h, k)$, the equation is

$$\frac{(y - k)^2}{a^2} - \frac{(x - h)^2}{b^2} = 1 \quad \text{where } b^2 = c^2 - a^2 \tag{66.4}$$

**Example 66.4**
Discuss and sketch

$$\frac{(x - 1)^2}{9} - \frac{(y - 2)^2}{16} = 1$$

Solution.   The center of the hyperbola is $(1, 2)$ and its branches open to the

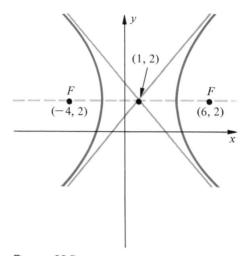

Figure 66.5

right and left. $a^2 = 9$, $b^2 = 16$, and $c^2 = a^2 + b^2 = 25$. Thus the vertices are $(-2, 2)$ and $(4, 2)$ and the foci are $(-4, 2)$ and $(6, 2)$. The equations of the asymptotes are $y - 2 = \pm\frac{4}{3}(x - 1)$. These were obtained by replacing 1 by 0 in the original equation (See Remark 66.1). The sketch is shown in Figure 66.5. ●

**Example 66.5**
Discuss and sketch $y^2 - 4x^2 + 2y + 8x - 7 = 0$.

Solution. We complete the square in $x$ and $y$ in order to reduce the equation to standard form:

$$(y^2 + 2y) - 4(x^2 - 2x) = 7$$
$$(y^2 + 2y + 1) - 4(x^2 - 2x + 1) = 7 + 1 - 4(1)$$
$$(y + 1)^2 - 4(x - 1)^2 = 4$$

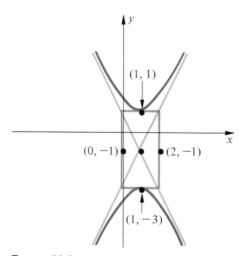

Figure 66.6

or

$$\frac{(y + 1)^2}{4} - \frac{(x - 1)^2}{1} = 1$$

The center of the hyperbola is $(1, -1)$ and its branches open upward and downward. $a^2 = 4$, $b^2 = 1$, and $c^2 = 5$. The vertices are $(1, -3)$ and $(1, 1)$ and the foci are $(1, -1 - \sqrt{5})$ and $(1, -1 + \sqrt{5})$. The equation of the asymptotes are $y + 1 = \pm 2(x - 1)$ or $2x + y - 1 = 0$ and $2x - y - 3 = 0$. The sketch is shown in Figure 66.6.  ●

## EXERCISES FOR SECTION 66

Discuss and sketch each of the following. Label the vertices and foci and sketch the asymptotes.

**1.** $\dfrac{x^2}{16} - \dfrac{y^2}{9} = 1$

**2.** $\dfrac{y^2}{25} - \dfrac{x^2}{16} = 1$

**3.** $\dfrac{x^2}{225} - \dfrac{y^2}{49} = 1$

**4.** $x^2 - y^2 = 1$ (equilateral hyperbola)

**5.** $\dfrac{y^2}{4} - \dfrac{x^2}{1} = 1$

**6.** $x^2 - 9y^2 = -36$

**7.** $9y^2 - 4x^2 = 9$

**8.** $9x^2 - 16y^2 - 9 = 0$

**9.** $\dfrac{(y - 4)^2}{9} - \dfrac{(x + 2)^2}{4} = 1$

**10.** $\dfrac{(x - 3)^2}{16} - \dfrac{(y - 1)^2}{25} = 1$

**11.** $\dfrac{(x - 1)^2}{16} - \dfrac{(y + 3)^2}{9} = 1$

**12.** $x^2 - y^2 - 2x + 2y - 2 = 0$

**13.** $y^2 - 4x^2 + 4y - 16x - 28 = 0$

**14.** $5x^2 - 4y^2 + 20x + 8y = 4$

**15.** $x^2 - 4y - 9y^2 = 0$

**16.** $36x^2 - 36y^2 + 24x + 36y + 31 = 0$

**17.** $4y^2 - x^2 - 2y - x - 16 = 0$

**18.** $4x^2 - y^2 - 2x - y - 16 = 0$

**19.** $4x^2 = y^2 - 4y + 8$

**20.** $9x^2 - 16y^2 - 18x + 32y - 151 = 0$

In Exercises 21 through 38, find an equation(s) of the hyperbola(s) described.

**21.** Center $(0, 0)$, focus $(8, 0)$, vertex $(6, 0)$.

**22.** Center $(0, 0)$, vertex $(3, 0)$, asymptotes $y = \pm\frac{2}{3}x$.

**23.** Vertices $(\pm 2, 0)$, focus $(4, 0)$.

**24.** Asymptotes $y = \pm\frac{2}{3}x$, vertex $(9, 0)$.

**25.** Vertices $(0, \pm 3)$, foci $(0, \pm 5)$.

**26.** Vertices $(\pm 3, 0)$, foci $(\pm 5, 0)$.

**27.** Center $(2, 5)$, vertex $(2, 7)$, focus $(2, 0)$.

**28.** Vertices $(6, 1)$ and $(0, 1)$, focus $(8, 1)$.

**29.** Asymptotes $4x - 3y + 13 = 0$ and $4x + 3y - 5 = 0$, focus $(-1, -2)$.

**30.** Transverse axis of length 8, foci $(0, \pm 5)$.

**31.** Center $(-4, 1)$, vertex $(2, 1)$, conjugate axis of length 8.

**32.** Center $(0, 0)$, passing through $(1, 5)$ and $(2, 7)$.

**33.** Passing through $(2, -2)$, $(-3, 8)$, $(-1, 1)$, and $(2, 8)$.

**34.** Vertices $(\pm 3, 0)$ and passing through $(4, 2)$.

**35.** Asymptotes $y = \pm 2x/3$ and passing through $(6, 1)$.

**36.** Foci $(0, \pm 4)$, $e = \frac{5}{4}$.

**37.** Ends of the conjugate axis at $(-2, 3)$ and $(-2, -1)$ and a vertex at $(1, 1)$.

**38.** An end of a conjugate axis at $(0, 3)$ and vertices at $(-4, 5)$ and $(-4, 1)$.

**39.** The chord passing through a focus and perpendicular to the transverse axis extended is the latus rectum of the hyperbola.
    (a) Show that the length of the latus rectum is $2b^2/a$.
    (b) Show that both latera recta of a hyperbola are the same length.
    (c) Find the length of the latus rectum in Exercises 1–5.

**40.** Discuss and sketch $x|x| + y|y| = 9$.

## SECTION 67
## FOCUS-DIRECTRIX PROPERTIES

The conics may be defined in the following manner:

**Definition 67.1** Let $l$ be a given line called the **directrix** and $F$ (the **focus**) a given point not on $l$. Let $e$ be a given positive number called the **eccentricity.** The set of all points $P(x, y)$ such that the ratio of the distance $|PQ|$ to $|PF|$ is equal to $e$ is called a **conic.** (Figure 67.1). The curve is an ellipse if $0 < e < 1$, a parabola if $e = 1$, and a hyperbola if $e > 1$.

> **Remark 67.1**
> In terms of Definition 67.1, the circle is considered to be a degenerate form of the ellipse with $e = 0$.

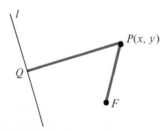

Figure 67.1

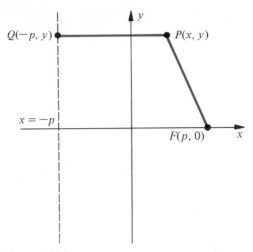

Figure 67.2

In order to illustrate the consistency of this definition with our previous definitions, let us consider the derivation of the equation of the parabola. We take the focus $F$ at the point $(p, 0)$ and the directrix to be $x = -p$. Then $P(x, y)$ lies on the parabola if and only if

$$\frac{|PF|}{|PQ|} = e \text{ or } |PF| = |PQ|e \text{ (Figure 67.2). For the parabola, } e = 1 \text{ and hence } |PF| = |PQ|$$

Thus $\sqrt{(x - p)^2 + y^2} = |x + p|$ or $y^2 = 4px$, as before.

For the ellipse and the hyperbola, the choice of focus and directrix is not so obvious. If we take $F(c, 0)$ as our focal point, then $x = a^2/c$ is the appropriate choice for the directrix line.

With these choices, the equation for the ellipse is

$$\frac{x^2}{c^2/e^2} + \frac{y^2}{c^2(1 - e^2)/e^2} = 1$$

and for the hyperbola is

$$\frac{x^2}{c^2/3^2} - \frac{y^2}{c^2(1 - e^2)/e^2} = 1$$

If $e = c/a$, then the equations reduce to the standard forms.

# SECTION 68
# THE GENERAL SECOND-DEGREE EQUATION:

$$Ax^2 + Bxy + Cy^2 + Dx + Ey + F = 0$$

In previous sections we have seen that the circle, parabola, ellipse, and hyperbola can be described by second-degree equations in two variables. Now we consider the converse: What is the graph of a given second-degree equation in two unknowns? We shall show that the graph of such an equation is always one of the conics or a degenerate form of a conic.

We can write the general second-degree equation in $x$ and $y$ as

$$Ax^2 + Bxy + Cy^2 + Dx + Ey + F = 0 \qquad (68.1)$$

where $A$, $B$, and $C$ are not all zero. For the sake of convenience, let us consider separately the cases $B = 0$ and $B \neq 0$.

In the case where $B = 0$, we have that $Ax^2 + Cy^2 + Dx + Ey + F = 0$. Assuming that the equation does *not* represent a degenerate form, it follows from preceding sections that after completing the squares, we must obtain the equation of a circle, parabola, ellipse, or hyperbola. As we mentioned at the beginning of the chapter, these curves are called conics because they represent the intersection of a plane with a cone. The situations in which the plane and the cone intersect in a single point, a line, two intersecting lines, or do not intersect are called **degenerate conics.**

We may summarize the results of our previous discussions by the following theorem.

**Theorem 68.1**  If the graph of $Ax^2 + Cy^2 + Dx + Ey + F = 0$ is not degenerate and if not both $A$ and $C$ are zero, the graph is

1. A circle when $AC > 0$ and $A = C$.
2. An ellipse when $AC > 0$ and $A \neq C$.
3. A parabola when $AC = 0$.
4. A hyperbola when $AC < 0$.

**Example 68.1**

Identify the conic whose equation is $4y^2 - 5x^2 - 8y - 10x - 21 = 0$.

Solution.  Assume that the graph is not degenerate, and apply Theorem 68.1. We find that since $A = -5$ and $C = 4$, $AC = -20 < 0$. Thus we must have the equation of a hyperbola. This is easy to verify by completing the squares in $x$ and $y$. In so doing, we obtain $4(y^2 - 2y + 1) - 5(x^2 + 2x + 1) = 21 + 4 - 5$, which may be put into the form

$$\frac{(y - 1)^2}{5} - \frac{(x + 1)^2}{4} = 1$$

We have a hyperbola with its branches opening up and down and its center at $(-1, 1)$.  ●

**Example 68.2**

The equation $x^2 + 4y^2 = 0$ is a degenerate conic whose graph consists of only one point, $(0, 0)$.

**Example 68.3**

The equation $x^2 + 2y^2 + 3 = 0$ is a degenerate conic that has no graph since there are no real solutions.  ●

In many instances it is not easy to recognize the degenerate types shown in Examples 68.2 and 68.3.

**Example 68.4**

Identify the equation $x^2 - 4y^2 - 4x - 8y = 0$.

Solution. Theorem 68.1 says that since $AC = -4 < 0$, the equation must be a hyperbola if it is not degenerate. Completing the square in $x$ and $y$ yields $(x^2 - 4x + 4) - 4(y^2 + 2y + 1) = 4 - 4$. Thus $(x - 2)^2 - 4(y + 1)^2 = 0$. Since the left-hand side of the equation is the difference of two squares, its factors as the product of the sum and difference of the square roots:

$$[(x - 2) + 2(y + 1)][(x - 2) - 2(y + 1)] = 0$$

The graph of the equation is the union of the graphs of $(x - 2) + 2(y + 1) = 0$ and $(x - 2) - 2(y + 1) = 0$ (two straight lines) and is shown in Figure 68.1.  ●

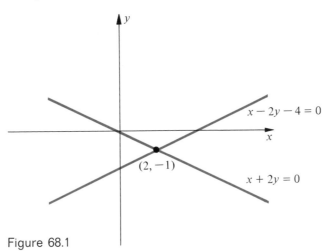

Figure 68.1

## EXERCISES FOR SECTION 68

Use Theorem 68.1 to identify each of the following.

1. $x^2 + y^2 + 6x + 4y = 12$

2. $x^2 + 4y^2 + 6x + 4y + 6 = 0$

3. $x^2 - 4y^2 + 6x + 4y + 6 = 0$

4. $y^2 = 4x + 8$

5. $3x^2 + 3y^2 + 6x + 9y + 100 = 0$

6. $x^2 + y^2 + 8x = 6$

7. $4x^2 - y^2 + 4x + 6y - 11 = 0$

8. $x^2 + y^2 + 3x - 2y = 10$

9. $x^2 + 3y^2 - 4x = 5$

10. $x^2 - 3y^2 = 2$

# SECTION **69**
# ROTATION OF AXES

The method of completing the square, although simple to use, is of no help in the case where the second-degree equation has an $xy$ term (that is, $B \neq 0$). If we could transform the equation $Ax^2 + Bxy + Cy^2 + Dx + Ey + F = 0$ to $A'x^2 + C'y^2 + D'x + E'y + F' = 0$, we could apply Theorem 68.1 and readily identify the conic. Before proceeding let us make the following observation. In each of the previous cases, $B = 0$, and thus no $xy$ term appeared. We saw that, in these situations, the graphs appeared with their axes parallel to or coinciding with the coordinate axes. On this basis, we might suspect that when the axes of the conics are not parallel to the coordinate axes, the equation will have an $xy$ term. This is indeed the case and we will show that by a suitable transformation called a **rotation,** through an angle $\theta$, we can eliminate the $xy$ term from the general second-degree equation.

Suppose that we rotate the coordinate axes through an angle $\theta$ in a counterclockwise direction (Figure 69.1). Then every point in the plane may be represented in two ways: $(x, y)$ in the given coordinate system and $(x', y')$ in the new coordinate system. We note from Figure 69.1 that

$$x = \overline{OR} = \overline{OP} \cos (\theta + \alpha)$$
$$y = \overline{RP} = \overline{OP} \sin (\theta + \alpha) \tag{A}$$

and

$$x' = \overline{OR'} = \overline{OP} \cos \alpha$$
$$y' = \overline{R'P} = \overline{OP} \sin \alpha \tag{B}$$

Using the equation

$$\cos (\theta + \alpha) = \cos \theta \cos \alpha - \sin \theta \sin \alpha$$
$$\sin (\theta + \alpha) = \sin \theta \cos \alpha + \cos \theta \sin \alpha$$

in (A) yields

$$x = \overline{OP} \cos \theta \cos \alpha - \overline{OP} \sin \theta \sin \alpha$$
$$y = \overline{OP} \sin \theta \cos \alpha + \overline{OP} \cos \theta \sin \alpha$$

Substituting from (B) we find that

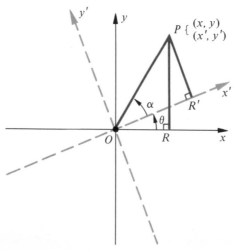

Figure 69.1

$$x = x' \cos \theta - y' \sin \theta$$
$$y = x' \sin \theta + y' \cos \theta \tag{69.1}$$

which are called the **general transformation equations for rotation of axes.**

If we solve equation (69.1) for $x'$ and $y'$ we obtain the inverse transformation equation for rotation of axes:

$$x' = x \cos \theta + y \sin \theta$$
$$y' = -x \sin \theta + y \cos \theta \tag{69.2}$$

Our primary interest in deriving these equations is to see whether we can eliminate the $xy$ term from the general second-degree equation. The following theorem shows how this is done.

**Theorem 69.1**   The general second-degree equation $Ax^2 + Bxy + Cy^2 + Dx + Ey + F = 0$ can be transformed into $A'x'^2 + C'y'^2 + D'x' + E'y' + F' = 0$ by a rotation of axes through an angle $\theta$, where

$$\tan 2\theta = \frac{B}{A - C} \quad \text{if } A \neq C \tag{69.3}$$

or

$$\theta = \pi/4 \quad \text{if } A = C \tag{69.4}$$

Proof.   If we use these equations in the general quadratic equation, we obtain a new quadratic equation of the form $A'x'^2 + B'x'y' + C'y'^2 + D'x' + E'y' + F' = 0$, where the new coefficients and the old are related as follows:

$$A' = A \cos^2 \theta + B \cos \theta \sin \theta + C \sin^2 \theta$$
$$B' = B(\cos^2 \theta - \sin^2 \theta) + 2(C - A) \sin \theta \cos \theta$$
$$C' = A \sin^2 \theta - B \sin \theta \cos \theta + C \cos^2 \theta$$
$$D' = D \cos \theta + E \sin \theta \tag{69.5}$$
$$E' = -D \sin \theta + E \cos \theta$$
$$F' = F$$

If our original equation contains an $xy$ term, we can always find an angle of rotation $\theta$ such that the $xy$ term is eliminated. To find this angle, we simply set $B' = 0$ in (69.5) and solve for $\theta$. Thus

$$B' = B(\cos^2 \theta - \sin^2 \theta) + 2(C - A) \sin \theta \cos \theta = 0$$

We note that

$$\cos^2 \theta - \sin^2 \theta = \cos 2\theta$$
$$2 \sin \theta \cos \theta = \sin 2\theta$$

and therefore,

$$B' = B \cos 2\theta + (C - A) \sin 2\theta = 0$$

Finally,

$$\frac{\sin 2\theta}{\cos 2\theta} = \frac{B}{A - C}$$

or

$$\tan 2\theta = \frac{B}{A - C} \quad \text{if } A \neq C$$

If $A = C$, then $\cos 2\theta = 0$ and $\theta = \pi/4$.  ●

### Remark 69.1
As with a translation, the transformation of an equation by means of a rotation does not change the graph.

### Example 69.1
Rotate the axes to remove the $xy$ term from the equation $5x^2 + 8xy + 5y^2 = 36$. Identify and sketch the graph, showing both sets of axes.

Solution.  Theorem 69.1 tells us that $\theta = \pi/4$ since $A = C = 5$. The equations of the rotation are

$$x = \frac{\sqrt{2}}{2}x' - \frac{\sqrt{2}}{2}y'$$

$$y = \frac{\sqrt{2}}{2}x' + \frac{\sqrt{2}}{2}y'$$

Substituting these in the given equation yields

$$5\left(\frac{\sqrt{2}}{2}x' - \frac{\sqrt{2}}{2}y'\right)^2 + 8\left(\frac{\sqrt{2}}{2}x' - \frac{\sqrt{2}}{2}y'\right)\left(\frac{\sqrt{2}}{2}x' + \frac{\sqrt{2}}{2}y'\right)$$

$$+ 5\left(\frac{\sqrt{2}}{2}x' + \frac{\sqrt{2}}{2}y'\right) = 36$$

This simplifies to $\dfrac{x'^2}{4} + \dfrac{y'^2}{36} = 1$. The graph is the ellipse shown in Figure 69.2.  ●

### Example 69.2
By means of a suitable rotation, reduce the given equation to standard form, and identify and sketch the graph of $4x^2 + 3\sqrt{3}xy + y^2 = 22$.

Solution.  From Theorem 69.1 we have

$$\tan 2\theta = \frac{B}{A - C} = \frac{3\sqrt{3}}{4 - 1} = \frac{3\sqrt{3}}{3} = \sqrt{3}$$

We choose $2\theta$ to be in the first quadrant. Thus $\tan 2\theta = \sqrt{3}$, $2\theta = \pi/3$, $\theta = \pi/6$, and

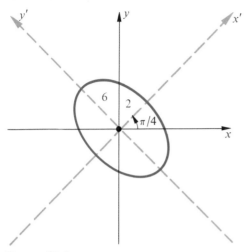

Figure 69.2                                    Figure 69.3

$$\sin\theta = \sin\frac{\pi}{6} = \frac{1}{2}$$

$$\cos\theta = \cos\frac{\pi}{6} = \frac{\sqrt{3}}{2}$$

The equations of the rotation are

$$x = \frac{\sqrt{3}}{2}x' - \frac{y'}{2}$$

$$y = \frac{x'}{2} + \frac{\sqrt{3}}{2}y'$$

Substitution in the original equation yields

$$4\left(\frac{\sqrt{3}}{2}x' - \frac{y'}{2}\right)^2 + 3\sqrt{3}\left(\frac{\sqrt{3}}{2}x' - \frac{y'}{2}\right)\left(\frac{x'}{2} + \frac{\sqrt{3}}{2}y'\right)$$

$$+ \left(\frac{x'}{2} + \frac{\sqrt{3}}{2}y'\right)^2 = 22$$

which simplifies to $x'^2/4 - y'^2/44 = 1$. The graph is the hyperbola shown in Figure 69.3.   ●

**Example 69.3**
Rotate the axes to remove the $xy$ term from the equation $16x^2 + 24xy + 9y^2 - 130x + 90y = 0$. Write the transformed equation in standard form, and identify and sketch the graph.

Solution.   From Theorem 69.1 we have

$$\tan 2\theta = \frac{B}{A - C} = \frac{24}{16 - 9} = \frac{24}{7}$$

We choose $2\theta$ to be in the first quadrant (see Figure 69.4).

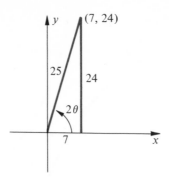

Figure 69.4

We use the identities

$$\sin \theta = \sqrt{\frac{1 - \cos 2\theta}{2}}$$

$$\cos \theta = \sqrt{\frac{1 + \cos 2\theta}{2}}$$

along with the fact that $\cos 2\theta = 7/25$ (see Figure 69.4), to find

$$\sin \theta = \sqrt{\frac{1 - 7/25}{2}} = \frac{3}{5}$$

$$\cos \theta = \sqrt{\frac{1 + 7/25}{2}} = \frac{4}{5}$$

Thus the equations of rotation are

$$x = \frac{4x'}{5} - \frac{3y'}{5}$$

$$y = \frac{3x'}{5} + \frac{4y'}{5}$$

Substituting in the original equation yields

$$16\left(\frac{4x'}{5} - \frac{3y'}{5}\right)^2 + 24\left(\frac{4x'}{5} - \frac{3y'}{5}\right)\left(\frac{3x'}{5} + \frac{4y'}{5}\right) + 9\left(\frac{3x'}{5} + \frac{4y'}{5}\right)^2$$
$$- 130\left(\frac{4x'}{5} - \frac{3y'}{5}\right) + 90\left(\frac{3x'}{5} + \frac{4y'}{5}\right) = 0$$

which simplifies to $y'^2 - 2y' - 6x' = 0$. In terms of the $x'y'$ coordinate system we have

$$y'^2 - 2y' + 1 = 6x' + 1$$

or

$$(y' - 1)^2 = 6\left(x' + \frac{1}{6}\right)$$

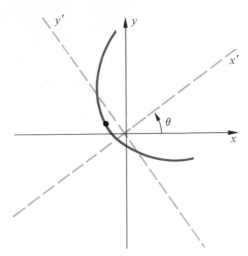

Figure 69.5

This equation is the standard form of a parabola. The graph of the parabola is shown in Figure 69.5, where $\theta = \sin^{-1}\left(\frac{3}{5}\right) \approx 37°$. •

We now ask if we can tell whether the curve is a parabola, an ellipse, or a hyperbola without first rotating the axes to eliminate the $xy$ term. To determine this, let us consider the general quadratic equation with the $xy$ terms removed. For $A'x'^2 + C'y'^2 + D'x' + E'y' + F' = 0$ we have, by Theorem 68.1,

$$\begin{aligned}
&\text{a parabola if } A'C' = 0\\
&\text{an ellipse if } A'C' > 0\\
&\text{a hyperbola if } A'C' < 0
\end{aligned} \tag{69.6}$$

It can be shown by the use of equation (69.5) (see Exercise 22) that the coefficients $A$, $B$, and $C$ and $A'$, $B'$, and $C'$ satisfy the relation

$$B^2 - 4AC = B'^2 - 4A'C' \tag{69.7}$$

for any rotations.

$B^2 - 4AC$ is called the **discriminant** of the general second-degree equation and is invariant under a rotation of axes. When we transform a given equation by a rotation, we determine the value of $\theta$ for which $B' = 0$. Thus, for this computed value of $\theta$, equation (69.7) becomes $B^2 - 4AC = -4A'C'$.

Because the graph of $A'x'^2 + C'y'^2 + D'x' + E'y' + F' = 0$ is easily identified and because $B^2 - 4AC$ is invariant, we will have little trouble distinguishing the graph of $Ax^2 + Bxy + Cy^2 + Dx + Ey + F = 0$. The method is given in Theorem 69.2.

**Theorem 69.2**  If the graph of $Ax^2 + Bxy + Cy^2 + Dx + Ey + F = 0$ is not degenerate and not all $A$, $B$ and $C$ are equal to zero, then the graph is

1. An ellipse if $B^2 - 4AC < 0$.

2. A parabola if $B^2 - 4AC = 0$.
3. A hyperbola if $B^2 - 4AC > 0$.

**Example 69.4**

Use the discriminant to identify the equation $4x^2 + 4xy + y^2 - 3x = 6$.

Solution. Here $A = 4, B = 4$, and $C = 1$. Thus $B^2 - 4AC = 16 - 16 = 0$, and the graph of the equation, if not degenerate, is a parabola.   ●

## EXERCISES FOR SECTION 69

In Exercises 1–10, discuss and sketch the graph of each of the following equations.

1. $x^2 + y^2 + 6x + 4y = 12$        6. $x^2 + y^2 + 8x = 6$

2. $x^2 + 4y^2 + 6x + 4y + 6 = 0$       7. $4x^2 - y^2 + 4x + 6y - 11 = 0$

3. $x^2 - 4y^2 + 6x + 4y + 6 = 0$       8. $x^2 + y^2 + 3x - 2y = 10$

4. $y^2 = 4x + 8$                9. $x^2 + 3y^2 - 4x = 5$

5. $3x^2 + 3y^2 + 6x + 9y + 100 = 0$     10. $x^2 - 3y^2 = 2$

In Exercises 11–20, identify the curve, reduce to standard form, and sketch, showing both sets of axes.

11. $5x^2 + 6xy + 5y^2 = 8$

12. $5x^2 + 8xy + 5y^2 = 0$

13. $xy - 2y - 4x = 0$

14. $9x^2 + 24xy + 16y^2 + 90x - 130y = 0$

15. $2x^2 + 5xy + 2y^2 = 0$

16. $2x^2 + 5xy + 2y^2 = -18$

17. $13x^2 - 10xy + 13y^2 - 82x + 98y + 157 = 0$

18. $x^2 + xy + y^2 + 4\sqrt{2}x - 4\sqrt{2}y = 0$

19. $9x^2 - 6xy + y^2 - 12\sqrt{10}x - 36\sqrt{10}y = 0$

20. $9x^2 - 24xy + 16y^2 - 50x - 100y - 175 = 0$

21. (a) Find the equation of the locus (path) of the point that moves so that the sum of its distances from $(4, -2)$ and $(5, 3)$ is 6.
   (b) Use Theorem 69.2 to verify that the equation obtained in part (a) is the equation of an ellipse.

22. (a) Find the equation of the locus (path) of the point that moves so that the sum of its distance from $(3, 4)$ and $(-3, 4)$ is 12.
   (b) Use Theorem 69.2 to verify that the equation obtained in part (a) is the equations of an ellipse.

23. (a) Find the equation of the locus of the point that moves so that the difference of its distances from $(4, 3)$ and $(-4, -3)$ is 6.

(b) Use Theorem 69.2 to verify that the equation obtained in part (a) is the equation of a hyperbola.

24. (a) Find the equation of the locus of the point that moves so that the difference of its distances from $(2, 1)$ and $(-2, -1)$ is 4.
    (b) Use Theorem 69.2 to verify that the equation obtained in part (a) is the equation of a hyperbola.

25. Show that the rotation equations of (69.1) produce the relationships shown in (69.2).

26. Use the equations of (69.5) to show that $B^2 - 4AC = B'^2 - 4A'C'$ when the general second-degree equation has been transformed by a rotation.

27. Show that the graph of $Ax^2 + Cy^2 + Dx + Ey + F = 0$ consists of two intersecting straight lines if $AC < 0$ and $F = \dfrac{D^2}{4A} + \dfrac{E^2}{4C}$.

28. Show that the equation $x^2 + y^2 = r^2$ will be transformed into $x'^2 + y'^2 = r^2$ for any angle $\theta$ in the equations for rotation of axes.

29. Suppose that we are given the transformation equations

$$x' = \tfrac{3}{5}x + \tfrac{4}{5}y$$
$$y' = -\tfrac{4}{5}x + \tfrac{3}{5}y$$

What is the angle $\theta$ of rotation?

30. (a) Show that the transformation equations generated by a translation of origin to $(h, k)$ followed by a rotation through an angle $\theta$ are given by

$$x'' = (x - h)\cos\theta + (y - k)\sin\theta$$

$$y'' = -(x - h)\sin\theta + (y - k)\cos\theta$$

(b) Show that the transformation equations generated by a rotation through an angle $\theta$ followed by a translation of origin to $(h, k)$ are given by

$$x'' = x\cos\theta + y\sin\theta - h$$
$$y'' = -x\sin\theta + y\cos\theta - k$$

(c) What conclusion may be drawn from the results of parts (a) and (b)?

# SECTION 70
# GRAPHS OF POLAR EQUATIONS

We have studied the straight line, the conic sections, and their corresponding graphs. Now let us consider curves in the plane defined by an equation in polar coordinates. If the given equation is of the form $F(r, \theta) = 0$, the set of all points $P(r, \theta)$ that satsify the equation will be called the **polar graph.** Keep in mind that the point $P(r, \theta)$ has many pairs of coordinates. However, $P$ is on the graph if only one pair satisfies the equation. If the given equation is of the form $r = f(\theta)$, the previous remarks still hold. In this instance we usually think of $\theta$ as the independent variable and $r$ the dependent variable.

For the sake of reference and convenience, we again list the relationships between

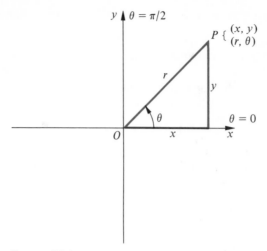

Figure 70.1

rectangular and polar coordinates (Figure 70.1):

$$x = r \cos \theta$$
$$y = r \sin \theta$$

(57.1)

$$r = \sqrt{x^2 + y^2}$$

$$\tan \theta = \frac{y}{x} \quad \text{or} \quad \theta = \tan^{-1}\left(\frac{y}{x}\right), \quad x \neq 0$$

(57.2)

These equations enable us to change from one coordinate system to the other.

If we hold $r$ fixed at $r = r_1$ and let $\theta$ vary, we have a circle with center at the pole and radius of $r_1$. If we hold $\theta$ fixed at $\theta = \theta_1$ and let $r$ vary, we have a line through the pole that makes an angle of $\theta$ with the polar axis. Thus we describe the horizontal or polar axis in terms of the equation $\theta = 0$ and the line through the pole and perpendicular to the polar axis (the $y$-axis in rectangular coordinates) in terms of the equation $\theta = \pi/2$.

If we begin with an equation that gives $r$ explicitly in terms of $\theta$ [$r = f(\theta)$], we can arrive at a sketch by plotting points. We may obtain as many points as we wish simply by substituting values of $\theta$ and then computing the corresponding values for $r$. This method may be cumbersome and ineffective at times, so we attempt to simplify and order our work by using the following procedures in dealing with the properties of polar graphs.

These properties include tests for symmetry, extent (analogous to range and domain in rectangular coordinates), points at the pole, and a table of values.

1. Test for *Symmetry*
   (a) The curve is *symmetric with respect to the pole* if the equation is unaltered when $r$ is replaced by $-r$ (Figure 70.2).
   (b) The curve is *symmetric with respect to the polar axis* if the equation is unaltered when $\theta$ is replaced by $-\theta$ (Figure 70.3).
   (c) The curve is *symmetric with respect to the line* $\theta = \pi/2$ if the equation is unaltered when $\theta$ is replaced by $\pi - \theta$ (Figure 70.4).

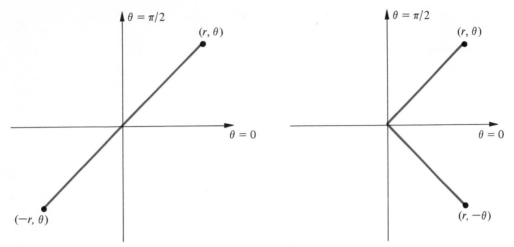

Figure 70.2                                      Figure 70.3

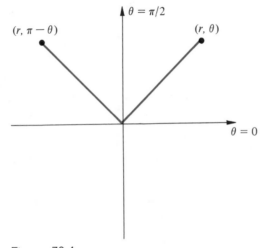

Figure 70.4

If our given expression is $F(r, \theta) = 0$, we may formulate our symmetry tests as follows:

(a) If $F(r, \theta) = F(-r, \theta)$, the curve is symmetric with respect to the pole.
(b) If $F(r, \theta) = F(r, -\theta)$, the curve is symmetric with respect to the polar axis.
(c) If $F(r, \theta) = F(r, \pi - \theta)$, the curve is symmetric with respect to the line $\theta = \pi/2$.

Our future discussion dealing with possible symmetries will be restricted to line symmetry with respect to the polar axis and the line $\theta = \pi/2$ and point symmetry with respect to the pole. It should be noted that these tests *are not unique* (see Exercises 29 and 30).

Test (b) and (c) can often be used quickly and simply. Since $\cos(-\theta) = \cos \theta$ and $\sin(\pi - \theta) = \sin \theta$, it follows that any equation in which $\theta$ appears only as a function of $\cos \theta$ or only as a function of $\sin \theta$ must be symmetric with respect to the polar axis and the line $\theta = \pi/2$, respectively.

2. Determine the *extent* of the function for $\theta$ and $r$, noting especially the maximum and minimum values of $r$.
3. Find the values of $\theta$ for which the curve passes through the pole (that is, set $r = 0$).
4. Construct a table of values.

The procedure outlined in steps 1–4 will normally enable us to graph the given equation.

### Example 70.1
Discuss and sketch the graph of $r = 4 \cos \theta$.

### Solution

**Symmetry:** We observe that the graph is symmetric with respect to the polar axis since the equation is unaltered when $\theta$ is replaced by $-\theta$.

**Extent:** There are no restrictions on $\theta$, but since $\cos \theta$ has a period of $2\pi$, we need only consider the interval $0 \leq \theta \leq 2\pi$. In this interval $-1 \leq \cos \theta \leq 1$ (see Table 70.1). Hence $r$ varies from $-4$ to 4. The maximum value 4, occurs when $\theta = 0$ and the minimum value, $-4$, occurs when $\theta = \pi$.

**Table 70.1**

| $\theta$ | 0° | 20° | 45° | 60° | 90° |
|---|---|---|---|---|---|
| $r$ | 4 | 3.48 | 2.84 | 2 | 0 |

**Points at the Pole:** The curve passes through the pole since $r = 0$ when $\theta = \pi/2$.

**Graphing Comments:** As $\theta$ varies from 0 to $\pi/2$, $\cos \theta$ decreases from 1 to 0 and $r$ decreases from 4 to 0. The point $P(r, \theta)$ thus traces a continuous arc of the curve that lies above the polar axis and connects the point $(4, 0)$ with the origin. We make use of our symmetry to graph the corresponding arc of the curve that lies below the axis (Figure 70.5).

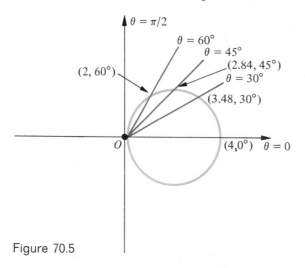

Figure 70.5

For $\pi/2 < \theta \le \pi$, $r$ becomes negative, achieving a minimum value of $-4$ when $\theta = \pi$. But the use of our symmetry has already given us the points in this interval, and we have no need to plot them a second time. For this reason Table 70.1 ends at $\theta = \pi/2$

As $\theta$ varies from $\pi$ to $2\pi$, the curve is traced a *second time*. The graph is a circle. This can be verified by converting our given equation to rectangular form as follows. Since $\cos \theta = x/r$, $r = 4 \cos \theta$ is equivalent to $r = 4x/r$. Multiplying by $r$ we have $r^2 = 4x$. But $r^2 = x^2 + y^2$, and thus $x^2 + y^2 = 4x$. Completing the square of the terms in $x$ we have $(x - 2)^2 + y^2 = 4$, which is the equation of a circle with center $(2, 0)$ and radius 2. ●

### Remark 70.1

The graph of $r = a \cos \theta$, $a > 0$, will always be a circle with center $(a/2, 0)$, radius $a/2$, and passing through the pole. The graph of $r = a \sin \theta$, $a > 0$, is the same Figure rotated through an angle of $\pi/2$.

### Example 70.2

Discuss and sketch the graph of $r = 2(1 + \cos \theta)$.

### Solution

**Symmetry:** The graph is symmetric with respect to the polar axis since the equation is unaltered when $\theta$ is replaced by $-\theta$.

**Extent:** Here, again, there is no restriction on $\theta$. As before, we consider $\theta$ only in the interval $0 \le \theta \le 2\pi$, owing to its periodicity. Since $-1 \le \cos \theta \le 1$, the values for $r$ vary from 0 to 4 (see Table 70.2). The maximum value, 4, occurs when $\theta = 0$ and the minimum value, 0, occurs when $\theta = \pi$.

### Table 70.2

| $\theta$ | 0 | 30° | 45° | 60° | 90° | 120° | 135° | 150° | 180° |
|---|---|---|---|---|---|---|---|---|---|
| $r$ | 4 | 3.74 | 3.42 | 3 | 2 | 1 | 0.58 | 0.26 | 0 |

**Points at the Pole:** The curve passes through the pole since $r = 0$ when $\theta = \pi$.

**Graphing Comments:** As $\theta$ varies from 0 to $\pi$, $\cos \theta$ decreases from 1 to $-1$ and $r$ decreases from 4 to 0. Since the curve is symmetric with respect to the polar axis, Table 70.2 ends at $\theta = \pi$. We obtain the lower half of the curve by reflection in the polar axis (Figure 70.6). The curve is called a **cardioid** because of its heart-shaped appearance.

The rectangular form of the equation that is satisfied by the points on this curve can be found as follows. Since $\cos \theta = x/r$, we have $r = 2\left(1 + \dfrac{x}{r}\right)$.

Multiplying by $r$ yields $r^2 = 2(r + x)$ or $r^2 - 2x = 2r$. Squaring both sides we have

$$(r^2 - 2x)^2 = 4r^2$$

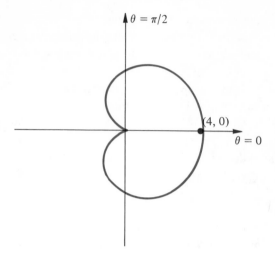

Figure 70.6

Thus

$$(x^2 + y^2 - 2x)^2 = 4(x^2 + y^2)$$

The last expression gives us some idea of why we might choose to work with the polar form rather than the rectangular form of the curve.   ●

**Example 70.3**
Discuss and sketch $r = 2 + \sin \theta$.

Solution

**Symmetry:** If we replace $\theta$ by $\pi - \theta$, the equation is unchanged and thus the graph is symmetric with respect to the line $\theta = \pi/2$.

**Extent:** There is no restriction on $\theta$, so we choose the interval $0 \leq \theta \leq 2\pi$. In this interval $-1 \leq \sin \theta \leq 1$ and thus $r$ varies from 1 to 3 (see Table 70.3). The maximum value, 3, occurs when $\theta = \pi/2$ and the minimum value, 1, occurs when $\theta = 3\pi/2$ or, if we wish, when $\theta = \pi/2$.

**Table 70.3**

| $\theta$ | 0 | 30° | 45° | 60° | 90° | −30° | −45° | −60° | −90° |
|---|---|---|---|---|---|---|---|---|---|
| $r$ | 2 | 2.5 | 2.71 | 2.87 | 3 | 1.5 | 1.29 | 1.13 | 1 |

**Points at the Pole:** The curve does not pass through the pole because $\sin \theta$ is never equal to $-2$.   ●

**Graphing Comments:** As $\theta$ varies from 0 to $\pi/2$, $\sin \theta$ increases from 0 to 1 and $r$ increases continuously from 2 to 3. Since the curve is symmetric with respect to the line $\theta = \pi/2$, we consider some special values for $\theta$ in the interval $-\pi/2 \leq \theta < 0$ rather than in the interval $\pi/2 < \theta \leq \pi$. Once we have a sketch of the curve to the right of $\theta = \pi/2$, the remaining part of the

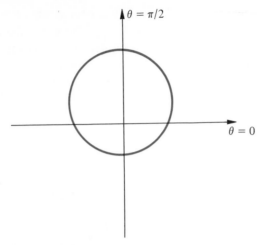

Figure 70.7

curve is obtained by a reflection in the line $\theta = \pi/2$ (see Figure 70.7). This curve is called a **limacon.**

The rectangular form of the equation is

$$(x^2 + y^2 - y)^2 = 4(x^2 + y^2)$$

The details are left to the Exercises. ●

**Example 70.4**
Discuss and sketch $r = 1 - 2 \cos \theta$.

Solution

**Symmetry:** The curve is symmetric with respect to the polar axis since the equation is unaltered when $\theta$ is replaced by $-\theta$.

**Extent:** There is no restriction on $\theta$, so we consider $\theta$ only in the interval $0 \le \theta \le 2\pi$. Since $-1 \le \cos \theta \le 1$, $r$ varies from $-1$ to 3 (see Table 70.4). The maximum value, 3, occurs when $\theta = \pi$ and the minimum value, $-1$, occurs when $\theta = 0$.

**Table 70.4**

| $\theta$ | 0 | 30° | 45° | 60° | 90° | 120° | 135° | 150° | 180° |
|---|---|---|---|---|---|---|---|---|---|
| $r$ | $-1$ | $-0.74$ | $-0.42$ | 0 | 1 | 2 | 2.42 | 2.74 | 3 |

**Points at the Pole:** $r$ will be 0 when $1 - 2 \cos \theta = 0$. Solving this equation, we have $\cos \theta = 1/2$ and thus $\theta = \pi/3$ and $\theta = 5\pi/3$. The curve will pass through the pole twice, once when $\theta = \pi/3$ and again when $\theta = 5\pi/3$. In plotting these points we must be careful to note that the quadrant in which the points lie is *not* determined by the signs of the polar coordinates (as is the case with rectangular coordinates) but by the *size* of $\theta$ and the *sign* of $r$.

**Graphing Comments:** As $\theta$ varies from 0 to $\pi$, $\cos \theta$ decreases from 1 to $-1$

and $r$ increases from $-1$ to $3$. We plot the points from Table 70.4 and carefully connect them with a smooth curve. The curve for the interval $\pi < \theta \leq 2\pi$ is obtained by making use of our symmetry (Figure 70.8). The curve is a limaçon.

The rectangular form of the equation is $(x^2 + y^2 + 2x)^2 = x^2 + y^2$. ●

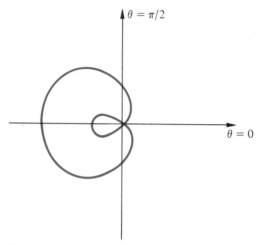

Figure 70.8

The graphs of the equation shown in Examples 70.2, 70.3, and 70.4 are specific examples of the graphs of equations of the form $r = a \pm b \cos \theta$ or $r = a \pm b \sin \theta$, $a > 0$. Such graphs are, in general, limaçons. In the case where $a = b$, we have a special type of limaçon, the cardiod. If $a > b$, we have a limaçon that does not pass through the pole (see Example 70.3). If $a < b$, the limaçon passes through the pole twice and has an interior loop, as in Example 70.4.

### Example 70.5
Discuss and sketch the graph of $r^2 = 4 \cos 2\theta$.

### Solution

**Symmetry:** We observe that the equation is unchanged when $r$ is replaced by $-r$, and therefore we have symmetry with respect to the pole. In addition, since $\cos 2(-\theta) = \cos 2\theta$, we have symmetry with respect to the polar axis.

**Extent:** Since $2\theta$ varies from $0$ to $2\pi$ as $\theta$ varies from $0$ to $\pi$, it appears that the entire curve can be obtained by examining values of $\theta$ in the interval $0 \leq \theta \leq \pi$. But $r^2$ can never be negative and $\cos 2\theta$ is negative for $\pi/4 < \theta < 3\pi/4$. Thus we consider only those values for $\theta$ in the interval $0 \leq \theta \leq \pi/4$ (see Table 70.5). For values of $\theta$ in this interval, $r$ has a maximum value of $2$ when $\theta = 0$ and a minimum value of $0$ when $\theta = \pi/4$.

**Points at the Pole:** $r = 0$ when $4 \cos 2\theta = 0$ or $\cos 2\theta = 0$. Thus $2\theta = \pi/2$ or

**Table 70.5**

| $\theta$ | 0 | 15° | $22\frac{1}{2}$° | 30° | 45° |
|---|---|---|---|---|---|
| $2\theta$ | 0 | 30° | 45° | 60° | 90° |
| $r$ | 2 | 1.86 | 1.68 | 1.42 | 0 |

$3\pi/2$ and $\theta = \pi/4$ and $3\pi/4$. The curve passes through the pole twice, once when $\theta = \pi/4$ and again when $\theta = 3\pi/4$. •

**Graphing Comments:** As $\theta$ varies from 0 to $\pi/4$, $\cos 2\theta$ decreases from 1 to 0 and $r$ decreases from 2 to 0. With these considerations and the points from Table 70.5, we have the quarter of the curve that lies in the first quadrant. The remaining curve is obtained through the use of symmetry (Figure 70.9). This curve is called a **lemniscate.** The rectangular form of the given equation is $(x^2 + y^2)^2 = 4(x^2 - y^2)$. •

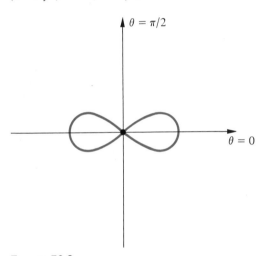

Figure 70.9

Graphs of equations of the form $r^2 = a^2 \cos 2\theta$ or $r^2 = a^2 \sin 2\theta$, $a > 0$, are called **lemniscates.** The graph of $r^2 = a^2 \sin 2\theta$ is a rotation through an angle of 45° of the graph of $r^2 = a^2 \cos 2\theta$.

### Example 70.6
Discuss and sketch $r = 2 \sin 3\theta$.

### Solution

**Symmetry:** The equation is unaltered when $\theta$ is replaced by $\pi - \theta$ and thus the curve is symmetric with respect to the line $\theta = \pi/2$.

**Extent:** There is no restriction on $\theta$, so we might choose the interval $0 \le \theta \le 2\pi$. But $\sin 3\theta$ has a period of $2\pi/3$. Thus we need only consider $\theta$ in the interval $0 \le \theta \le 2\pi/3$. We further note from the period of $\sin 3\theta$ that whatever figure appears for the interval $0 \le \theta \le 2\pi/3$ is repeated in the intervals $2\pi/3 \leq \theta \leq 4\pi/3$ and $4\pi/3 \leq \theta \leq 2\pi$. Thus the remaining parts of the curve are obtained by a rotation through an angle of $2\pi/3$. In the interval $0 \leq \theta \leq 2\pi/3$, the maximum value for $r$ is 2 and occurs when $3\theta = \pi/2$ or $\theta = \pi/6$. The minimum value is $-2$ and occurs when $\theta = \pi/2$.

**Points at the Pole:** $r = 0$ when $\sin 3\theta = 0$. Thus $\theta = 0, 60°, 120°, 180°$ and $240°$. The curve passes through the pole four times.   ●

**Table 70.6**

| $\theta$ | 0° | 10° | 15° | 20° | 30° | 40° | 45° | 50° | 60° |
|---|---|---|---|---|---|---|---|---|---|
| $3\theta$ | 0° | 30° | 45° | 60° | 90° | 120° | 135° | 150° | 180° |
| $r$ | 0 | 1 | 1.42 | 1.74 | 2 | 1.74 | 1.42 | 1 | 0 |

**Graphing Comments:** The points from Table 70.6 give us only the first loop of the curve shown in Figure 70.10, but this is sufficient, since the remaining two

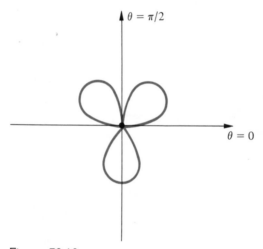

Figure 70.10

loops may be obtained by a rotation of this initial loop through an angle of $2\pi/3$. If our table of values had included special values of $\theta$ for $0 \leq \theta \leq 2\pi/3$, we would have obtained the first and third loops of the curve. The second loop could be graphed by then using the symmetry with respect to the line $\theta = \pi/2$. This type of curve is called a **petal curve.** In Example 70.6 we have a curve of three petals. Its equation in rectangular form is given by $(x^2 + y^2)^2 = 2y(x^2 - y^2)$.

The graph of $r = 3 \sin 2\theta$ is shown in Figure 70.11. The details are left until Exercise 9.   ●

Curves of equations of the form $r = a \sin n\theta$ or $r = a \cos n\theta$, $a > 0$, $n$ a positive integer, are generally called petal curves. If $n$ is odd, the curve has $n$ petals; if $n$ is even, the curve has $2n$ petals. In the special case where $n = 1$, we have a circle passing through the pole.

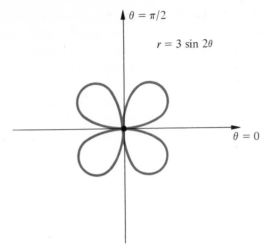

Figure 70.11

Suppose that we wish to find the points of intersection of two curves whose equations are in polar form. We solve the equations simultaneously to obtain pairs of numbers that satisfy both equations. Although these points must be points of intersection of the curves, some points of intersection cannot be found in this way. The following example illustrates the problem.

### Example 70.7
Find the points of intersection of the circle $r = \cos \theta$ and the cardioid $r = 1 - \cos \theta$.

Solution.   Equating $r$ in the given pair of equations yields $\cos \theta = 1 - \cos \theta$ or $\cos \theta = 1/2$. Thus $\theta = \pi/3, 5\pi/3$ for $0 \le \theta \le 2\pi$. This gives us the points $(1/2, \pi/3)$ and $(1/2, 5\pi/3)$. The graphs of the two curves as shown in Figure 70.12.   ●

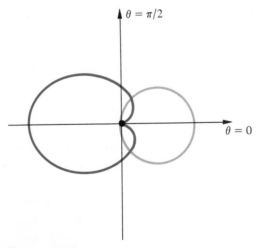

Figure 70.12

As we see from Figure 70.12 (page 383), there are three points of intersection. Both curves pass through the pole, but this point was not obtained by algebraic means. The reason for this lies in the fact that the pole has many different representations. On the curve $r = \cos \theta$ it is represented by $(0, \pi/2)$, and on the curve $r = 1 - \cos \theta$ it is represented by $(0, 0)$. Thus, although the pole does lie on both curves, it does not have a common representation that satisfies both equations. For this reason the intersections of polar equations are generally determined not only by algebraic methods, but also by graphing the curves. In this way we can be sure that we have found all points of intersection.

## EXERCISES FOR SECTION 70

In Exercises 1–25, discuss and sketch the equations.

1. $r = 2$

2. $r = \dfrac{\pi}{2}$

3. $\theta = \dfrac{\pi}{6}$

4. $\theta = 2$

5. $r = \cos \theta$

6. $r = \sin \theta$

7. $r = 1 + \sin \theta$

8. $r = \cos 2\theta$

9. $r = 3 \sin 2\theta$

10. $r = 2 \cos 3\theta$

11. $r = 4 \cos 4\theta$

12. $r = a(1 - \cos \theta), \ a > 0$

13. $r = a(1 + \cos \theta), \ a > 0$

14. $r = 1 - 2 \sin \theta$

15. $r = 2 - \sin \theta$

16. $r = 2 - 2 \sin \theta$

17. $r = 1 + 3 \cos \theta$

18. $r = 2 - \cos \theta$

19. $r^2 = 4 \cos 2\theta$

20. $r^2 = 6 \sin 2\theta$

21. $r^2 = a^2 \cos 2\theta, \ a > 0$

22. $r^2 = a^2 \sin 2\theta, \ a > 0$

23. $r = \dfrac{1}{\sin \theta}$

24. $r = a\theta, \ a > 0$ (spiral of Archimedes)

25. $r^2 = \sin^2 \theta$

26. Determine the polar equation of each of the following, and sketch $a > 0$.
    (a) $x^2 + y^2 - 2ay = 0$        (c) $(x^2 + y^2)^2 = 2a^2xy$
    (b) $(x^2 + y^2)^2 = x^2 - y^2$    (d) $(x^2 + y^2)^3 = 4a^2x^2y^2$

27. Determine the Cartesian equation of each of the following, and sketch.
    (a) $r = \sin \theta + \cos \theta$        (d) $r^2(16 \cos^2 \theta + 9 \sin^2 \theta) = 144$        (c) $\theta = \dfrac{\pi}{3}$

    (b) $r = 4 \cos \theta$        (e) $r = \dfrac{9}{5 - 4 \cos \theta}$

28. Sketch the following pairs of equations and find all points of intersection. Use the results of Exercises 1–25 where appropriate.

(a) $r = \cos 2\theta$ and $r = \sin \theta$  
(b) $r = 1 - \sin \theta$ and $r = 1 - \cos \theta$  
(c) $r = \cos \theta$ and $r = 1 + \sin \theta$  
(d) $r = 1 + 2 \cos \theta$ and $r = 1 - 2 \cos \theta$  

(e) $r = 2a \cos \theta$ and $r = a$  
(f) $r = \cos \theta$ and $r^2 = \cos^2(2\theta)$  
(g) $r = 2(1 + \cos \theta)$ and $r = 6 \cos \theta$  

**29.** Show that the graph of the equation $F(r, \theta) = 0$ is symmetric with respect to
(a) The pole, if the point $(r, \pi + \theta)$ satisfies the equation whenever the point $(r, \theta)$ does.
(b) The polar axis, if the point $(-r, \pi - \theta)$ satisfies the equation whenever the point $(r, \theta)$ does.
(c) The line $\theta = \pi/2$, if the point $(-r, -\theta)$ satisfies the equation whenever the point $(r, \theta)$ does.

**30.** Using symmetry tests a, b, and c, and the results of Exercise 29, show that the graphs of the following equations are symmetric with respect to the pole, the polar axis, and the line $\theta = \pi/2$.
(a) $r^2 = \sin \theta$　　(b) $r^2 = \cos \theta$

**31.** Show that $r(A \cos + B \sin \theta) = 0$ is always a straight line provided that $A$ and $B$ are not both 0.

**32.** Show that $r = 2A \sin \theta + 2B \cos \theta$ is either a circle through the origin or a single point.

**33.** Show that $r^2 \cos 2\theta = A^2$ is an equilateral hyperbola.

**34.** The polar equation of a conic section with its focus at the pole and its directrix perpendicular to the polar axis and $p$ units to the left of the pole is given by

$$r = \frac{ep}{1 - e \cos \theta} \tag{1}$$

If the directrix is $p$ units to the right of the pole, the equation becomes

$$r = \frac{ep}{1 + e \cos \theta} \tag{2}$$

The polar equation of the conic with its focus at the pole and directrix parallel to the polar axis and $p$ units from it is given by

$$r = \frac{ep}{1 + e \sin \theta} \tag{3}$$

if the directrix is above the polar axis, and

$$r = \frac{ep}{1 - e \sin \theta} \tag{4}$$

if the directrix is below the polar axis. Show that equations (1)–(4) represent an ellipse if $0 < e < 1$, a parabola if $e = 1$, a hyperbola if $e > 1$.

**35.** Identify and sketch the following conics using the results of Exercise 34.

(a) $r = \dfrac{6}{1 - \cos \theta}$　　(d) $r = \dfrac{6}{2 + 3 \sin \theta}$

(b) $r = \dfrac{1}{1 - \cos \theta}$　　(e) $r = \dfrac{4}{2 - \sin \theta}$

(c) $r = \dfrac{12}{3 + 4 \cos \theta}$

# CHAPTER 11 SEQUENCES AND SERIES

## SECTION 71
### SEQUENCES

**Definition 71.1** An **infinite sequence** is a function whose domain is the set of natural numbers $1, 2, 3, \ldots$. A **finite sequence** is a function whose domain is a subset of the natural numbers $1, 2, 3, \ldots, n$.

If we represent the function by the letter $F$, the infinite sequence could be described by its ordered pairs as follows:

$$F = \{(1, F(1)), (2, F(2)), \ldots, (n, F(n)), \ldots\} \tag{A}$$

Since the domain of the sequence is always the ordered set of natural numbers, we can describe such a sequence as (A) by listing only its range elements,

$$F(1), F(2), \ldots, F(n), \ldots \tag{B}$$

in the order of the natural numbers with which they are associated. A simpler notation for the sequence in form (A) is achieved by letting

$$F(1) = a_1, F(2) = a_2, \ldots, F(n) = a_n, \ldots$$

Thus the infinite sequence in (A) can be expressed as

$$a_1, a_2, \ldots, a_n, \ldots$$

where $a_1$ is called the **first term** of the sequence, $a_2$ the **second term,** $a_3$ the **third term,** $a_n$ the **$n$th term** or the **general term.**

**387**

**Example 71.1**

The infinite sequence that has $a_n = 1/n$ as its general term can be described as

$$a_1, a_2, \ldots, a_n, \ldots$$

where the terms of this sequence are found by successively substituting the numbers 1, 2, 3, ... for $n$ as follows:

$$a_1 = \tfrac{1}{1} = 1$$
$$a_2 = \tfrac{1}{2}$$
$$a_3 = \tfrac{1}{3}$$

Thus the sequence can be expressed as $1, 1/2, 1/3, \ldots, 1/n, \ldots$.

**Example 71.2**

Find the $k$th term and the $(k + 1)$th term of the sequence whose $n$th term is $a_n = n(n + 2)$.

**Solution.** If we replace $n$ by $k$, we obtain $a_k = k(k + 2)$, and if we replace $n$ by $k + 1$, we obtain $a_{k+1} = (k + 1)(k + 1 + 2) = (k + 1)(k + 3)$. ●

**Example 71.3**

Find the first five terms of the infinite sequence whose $n$th term is given by $a_n = n^2$.

**Solution.** We need to compute $a_1, a_2, a_3, a_4$, and $a_5$. We do so by successively substituting the numbers 1, 2, 3, 4, and 5 for $n$ in the formula $a_n = n^2$. Thus we have $a_1 = 1^2 = 1$, $a_2 = 2^2 = 4$, $a_3 = 3^2 = 9$, $a_4 = 4^2 = 16$, and $a_5 = 5^2 = 25$. The first five terms of this sequence are 1, 4, 9, 16, and 25. We can describe the entire sequence as

$$1, 4, 9, 16, 25, \ldots, n^2, \ldots \quad ●$$

**Example 71.4**

The first twenty-five terms of the infinite sequence whose $n$th term is $a_n = n + 2$ constitutes a finite sequence that can be described as

$$3, 4, 5, 6, 7, \ldots, n + 2, \ldots, 27$$

When we deal with sequences order is most important. Thus if two different terms in the sequence are equal, both terms must be written in their appropriate positions. For example, suppose that the $n$th term of a sequence is $a_n = n^2 - 5n + 6$. Substitution of 1 and 4 for $n$ shows us that $a_1 = a_4 = 2$, and substitution of 2 and 3 for $n$ shows us that $a_2 = a_3 = 0$. We must express the sequence as

$$2, 0, 0, 2, 6, \ldots, n^2 - 5n + 6, \ldots$$

If we are given several terms of a sequence, it is sometimes possible to find an expression for a general term by inspection or by trial and error. For example, if the first three terms of a sequence are 1, 3, 5, ..., then a general term for this sequence could be $a_n = 2n - 1$.

If, however, we are given only a finite number of successive terms of a sequence and we do find a general term, it is *not unique*.

### Example 71.5

Suppose that the first three terms in an infinite sequence are 1, 4, 9, .... A general term for this sequence could be $a_n = n^2$, since $a_1 = 1$, $a_2 = 4$, and $a_3 = 9$. However, this is not the only possibility. Consider $a_n = n^2 + (n - 1)(n - 2)(n - 3)$. Here $a_1 = 1$, $a_2 = 4$, and $a_3 = 9$. As a matter of fact, we could generate an infinite number of general terms for this sequence by letting

$$a_n = n^2 + (n - 1)(n - 2)(n - 3)f(n)$$

where $f(n)$ is *any* function whose domain is $N$.

## EXERCISES FOR SECTION 71

1. Find the first five terms, the twelfth term, the $k$th term, and the $(k + 1)$th term of the sequence whose general term is

   (a) $a_n = 2n + 1$         (d) $a_n = \left(-\dfrac{1}{5}\right)^n$         (g) $a_n = 3n + 2$

   (b) $a_n = \dfrac{1}{n + 1}$         (e) $a_n = \dfrac{(-1)^n}{n^2}$         (h) $a_n = \dfrac{3^n}{n^2 + 1}$

   (c) $a_n = \dfrac{1}{n(n + 1)}$         (f) $a_n = \dfrac{(-1)^n}{n + 1}$         (i) $a_n = (-1)^n 3^n$

2. Find two different general or $n$th terms for each of the following sequences when the first four terms are as given.

   (a) $4, 7, 10, 13, \ldots$         (g) $\frac{1}{3}, \frac{1}{9}, \frac{1}{27}, \frac{1}{81}, \ldots$
   (b) $1, 8, 27, 64, \ldots$         (h) $1, -1, 1, -1, \ldots$
   (c) $-1, \frac{1}{2}, -\frac{1}{3}, \frac{1}{4}, \ldots$         (i) $\frac{1}{2}, \frac{2}{3}, \frac{3}{4}, \frac{4}{5}, \ldots$
   (d) $4, 8, 12, 16, \ldots$         (j) $x, x^2, x^3, x^4, \ldots$
   (e) $5, 8, 11, 14, \ldots$         (k) $x^2, x^4, x^6, x^8, \ldots$
   (f) $2, 5, 10, 17, \ldots$         (l) $-x, x^3, -x^5, x^7, \ldots$

3. Find the first four terms of each of the sequences when

   (a) $a_1 = 2, a_n = a_{n-1} + 1$         (d) $a_1 = 5, a_n = \dfrac{a_{n-1}}{2}$

   (b) $a_1 = 2, a_n = a_{n-1}^2$         (e) $a_1 = 5, a_n = 4a_{n-1} - 1$
   (c) $a_1 = 1, a_n = na_{n-1}$

4. A ball is dropped from a height of 10 feet. If it rebounds one-half of the distance it has fallen after each fall, how far will it rebound the fifth time? The $k$th time?

5. An automobile purchased for $P$ dollars depreciates in value $r$ per cent every year. Find its value at the end of $k$ years.

# SECTION 72
## SERIES

Associated with every sequence is a **series** that can be formed by taking the algebraic sum of the corresponding terms of the given sequence.

> **Example 72.1**
> With the finite sequence 1, 3, 5, ..., $2n - 1$, there is associated the finite series
>
> $$S_n = 1 + 3 + 5 + \cdots + (2n - 1)$$
>
> With the finite sequence 1, 1/2, 1/4, ..., $1/(2)^{n-1}$, there is associated the finite series
>
> $$S_n = 1 + 1/2 + 1/4 + \cdots + 1/(2)^{n-1} \quad \bullet$$

In general, the finite series

$$S_n = a_1 + a_2 + a_3 + \cdots + a_n$$

is obtained by summing the terms of the finite sequence

$$a_1, a_2, a_3, \ldots, a_n$$

The infinite series

$$a_1 + a_2 + a_3 + \cdots + a_n + \cdots$$

is obtained by summing the terms of the infinite sequence $a_1, a_2, a_3, \ldots, a_n \ldots$. The $n$th or general term of a series is the $n$th or general term of the corresponding sequence.

A finite series will always have for its sum a finite number. The meaning, if any, of the "sum" of an infinite series will be discussed later.

A series for which the general term is known can be represented in a compact form by means of **sigma** or **summation notation.** The symbol

$$\sum_{k=1}^{n} a_k$$

represents the sum of the first $n$ terms of the finite series $a_1 + a_2 + \cdots + a_n$; that is,

$$\sum_{k=1}^{n} a_k = a_1 + a_2 + a_3 + \cdots + a_n \tag{72.1}$$

The Greek capital letter $\Sigma$ (sigma) is used to denote the sum of the series, the symbol $a_k$ represents the $k$th or general term of the series, the letter $k$ is called the **summation variable,** and the numbers 1 and $n$ indicate the extremes of the set of natural numbers that we sum over.

**Example 72.2**
Evaluate

$$\sum_{k=1}^{4} k(k + 1)$$

Solution. We have $a_k = k(k + 1)$ and the summation variable $k$ ranges over the natural numbers 1, 2, 3, and 4. To obtain the indicated sum, we substitute in succession the values of $k$ and then add the resulting terms as follows:

$$\sum_{k=1}^{4} k(k + 1) = 1(1 + 1) + 2(2 + 1) + 3(3 + 1) + 4(4 + 1)$$
$$= \quad 2 \quad + \quad 6 \quad + \quad 12 \quad + \quad 20 \quad = 40 \quad \bullet$$

**Example 72.3**
Evaluate

$$\sum_{j=3}^{7} (j^2 + 1)$$

Solution. We have $a_j = j^2 + 1$ and the summation variable $j$ ranges over the natural numbers 3, 4, 5, 6, and 7. To obtain the indicated sum, we substitute in succession the values of $j$ and then add the resulting terms as follows:

$$\sum_{j=3}^{7} (j^2 + 1) = (3^2 + 1) + (4^2 + 1) + (5^2 + 1) + (6^2 + 1) + (7^2 + 1)$$
$$= 10 + 17 + 26 + 37 + 50 = 140$$

(*Note:* The letter used for the summation variable is arbitrary, and the summation range may begin with a value other than 1.)  $\bullet$

**Theorem 72.1**   If $a_k = c$, where $c$ is any real number, then

$$\sum_{k=1}^{n} a_k = \sum_{k=1}^{n} c = cn$$

Proof.   We have

$$\sum_{k=1}^{n} a_k = a_1 + a_2 + a_3 + \cdots + a_n$$
$$= c + c + c + \cdots + c = cn \quad \bullet$$

**Remark 72.1**

If we apply Theorem 72.1 when $a_k = c = 1$, we have

$$\sum_{k=1}^{n} 1 = n$$

**Example 72.4**

Use sigma notation to describe a finite series associated with each of the following finite sequences.

(a) $1, \dfrac{1}{2}, \dfrac{1}{3}, \ldots, \dfrac{1}{n}$            (c) $-1, \dfrac{1}{2}, -\dfrac{1}{4}, \ldots, \dfrac{(-1)^n}{2^{n-1}}$

(b) $\dfrac{1}{2}, \dfrac{2}{3}, \dfrac{3}{4}, \ldots, \dfrac{n}{n+1}$       (d) $3^2, 4^2, \ldots, 100^2$

**Solution**

(a) We have $a_k = \dfrac{1}{k}$, thus the associated series

$$1 + \frac{1}{2} + \frac{1}{3} + \cdots + \frac{1}{n} = \sum_{k=1}^{n} \frac{1}{k}$$

(b) We have $a_k = \dfrac{k}{k+1}$, thus the associated series

$$\frac{1}{2} + \frac{2}{3} + \frac{3}{4} + \cdots + \frac{n}{n+1} = \sum_{k=1}^{n} \frac{k}{k+1}$$

(c) We have $a_k = \dfrac{(-1)^k}{2^{k-1}}$, thus the associated series

$$-1 + \frac{1}{2} - \frac{1}{4} + \cdots + \frac{(-1)^n}{2^{n-1}} = \sum_{k=1}^{n} \frac{(-1)^k}{2^{k-1}}$$

(d) We have $a_j = j^2$, thus the associated series

$$3^2 + 4^2 + \cdots + 100^2 = \sum_{j=3}^{100} j^2 \quad \bullet$$

We now present another very useful theorem about sums.

**Theorem 72.2**   If $\displaystyle\sum_{k=1}^{n} a_k = S$ and if $\displaystyle\sum_{k=1}^{n} b_k = R$, then

1. $\displaystyle\sum_{k=1}^{n} (a_k + b_k) = \sum_{k=1}^{n} a_k + \sum_{k=1}^{n} b_k = S + R.$

2. $\displaystyle\sum_{k=1}^{n} ca_k = c \sum_{k=1}^{n} a_k = cS$, where $c$ is any real number.

We leave the proof of this theorem as an exercise.

We conclude this discussion with an example of the usefulness of Theorem 72.2.

**Example 72.5**

Prove that

$$\sum_{k=1}^{n} k = 1 + 2 + \cdots + n = \frac{n(n+1)}{2}$$

**Solution.** We first observe that $\displaystyle\sum_{k=1}^{n} [(k+1)^2 - k^2]$ can be expressed in terms of $n$ as follows:

$$\sum_{k=1}^{n} [(k+1)^2 - k^2] = [2^2 - 1^2] + [3^2 - 2^2] + [4^2 - 3^2]$$

$$+ [5^2 - 4^2] + \cdots + [(n+1)^2 - n^2]$$

or

$$\sum_{k=1}^{n} [(k+1)^2 - k^2] = (n+1)^2 - 1 \tag{A}$$

Form (A) was obtained by simply observing that all but two of the terms on the right side of the equation cancel each other out.

Next we note that $\displaystyle\sum_{k=1}^{n} [(k+1)^2 - k^2]$ can be simplified by using algebra and Theorem 72.2 as follows.

$$\sum_{k=1}^{n} [(k+1)^2 - k^2] = \sum_{k=1}^{n} [k^2 + 2k + 1 - k^2] = \sum_{k=1}^{n} (2k + 1)$$

$$= 2 \sum_{k=1}^{n} k + \sum_{k=1}^{n} 1 \tag{B}$$

We now equate forms (A) and (B) and solve for $\displaystyle\sum_{k=1}^{n} k$ as follows to obtain the desired result:

$$2 \sum_{k=1}^{n} k + \sum_{k=1}^{n} 1 = (n+1)^2 - 1$$

$$2 \sum_{k=1}^{n} k + n = n^2 + 2n + 1 - 1$$

$$\sum_{k=1}^{n} k = \frac{n^2 + n}{2} \quad \text{or} \quad \frac{n(n+1)}{2} \quad \bullet$$

## EXERCISES FOR SECTION 72

Express each of the following in summation notation.

1. $1 + 2 + 3 + \cdots + 10$

2. $1^2 + 2^2 + 3^2 + \cdots + (t-1)^2$

3. $1 + 2 + 3 + \cdots + (n-1)$

4. $1^2 + 2^3 + 3^3 + \cdots + (t+1)^3$

5. $3 + 6 + 9 + \cdots + 300$

6. $1/3 + 1/9 + 1/27 + \cdots + 1/3^k$

7. $2 + 4 + 8 + \cdots + 128$

8. $1 + 1/3 + 1/9 + 1/27 + \cdots + 1/3^{k-1}$

9. $2 + 4 + 8 + \cdots + 2^n$

10. $2 + 2 + 2 + 2 + 2 + 2$

11. $1 + 4 + 9 + 16 + 25 + 36$

12. $1 - 1 + 1 - 1 + 1 - 1$

13. $1 + 4 + 9 + 16 + \cdots + n^2$

14. $4 + 7 + 10 + 13 + \cdots + (3n+1)$

Write each of the following in expanded form.

15. $\sum_{k=2}^{6} k$

16. $\sum_{t=1}^{4} \frac{1}{t}$

17. $\sum_{i=2}^{100} i(i-1)$

18. $\sum_{k=1}^{n} (2k)^k$

19. $\sum_{k=1}^{n} 2k^k$

20. $\sum_{k=3}^{8} 3$

21. $\sum_{k=1}^{4} \frac{(-1)^k}{3^k}$

22. $\sum_{i=1}^{4} \frac{1}{2^i}$

23. $\sum_{k=1}^{n-1} \frac{1}{k(k+1)}$

24. $\sum_{k=1}^{3} \frac{1}{2^{k-1}}$

25. $\sum_{k=1}^{n} \frac{x^{2k}}{2k}$

26. $\sum_{k=1}^{n} \frac{x^{2k-1}}{2k-1}$

27. Show that

$$\sum_{k=1}^{n} a_k b_k \neq \left( \sum_{k=1}^{n} a_k \right)\left( \sum_{k=1}^{n} b_k \right)$$

28. Prove Theorem 72.2.

29. Use the result of Example 72.5 and the fact that

$$\sum_{k=1}^{n} (k + 1)^3 - k^3 = (n + 1)^3 - 1$$

to prove that

$$\sum_{k=1}^{n} k^2 = \frac{n(n + 1)(2n + 1)}{6}$$

# SECTION 73
## ARITHMETIC SEQUENCES AND SERIES

**Definition 73.1** An **arithmetic progression** is a sequence of numbers in which each term after the first is obtained by adding a fixed number $d$ to the preceding term. $d$ is called the **common difference.**

If $a$ is the first term of an arithmetic progression, with a common difference $d$, then the first $n$ terms of this sequence are

$$a, a + d, a + 2d, a + 3d, \ldots, [a + (n - 1)d]$$

where the $n$th or last term, denoted $l$, is given by the formula

$$l = a + (n - 1)d \tag{73.1}$$

### Example 73.1
The first four terms of the arithmetic progression with a first term equal to 2 and a common difference equal to 3 are 2, 5, 8, 11. The $n$th or last term is $l = 2 + (n - 1)3$ or $l = 3n - 1$.

### Example 73.2
If the first four terms of an arithmetic progression are $-11, -8, -5$, and $-2$, find the twentieth term.

Solution. We can compute the common difference $d$ by subtracting the first term from the second term. Thus $d = -8 - (-11) = 3$. We then use formula (73.1), where $n = 20$, to find that the twentieth term is $l = -11 + (20 - 1)3 = -11 + 57$ or 46. ●

Associated with the finite arithmetic sequence $a, a + d, a + 2d, \ldots, a + (n - 1)d$ is the finite **arithmetic series.**

$$S_n = a + (a + d) + (a + 2d) + \cdots + [a + (n - 1)d] = \sum_{k=1}^{n} [a + (k - 1)d]$$

The sum of this arithmetic progression or series denoted $S_n$ can be expressed in terms of $n$, the number of terms; $a$, the first term; and $l$, the last term, as follows.

We can express $S_n$ in terms of $a$, $l$, and $d$ as

$$S_n = a + (a + d) + (a + 2d) + \cdots + l \qquad \textbf{(A)}$$

We can also express $S_n$ in reverse order in terms of $a$, $l$ and $d$ as

$$S_n = l + (l - d) + (l - 2d) + \cdots + a \qquad \textbf{(B)}$$

We now add forms (A) and (B) to obtain

$$2S_n = (a + l) + [(a + d) + (l - d)] + \cdots + (l + a)$$
$$2S_n = (a + l) + (a + l) + \cdots + (a + l)$$
$$2S_n = n(a + l)$$

$$\boxed{S_n = \frac{n(a + l)}{2}} \qquad (73.2)$$

If we wish to express $S_n$ in terms of $a$, $n$, and $d$ we can substitute (73.1) into (73.2) to obtain

$$\boxed{S_n = \frac{n}{2}[2a + (n - 1)d]} \qquad (73.3)$$

**Example 73.3**

Find the sum of the first 15 terms of the arithmetic progression $4, 7, 10, \ldots$.

**Solution.** We first compute $d = 7 - 4 = 3$ and then use (73.1) to find the fifteenth term $l = 4 + (15 - 1)3 = 46$. Finally, we use (73.2) to obtain

$$S_{15} = \frac{n(a + l)}{2} = \frac{15(4 + 46)}{2} = 375 \quad \bullet$$

**Example 73.4**

Find the sum of the first $n$ consecutive odd natural numbers.

**Solution.** This set is an arithmetic progression with $l = 2n - 1$ and $a = 1$. We apply (73.2) to obtain

$$S_n = \frac{n(1 + 2n - 1)}{2} = \frac{2n^2}{2} = n^2 \quad \bullet$$

## EXERCISES FOR SECTION 73

For each of the following arithmetic progressions, find an $n$th term and the sum.

**1.** $2, 5, 8, \ldots$ to 10 terms

**2.** $11, 5, -1, \ldots$ to 9 terms

**3.** $-\frac{2}{3}, \frac{2}{3}, 2, \ldots$ to 15 terms

**4.** $32, 29, 26, \ldots$ to 10 terms

**5.** $x - 2y, x - y, x, \ldots$ to 20 terms

6. Find the sum of the first $n$ terms of the following arithmetic progressions.
   (a) $3, 5, 7, \ldots$      (c) $1, -1, -3, \ldots$
   (b) $5, 9, 13, \ldots$      (d) $\frac{3}{2}, 5, \frac{17}{2}, \ldots$

7. If $l = 43$, $n = 10$, and $S_n = 250$, find $a$ and $d$.

8. If $S_n = 48$, $a = 3$, and $l = 5$, find $n$.

9. If $l = 27$, $n = 11$, and $d = 6$, find $a$ and $S_n$.

10. Find the sum of all the even integers between 27 and 31.

11. Find the sum of all the odd integers between 28 and 462.

12. Find the sum of all integers divisible by 5 between 17 and 119.

13. If the sum of the first 20 terms of an arithmetic progression is 590 and if the sum of the first 40 terms of the same progression is 2380, find the $n$th term of the progression.

14. The terms between any two given terms of a progression are called the **means** between these two terms. For example, in the progression $3, 7, 11, 15, 19, \ldots$, 7 is the arithmetic mean between 3 and 11 and 11 and 15 are the arithmetic means between 7 and 19.
    (a) Find the arithmetic mean between 7 and 19.
    (b) Insert three arithmetic means between 7 and 19.
    (c) Insert four arithmetic means between 7 and 19.

15. A **harmonic progression** is a sequence of numbers whose reciprocals form an arithmetic progression. Determine which of the following sequences are harmonic.
    (a) $\frac{1}{2}, \frac{1}{5}, \frac{1}{8}, \ldots$
    (b) $3, 6, 9, \ldots$
    (c) $\frac{2}{3}, \frac{1}{5}, \frac{2}{17}, \ldots$
    (d) Insert two harmonic means between 5 and 10.
    (e) Find the twentieth term of the harmonic progression $\frac{1}{4}, \frac{1}{9}, \frac{1}{14}, \ldots$

16. Prove that the following is an alternative definition for an arithmetic progression by using formula (73.1): An **arithmetic progression** is any sequence whose sequence function is linear.

SECTION **74**
## GEOMETRIC SEQUENCES AND SERIES

**Definition 74.1**  A **geometric progression** is a sequence of numbers in which each term after the first is obtained from the preceding one by multiplying the preceding term by a fixed number $r$ called the **common ratio.**

If $a$ is the first term of a geometric progression, with a common ratio $r$, then the first $n$ terms of this sequence are

$$a, ar, ar^2, ar^3, \ldots, ar^{n-1}$$

where the $n$th or last term, denoted $l$, is

$$l = ar^{n-1} \tag{74.1}$$

**Example 74.1**

Find the first four terms of a geometric progression whose first term is 3 and whose common ratio is $\frac{1}{3}$. Then find the $n$th term.

**Solution.** We are given that $a = 3$ and $r = \frac{1}{3}$. Thus the first four terms of the geometric progression are $3, 1, \frac{1}{3}, \frac{1}{9}$, and the $n$th term is $l = 3(\frac{1}{3})^{n-1}$.   ●

**Example 74.2**

If the first three terms of a geometric progression are 18, 6, and 2, find the common ratio $r$ and then find the fifteenth term.

**Solution.** We compute the common ratio $r$ by dividing the second term 6 by the first term 18 to obtain $r = \frac{6}{18} = \frac{1}{3}$. We now compute the fifteenth term by substituting $r = \frac{1}{3}$, $a = 18$, and $n = 15$ into formula (74.1) to obtain $l = 18(\frac{1}{3})^{15-1} = 18(\frac{1}{3})^{14}$   ●

Associated with the finite geometric sequence $a, ar, ar^2, \ldots, ar^{n-1}$ is the finite **geometric series**

$$S_n = a + ar + ar^2 + \cdots + ar^{n-1} = \sum_{k=1}^{n} ar^{k-1} \qquad \text{(A)}$$

The sum of this geometric series, denoted by $S_n$ or $\displaystyle\sum_{k=1}^{n} ar^{k-1}$, can be expressed in terms of $a$, the first term; $r$, the common ratio; $n$, the number of terms, as follows. We multiply $S_n$ by $r$ and obtain

$$rS = ar + ar^2 + ar^3 + \cdots + ar^{n-1} + ar^n \qquad \text{(B)}$$

Next we subtract (B) from (A) to obtain

$$S_n - rS_n = a - ar^n$$
$$S_n(1 - r) = a(1 - r^n)$$

$$S_n = \frac{a(1 - r^n)}{1 - r}, \quad r \neq 1 \qquad (74.2)$$

If we wish to express $S_n$ in terms of $a$, $r$, and $l$, we write (74.2) as

$$S_n = \frac{a - r(ar^{n-1})}{1 - r} = \frac{a - rl}{1 - r}, \quad r \neq 1 \qquad (74.3)$$

It should be noted from (A) that when $r = 1$, then $S_n = \displaystyle\sum_{k=1}^{n} a = na$.

If any three of the elements $a$, $n$, $r$, $l$, and $S_n$ are given, the remaining two can be found by use of formulas (74.1), (74.2), and (74.3).

**Example 74.3**
Find the sixth term and the sum of the first six terms of the geometric progression 6, 12, 24, . . . .

Solution.  Here we have $a = 6, r = \frac{12}{6} = 2$, and $n = 6$. We obtain the sixth term, $l = 6(2)^5 = 192$, by using (74.1) and then we use (74.3) to compute

$$S_6 = \frac{6 - (2)192}{1 - 2} = 378 \quad \bullet$$

## EXERCISES FOR SECTION 74

For each of the following geometric progressions, find an $n$th term and the sum.

1. $2, 1, \frac{1}{2}, \ldots$ to 7 terms

2. $2, 6, 18, \ldots$ to 10 terms

3. $\frac{1}{2}, -\frac{5}{2}, \ldots$ to 5 terms

4. $\frac{1}{36}, \frac{1}{30}, \ldots$ to 5 terms

5. $1, 2, 4, \ldots$ to 10 terms

6. $\dfrac{x}{y}, \dfrac{x}{yz}, \ldots$ to 8 terms

7. $\dfrac{a}{b}, -1, \ldots$ to 5 terms

8. Find the first term of a geometric progression if the sixth term is 729 and the common ratio is 3.

9. Find a value for $b$ such that $-\frac{3}{2}, b, -\frac{2}{27}$ forms a geometric progression.

10. If $a = 16, n = 6$, and $l = \frac{1}{2}$, find $S_n$ and $r$.

11. If the second term of a geometric progression is 3 and the fifth term is $\frac{81}{8}$, find the tenth term.

12. Prove that $x, x + 2, x + 4$ cannot be a geometric progression.

13. Evaluate each of the following geometric series.

(a) $\displaystyle\sum_{k=1}^{5} \left(\frac{1}{2}\right)^{k-1}$     (c) $\displaystyle\sum_{k=1}^{10} \left(\frac{1}{e}\right)^{k}$

(b) $\displaystyle\sum_{k=1}^{5} \left(\frac{1}{2}\right)^{k}$     (d) $\displaystyle\sum_{k=1}^{6} \left(\frac{1}{5}\right)^{k}$

14. A ball is dropped from a height of 20 feet. If it rebounds one-half the distance it has fallen after each fall, how far will it rebound the fifth time? Through what distance has it traveled when it strikes the ground the sixth time?

15. An automobile purchased for $5,000 depreciates in value 15 per cent every year. Find its value after 4 years.

16. The population of a certain city is $P$. If it increases 5 per cent every year, what will the population be at the end of 10 years?

**17.** A sum of $P$ dollars is invested today at an interest rate of $r$ per cent per year.
   (a) Find what amount will accumulate in 5 years if interest is compounded annually.
   (b) Suppose that the amount is $400 and the interest rate is 6 per cent compounded semiannually; what amount will be accumulated in 5 years?

**18.** (a) Find the geometric means between 9 and 4. (See Exercise 14 in Section 73.)
   (b) Find the geometric means between 9 and 6.
   (c) Insert three geometric means between 18 and 288.

**19.** Prove that the following is an alternative definition for a geometric progression by using formula (74.1): A **geometric progression** is any sequence whose sequence function is of the form $ka^y$, where $y$ is a linear function of $x$.

**20.** Suppose that $a > 0, r > 0$, in the geometric progression $a, ar, ar^2, \ldots, ar^{n-1}$. Show that $\log a$, $\log ar$, $\log ar^2$, $\ldots$, $\log ar^{n-1}$ forms an arithmetic progression.

**21.** Show that the reciprocals of a geometric progression also form a geometric progression.

**22.** Prove that the arithmetic mean between $a$ and $b$ is greater than the geometric mean between $a$ and $b$, $a \neq b$.

**23.** Show that if $A$, $G$, and $H$ are the respective arithmetic, geometric, and harmonic means between $a$ and $b$, then $G^2 = AH$.

# SECTION 75
## INFINITE GEOMETRIC SERIES

We would now like to investigate the **infinite geometric series** associated with the infinite geometric sequence

$$a, ar, ar^2, \ldots, ar^{n-1}, \ldots.$$

This infinite series can be described as

$$S = a + ar + ar^2 + \cdots + ar^{n-1} + \cdots$$

where the symbol $S$ stands for the "sum" of this series, if it exists. It should be carefully noted that the word "sum" is not being used in the ordinary sense but to imply an entirely new concept. The following discussion and examples will illustrate this new concept.

In Section 74 we found that the sum of a finite geometric series could be described as

$$S_n = \frac{a - ar^n}{1 - r}, \quad r \neq 1$$

We can rewrite this as

$$S_n = \frac{a}{1 - r} - \left(\frac{a}{1 - r}\right)r^n, \quad r \neq 1 \tag{75.1}$$

Now, suppose that we consider all $-1 < r < 1$, or, equivalently, all $r$ such that

$|r| < 1$. When $r$ is raised to positive integral powers, $n$, we can see that $r^n$ grows numerically smaller and smaller. For example, if $r = \frac{1}{2}$, then

$$r^2 = (\tfrac{1}{2})^2 = \tfrac{1}{4} \quad r^3 = (\tfrac{1}{2})^3 = \tfrac{1}{8} \quad r^4 = (\tfrac{1}{2})^4 = \tfrac{1}{16}$$

and so on. If $r = -\frac{1}{2}$, then

$$r^2 = (-\tfrac{1}{2})^2 = \tfrac{1}{4} \quad r^3 = (-\tfrac{1}{2})^3 = -\tfrac{1}{8} \quad r^4 = (-\tfrac{1}{2})^4 = \tfrac{1}{16}$$

and so on. We would like to say that if $|r| < 1$ and $n$ is a positive integer, then $r^n$ can be made as close to zero as desired by selecting large enough values of $n$. Since $r^n$ can be made as numerically close to zero as desired, the expression $\left(\dfrac{a}{1-r}\right) r^n$ can also be made as numerically close to zero as desired. These observations and formula (75.1) enable us to define the **sum of an infinite geometric series** as follows:

$$S_\infty = \frac{a}{1-r} \quad \text{if } |r| < 1 \tag{75.2}$$

### Example 75.1

Find the sum (if it exists) for the infinite geometric series $1 + \frac{1}{3} + \frac{1}{9} + \cdots$.

Solution. Here $a = 1$ and $r = \frac{1}{3} \div 1 = \frac{1}{3}$. Since $|r| < 1$ we can apply (75.2) to obtain

$$S_\infty = \frac{1}{1 - (\frac{1}{3})} = \frac{1}{\frac{2}{3}} = \frac{3}{2}$$

*Note:* Using summation notation, we can say that

$$\sum_{k=1}^{\infty} \left(\frac{1}{3}\right)^{k-1} = \frac{3}{2} \quad \bullet$$

### Example 75.2

Find the sum (if it exists) for the infinite geometric series

$$\sum_{k=1}^{\infty} 3\left(\frac{1}{5}\right)^{k-1}$$

Solution. The given series in expanded form is

$$\sum_{k=1}^{\infty} 3\left(\frac{1}{5}\right)^{k-1} = 3 + \frac{3}{5} + \frac{3}{25} + \cdots$$

Here $a = 3$ and $r = \frac{1}{5}$. Since $|r| < 1$, we can apply (75.2) to obtain

$$S_\infty = \frac{3}{1 - (\frac{1}{5})} = \frac{3}{(\frac{4}{5})} = \frac{15}{4}$$

or

$$\sum_{k=1}^{\infty} 3 \left(\frac{1}{5}\right)^{k-1} = \frac{15}{4} \quad \bullet$$

**Example 75.3**

Find the sum (if it exists) for the infinite geometric series $1 + 2 + 4 + \cdots$.

Solution.   Here $a = 1$ and $r = 2$. Since $|r| > 1$, there is *no sum*.   $\bullet$

**Example 75.4**

Express $0.12\overline{12}$ in the form $p/q$, where $p$ and $q$ are integers.

Solution.   We note that $0.12\overline{12}$ can be written as the infinite geometric series $0.12 + 0.0012 + 0.000012 + \cdots$. Here $a = 0.12$ and $r = 0.0012/0.12 = 0.01$. We now apply (75.2) to obtain

$$S_{\infty} = \frac{0.12}{1 - 0.01} = \frac{0.12}{0.99} = \frac{4}{33}$$

Thus $0.12\overline{12} = \dfrac{4}{33}$.   $\bullet$

If we use sigma notation to express the infinite geometric series

$$a + ar + ar^2 + \cdots + ar^{k-1} + \cdots, |r| < 1,$$

we can express the sum of this series as

$$\sum_{k=1}^{\infty} ar^{k-1} = \frac{a}{1 - r}; |r| < 1$$

## EXERCISES FOR SECTION 75

Find the sum for each of the following infinite geometric series if it exists. If the sum does not exist, state why.

**1.** $1 + \frac{1}{2} + \frac{1}{4} + \cdots$

**2.** $1 + \frac{1}{5} + \frac{1}{25} + \cdots$

**3.** $4 + 2 + 1 + \cdots$

**4.** $1 - \frac{1}{2} + \frac{1}{4} - \cdots$

**5.** $6 + 2 + \frac{2}{3} + \cdots$

**6.** $\frac{1}{3} - \frac{2}{9} + \frac{4}{27} - \cdots$

**7.** $2 + 0.2 + 0.02 + \cdots$

**8.** $\sum_{k=1}^{\infty} 2 \left(\frac{4}{5}\right)^{k-1}$

**9.** $\sum_{k=1}^{\infty} \left(\frac{4}{5}\right)^{k}$

**10.** $\sum_{k=1}^{\infty} 3^{k-1}$

**11.** $\sum_{k=1}^{\infty} 2^{k}$

**12.** $\sum_{k=1}^{\infty} \frac{1}{e^{k}}$

**13.** $\sum_{k=1}^{\infty} \left(\frac{1}{e}\right)^{k-1}$

Express each of the following in the form $p/q$, where $p$ and $q$ are integers.

**14.** $0.13\overline{13}$    **17.** $0.99\overline{9}$    **20.** $1.21\overline{21}$

**15.** $0.060\overline{6}$    **18.** $2.\overline{142857}$    **21.** $3.123\overline{123}$

**16.** $0.160\overline{160}$    **19.** $2.11\overline{1}$    **22.** $0.59\overline{9}$

**23.** Suppose that a ball rebounds two-thirds the distance it falls. If we assume infinitely many bounces, find the distance a ball travels in coming to rest if it is dropped from a height of 30 feet.

**24.** Suppose that each swing of a pendulum bob is three-fourths as long as the preceding swing. If the first swing is 40 inches long and if we assume an infinite number of swings, how far does the bob travel before it comes to rest?

SECTION **76**
## MATHEMATICAL INDUCTION

**Mathematical induction** is the process of proving a general theorem by examining at least one special case. Mathematical induction allows us to prove a general theorem or formula for an infinity of cases without examining each case separately. There are two distinct steps in a proof by mathematical induction. They are often called the **basis of induction** and the **induction step,** respectively.

Given the statement $P(n)$ where $n \in N$. If

(1) $P(1)$ is true

and

(2) If $P(k)$ is true, then $P(k + 1)$ is true.

Then

$P(n)$ is true for all $n \in N$

**Example 76.1**
Prove that if $n$ is any natural number, then

$$4 + 8 + 12 + \cdots + 4n = 2n(n + 1)$$

**Proof.** We must establish both steps of the mathematical induction process. If we let $n = 1$ in the given formula, we have $4 = 2(1)(1 + 1)$. Thus step 1 is satisfied.

Next we assume that the formula is true for $n = k$. That is,

$$4 + 8 + 12 + \cdots + 4k = 2k(k + 1) \tag{A}$$

With this assumption we hope to show that the formula is true for $n = k + 1$. That is,

$$4 + 8 + 12 + \cdots + 4k + 4(k + 1) = 2(k + 1)(k + 2) \tag{B}$$

We now add $4(k + 1)$ to both sides of (A) to obtain

$$4 + 8 + 12 + \cdots + 4k + 4(k + 1) = 2k(k + 1) + 4(k + 1)$$

or

$$4 + 8 + 12 + \cdots + 4k + 4(k + 1) = (k + 1)(2k + 4)$$

or

$$4 + 8 + 12 + \cdots + 4k + 4(k + 1) = 2(k + 1)(k + 2)$$

This last form is (B). Thus step 2 has also been established. We can now conclude that the formula is valid for *all n*.  ●

**Example 76.2**

Prove that if $n$ is any natural number, then

$$1^2 + 2^2 + 3^2 + \cdots + n^2 = \frac{n(n + 1)(2n + 1)}{6}$$

**Proof.**  We see that the formula is valid when we replace $n$ by 1, as follows:

$$1^2 = \frac{1(1 + 1)(2 + 1)}{6} = \frac{6}{6} = 1$$

Thus step 1 is established. Next, we assume that the formula is true for $n = k$. That is,

$$1^2 + 2^2 + 3^2 + \cdots + k^2 = \frac{k(k + 1)(2k + 1)}{6} \tag{A}$$

Now, we must show, with this assumption (A), that the formula is also true for $n = k + 1$. We can do this by adding the term $(k + 1)^2$ to both sides of (A), as follows, to obtain result B:

$$1^2 + 2^2 + 3^2 + \cdots + k^2 + (k + 1)^2 = \frac{k(k + 1)(2k + 1)}{6} + (k + 1)^2$$

or

$$1^2 + 2^2 + 3^2 + \cdots + k^2 + (k + 1)^2 = \frac{k(k + 1)(2k + 1) + 6(k + 1)^2}{6}$$

or

$$1^2 + 2^2 + 3^2 + \cdots + k^2 + (k + 1)^2 = \frac{(k + 1)(2k^2 + 7k + 6)}{6}$$

or

$$1^2 + 2^2 + 3^2 + \cdots + k^2 + (k + 1)^2 = \frac{(k + 1)(k + 2)(2k + 3)}{6} \tag{B}$$

Form (B) is the original formula when we replace $n$ by $k + 1$. Thus step (2) is also established and the induction process is completed.  ●

Neither step 1 nor step 2 alone is sufficient in a proof by mathematical induction (see Examples 76.3 and 76.4).

**Example 76.3**

Consider the false formula

$$2 + 4 + 6 + \cdots + 2n = n(n + 1) + (n - 1)$$

We see that if we let $n = 1$, the formula is true, since $2 = 1(1 + 1) + (1 - 1)$. Thus step 1 is true. However, if we assume that this formula is true for $n = k$, we cannot establish that the formula is also true for $n = k + 1$. We leave the details of this to the reader. The formula is false since we cannot establish *both* steps of the induction process.

**Example 76.4**

Consider the false formula

$$2 + 4 + 6 + \cdots + 2n = n(n + 1) + 1$$

If we assume that this formula is true for $n = k$, we have

$$2 + 4 + 6 + \cdots + 2k = k(k + 1) + 1 \tag{A}$$

If we now add the term $2(k + 1)$ to both sides of (A), we obtain

$$2 + 4 + 6 + \cdots + 2k + 2(k + 1) = k(k + 1) + 1 + 2(k + 1)$$

or

$$2 + 4 + 6 + \cdots + 2k + 2(k + 1) = (k^2 + 3k + 2) + 1$$

or

$$2 + 4 + 6 + \cdots + 2k + 2(k + 1) = (k + 1)(k + 2) + 1 \tag{B}$$

Result (B) is the original formula when $n$ is replaced by $k + 1$. Thus step 2 is established. However, if we let $n = 1$ in the given formula, we see that step 1 fails since

$$2 \neq 1(1 + 1) + 1$$

Thus the formula is false since we cannot establish *both* steps of the induction process.

## EXERCISES FOR SECTION 76

Prove each of the following by mathematical induction. Assume that $n$ is a natural number.

**1.** $1 + 2 + 3 + \cdots + n = \dfrac{n(n + 1)}{2}$

**2.** $1^3 + 2^3 + 3^3 + \cdots + n^3 = \left[ \dfrac{n(n + 1)}{2} \right]^2$

**3.** $1 + 3 + 5 + \cdots + (2n - 1) = n^2$

**4.** $5 + 8 + 11 + \cdots + (3n + 2) = \dfrac{n(3n + 7)}{2}$

**5.** $6 + 10 + 14 + \cdots + (4n + 2) = n(2n + 4)$

**6.** $\dfrac{1}{1(2)} + \dfrac{1}{2(3)} + \dfrac{1}{3(4)} + \cdots + \dfrac{1}{n(n + 1)} = \dfrac{n}{n + 1}$

**7.** $a + ar + ar^2 + \cdots + ar^{n-1} = \dfrac{a - ar^n}{1 - r}, \quad r \neq 1$

**8.** $3 \cdot 1^2 + 3 \cdot 2^2 + 3 \cdot 3^2 + \cdots + 3n^2 = \dfrac{n(n + 1)(2n + 1)}{2}$

**9.** $2 + 2^2 + 2^3 + \cdots + 2^n = 2^{n+1} - 2$

**10.** $1 + \dfrac{1}{3} + \left(\dfrac{1}{3}\right)^2 + \cdots + \left(\dfrac{1}{3}\right)^{n-1} = \dfrac{3}{2}\left[1 - \left(\dfrac{1}{3}\right)^n\right]$

**11.** $a + (a + d) + (a + 2d) + \cdots + [a + (n - 1)d] = \dfrac{n[2a + (n - 1)d]}{2}$

**12.** $2 + 5 + 8 + \cdots + (3n - 1) = \dfrac{n(3n + 1)}{2}$

**13.** $\left(1 + \dfrac{1}{1}\right)\left(1 + \dfrac{1}{2}\right)\left(1 + \dfrac{1}{3}\right) \cdots \left(1 + \dfrac{1}{n}\right) = n + 1$

**14.** $3 + 6 + 9 + \cdots + 3n = \dfrac{3n(n + 1)}{2}$

**15.** $3^n < 3^{n+1}$

**16.** $3n \le 3^n$

# SECTION 77
## THE BINOMIAL THEOREM

The question arises: Is there a formula for expanding the binomial $(a + b)$ to any power? The answer is yes, and in this section we shall develop what is known as the **binomial formula.** This formula will have some restrictions on the powers of $n$.

    Before we can develop the binomial formula, we wish to establish a convenient notation known as **factorial notation.**

**Definition 77.1**   The symbol $n!$ (read "$n$ factorial") is the product of all positive integers between 1 and $n$ inclusive. That is,

$$n! = 1 \times 2 \times 3 \times \cdots \times n \tag{77.1}$$

**Example 77.1**

$4! = 1 \times 2 \times 3 \times 4 = 24$

$5! = 1 \times 2 \times 3 \times 4 \times 5 = 120$

$\dfrac{5!}{4!} = \dfrac{1 \times 2 \times 3 \times 4 \times 5}{1 \times 2 \times 3 \times 4} = 5$

$n!$ can also be expressed as

$$n! = n(n - 1)(n - 2) \cdots 3 \cdot 2 \cdot 1 \tag{77.2}$$

We can also express $n!$ as

$$n! = n(n - 1)! \qquad (77.3)$$

Since $1! = 1$, we can substitute 1 for $n$ in (77.3) to obtain

$$1! = 1(1 - 1)!$$

or

$$1! = 1 \cdot 0!$$

Now, if we wish to define $0!$ and retain the fact that $1! = 1$, we must insist that $1 \cdot 0! = 1$. The only possible real value for $0!$ is 1. Thus we define $0!$ to be equal to 1.

**Definition 77.2** The symbol $\binom{n}{r}$, where both $n$ and $r$ are positive integers and $n \geq r$, is called the **binomial coefficient symbol.** We define it as

$$\binom{n}{r} = \frac{n!}{(n - r)!r!} \qquad (77.4)$$

**Example 77.2**

Evaluate $\binom{5}{2}$, $\binom{3}{3}$, and $\binom{4}{0}$.

Solution. We use formula (77.4) as follows:

$$\binom{5}{2} = \frac{5!}{(5 - 2)!2!} = \frac{5!}{3!2!} = \frac{5 \cdot 4 \cdot 3!}{3! \cdot 2} = 10$$

$$\binom{3}{3} = \frac{3!}{(3 - 3)!3!} = \frac{3!}{0!3!} = \frac{3!}{3!} = 1$$

$$\binom{4}{0} = \frac{4!}{(4 - 0)!0!} = \frac{4!}{4!0!} = \frac{4!}{4!} = 1 \quad \bullet$$

**Example 77.3**

Evaluate $\binom{n}{n}$, $\binom{n}{0}$, and $\binom{n}{n - r}$.

Solution. We again use formula (77.4):

$$\binom{n}{n} = \frac{n!}{(n - n)!n!} = \frac{n!}{0!n!} = 1$$

$$\binom{n}{0} = \frac{n!}{(n - 0)!0!} = \frac{n!}{n!0!} = 1$$

$$\binom{n}{n - r} = \frac{n!}{[n - (n - r)]!(n - r)!} = \frac{n!}{r!(n - r)!}$$

Note that

$$\binom{n}{r} = \binom{n}{n - r} \quad \bullet$$

We are now prepared to prove the following theorem.

### Theorem 77.1  Binomial Theorem.   If $n$ is a positive integer, then

$$(a + b)^n = \binom{n}{0}a^n + \binom{n}{1}a^{n-1}b + \binom{n}{2}a^{n-2}b^2 + \cdots$$

$$+ \binom{n}{r}a^{n-r}b^r + \cdots + \binom{n}{n}b^n \qquad (77.5)$$

Proof:   We will use mathematical induction to prove this theorem. Examination of (77.6) clearly indicates that step 1 of the induction process is satisfied for $n = 1, 2, 3, 4,$ or $5$.

$$(a + b)^0 = 1$$
$$(a + b)^1 = a + b$$
$$(a + b)^2 = a^2 + 2ab + b^2$$
$$(a + b)^3 = a^3 + 3a^2b + 3ab^2 + b^3 \qquad\qquad (77.6)$$
$$(a + b)^4 = a^4 + 4a^3b + 6a^2b^2 + 4ab^3 + b^4$$
$$(a + b)^5 = a^5 + 5a^4b + 10a^3b^2 + 10a^3b^3 + 5ab^4 + b^5$$

We will assume that 77.5 is true for $n = k$. That is,

$$(a + b)^k = \binom{k}{0}a^k + \binom{k}{1}a^{k-1}b + \binom{k}{2}a^{k-2}b^2 + \cdots$$

$$+ \binom{k}{r-1}a^{k-r+1}b^{r-1} + \binom{k}{r}a^{k-r}b^r + \cdots + \binom{k}{k}b^k \qquad \textbf{(A)}$$

Now, we multiply formula (A) by $(a + b)$ to obtain

$$(a + b)^k(a + b) = a(a + b)^k + b(a + b)^k$$

or

$$(a + b)^{k+1} = a\left[ \binom{k}{0}a^k + \binom{k}{1}a^{k-1}b + \cdots + \binom{k}{r-1}a^{k-r+1}b^{r-1} \right.$$

$$+ \binom{k}{r}a^{k-r}b^r + \cdots + \binom{k}{k}b^k \right] + b\left[ \binom{k}{0}a^k + \binom{k}{1}a^{k-1}b + \cdots \right.$$

$$+ \binom{k}{r-1}a^{k-r+1}b^{r-1} + \binom{k}{r}a^{k-r}b^r + \cdots + \binom{k}{k}b^k \right]$$

or

$$(a + b)^{k+1} = \left[ \binom{k}{0}a^{k+1} + \binom{k}{1}a^kb + \cdots + \binom{k}{r-1}a^{k-r+2}b^{r-1} \right.$$

$$+ \binom{k}{r}a^{k-r+1}b^r + \cdots + \binom{k}{k}ab^k \right] + \left[ \binom{k}{0}a^kb + \binom{k}{1}a^{k-1}b^2 + \cdots \right.$$

$$+ \binom{k}{r-1}a^{k-r+1}b^r + \binom{k}{r}a^{k-r}b^{r+1} + \cdots + \binom{k}{k}b^{k+1} \right] \qquad \textbf{(B)}$$

We now note in (B) that the two terms involving $a^{k-r+1}$ and $b^r$ represent the

$(r + 1)$th term in the expansion of (B). A closer look at these two terms reveals the following:

$$\binom{k}{r}a^{k-r+1}b^r + \binom{k}{r-1}a^{k-r+1}b^r = \left[\binom{k}{r} + \binom{k}{r-1}\right]a^{k-r+1}b^r$$

We can now simplify the coefficient of $a^{k-r+1}b^r$ as follows:

$$
\begin{aligned}
\left[\binom{k}{r} + \binom{k}{r-1}\right] &= \frac{k!}{(k-r)!r!} + \frac{k!}{(k-r+1)!(r-1)!} \\
&= \frac{(k-r+1)k!}{(k-r+1)(k-r)!r!} + \frac{(r)k!}{(k-r+1)!(r)(r-1)!} \\
&= \frac{(k-r+1)k!}{(k-r+1)!r!} + \frac{(r)k!}{(k-r+1)!r!} \\
&= \frac{(k-r+1+r)k!}{(k-r+1)!r!} \\
&= \frac{(k+1)k!}{(k+1-r)!r!} \\
&= \frac{(k+1)!}{(k+1-r)!r!} = \binom{k+1}{r}
\end{aligned}
$$

Thus we have

$$\left[\binom{k}{r} + \binom{k}{r-1}\right]a^{k-r+1}b^r = \binom{k+1}{r}a^{k+1-r}b^r$$

But, this is precisely what we would obtain for the $(r + 1)$th term in (77.5) if we let $n = k + 1$. Thus form (B) is equivalent to

$$(a + b)^{k+1} = \binom{k+1}{0}a^{k+1} + \binom{k+1}{1}a^k b + \cdots + \binom{k+1}{r}a^{k+1-r}b^r$$
$$+ \cdots + \binom{k+1}{k+1}b^{k+1}$$

and step 2 of the induction process is completed. If we use the definition for $\binom{n}{r}$ to expand the coefficients in (77.5) we obtain

$$
\begin{aligned}
(a + b)^n = a^n + na^{n-1}b + \frac{n(n-1)}{2}a^{n-2}b^2 \\
+ \frac{n(n-1)(n-2)}{2\cdot3}a^{n-3}b^3 + \cdots + b^n \qquad (77.7) \quad \bullet
\end{aligned}
$$

### Remark 77.1

The first term in (77.7) is $a^n$ and the last term is $b^n$. There are $(n + 1)$ terms in the expansion. In successive terms after the first, the exponent of $a$ decreases by 1 and the exponent of $b$ increases by 1. The sum of the exponents of any one term is always $n$. The coefficients of the first and last terms are 1. The coefficient of any other term can be found by multiplying the coefficient of the preceding term by the exponent of $a$ in the preceding term and then dividing by the number of terms that precede it. The $r$th term in the binomial expansion is

$$\binom{n}{r-1} a^{n-r+1} b^{r-1}$$

**Example 77.4**

Expand and simplify $(x + 2y)^4$.

Solution.   By (77.7) we have

$$(x + 2y)^4 = x^4 + 4x^3(2y) + \frac{4(3)}{2} x^2(2y)^2 + \frac{4(3)(2)}{2(3)} x(2y)^3 + (2y)^4$$

or

$$(x + 2y)^4 = x^4 + 8x^3y + 24x^2y^2 + 32xy^3 + 16y^4 \quad \bullet$$

**Example 77.5**

Expand and simplify $(2x - y^2)^5$.

Solution.   By (77.7) we have

$$(2x - y^2)^5 = [2x + (-y^2)]^5$$

or

$$(2x - y^2)^5 = (2x)^5 + 5(2x)^4(-y^2) + 10(2x)^3(-y^2)^2$$
$$+ 10(2x)^2(-y^2)^3 + 5(2x)(-y^2)^4 + (-y^2)^5$$

or

$$(2x - y)^5 = 32x^5 - 80x^4y^2 + 80x^3y^4 - 40x^2y^6 + 10xy^8 - y^{10} \quad \bullet$$

**Example 77.6**

Find and simplify the sixth term in the expansion of $(x^2 - 2y)^8$.

Solution.   In the expansion the sixth term will involve $b^5$ where $b = -2y$. Thus the sixth term will be

$$\binom{8}{5}(x^2)^3(-2y)^5 = 56(x^6)(-32y^5) = -1792x^6y^5 \quad \bullet$$

**Example 77.7**

Find and simplify the term involving $h^6$ in the expansion of $(x^3 + h^2)^6$.

Solution.   The term in the given expansion that involves $h^6$ must be the fourth term, since we must have $b = h^2$ or $b^3 = (h^2)^3 = h^6$. Thus we have

$$\binom{6}{3}(x^3)^3(h^2)^3 = 20x^9h^6 \quad \bullet$$

It is of interest to note that if we lift the restriction that $n$ be a positive integer, the expansion is not restricted to $(n + 1)$ terms. If $n$ is any real number, the binomial formula generates an infinite series known as a **binomial series.** There are certain restrictions on this infinite expansion that are beyond the scope of this course. We will, however, state the binomial series and give an example of its usefulness.

**Binomial Series**

$$(x + y)^n = x^n + nx^{n-1}y + \frac{n(n-1)}{2!}x^{n-2}y^2 + \frac{n(n-1)(n-2)}{3!}x^{n-3}y^3$$

$$+ \cdots \quad \text{where } n \in R, |y| < |x| \qquad (77.8)$$

### Example 77.8
Use binomial series expansion to express $(1 + x)^{1/4}$ as an infinite series, where $|x| < 1$.

Solution.   We will use (77.8), where $x = 1$, $y = x$ and $n = \frac{1}{4}$, to obtain

$$(1 + x)^{1/4} = 1 + \tfrac{1}{4}x + \frac{\frac{1}{4}(-\frac{3}{4})}{2!}x^2 + \frac{\frac{1}{4}(-\frac{3}{4})(-\frac{7}{4})}{3!}x^3 + \cdots$$

or

$$(1 + x)^{1/4} = 1 + \tfrac{1}{4}x - \tfrac{3}{32}x^2 + \tfrac{7}{128}x^3 + \cdots \quad \bullet$$

### Example 77.9
Use the results of Example 77.8 to find the value of $\sqrt[4]{1.02}$ to three decimal places.

Solution.   $\sqrt[4]{1.02}$ can be written as $(1 + 0.02)^{1/4}$. We now use the results of Example 77.8 (letting $x = 0.02$) to obtain

$$\sqrt[4]{1.02} = (1 + 0.02)^{1/4} = 1 + \tfrac{1}{4}(0.02) - \tfrac{3}{32}(0.02)^2 + \tfrac{7}{128}(0.02)^3 + \cdots$$
or
$$(1 + 0.02)^{1/4} = 1 + 0.005 - 0.00004 + \cdots$$

We can readily see that no term after the third has any effect on the first three decimal places. Thus we say that

$$\sqrt[4]{1.02} = 1 + 0.005 - 0.00004 = 1.005 \quad \bullet$$

### Example 77.10
Use the binomial expansion to derive an infinite series for $(1 + x^2)^{-1}$, where $x^2 < 1$.

Solution.   In (77.8) we let $x = 1$, $y = x^2$, and $n = -1$, to obtain

$$(1 + x^2)^{-1} = 1 + (-1)x^2 + \frac{(-1)(-2)}{2!}(x^2)^2 + \frac{(-1)(-2)(-3)}{3!}(x^2)^3 + \cdots$$

$$+ \frac{(-1)(-2)(-3)\cdots(-k)}{k!}(x^2)^k + \cdots$$

or

$$(1 + x^2)^{-1} = 1 - x^2 + x^4 - x^6 + \cdots + (-1)^k x^{2k} + \cdots$$

This special result is actually an infinite geometric series whose first term $a = 1$ and whose common ratio $x = -x^2$. If we use (75.2), we would obtain

$$S_\infty = \frac{1}{1 - (-x^2)} = \frac{1}{1 + x^2} \quad \text{where } |x^2| < 1$$

This is precisely what we have obtained by the binomial expansion. That is, the "sum" of this infinite series is $(1 + x^2)^{-1}$ or $\dfrac{1}{1 + x^2}$. In sigma notation we have

$$\sum_{k=0}^{\infty} (-1)^k x^{2k} = \frac{1}{1 + x^2}, \quad x^2 < 1 \quad \bullet$$

## EXERCISES FOR SECTION 77

**1.** Evaluate the following.

(a) $\binom{6}{2}$    (b) $\binom{6}{4}$    (c) $\binom{3}{3}$    (d) $\binom{n}{n-1}$    (e) $\binom{10}{0}$

Expand and simplify each of the following by the binomial formula.

**2.** $(x + y)^6$        **6.** $\left[ \dfrac{1}{x} + x \right]^5$        **10.** $(2x - 3y)^6$

**3.** $(x - y)^6$        **7.** $(x^3 - y^3)^5$        **11.** $(x^{1/2} + x^{-1/2})^4$

**4.** $(x - 3)^5$        **8.** $(x - 3y)^6$        **12.** $(1 + \Delta x)^5$

**5.** $\left( \dfrac{x}{y} - \dfrac{y}{x} \right)^4$        **9.** $(x^{1/3} - y^{1/3})^4$        **13.** $(1 + 0.01)^4$

Do the following exercises without expanding completely.

**14.** Find and simplify the fifth term in the expansion of $(x^2 + y^3)^8$.

**15.** Find and simplify the seventh term in the expansion of $(a - 2b)^9$.

**16.** Find and simplify the fourth term in the expansion of $(\sqrt{x} + y)^6$.

**17.** Find and simplify the eighth term in the expansion of $(x^3 - 2y)^{12}$.

**18.** Find and simplify the term involving $b^8$ in the expansion of $(2a^2 - 3b^2)^6$.

**19.** Find and simplify the term involving $y^6$ in the expansion of $\left( 3x^2 + \dfrac{y}{3} \right)^8$.

**20.** Find and simplify the term involving $x^2$ in the expansion of $\left( \dfrac{x}{2} - \dfrac{4}{x} \right)^8$.

**21.** Find the middle term in the expansion of $(x^2 - y^2)^{10}$.

**22.** Find the middle term in the expansion of $(2x + y^3)^{12}$.

Use the binomial theorem to find and simplify the first four terms of each of the following. (Assume that all restrictions have been made.)

**23.** $(x - y)^{-2}$      **26.** $(1 + 2x)^{1/2}$

**24.** $(2x + 3y)^{-3}$      **27.** $(1 + 2x)^{-1/2}$

**25.** $(1 + x)^{1/3}$      **28.** $\sqrt[4]{1 - x}$

Use the binomial series to compute the value of each of the following to an accuracy of three decimal places.

**29.** $(1.01)^6 = (1 + 0.01)^6$      **31.** $(0.98)^8 = (1 - 0.02)^8$

**30.** $\sqrt[3]{1.02} = (1 + 0.02)^{1/3}$      **32.** $\sqrt[4]{0.98} = (1 - 0.02)^{1/4}$

**33.** Expand $(x + y + z)^3$ by the binomial theorem. (*Hint:* $(x + y + z)^3 = [(x + y) + z]^3$.)

# TABLES

**Table I.** Common Logarithms (Base 10)

| N | 0 | 1 | 2 | 3 | 4 | 5 | 6 | 7 | 8 | 9 |
|---|---|---|---|---|---|---|---|---|---|---|
| 10 | 0000 | 0043 | 0086 | 0128 | 0170 | 0212 | 0253 | 0294 | 0334 | 0374 |
| 11 | 0414 | 0453 | 0492 | 0531 | 0569 | 0607 | 0645 | 0682 | 0719 | 0755 |
| 12 | 0792 | 0828 | 0864 | 0899 | 0934 | 0969 | 1004 | 1038 | 1072 | 1106 |
| 13 | 1139 | 1173 | 1206 | 1239 | 1271 | 1303 | 1335 | 1367 | 1399 | 1430 |
| 14 | 1461 | 1492 | 1523 | 1553 | 1584 | 1614 | 1644 | 1673 | 1703 | 1732 |
| 15 | 1761 | 1790 | 1818 | 1847 | 1875 | 1903 | 1931 | 1959 | 1987 | 2014 |
| 16 | 2041 | 2068 | 2095 | 2122 | 2148 | 2175 | 2201 | 2227 | 2253 | 2279 |
| 17 | 2304 | 2330 | 2355 | 2380 | 2405 | 2430 | 2455 | 2480 | 2504 | 2529 |
| 18 | 2553 | 2577 | 2601 | 2625 | 2648 | 2672 | 2695 | 2718 | 2742 | 2765 |
| 19 | 2788 | 2810 | 2833 | 2856 | 2878 | 2900 | 2923 | 2945 | 2967 | 2989 |
| 20 | 3010 | 3032 | 3054 | 3075 | 3096 | 3118 | 3139 | 3160 | 3181 | 3201 |
| 21 | 3222 | 3243 | 3263 | 3284 | 3304 | 3324 | 3345 | 3365 | 3385 | 3404 |
| 22 | 3424 | 3444 | 3464 | 3483 | 3502 | 3522 | 3541 | 3560 | 3579 | 3598 |
| 23 | 3617 | 3636 | 3655 | 3674 | 3692 | 3711 | 3729 | 3747 | 3766 | 3784 |
| 24 | 3802 | 3820 | 3838 | 3856 | 3874 | 3892 | 3909 | 3927 | 3945 | 3962 |
| 25 | 3979 | 3997 | 4014 | 4031 | 4048 | 4065 | 4082 | 4099 | 4116 | 4133 |
| 26 | 4150 | 4166 | 4183 | 4200 | 4216 | 4232 | 4249 | 4265 | 4281 | 4298 |
| 27 | 4314 | 4330 | 4346 | 4362 | 4378 | 4393 | 4409 | 4425 | 4440 | 4456 |
| 28 | 4472 | 4487 | 4502 | 4518 | 4533 | 4548 | 4564 | 4579 | 4594 | 4609 |
| 29 | 4624 | 4639 | 4654 | 4669 | 4683 | 4698 | 4713 | 4728 | 4742 | 4757 |
| 30 | 4771 | 4786 | 4800 | 4814 | 4829 | 4843 | 4857 | 4871 | 4886 | 4900 |
| 31 | 4914 | 4928 | 4942 | 4955 | 4969 | 4983 | 4997 | 5011 | 5024 | 5038 |
| 32 | 5051 | 5065 | 5079 | 5092 | 5105 | 5119 | 5132 | 5145 | 5159 | 5172 |
| 33 | 5185 | 5198 | 5211 | 5224 | 5237 | 5250 | 5263 | 5276 | 5289 | 5302 |
| 34 | 5315 | 5328 | 5340 | 5353 | 5366 | 5378 | 5391 | 5403 | 5416 | 5428 |
| 35 | 5441 | 5453 | 5465 | 5478 | 5490 | 5502 | 5514 | 5527 | 5539 | 5551 |
| 36 | 5563 | 5575 | 5587 | 5599 | 5611 | 5623 | 5635 | 5647 | 5658 | 5670 |
| 37 | 5682 | 5694 | 5705 | 5717 | 5729 | 5740 | 5752 | 5763 | 5775 | 5786 |
| 38 | 5798 | 5809 | 5821 | 5832 | 5843 | 5855 | 5866 | 5877 | 5888 | 5899 |
| 39 | 5911 | 5922 | 5933 | 5944 | 5955 | 5966 | 5977 | 5988 | 5999 | 6010 |
| 40 | 6021 | 6031 | 6042 | 6053 | 6064 | 6075 | 6085 | 6096 | 6107 | 6117 |
| 41 | 6128 | 6138 | 6149 | 6160 | 6170 | 6180 | 6191 | 6201 | 6212 | 6222 |
| 42 | 6232 | 6243 | 6253 | 6263 | 6274 | 6284 | 6294 | 6304 | 6314 | 6325 |
| 43 | 6335 | 6345 | 6355 | 6365 | 6375 | 6385 | 6395 | 6405 | 6415 | 6425 |
| 44 | 6435 | 6444 | 6454 | 6464 | 6474 | 6484 | 6493 | 6503 | 6513 | 6522 |
| 45 | 6532 | 6542 | 6551 | 6561 | 6571 | 6580 | 6590 | 6599 | 6609 | 6618 |
| 46 | 6628 | 6637 | 6646 | 6656 | 6665 | 6675 | 6684 | 6693 | 6702 | 6712 |
| 47 | 6721 | 6730 | 6739 | 6749 | 6758 | 6767 | 6776 | 6785 | 6794 | 6803 |
| 48 | 6812 | 6821 | 6830 | 6839 | 6848 | 6857 | 6866 | 6875 | 6884 | 6893 |
| 49 | 6902 | 6911 | 6920 | 6928 | 6937 | 6946 | 6955 | 6964 | 6972 | 6981 |
| 50 | 6990 | 6998 | 7007 | 7016 | 7024 | 7033 | 7042 | 7050 | 7059 | 7067 |
| 51 | 7076 | 7084 | 7093 | 7101 | 7110 | 7118 | 7126 | 7135 | 7143 | 7152 |
| 52 | 7160 | 7168 | 7177 | 7185 | 7193 | 7202 | 7210 | 7218 | 7226 | 7235 |
| 53 | 7243 | 7251 | 7259 | 7267 | 7275 | 7284 | 7292 | 7300 | 7308 | 7316 |
| 54 | 7324 | 7332 | 7340 | 7348 | 7356 | 7364 | 7372 | 7380 | 7388 | 7396 |
| N | 0 | 1 | 2 | 3 | 4 | 5 | 6 | 7 | 8 | 9 |

**Table I.** Common Logarithms (Continued)

| N | 0 | 1 | 2 | 3 | 4 | 5 | 6 | 7 | 8 | 9 |
|---|---|---|---|---|---|---|---|---|---|---|
| 55 | 7404 | 7412 | 7419 | 7427 | 7435 | 7443 | 7451 | 7459 | 7466 | 7474 |
| 56 | 7482 | 7490 | 7497 | 7505 | 7513 | 7520 | 7528 | 7536 | 7543 | 7551 |
| 57 | 7559 | 7566 | 7574 | 7582 | 7589 | 7597 | 7604 | 7612 | 7619 | 7627 |
| 58 | 7634 | 7642 | 7649 | 7657 | 7664 | 7672 | 7679 | 7686 | 7694 | 7701 |
| 59 | 7709 | 7716 | 7723 | 7731 | 7738 | 7745 | 7752 | 7760 | 7767 | 7774 |
| 60 | 7782 | 7789 | 7796 | 7803 | 7810 | 7818 | 7825 | 7832 | 7839 | 7846 |
| 61 | 7853 | 7860 | 7868 | 7875 | 7882 | 7889 | 7896 | 7903 | 7910 | 7917 |
| 62 | 7924 | 7931 | 7938 | 7945 | 7952 | 7959 | 7966 | 7973 | 7980 | 7987 |
| 63 | 7993 | 8000 | 8007 | 8014 | 8021 | 8028 | 8035 | 8041 | 8048 | 8055 |
| 64 | 8062 | 8069 | 8075 | 8082 | 8089 | 8096 | 8102 | 8109 | 8116 | 8122 |
| 65 | 8129 | 8136 | 8142 | 8149 | 8156 | 8162 | 8169 | 8176 | 8182 | 8189 |
| 66 | 8195 | 8202 | 8209 | 8215 | 8222 | 8228 | 8235 | 8241 | 8248 | 8254 |
| 67 | 8261 | 8267 | 8274 | 8280 | 8287 | 8293 | 8299 | 8306 | 8312 | 8319 |
| 68 | 8325 | 8331 | 8338 | 8344 | 8351 | 8357 | 8363 | 8370 | 8376 | 8382 |
| 69 | 8388 | 8395 | 8401 | 8407 | 8414 | 8420 | 8426 | 8432 | 8439 | 8445 |
| 70 | 8451 | 8457 | 8463 | 8470 | 8476 | 8482 | 8488 | 8494 | 8500 | 8506 |
| 71 | 8513 | 8519 | 8525 | 8531 | 8537 | 8543 | 8549 | 8555 | 8561 | 8567 |
| 72 | 8573 | 8579 | 8585 | 8591 | 8597 | 8603 | 8609 | 8615 | 8621 | 8627 |
| 73 | 8633 | 8639 | 8645 | 8651 | 8657 | 8663 | 8669 | 8675 | 8681 | 8686 |
| 74 | 8692 | 8698 | 8704 | 8710 | 8716 | 8722 | 8727 | 8733 | 8739 | 8745 |
| 75 | 8751 | 8756 | 8762 | 8768 | 8774 | 8779 | 8785 | 8791 | 8797 | 8802 |
| 76 | 8808 | 8814 | 8820 | 8825 | 8831 | 8837 | 8842 | 8848 | 8854 | 8859 |
| 77 | 8865 | 8871 | 8876 | 8882 | 8887 | 8893 | 8899 | 8904 | 8910 | 8915 |
| 78 | 8921 | 8927 | 8932 | 8938 | 8943 | 8949 | 8954 | 8960 | 8965 | 8971 |
| 79 | 8976 | 8982 | 8987 | 8993 | 8998 | 9004 | 9009 | 9015 | 9020 | 9025 |
| 80 | 9031 | 9036 | 9042 | 9047 | 9053 | 9058 | 9063 | 9069 | 9074 | 9079 |
| 81 | 9085 | 9090 | 9096 | 9101 | 9106 | 9112 | 9117 | 9122 | 9128 | 9133 |
| 82 | 9138 | 9143 | 9149 | 9154 | 9159 | 9165 | 9170 | 9175 | 9180 | 9186 |
| 83 | 9191 | 9196 | 9201 | 9206 | 9212 | 9217 | 9222 | 9227 | 9232 | 9238 |
| 84 | 9243 | 9248 | 9253 | 9258 | 9263 | 9269 | 9274 | 9279 | 9284 | 9289 |
| 85 | 9294 | 9299 | 9304 | 9309 | 9315 | 9320 | 9325 | 9330 | 9335 | 9340 |
| 86 | 9345 | 9350 | 9355 | 9360 | 9365 | 9370 | 9375 | 9380 | 9385 | 9390 |
| 87 | 9395 | 9400 | 9405 | 9410 | 9415 | 9420 | 9425 | 9430 | 9435 | 9440 |
| 88 | 9445 | 9450 | 9455 | 9460 | 9465 | 9469 | 9474 | 9479 | 9484 | 9489 |
| 89 | 9494 | 9499 | 9504 | 9509 | 9513 | 9518 | 9523 | 9528 | 9533 | 9538 |
| 90 | 9542 | 9547 | 9552 | 9557 | 9562 | 9566 | 9571 | 9576 | 9581 | 9586 |
| 91 | 9590 | 9595 | 9600 | 9605 | 9609 | 9614 | 9619 | 9624 | 9628 | 9633 |
| 92 | 9638 | 9643 | 9647 | 9652 | 9657 | 9661 | 9666 | 9671 | 9675 | 9680 |
| 93 | 9685 | 9689 | 9694 | 9699 | 9703 | 9708 | 9713 | 9717 | 9722 | 9727 |
| 94 | 9731 | 9736 | 9741 | 9745 | 9750 | 9754 | 9759 | 9763 | 9768 | 9773 |
| 95 | 9777 | 9782 | 9786 | 9791 | 9795 | 9800 | 9805 | 9809 | 9814 | 9818 |
| 96 | 9823 | 9827 | 9832 | 9836 | 9841 | 9845 | 9850 | 9854 | 9859 | 9863 |
| 97 | 9868 | 9872 | 9877 | 9881 | 9886 | 9890 | 9894 | 9899 | 9903 | 9908 |
| 98 | 9912 | 9917 | 9921 | 9926 | 9930 | 9934 | 9939 | 9943 | 9948 | 9952 |
| 99 | 9956 | 9961 | 9965 | 9969 | 9974 | 9978 | 9983 | 9987 | 9991 | 9996 |
| N | 0 | 1 | 2 | 3 | 4 | 5 | 6 | 7 | 8 | 9 |

**Table II.** Natural Logarithms (Base *e*)

| | .00 | .01 | .02 | .03 | .04 | .05 | .06 | .07 | .08 | .09 |
|---|---|---|---|---|---|---|---|---|---|---|
| **1.0** | 0.0000 | 0.0100 | 0.0198 | 0.0296 | 0.0392 | 0.0488 | 0.0583 | 0.0677 | 0.0770 | 0.0862 |
| **1.1** | 0.0953 | 0.1044 | 0.1133 | 0.1222 | 0.1310 | 0.1398 | 0.1484 | 0.1570 | 0.1655 | 0.1740 |
| **1.2** | 0.1823 | 0.1906 | 0.1989 | 0.2070 | 0.2151 | 0.2231 | 0.2311 | 0.2390 | 0.2469 | 0.2546 |
| **1.3** | 0.2624 | 0.2700 | 0.2776 | 0.2852 | 0.2927 | 0.3001 | 0.3075 | 0.3148 | 0.3221 | 0.3293 |
| **1.4** | 0.3365 | 0.3436 | 0.3507 | 0.3577 | 0.3646 | 0.3716 | 0.3784 | 0.3853 | 0.3920 | 0.3988 |
| **1.5** | 0.4055 | 0.4121 | 0.4187 | 0.4253 | 0.4318 | 0.4383 | 0.4447 | 0.4511 | 0.4574 | 0.4637 |
| **1.6** | 0.4700 | 0.4762 | 0.4824 | 0.4886 | 0.4947 | 0.5008 | 0.5068 | 0.5128 | 0.5188 | 0.5247 |
| **1.7** | 0.5306 | 0.5365 | 0.5423 | 0.5481 | 0.5539 | 0.5596 | 0.5653 | 0.5710 | 0.5766 | 0.5822 |
| **1.8** | 0.5878 | 0.5933 | 0.5988 | 0.6043 | 0.6098 | 0.6152 | 0.6206 | 0.6259 | 0.6313 | 0.6366 |
| **1.9** | 0.6419 | 0.6471 | 0.6523 | 0.6575 | 0.6627 | 0.6678 | 0.6729 | 0.6780 | 0.6831 | 0.6881 |
| **2.0** | 0.6932 | 0.6981 | 0.7031 | 0.7080 | 0.7129 | 0.7178 | 0.7227 | 0.7275 | 0.7324 | 0.7372 |
| **2.1** | 0.7419 | 0.7467 | 0.7514 | 0.7561 | 0.7608 | 0.7655 | 0.7701 | 0.7747 | 0.7793 | 0.7839 |
| **2.2** | 0.7885 | 0.7930 | 0.7975 | 0.8020 | 0.8065 | 0.8109 | 0.8154 | 0.8198 | 0.8242 | 0.8286 |
| **2.3** | 0.8329 | 0.8373 | 0.8416 | 0.8459 | 0.8502 | 0.8544 | 0.8587 | 0.8629 | 0.8671 | 0.8713 |
| **2.4** | 0.8755 | 0.8796 | 0.8838 | 0.8879 | 0.8920 | 0.8961 | 0.9002 | 0.9042 | 0.9083 | 0.9123 |
| **2.5** | 0.9163 | 0.9203 | 0.9243 | 0.9282 | 0.9322 | 0.9361 | 0.9400 | 0.9439 | 0.9478 | 0.9517 |
| **2.6** | 0.9555 | 0.9594 | 0.9632 | 0.9670 | 0.9708 | 0.9746 | 0.9783 | 0.9821 | 0.9858 | 0.9895 |
| **2.7** | 0.9933 | 0.9969 | 1.0006 | 1.0043 | 1.0080 | 1.0116 | 1.0152 | 1.0188 | 1.0225 | 1.0260 |
| **2.8** | 1.0296 | 1.0332 | 1.0367 | 1.0403 | 1.0438 | 1.0473 | 1.0508 | 1.0543 | 1.0578 | 1.0613 |
| **2.9** | 1.0647 | 1.0682 | 1.0716 | 1.0750 | 1.0784 | 1.0818 | 1.0852 | 1.0886 | 1.0919 | 1.0953 |
| **3.0** | 1.0986 | 1.1019 | 1.1053 | 1.1086 | 1.1119 | 1.1151 | 1.1184 | 1.1217 | 1.1249 | 1.1282 |
| **3.1** | 1.1314 | 1.1346 | 1.1378 | 1.1410 | 1.1442 | 1.1474 | 1.1506 | 1.1537 | 1.1569 | 1.1600 |
| **3.2** | 1.1632 | 1.1663 | 1.1694 | 1.1725 | 1.1756 | 1.1787 | 1.1817 | 1.1848 | 1.1878 | 1.1909 |
| **3.3** | 1.1939 | 1.1969 | 1.2000 | 1.2030 | 1.2060 | 1.2090 | 1.2119 | 1.2149 | 1.2179 | 1.2208 |
| **3.4** | 1.2238 | 1.2267 | 1.2296 | 1.2326 | 1.2355 | 1.2384 | 1.2413 | 1.2442 | 1.2470 | 1.2499 |
| **3.5** | 1.2528 | 1.2556 | 1.2585 | 1.2613 | 1.2641 | 1.2669 | 1.2698 | 1.2726 | 1.2754 | 1.2782 |
| **3.6** | 1.2809 | 1.2837 | 1.2865 | 1.2892 | 1.2920 | 1.2947 | 1.2975 | 1.3002 | 1.3029 | 1.3056 |
| **3.7** | 1.3083 | 1.3110 | 1.3137 | 1.3164 | 1.3191 | 1.3218 | 1.3244 | 1.3271 | 1.3297 | 1.3324 |
| **3.8** | 1.3350 | 1.3376 | 1.3403 | 1.3429 | 1.3455 | 1.3481 | 1.3507 | 1.3533 | 1.3558 | 1.3584 |
| **3.9** | 1.3610 | 1.3635 | 1.3661 | 1.3686 | 1.3712 | 1.3737 | 1.3762 | 1.3788 | 1.3813 | 1.3838 |
| **4.0** | 1.3863 | 1.3888 | 1.3913 | 1.3938 | 1.3962 | 1.3987 | 1.4012 | 1.4036 | 1.4061 | 1.4085 |
| **4.1** | 1.4110 | 1.4134 | 1.4159 | 1.4183 | 1.4207 | 1.4231 | 1.4255 | 1.4279 | 1.4303 | 1.4327 |
| **4.2** | 1.4351 | 1.4375 | 1.4398 | 1.4422 | 1.4446 | 1.4469 | 1.4493 | 1.4516 | 1.4540 | 1.4563 |
| **4.3** | 1.4586 | 1.4609 | 1.4633 | 1.4656 | 1.4679 | 1.4702 | 1.4725 | 1.4748 | 1.4771 | 1.4793 |
| **4.4** | 1.4816 | 1.4839 | 1.4861 | 1.4884 | 1.4907 | 1.4929 | 1.4951 | 1.4974 | 1.4996 | 1.5019 |
| **4.5** | 1.5041 | 1.5063 | 1.5085 | 1.5107 | 1.5129 | 1.5151 | 1.5173 | 1.5195 | 1.5217 | 1.5239 |
| **4.6** | 1.5261 | 1.5282 | 1.5304 | 1.5326 | 1.5347 | 1.5369 | 1.5390 | 1.5412 | 1.5433 | 1.5454 |
| **4.7** | 1.5476 | 1.5497 | 1.5518 | 1.5539 | 1.5560 | 1.5581 | 1.5602 | 1.5623 | 1.5644 | 1.5665 |
| **4.8** | 1.5686 | 1.5707 | 1.5728 | 1.5748 | 1.5769 | 1.5790 | 1.5810 | 1.5831 | 1.5851 | 1.5872 |
| **4.9** | 1.5892 | 1.5913 | 1.5933 | 1.5953 | 1.5974 | 1.5994 | 1.6014 | 1.6034 | 1.6054 | 1.6074 |
| **5.0** | 1.6094 | 1.6114 | 1.6134 | 1.6154 | 1.6174 | 1.6194 | 1.6214 | 1.6233 | 1.6253 | 1.6273 |
| **5.1** | 1.6292 | 1.6312 | 1.6332 | 1.6351 | 1.6371 | 1.6390 | 1.6409 | 1.6429 | 1.6448 | 1.6467 |
| **5.2** | 1.6487 | 1.6506 | 1.6525 | 1.6544 | 1.6563 | 1.6582 | 1.6601 | 1.6620 | 1.6639 | 1.6658 |
| **5.3** | 1.6677 | 1.6696 | 1.6715 | 1.6734 | 1.6752 | 1.6771 | 1.6790 | 1.6808 | 1.6827 | 1.6845 |
| **5.4** | 1.6864 | 1.6882 | 1.6901 | 1.6919 | 1.6938 | 1.6956 | 1.6974 | 1.6993 | 1.7011 | 1.7029 |

**Table II.** Natural Logarithms (Continued)

| | .00 | .01 | .02 | .03 | .04 | .05 | .06 | .07 | .08 | .09 |
|---|---|---|---|---|---|---|---|---|---|---|
| **5.5** | 1.7047 | 1.7066 | 1.7084 | 1.7102 | 1.7120 | 1.7138 | 1.7156 | 1.7174 | 1.7192 | 1.7210 |
| **5.6** | 1.7228 | 1.7246 | 1.7263 | 1.7281 | 1.7299 | 1.7317 | 1.7334 | 1.7352 | 1.7370 | 1.7387 |
| **5.7** | 1.7405 | 1.7422 | 1.7440 | 1.7457 | 1.7475 | 1.7492 | 1.7509 | 1.7527 | 1.7544 | 1.7561 |
| **5.8** | 1.7579 | 1.7596 | 1.7613 | 1.7630 | 1.7647 | 1.7664 | 1.7681 | 1.7699 | 1.7716 | 1.7733 |
| **5.9** | 1.7750 | 1.7766 | 1.7783 | 1.7800 | 1.7817 | 1.7834 | 1.7851 | 1.7868 | 1.7884 | 1.7901 |
| **6.0** | 1.7918 | 1.7934 | 1.7951 | 1.7967 | 1.7984 | 1.8001 | 1.8017 | 1.8034 | 1.8050 | 1.8066 |
| **6.1** | 1.8083 | 1.8099 | 1.8116 | 1.8132 | 1.8148 | 1.8165 | 1.8181 | 1.8197 | 1.8213 | 1.8229 |
| **6.2** | 1.8245 | 1.8262 | 1.8278 | 1.8294 | 1.8310 | 1.8326 | 1.8342 | 1.8358 | 1.8374 | 1.8390 |
| **6.3** | 1.8405 | 1.8421 | 1.8437 | 1.8453 | 1.8469 | 1.8485 | 1.8500 | 1.8516 | 1.8532 | 1.8547 |
| **6.4** | 1.8563 | 1.8579 | 1.8594 | 1.8610 | 1.8625 | 1.8641 | 1.8656 | 1.8672 | 1.8687 | 1.8703 |
| **6.5** | 1.8718 | 1.8733 | 1.8749 | 1.8764 | 1.8779 | 1.8795 | 1.8810 | 1.8825 | 1.8840 | 1.8856 |
| **6.6** | 1.8871 | 1.8886 | 1.8901 | 1.8916 | 1.8931 | 1.8946 | 1.8961 | 1.8976 | 1.8991 | 1.9006 |
| **6.7** | 1.9021 | 1.9036 | 1.9051 | 1.9066 | 1.9081 | 1.9095 | 1.9110 | 1.9125 | 1.9140 | 1.9155 |
| **6.8** | 1.9169 | 1.9184 | 1.9199 | 1.9213 | 1.9228 | 1.9242 | 1.9257 | 1.9272 | 1.9286 | 1.9301 |
| **6.9** | 1.9315 | 1.9330 | 1.9344 | 1.9359 | 1.9373 | 1.9387 | 1.9402 | 1.9416 | 1.9430 | 1.9445 |
| **7.0** | 1.9459 | 1.9473 | 1.9488 | 1.9502 | 1.9516 | 1.9530 | 1.9544 | 1.9559 | 1.9573 | 1.9587 |
| **7.1** | 1.9601 | 1.9615 | 1.9629 | 1.9643 | 1.9657 | 1.9671 | 1.9685 | 1.9699 | 1.9713 | 1.9727 |
| **7.2** | 1.9741 | 1.9755 | 1.9769 | 1.9782 | 1.9796 | 1.9810 | 1.9824 | 1.9838 | 1.9851 | 1.9865 |
| **7.3** | 1.9879 | 1.9892 | 1.9906 | 1.9920 | 1.9933 | 1.9947 | 1.9961 | 1.9974 | 1.9988 | 2.0001 |
| **7.4** | 2.0015 | 2.0028 | 2.0042 | 2.0055 | 2.0069 | 2.0082 | 2.0096 | 2.0109 | 2.0122 | 2.0136 |
| **7.5** | 2.0149 | 2.0162 | 2.0176 | 2.0189 | 2.0202 | 2.0215 | 2.0229 | 2.0242 | 2.0255 | 2.0268 |
| **7.6** | 2.0281 | 2.0295 | 2.0308 | 2.0321 | 2.0334 | 2.0347 | 2.0360 | 2.0373 | 2.0386 | 2.0399 |
| **7.7** | 2.0412 | 2.0425 | 2.0438 | 2.0451 | 2.0464 | 2.0477 | 2.0490 | 2.0503 | 2.0516 | 2.0528 |
| **7.8** | 2.0541 | 2.0554 | 2.0567 | 2.0580 | 2.0592 | 2.0605 | 2.0618 | 2.0631 | 2.0643 | 2.0656 |
| **7.9** | 2.0669 | 2.0681 | 2.0694 | 2.0707 | 2.0719 | 2.0732 | 2.0744 | 2.0757 | 2.0769 | 2.0782 |
| **8.0** | 2.0794 | 2.0807 | 2.0819 | 2.0832 | 2.0844 | 2.0857 | 2.0869 | 2.0882 | 2.0894 | 2.0906 |
| **8.1** | 2.0919 | 2.0931 | 2.0943 | 2.0956 | 2.0968 | 2.0980 | 2.0992 | 2.1005 | 2.1017 | 2.1029 |
| **8.2** | 2.1041 | 2.1054 | 2.1066 | 2.1078 | 2.1090 | 2.1102 | 2.1114 | 2.1126 | 2.1138 | 2.1150 |
| **8.3** | 2.1163 | 2.1175 | 2.1187 | 2.1199 | 2.1211 | 2.1223 | 2.1235 | 2.1247 | 2.1259 | 2.1270 |
| **8.4** | 2.1282 | 2.1294 | 2.1306 | 2.1318 | 2.1330 | 2.1342 | 2.1353 | 2.1365 | 2.1377 | 2.1389 |
| **8.5** | 2.1401 | 2.1412 | 2.1424 | 2.1436 | 2.1448 | 2.1459 | 2.1471 | 2.1483 | 2.1494 | 2.1506 |
| **8.6** | 2.1518 | 2.1529 | 2.1541 | 2.1552 | 2.1564 | 2.1576 | 2.1587 | 2.1599 | 2.1610 | 2.1622 |
| **8.7** | 2.1633 | 2.1645 | 2.1656 | 2.1668 | 2.1679 | 2.1691 | 2.1702 | 2.1713 | 2.1725 | 2.1736 |
| **8.8** | 2.1748 | 2.1759 | 2.1770 | 2.1782 | 2.1793 | 2.1804 | 2.1815 | 2.1827 | 2.1838 | 2.1849 |
| **8.9** | 2.1861 | 2.1872 | 2.1883 | 2.1894 | 2.1905 | 2.1917 | 2.1928 | 2.1939 | 2.1950 | 2.1961 |
| **9.0** | 2.1972 | 2.1983 | 2.1994 | 2.2006 | 2.2017 | 2.2028 | 2.2039 | 2.2050 | 2.2061 | 2.2072 |
| **9.1** | 2.2083 | 2.2094 | 2.2105 | 2.2116 | 2.2127 | 2.2138 | 2.2148 | 2.2159 | 2.2170 | 2.2181 |
| **9.2** | 2.2192 | 2.2203 | 2.2214 | 2.2225 | 2.2235 | 2.2246 | 2.2257 | 2.2268 | 2.2279 | 2.2289 |
| **9.3** | 2.2300 | 2.2311 | 2.2322 | 2.2332 | 2.2343 | 2.2354 | 2.2364 | 2.2375 | 2.2386 | 2.2396 |
| **9.4** | 2.2407 | 2.2418 | 2.2428 | 2.2439 | 2.2450 | 2.2460 | 2.2471 | 2.2481 | 2.2492 | 2.2502 |
| **9.5** | 2.2513 | 2.2523 | 2.2534 | 2.2544 | 2.2555 | 2.2565 | 2.2576 | 2.2586 | 2.2597 | 2.2607 |
| **9.6** | 2.2618 | 2.2628 | 2.2628 | 2.2649 | 2.2659 | 2.2670 | 2.2680 | 2.2690 | 2.2701 | 2.2711 |
| **9.7** | 2.2721 | 2.2732 | 2.2742 | 2.2752 | 2.2762 | 2.2773 | 2.2783 | 2.2793 | 2.2803 | 2.2814 |
| **9.8** | 2.2824 | 2.2834 | 2.2844 | 2.2854 | 2.2865 | 2.2875 | 2.2885 | 2.2895 | 2.2905 | 2.2915 |
| **9.9** | 2.2925 | 2.2935 | 2.2946 | 2.2956 | 2.2966 | 2.2976 | 2.2986 | 2.2996 | 2.3006 | 2.3016 |

**Table III.** Values of the Exponential Function

| x | $e^x$ | $e^{-x}$ | x | $e^x$ | $e^{-x}$ |
|---|---|---|---|---|---|
| 0.00 | 1.0000 | 1.0000 | 2.5 | 12.182 | 0.0821 |
| 0.05 | 1.0513 | 0.9512 | 2.6 | 13.464 | 0.0743 |
| 0.10 | 1.1052 | 0.9048 | 2.7 | 14.880 | 0.0672 |
| 0.15 | 1.1618 | 0.8607 | 2.8 | 16.445 | 0.0608 |
| 0.20 | 1.2214 | 0.8187 | 2.9 | 18.174 | 0.0550 |
| 0.25 | 1.2840 | 0.7788 | 3.0 | 20.086 | 0.0498 |
| 0.30 | 1.3499 | 0.7408 | 3.1 | 22.198 | 0.0450 |
| 0.35 | 1.4191 | 0.7047 | 3.2 | 24.533 | 0.0408 |
| 0.40 | 1.4918 | 0.6703 | 3.3 | 27.113 | 0.0369 |
| 0.45 | 1.5683 | 0.6376 | 3.4 | 29.964 | 0.0334 |
| 0.50 | 1.6487 | 0.6065 | 3.5 | 33.115 | 0.0302 |
| 0.55 | 1.7333 | 0.5769 | 3.6 | 36.598 | 0.0273 |
| 0.60 | 1.8221 | 0.5488 | 3.7 | 40.447 | 0.0247 |
| 0.65 | 1.9155 | 0.5220 | 3.8 | 44.701 | 0.0224 |
| 0.70 | 2.0138 | 0.4966 | 3.9 | 49.402 | 0.0202 |
| 0.75 | 2.1170 | 0.4724 | 4.0 | 54.598 | 0.0183 |
| 0.80 | 2.2255 | 0.4493 | 4.1 | 60.340 | 0.0166 |
| 0.85 | 2.3396 | 0.4274 | 4.2 | 66.686 | 0.0150 |
| 0.90 | 2.4596 | 0.4066 | 4.3 | 73.700 | 0.0136 |
| 0.95 | 2.5857 | 0.3867 | 4.4 | 81.451 | 0.0123 |
| 1.0 | 2.7183 | 0.3679 | 4.5 | 90.017 | 0.0111 |
| 1.1 | 3.0042 | 0.3329 | 4.6 | 99.484 | 0.0101 |
| 1.2 | 3.3201 | 0.3012 | 4.7 | 109.95 | 0.0091 |
| 1.3 | 3.6693 | 0.2725 | 4.8 | 121.51 | 0.0082 |
| 1.4 | 4.0552 | 0.2466 | 4.9 | 134.29 | 0.0074 |
| 1.5 | 4.4817 | 0.2231 | 5.0 | 148.41 | 0.0067 |
| 1.6 | 4.9530 | 0.2019 | 5.5 | 244.69 | 0.0041 |
| 1.7 | 5.4739 | 0.1827 | 6.0 | 403.43 | 0.0025 |
| 1.8 | 6.0496 | 0.1653 | 6.5 | 665.14 | 0.0015 |
| 1.9 | 6.6859 | 0.1496 | 7.0 | 1,096.6 | 0.0009 |
| 2.0 | 7.3891 | 0.1353 | 7.5 | 1,808.0 | 0.0006 |
| 2.1 | 8.1662 | 0.1225 | 8.0 | 2,981.0 | 0.0003 |
| 2.2 | 9.0250 | 0.1108 | 8.5 | 4,914.8 | 0.0002 |
| 2.3 | 9.9742 | 0.1003 | 9.0 | 8,103.1 | 0.0001 |
| 2.4 | 11.023 | 0.0907 | 10.0 | 22,026 | 0.00005 |

**Table IV.** Trigonometric Functions (in radians)

| $S$ | $\sin S$ | $\cos S$ | $\tan S$ | $\cot S$ | $\sec S$ | $\csc S$ |
|------|----------|----------|----------|----------|----------|----------|
| .00 | .00000 | 1.00000 | .00000 | $\infty$ | 1.00000 | $\infty$ |
| .01 | .01000 | .99995 | .01000 | 99.9967 | 1.00005 | 100.0017 |
| .02 | .02000 | .99980 | .02000 | 49.9933 | 1.00020 | 50.0033 |
| .03 | .03000 | .99955 | .03001 | 33.3233 | 1.00045 | 33.3383 |
| .04 | .03999 | .99920 | .04002 | 24.9867 | 1.00080 | 25.0067 |
| .05 | .04998 | .99875 | .05004 | 19.9833 | 1.00125 | 20.0083 |
| .06 | .05996 | .99820 | .06007 | 16.6467 | 1.00180 | 16.6767 |
| .07 | .06994 | .99755 | .07011 | 14.2624 | 1.00246 | 14.2974 |
| .08 | .07991 | .99680 | .08017 | 12.4733 | 1.00321 | 12.5133 |
| .09 | .08988 | .99595 | .09024 | 11.0811 | 1.00406 | 11.1261 |
| .10 | .09983 | .99500 | .10033 | 9.9666 | 1.00502 | 10.0167 |
| .11 | .10978 | .99396 | .11045 | 9.0542 | 1.00608 | 9.1093 |
| .12 | .11971 | .99281 | .12058 | 8.2933 | 1.00724 | 8.3534 |
| .13 | .12963 | .99156 | .13074 | 7.6489 | 1.00851 | 7.7140 |
| .14 | .13954 | .99022 | .14092 | 7.0961 | 1.00988 | 7.1662 |
| .15 | .14944 | .98877 | .15114 | 6.6166 | 1.01136 | 6.6917 |
| .16 | .15932 | .98723 | .16138 | 6.1966 | 1.01294 | 6.2767 |
| .17 | .16918 | .98558 | .17166 | 5.8256 | 1.01463 | 5.9108 |
| .18 | .17903 | .98384 | .18197 | 5.4954 | 1.01642 | 5.5857 |
| .19 | .18886 | .98200 | .19232 | 5.1997 | 1.01833 | 5.2950 |
| .20 | .19867 | .98007 | .20271 | 4.9332 | 1.02034 | 5.0335 |
| .21 | .20846 | .97803 | .21314 | 4.6917 | 1.02246 | 4.7971 |
| .22 | .21823 | .97590 | .22362 | 4.4719 | 1.02470 | 4.5823 |
| .23 | .22798 | .97367 | .23414 | 4.2709 | 1.02705 | 4.3864 |
| .24 | .23770 | .97134 | .24472 | 4.0864 | 1.02951 | 4.2069 |
| .25 | .24740 | .96891 | .25534 | 3.9163 | 1.03209 | 4.0420 |
| .26 | .25708 | .96639 | .26602 | 3.7591 | 1.03478 | 3.8898 |
| .27 | .26673 | .96377 | .27676 | 3.6133 | 1.03759 | 3.7491 |
| .28 | .27636 | .96106 | .28755 | 3.4776 | 1.04052 | 3.6185 |
| .29 | .28595 | .95824 | .29841 | 3.3511 | 1.04358 | 3.4971 |
| .30 | .29552 | .95534 | .30934 | 3.2327 | 1.04675 | 3.3839 |
| .31 | .30506 | .95233 | .32033 | 3.1218 | 1.05005 | 3.2781 |
| .32 | .31457 | .94924 | .33139 | 3.0176 | 1.05348 | 3.1790 |
| .33 | .32404 | .94604 | .34252 | 2.9195 | 1.05704 | 3.0860 |
| .34 | .33349 | .94275 | .35374 | 2.8270 | 1.06072 | 2.9986 |
| .35 | .34290 | .93937 | .36503 | 2.7395 | 1.06454 | 2.9163 |
| .36 | .35227 | .93590 | .37640 | 2.6567 | 1.06849 | 2.8389 |
| .37 | .36162 | .93233 | .38786 | 2.5782 | 1.07258 | 2.7654 |
| .38 | .37092 | .92866 | .39941 | 2.5037 | 1.07682 | 2.6960 |
| .39 | .38019 | .92491 | .41105 | 2.4328 | 1.08119 | 2.6303 |
| .40 | .38942 | .92106 | .42279 | 2.3652 | 1.08570 | 2.5679 |

**Table IV.** Trigonometric Functions (Continued)

| $S$ | $\sin S$ | $\cos S$ | $\tan S$ | $\cot S$ | $\sec S$ | $\csc S$ |
|---|---|---|---|---|---|---|
| .40 | .38942 | .92106 | .42279 | 2.3652 | 1.0857 | 2.5679 |
| .41 | .39861 | .91712 | .43463 | 2.3008 | 1.0904 | 2.5087 |
| .42 | .40776 | .91309 | .44657 | 2.2393 | 1.0952 | 2.4524 |
| .43 | .41687 | .90897 | .45862 | 2.1804 | 1.1002 | 2.3988 |
| .44 | .42594 | .90475 | .47078 | 2.1241 | 1.1053 | 2.3478 |
| .45 | .43497 | .90045 | .48306 | 2.0702 | 1.1106 | 2.2990 |
| .46 | .44395 | .89605 | .49545 | 2.0184 | 1.1160 | 2.2525 |
| .47 | .45289 | .89157 | .50797 | 1.9686 | 1.1216 | 2.2081 |
| .48 | .46178 | .88699 | .52061 | 1.9208 | 1.1274 | 2.1655 |
| .49 | .47063 | .88233 | .53339 | 1.8748 | 1.1334 | 2.1248 |
| .50 | .47943 | .87758 | .54630 | 1.8305 | 1.1395 | 2.0858 |
| .51 | .48818 | .87274 | .55936 | 1.7878 | 1.1458 | 2.0484 |
| .52 | .49688 | .86782 | .57256 | 1.7465 | 1.1523 | 2.0126 |
| .53 | .50553 | .86281 | .58592 | 1.7067 | 1.1590 | 1.9781 |
| .54 | .51414 | .85771 | .59943 | 1.6683 | 1.1659 | 1.9450 |
| .55 | .52269 | .85252 | .61311 | 1.6310 | 1.1730 | 1.9132 |
| .56 | .53119 | .84726 | .62695 | 1.5950 | 1.1803 | 1.8826 |
| .57 | .53963 | .84190 | .64097 | 1.5601 | 1.1878 | 1.8531 |
| .58 | .54802 | .83646 | .65517 | 1.5263 | 1.1955 | 1.8247 |
| .59 | .55636 | .83094 | .66956 | 1.4935 | 1.2035 | 1.7974 |
| .60 | .56464 | .82534 | .68414 | 1.4617 | 1.2116 | 1.7710 |
| .61 | .57287 | .81965 | .69892 | 1.4308 | 1.2200 | 1.7456 |
| .62 | .58104 | .81388 | .71391 | 1.4007 | 1.2287 | 1.7211 |
| .63 | .58914 | .80803 | .72911 | 1.3715 | 1.2376 | 1.6974 |
| .64 | .59720 | .80210 | .74454 | 1.3431 | 1.2467 | 1.6745 |
| .65 | .60519 | .79608 | .76020 | 1.3154 | 1.2561 | 1.6524 |
| .66 | .61312 | .78999 | .77610 | 1.2885 | 1.2658 | 1.6310 |
| .67 | .62099 | .78382 | .79225 | 1.2622 | 1.2758 | 1.6103 |
| .68 | .62879 | .77757 | .80866 | 1.2366 | 1.2861 | 1.5903 |
| .69 | .63654 | .77125 | .82534 | 1.2116 | 1.2966 | 1.5710 |
| .70 | .64422 | .76484 | .84229 | 1.1872 | 1.3075 | 1.5523 |
| .71 | .65183 | .75836 | .85953 | 1.1634 | 1.3186 | 1.5341 |
| .72 | .65938 | .75181 | .87707 | 1.1402 | 1.3301 | 1.5166 |
| .73 | .66687 | .74517 | .89492 | 1.1174 | 1.3420 | 1.4995 |
| .74 | .67429 | .73847 | .91309 | 1.0952 | 1.3542 | 1.4830 |
| .75 | .68164 | .73169 | .93160 | 1.0734 | 1.3667 | 1.4671 |
| .76 | .68892 | .72484 | .95045 | 1.0521 | 1.3796 | 1.4515 |
| .77 | .69614 | .71791 | .96967 | 1.0313 | 1.3929 | 1.4365 |
| .78 | .70328 | .71091 | .98926 | 1.0109 | 1.4066 | 1.4219 |
| .79 | .71035 | .70385 | 1.0092 | .99084 | 1.4208 | 1.4078 |
| .80 | .71736 | .69671 | 1.0296 | .97121 | 1.4353 | 1.3940 |

**Table IV.** Trigonometric Functions (Continued)

| $S$ | $\sin S$ | $\cos S$ | $\tan S$ | $\cot S$ | $\sec S$ | $\csc S$ |
|------|----------|----------|----------|----------|----------|----------|
| .80 | .71736 | .69671 | 1.0296 | .97121 | 1.4353 | 1.3940 |
| .81 | .72429 | .68950 | 1.0505 | .95197 | 1.4503 | 1.3807 |
| .82 | .73115 | .68222 | 1.0717 | .93309 | 1.4658 | 1.3677 |
| .83 | .73793 | .67488 | 1.0934 | .91455 | 1.4818 | 1.3551 |
| .84 | .74464 | .66746 | 1.1156 | .89635 | 1.4982 | 1.3429 |
| .85 | .75128 | .65998 | 1.1383 | .87848 | 1.5152 | 1.3311 |
| .86 | .75784 | .65244 | 1.1616 | .86091 | 1.5327 | 1.3195 |
| .87 | .76433 | .64483 | 1.1853 | .84365 | 1.5508 | 1.3083 |
| .88 | .77074 | .63715 | 1.2097 | .82668 | 1.5695 | 1.2975 |
| .89 | .77707 | .62941 | 1.2346 | .80998 | 1.5888 | 1.2869 |
| .90 | .78333 | .62161 | 1.2602 | .79355 | 1.6087 | 1.2766 |
| .91 | .78950 | .61375 | 1.2864 | .77738 | 1.6293 | 1.2666 |
| .92 | .79560 | .60582 | 1.3133 | .76146 | 1.6507 | 1.2569 |
| .93 | .80162 | .59783 | 1.3409 | .74578 | 1.6727 | 1.2475 |
| .94 | .80756 | .58979 | 1.3692 | .73034 | 1.6955 | 1.2383 |
| .95 | .81342 | .58168 | 1.3984 | .71511 | 1.7191 | 1.2294 |
| .96 | .81919 | .57352 | 1.4284 | .70010 | 1.7436 | 1.2207 |
| .97 | .82489 | .56530 | 1.4592 | .68531 | 1.7690 | 1.2123 |
| .98 | .83050 | .55702 | 1.4910 | .67071 | 1.7953 | 1.2041 |
| .99 | .83603 | .54869 | 1.5237 | .65631 | 1.8225 | 1.1961 |
| 1.00 | .84147 | .54030 | 1.5574 | .64209 | 1.8508 | 1.1884 |
| 1.01 | .84683 | .53186 | 1.5922 | .62806 | 1.8802 | 1.1809 |
| 1.02 | .85211 | .52337 | 1.6281 | .61420 | 1.9107 | 1.1736 |
| 1.03 | .85730 | .51482 | 1.6652 | .60051 | 1.9424 | 1.1665 |
| 1.04 | .86240 | .50622 | 1.7036 | .58699 | 1.9754 | 1.1595 |
| 1.05 | .86742 | .49757 | 1.7433 | .57362 | 2.0098 | 1.1528 |
| 1.06 | .87236 | .48887 | 1.7844 | .56040 | 2.0455 | 1.1463 |
| 1.07 | .87720 | .48012 | 1.8270 | .54734 | 2.0828 | 1.1400 |
| 1.08 | .88196 | .47133 | 1.8712 | .53441 | 2.1217 | 1.1338 |
| 1.09 | .88663 | .46249 | 1.9171 | .52162 | 2.1622 | 1.1279 |
| 1.10 | .89121 | .45360 | 1.9648 | .50897 | 2.2046 | 1.1221 |
| 1.11 | .89570 | .44466 | 2.0143 | .49644 | 2.2489 | 1.1164 |
| 1.12 | .90010 | .43568 | 2.0660 | .48404 | 2.2952 | 1.1110 |
| 1.13 | .90441 | .42666 | 2.1198 | .47175 | 2.3438 | 1.1057 |
| 1.14 | .90863 | .41759 | 2.1759 | .45959 | 2.3947 | 1.1006 |
| 1.15 | .91276 | .40849 | 2.2345 | .44753 | 2.4481 | 1.0956 |
| 1.16 | .91680 | .39934 | 2.2958 | .43558 | 2.5041 | 1.0907 |
| 1.17 | .92075 | .39015 | 2.3600 | .42373 | 2.5631 | 1.0861 |
| 1.18 | .92461 | .38092 | 2.4273 | .41199 | 2.6252 | 1.0815 |
| 1.19 | .92837 | .37166 | 2.4979 | .40034 | 2.6906 | 1.0772 |
| 1.20 | .93204 | .36236 | 2.5722 | .38878 | 2.7597 | 1.0729 |

**Table IV.** Trigonometric Functions (Continued)

| $S$ | $\sin S$ | $\cos S$ | $\tan S$ | $\cot S$ | $\sec S$ | $\csc S$ |
|---|---|---|---|---|---|---|
| 1.20 | .93204 | .36236 | 2.5722 | .38878 | 2.7597 | 1.07292 |
| 1.21 | .93562 | .35302 | 2.6503 | .37731 | 2.8327 | 1.06881 |
| 1.22 | .93910 | .34365 | 2.7328 | .36593 | 2.9100 | 1.06485 |
| 1.23 | .94249 | .33424 | 2.8198 | .35463 | 2.9919 | 1.06102 |
| 1.24 | .94578 | .32480 | 2.9119 | .34341 | 3.0789 | 1.05732 |
| 1.25 | .94898 | .31532 | 3.0096 | .33227 | 3.1714 | 1.05376 |
| 1.26 | .95209 | .30582 | 3.1133 | .32121 | 3.2699 | 1.05032 |
| 1.27 | .95510 | .29628 | 3.2236 | .31021 | 3.3752 | 1.04701 |
| 1.28 | .95802 | .28672 | 3.3414 | .29928 | 3.4878 | 1.04382 |
| 1.29 | .96084 | .27712 | 3.4672 | .28842 | 3.6085 | 1.04076 |
| 1.30 | .96356 | .26750 | 3.6021 | .27762 | 3.7383 | 1.03782 |
| 1.31 | .96618 | .25785 | 3.7471 | .26687 | 3.8782 | 1.03500 |
| 1.32 | .96872 | .24818 | 3.9033 | .25619 | 4.0294 | 1.03230 |
| 1.33 | .97115 | .23848 | 4.0723 | .24556 | 4.1933 | 1.02971 |
| 1.34 | .97348 | .22875 | 4.2556 | .23498 | 4.3715 | 1.02724 |
| 1.35 | .97572 | .21901 | 4.4552 | .22446 | 4.5661 | 1.02488 |
| 1.36 | .97786 | .20924 | 4.6734 | .21398 | 4.7792 | 1.02264 |
| 1.37 | .97991 | .19945 | 4.9131 | .20354 | 5.0138 | 1.02050 |
| 1.38 | .98185 | .18964 | 5.1774 | .19315 | 5.2731 | 1.01848 |
| 1.39 | .98370 | .17981 | 5.4707 | .18279 | 5.5613 | 1.01657 |
| 1.40 | .98545 | .16997 | 5.7979 | .17248 | 5.8835 | 1.01477 |
| 1.41 | .98710 | .16010 | 6.1654 | .16220 | 6.2459 | 1.01307 |
| 1.42 | .98865 | .15023 | 6.5811 | .15195 | 6.6567 | 1.01148 |
| 1.43 | .99010 | .14033 | 7.0555 | .14173 | 7.1260 | 1.00999 |
| 1.44 | .99146 | .13042 | 7.6018 | .13155 | 7.6673 | 1.00862 |
| 1.45 | .99271 | .12050 | 8.2381 | .12139 | 8.2986 | 1.00734 |
| 1.46 | .99387 | .11057 | 8.9886 | .11125 | 9.0441 | 1.00617 |
| 1.47 | .99492 | .10063 | 9.8874 | .10114 | 9.9378 | 1.00510 |
| 1.48 | .99588 | .09067 | 10.9834 | .09105 | 11.0288 | 1.00414 |
| 1.49 | .99674 | .08071 | 12.3499 | .08097 | 12.3903 | 1.00327 |
| 1.50 | .99749 | .07074 | 14.1014 | .07091 | 14.1368 | 1.00251 |
| 1.51 | .99815 | .06076 | 16.4281 | .06087 | 16.4585 | 1.00185 |
| 1.52 | .99871 | .05077 | 19.6695 | .05084 | 19.6949 | 1.00129 |
| 1.53 | .99917 | .04079 | 24.4984 | .04082 | 24.5188 | 1.00083 |
| 1.54 | .99953 | .03079 | 32.4611 | .03081 | 32.4765 | 1.00047 |
| 1.55 | .99978 | .02079 | 48.0785 | .02080 | 48.0889 | 1.00022 |
| 1.56 | .99994 | .01080 | 92.6205 | .01080 | 92.6259 | 1.00006 |
| 1.57 | 1.00000 | .00080 | 1255.77 | .00080 | 1255.77 | 1.00000 |
| 1.58 | .99996 | −.00920 | −108.649 | −.00920 | −108.654 | 1.00004 |
| 1.59 | .99982 | −.01920 | −52.0670 | −.01921 | −52.0766 | 1.00018 |
| 1.60 | .99957 | −.02920 | −34.2325 | −.02921 | −34.2471 | 1.00043 |

**Table V.** Trigonometric Functions (in degrees)

| ↳ | sin | cos | tan | cot | sec | csc | |
|---|---|---|---|---|---|---|---|
| 0° | .0000 | 1.0000 | .0000 | . . . . | 1.000 | . . . . | 90° |
| 1° | .0175 | .9998 | .0175 | 57.29 | 1.000 | 57.30 | 89° |
| 2° | .0349 | .9994 | .0349 | 28.64 | 1.001 | 28.65 | 88° |
| 3° | .0523 | .9986 | .0524 | 19.08 | 1.001 | 19.11 | 87° |
| 4° | .0698 | .9976 | .0699 | 14.30 | 1.002 | 14.34 | 86° |
| 5° | .0872 | .9962 | .0875 | 11.43 | 1.004 | 11.47 | 85° |
| 6° | .1045 | .9945 | .1051 | 9.514 | 1.006 | 9.567 | 84° |
| 7° | .1219 | .9925 | .1228 | 8.144 | 1.008 | 8.206 | 83° |
| 8° | .1392 | .9903 | .1405 | 7.115 | 1.010 | 7.185 | 82° |
| 9° | .1564 | .9877 | .1584 | 6.314 | 1.012 | 6.392 | 81° |
| 10° | .1736 | .9848 | .1763 | 5.671 | 1.015 | 5.759 | 80° |
| 11° | .1908 | .9816 | .1944 | 5.145 | 1.019 | 5.241 | 79° |
| 12° | .2079 | .9781 | .2126 | 4.705 | 1.022 | 4.810 | 78° |
| 13° | .2250 | .9744 | .2309 | 4.331 | 1.026 | 4.445 | 77° |
| 14° | .2419 | .9703 | .2493 | 4.011 | 1.031 | 4.134 | 76° |
| 15° | .2588 | .9659 | .2679 | 3.732 | 1.035 | 3.864 | 75° |
| 16° | .2756 | .9613 | .2867 | 3.487 | 1.040 | 3.628 | 74° |
| 17° | .2924 | .9563 | .3057 | 3.271 | 1.046 | 3.420 | 73° |
| 18° | .3090 | .9511 | .3249 | 3.078 | 1.051 | 3.236 | 72° |
| 19° | .3256 | .9455 | .3443 | 2.904 | 1.058 | 3.072 | 71° |
| 20° | .3420 | .9397 | .3640 | 2.747 | 1.064 | 2.924 | 70° |
| 21° | .3584 | .9336 | .3839 | 2.605 | 1.071 | 2.790 | 69° |
| 22° | .3746 | .9272 | .4040 | 2.475 | 1.079 | 2.669 | 68° |
| 23° | .3907 | .9205 | .4245 | 2.356 | 1.086 | 2.559 | 67° |
| 24° | .4067 | .9135 | .4452 | 2.246 | 1.095 | 2.459 | 66° |
| 25° | .4226 | .9063 | .4663 | 2.145 | 1.103 | 2.366 | 65° |
| 26° | .4384 | .8988 | .4877 | 2.050 | 1.113 | 2.281 | 64° |
| 27° | .4540 | .8910 | .5095 | 1.963 | 1.122 | 2.203 | 63° |
| 28° | .4695 | .8829 | .5317 | 1.881 | 1.133 | 2.130 | 62° |
| 29° | .4848 | .8746 | .5543 | 1.804 | 1.143 | 2.063 | 61° |
| 30° | .5000 | .8660 | .5774 | 1.732 | 1.155 | 2.000 | 60° |
| 31° | .5150 | .8572 | .6009 | 1.664 | 1.167 | 1.942 | 59° |
| 32° | .5299 | .8480 | .6249 | 1.600 | 1.179 | 1.887 | 58° |
| 33° | .5446 | .8387 | .6494 | 1.540 | 1.192 | 1.836 | 57° |
| 34° | .5592 | .8290 | .6745 | 1.483 | 1.206 | 1.788 | 56° |
| 35° | .5736 | .8192 | .7002 | 1.428 | 1.221 | 1.743 | 55° |
| 36° | .5878 | .8090 | .7265 | 1.376 | 1.236 | 1.701 | 54° |
| 37° | .6018 | .7986 | .7536 | 1.327 | 1.252 | 1.662 | 53° |
| 38° | .6157 | .7880 | .7813 | 1.280 | 1.269 | 1.624 | 52° |
| 39° | .6293 | .7771 | .8098 | 1.235 | 1.287 | 1.589 | 51° |
| 40° | .6428 | .7660 | .8391 | 1.192 | 1.305 | 1.556 | 50° |
| 41° | .6561 | .7547 | .8693 | 1.150 | 1.325 | 1.524 | 49° |
| 42° | .6691 | .7431 | .9004 | 1.111 | 1.346 | 1.494 | 48° |
| 43° | .6820 | .7314 | .9325 | 1.072 | 1.367 | 1.466 | 47° |
| 44° | .6947 | .7193 | .9657 | 1.036 | 1.390 | 1.440 | 46° |
| 45° | .7071 | .7071 | 1.000 | 1.000 | 1.414 | 1.414 | 45° |
| | cos | sin | cot | tan | csc | sec | ↩ |

**Table VI.** Powers and Roots

| No. | Sq. | Sq. Root | Cube | Cube Root | No. | Sq. | Sq. Root | Cube | Cube Root |
|---|---|---|---|---|---|---|---|---|---|
| 1 | 1 | 1.000 | 1 | 1.000 | 51 | 2,601 | 7.141 | 132,651 | 3.708 |
| 2 | 4 | 1.414 | 8 | 1.260 | 52 | 2,704 | 7.211 | 140,608 | 3.733 |
| 3 | 9 | 1.732 | 27 | 1.442 | 53 | 2,809 | 7.280 | 148,877 | 3.756 |
| 4 | 16 | 2.000 | 64 | 1.587 | 54 | 2,916 | 7.348 | 157,464 | 3.780 |
| 5 | 25 | 2.236 | 125 | 1.710 | 55 | 3,025 | 7.416 | 166,375 | 3.803 |
| 6 | 36 | 2.449 | 216 | 1.817 | 56 | 3,136 | 7.483 | 175,616 | 3.826 |
| 7 | 49 | 2.646 | 343 | 1.913 | 57 | 3,249 | 7.550 | 185,193 | 3.849 |
| 8 | 64 | 2.828 | 512 | 2.000 | 58 | 3,364 | 7.616 | 195,112 | 3.871 |
| 9 | 81 | 3.000 | 729 | 2.080 | 59 | 3,481 | 7.681 | 205,379 | 3.893 |
| 10 | 100 | 3.162 | 1,000 | 2.154 | 60 | 3,600 | 7.746 | 216,000 | 3.915 |
| 11 | 121 | 3.317 | 1,331 | 2.224 | 61 | 3,721 | 7.810 | 226,981 | 3.936 |
| 12 | 144 | 3.464 | 1,728 | 2.289 | 62 | 3,844 | 7.874 | 238,328 | 3.958 |
| 13 | 169 | 3.606 | 2,197 | 2.351 | 63 | 3,969 | 7.937 | 250,047 | 3.979 |
| 14 | 196 | 3.742 | 2,744 | 2.410 | 64 | 4,096 | 8.000 | 262,144 | 4.000 |
| 15 | 225 | 3.873 | 3,375 | 2.466 | 65 | 4,225 | 8.062 | 274,625 | 4.021 |
| 16 | 256 | 4.000 | 4,096 | 2.520 | 66 | 4,356 | 8.124 | 287,496 | 4.041 |
| 17 | 289 | 4.123 | 4,913 | 2.571 | 67 | 4,489 | 8.185 | 300,763 | 4.062 |
| 18 | 324 | 4.243 | 5,832 | 2.621 | 68 | 4,624 | 8.246 | 314,432 | 4.082 |
| 19 | 361 | 4.359 | 6,859 | 2.668 | 69 | 4,761 | 8.307 | 328,509 | 4.102 |
| 20 | 400 | 4.472 | 8,000 | 2.714 | 70 | 4,900 | 8.367 | 343,000 | 4.121 |
| 21 | 441 | 4.583 | 9,261 | 2.759 | 71 | 5,041 | 8.426 | 357,911 | 4.141 |
| 22 | 484 | 4.690 | 10,648 | 2.802 | 72 | 5,184 | 8.485 | 373,248 | 4.160 |
| 23 | 529 | 4.796 | 12,167 | 2.844 | 73 | 5,329 | 8.544 | 389,017 | 4.179 |
| 24 | 576 | 4.899 | 13,824 | 2.884 | 74 | 5,476 | 8.602 | 405,224 | 4.198 |
| 25 | 625 | 5.000 | 15,625 | 2.924 | 75 | 5,625 | 8.660 | 421,875 | 4.217 |
| 26 | 676 | 5.099 | 17,576 | 2.962 | 76 | 5,776 | 8.718 | 438,976 | 4.236 |
| 27 | 729 | 5.196 | 19,683 | 3.000 | 77 | 5,929 | 8.775 | 456,533 | 4.254 |
| 28 | 784 | 5.292 | 21,952 | 3.037 | 78 | 6,084 | 8.832 | 474,552 | 4.273 |
| 29 | 841 | 5.385 | 24,389 | 3.072 | 79 | 6,241 | 8.888 | 493,039 | 4.291 |
| 30 | 900 | 5.477 | 27,000 | 3.107 | 80 | 6,400 | 8.944 | 512,000 | 4.309 |
| 31 | 961 | 5.568 | 29,791 | 3.141 | 81 | 6,561 | 9.000 | 531,441 | 4.327 |
| 32 | 1,024 | 5.657 | 32,768 | 3.175 | 82 | 6,724 | 9.055 | 551,368 | 4.344 |
| 33 | 1,089 | 5.745 | 35,937 | 3.208 | 83 | 6,889 | 9.110 | 571,787 | 4.362 |
| 34 | 1,156 | 5.831 | 39,304 | 3.240 | 84 | 7,056 | 9.165 | 592,704 | 4.380 |
| 35 | 1,225 | 5.916 | 42,875 | 3.271 | 85 | 7,225 | 9.220 | 614,125 | 4.397 |
| 36 | 1,296 | 6.000 | 46,656 | 3.302 | 86 | 7,396 | 9.274 | 636,056 | 4.414 |
| 37 | 1,369 | 6.083 | 50,653 | 3.332 | 87 | 7,569 | 9.327 | 658,503 | 4.431 |
| 38 | 1,444 | 6.164 | 54,872 | 3.362 | 88 | 7,744 | 9.381 | 681,472 | 4.448 |
| 39 | 1,521 | 6.245 | 59,319 | 3.391 | 89 | 7,921 | 9.434 | 704,969 | 4.465 |
| 40 | 1,600 | 6.325 | 64,000 | 3.420 | 90 | 8,100 | 9.487 | 729,000 | 4.481 |
| 41 | 1,681 | 6.403 | 68,921 | 3.448 | 91 | 8,281 | 9.539 | 753,571 | 4.498 |
| 42 | 1,764 | 6.481 | 74,088 | 3.476 | 92 | 8,464 | 9.592 | 778,688 | 4.514 |
| 43 | 1,849 | 6.557 | 79,507 | 3.503 | 93 | 8,649 | 9.644 | 804,357 | 4.531 |
| 44 | 1,936 | 6.633 | 85,184 | 3.530 | 94 | 8,836 | 9.695 | 830,584 | 4.547 |
| 45 | 2,025 | 6.708 | 91,125 | 3.557 | 95 | 9,025 | 9.747 | 857,375 | 4.563 |
| 46 | 2,116 | 6.782 | 97,336 | 3.583 | 96 | 9,216 | 9.798 | 884,736 | 4.579 |
| 47 | 2,209 | 6.856 | 103,823 | 3.609 | 97 | 9,409 | 9.849 | 912,673 | 4.595 |
| 48 | 2,304 | 6.928 | 110,592 | 3.634 | 98 | 9,604 | 9.899 | 941,192 | 4.610 |
| 49 | 2,401 | 7.000 | 117,649 | 3.659 | 99 | 9,801 | 9.950 | 970,299 | 4.626 |
| 50 | 2,500 | 7.071 | 125,000 | 3.684 | 100 | 10,000 | 10.000 | 1,000,000 | 4.642 |

# ANSWERS TO SELECTED EXERCISES

**Section 1, p. 3**

**1.** (b), (c), and (e) are well defined.

**3.** (a), (d)

**Section 2, p. 5**

**3.** (a), (d), and (e) are true

**7.** (a), (b), and (e)

**Section 3, p. 7**

**1.** (a) $U$      (g) $\{1, 2, 3, 4, 5, 6, 8, 10\}$

   (c) $C$      (i) $\{1, 3, 5\}$

   (e) $\varnothing$      (k) $\varnothing$

**3.** (a) $\{1, 2, 3, 4\}$

   (c) $A'$

**4.** (a) $\mathbb{R}$     (c) $J$     (e) $\varnothing$        (g) $H$     (i) $N$

   (b) $\varnothing$     (d) $Q$     (f) $J - \{0\}$     (h) $\mathbb{R}$     (j) $Q$

**Section 4, p. 11**

**1.** (a) $0.\overline{857142}$      (e) $0.375\overline{0}$

   (c) $-0.5\overline{0}$      (g) $0.4\overline{0}$

**2.** (a) $-\frac{2}{1}$     (e) $\frac{16}{9}$     (g) $4$

   (c) $\frac{2}{11}$     (f) $\frac{3,092}{990}$

**5.** (a) $x \in \mathbb{R},\ x \neq 0$      (e) none

   (c) $x = 0$     (g) none

**6.** (a) $x = 1$     (c) $x = \pm 2$     (e) $x = -1$     (g) $x = -1$

**7.** (a) $x = -1$     (c) $x = 0$     (e) $x = 3$     (g) $x = -\frac{1}{4}$

**8.** (a) $-1$; 0 and 1; none     (c) 1; $\pm 3$; none

   (e) $\frac{1}{4}$; $-2$, 0, and 1; none

## Section 5, p. 15

**1.** (a) and (g)

**2.** (a)
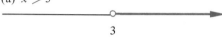

(c)

(e) ∅

(g)

(i)

**3.** (a) $(2, 5]$              (g) $[2, 2] \cup [4, 5]$

  (c) $[1, 3] \cup (7, \infty)$       (i) $(-\infty, 3] \cup (5, \infty)$

  (e) $(-\infty, 2) \cup (6, \infty)$

**4.** (a) Let $a = -3$ and $b = -2$

  (c) Let $a = -4$

## Section 6, p. 18

**1.** (a) $x > 3$

  (c) $x \geq 1$

  (e) $-\frac{5}{2} \leq x < -\frac{3}{2}$

  (g) ∅

  (i) $x < 3$ or $x > 11$

**2.** (a) $(-\infty, -2) \cup [3, \infty)$

  (c) $(-\infty, -2] \cup (2, \infty)$

  (e) $[-1, -1] \cup (1, \infty)$

**3.** (a) $3 < 2x + 3 < 7$      (c) $0 < 2 - x < 2$

**4.** (a) $2 < x + y < 5$

  (c) $0 < xy < 4$

  (e) $2 < x^2 + y < 5$

**5.** (a) Given interval contains 0. Hence lower bound for $x^2$ cannot be 4.

  (c) $0 \leq x^2 \leq 9$

  (e) $4 \leq x^2 \leq 9$

## Section 7, p. 25

**1.** (a) $x = \pm 6$

  (c) ∅

  (e) $x = \frac{3}{2}$

(g) $x = 1$ or $-2$

(i) $x = -1, 7$

**2.** (a) $-6 < x < 4$    (e) $-\frac{1}{2}(\delta + 3) < y < \frac{1}{2}(\delta - 3)$
  (c) $2.9 < x < 3.1$    (g) $x \geq a + b$ or $x \leq a - b$
                     (i)  $-2 < x^2 + 2x - 8 < 2$    or    $0 < x^2 + 2x - 6 < 4$

**3.**   (a) $-3 < x < 1$

  (c) $-3 < x < 2$

  (e) $-4 < x < 8$

  (g) $1 < x < 3$

  (i) $\mathbb{R}$

**4.** (a) $|x - 6| < 4$    (e) $|x + 1.95| < 0.05$
  (c) $|x| < 2$

**5.** (a) $2 < |x| < 6$    (e) $2 < |x| < 5$
  (c) $0 \leq |x| < 1$

**6.** (a) $0 \leq |x| < 4$    (c) $1 < |x + 3| < 7$
  (b) $0 \leq |x + 1| < 5$    (d) $0 \leq |x - 3| < 5$

**7.** (a) $|x| \geq \frac{1}{4}$

(c) $\{x \mid -\frac{1}{2} \leq x \leq \frac{3}{2}, x \neq \frac{1}{2}\}$

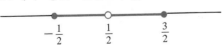

(e) $\{x \mid 0 \leq x \leq 2, x \neq 1\}$

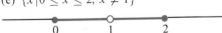

(g) $\{x \mid x < -1\} \cup \{x \mid x > 3\}$

**8.** (a) $\begin{cases} 2 - x, & x \geq 0 \\ 3x + 2, & -2 \leq x < 0 \\ x - 2, & x < -2 \end{cases}$

   (c) $\begin{cases} 2x + 6, & x \geq 3 \\ 6x - 6, & 0 \leq x < 3 \\ -2x - 6, & x < 0 \end{cases}$

**9.** (a) $x = -\frac{1}{3}, 1$

   (c) $x = -5, \frac{5}{3}$

**10.** (a) $\{x \mid x < -\frac{4}{3}\} \cup \{x \mid x > 4\}$

    (c) $\{x \mid -6 < x < 2\}$

## Section 8, p. 30

**1.** (a) 4      (e) 64      (i) 5      (m) 4

   (c) $\frac{1}{32}$     (g) $\frac{64}{27}$     (k) 3

**3.** (a) $-4 \leq x \leq 4$     (c) $-3 \leq x \leq 3$     (e) $-\frac{1}{2} < x$ or $x > \frac{1}{2}$     (g) $|x| \geq \frac{1}{2}$

   (i) $x < 1$ or $x > 3$

**5.** (a) $x \geq 3$     (c) $x \geq 4$     (e) $x = \varnothing$

**6.** (a) $2x - 4$     (b) 0

**7.** (a) $4 - 2x$     (b) 0

**8.** (a) $x \geq \frac{5}{3}$     (c) $x \geq 2$     (e) all $x$

   (g) $x \geq -3$     (i) $x \geq -1$

**9.** (a) $x = 4$     (e) $x = 2$     (i) $x = -3$     (m) $x = 1$ or $-3$

   (c) $x = 0$     (g) $x = -\frac{1}{2}$     (k) $x = \pm 2$

## Section 9, p. 37

**1.** (a) $A \times B = \{(2, -1), (2, 0), (2, 1), (2, 2), (3, -1), (3, 0), (3, 1), (3, 2), (4, -1), (4, 0), (4, 1),$
      $(4, 2)\}$

    $B \times A = \{(-1, 2), (0, 2), (1, 2), (2, 2), (-1, 3), (0, 3), (1, 3), (2, 3), (-1, 4), (0, 4), (1, 4),$
      $(2, 4)\}$

    $A \times A = \{(2, 2), (2, 3), (2, 4), (3, 2), (3, 3), (3, 4), (4, 2), (4, 3), (4, 4)\}$

    $B \times B = \{(-1, -1), (-1, 0), (-1, 1), (-1, 2), (0, -1), (0, 0), (0, 1), (0, 2), (1, -1), (1, 0),$
      $(1, 1), (1, 2), (2, -1), (2, 0), (2, 1), (2, 2)\}$

   (c) $D = \{(2, 2)\}$

   (e) $F = \{(2, -1), (2, 0), (2, 1), (3, -1), (3, 0), (3, 1), (3, 2), (4, -1), (4, 0), (4, 1), (4, 2)\}$

   (g) No

**3.** (a)         Rectangle                (b)         Right triangle

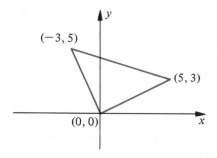

(c)                    Trapezoid                                    (e)            Isosceles triangle

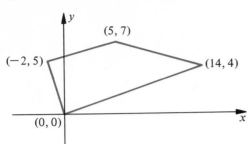

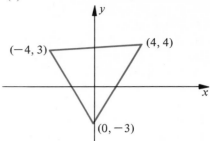

(d)                     Square

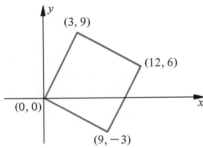

4. (a) quadrants I and III
   (c) quadrant II
   (e) quadrants I, II, and IV

5.

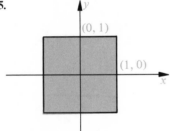

**Section 10, p. 40**
  1. (a) 5          (e) 1
     (c) $\sqrt{2}$
  7. $x = 2$
  8. $(0, \frac{31}{2})$
  9. $x = 1, 2$
 10. 4
 11. $x^2 + y^2 = r^2$
 12. $(x - 2)^2 + (y - 3)^2 = 16$
 13. $y^2 = 16x$

**Section 11, p. 46**
  1. $D = \{1, 2, 3\}.\ R = \{2, 3, 4, 5\}$
  3. $D = \{1, 2, 3, 6\},\ R = \{2\}$

**5.** $D = \{x \mid -6 \leq x \leq 6\}$, $R = \{y \mid -6 \leq y \leq 6\}$
**7.** $D = \{x \mid x \neq 3\}$, $R = \{y \mid y \neq 0\}$
**9.** $D = \{x \mid x \neq -3\}$, $R = \{y \mid y \neq 1\}$
**11.** $D = \{x \mid x \leq 4\}$, $R = \{y \mid y \in \mathbb{R}\}$
**13.** $D = \{x \mid x \neq 2\}$, $R = \{y \mid y \neq 4\}$
**15.** $D = \{x \mid x \in \mathbb{R}\}$, $R = \{y \mid y \leq 4\}$
**17.** $D = \{x \mid x \in \mathbb{R}\}$, $R = \{y \mid y \leq -2 \text{ or } y \geq 2\}$
**19.** $D = \{x \mid x > 0\}$, $R = \{y \mid y \neq 2\}$
**21.** $D = \{-1, 0, 1\}$, $R = \{0, 1\}$
**23.** $D = \{-1, 0, 1, 2, 3\}$, $R = \{0, 1, 2, 3\}$
**25.** $D = \{-1, 0, 1, 2, 3\}$, $R = \{3\}$
**27.** $D = \{x \mid x \geq 2$, $R = \{y \mid y \in \mathbb{R}\}$
**29.** $D = \{x \mid x \geq 0\}$, $R = \{y \mid y \geq 0\}$
**31.** $D = \{x \mid x \in \mathbb{R}\}$, $R = \{y \mid y \geq 0\}$
**33.** $D = \{x \mid -2 \leq x \leq 2\}$, $R = \{y \mid -2 \leq y \leq 2\}$
**35.** $D = \{x \mid -1 \leq x \leq 1\}$, $R = \{y \mid -3 \leq y \leq 3\}$
**37.** $D = \{x \mid x \in \mathbb{R}\}$, $R = \{y \mid y \geq 0\}$

## Section 12, p. 51

Problems 1, 2, 5, 6, and 8 are functions.
**11.** $D = \{x \mid x \neq 1\}$, $R = \{y \mid y \neq 0\}$
**13.** $D = \{x \mid x \geq -5\}$, $R = \{y \mid y \geq 0\}$
**15.** $D = \{x \mid x \neq 0\}$, $R = \{y \mid y \neq 4\}$
**17.** $D = \{x \mid x \in \mathbb{R}\}$, $R = \{y \mid y \in \mathbb{R}\}$
**19.** Same as 17.
**21.** $D = \{x \mid x \in \mathbb{R}\}$, $R = \{y \mid y \geq 1\}$

**23.** (a) $x = 8$      (c) $x = \dfrac{4}{3}$      (e) $\pm\sqrt{\dfrac{4}{3}}$

      (b) $x = 0$      (d) none
**25.** $\pm 2$; $\pm\sqrt{2}$, $\pm\sqrt{6}$
**27.** (a) $3x^2$, $5x^4$, $10x^9$
      (b) no
**29.** $A = \pi r^2$, $r \geq 0$
**31.** $P = 2\omega + 20$, $\omega \geq 0$
**33.** (a) $7h + h^2$
      (b) $7 + h$
      (c) $2x + h + 1$
**35.** $2x - 2 + h$
**37.** (a) $f(h) = h^4 + 3h^2$
        $f(-h) = h^4 + 3h^2$
      (b) $f(h) = f(-h)$
**39.** $0 < L < 0.05$

## Section 13, p. 57

**1.** (a) $(-2, 0), (0, 2)$

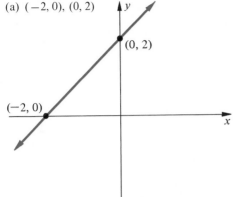

(c) $(0, 0)$

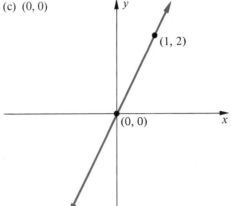

(e) none, $(0, 4)$

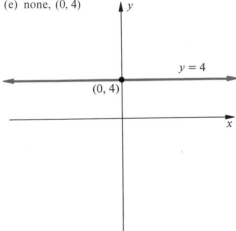

**3.** (a) 4     (d) 10
    (b) 6     (e) yes
    (c) 10     (f) no
           (g) yes

**5.** $f(3x) = 5(3x) = 15x$
   $3f(x) = 3(5x) = 15x$

**7.** $f(4x + 3) = 2(4x + 3) = 8x + 6$
   $f(4x) + f(3) = 2(4x) + 2(3) = 8x + 6$

**9.** (a) 90
   (b) 80

**11.** No, No

## Section 14, p. 62

**1.** $(-3, -3)$        **9.** $f(x) = -\frac{3}{2}x + 4$

**3.** $\emptyset$            **11.** $x^2 + y^2 - 8x - 2y + 12 = 0$

**5.** $(\frac{1}{4}, \frac{15}{16})$      **13.** $T = 52°$

**7.** $(\frac{7}{2}, 3, 1)$      **15.** $(3, -1), (-1, 2),$ and $(2, 5)$

**Section 15, p. 69**

1. $\dfrac{3 \pm \sqrt{2}}{2}$

3. none

5. $-1 \pm 2\sqrt{3}$

7. $1 \pm \sqrt{6}$

9. $\dfrac{-1 \pm \sqrt{y}}{y - 1}$

11. $x$-intercept at $(-1, 0)$
vertex at $(-1, 0)$
$y$-intercept at $(0, 1)$
range: $\{y \mid y \geq 0\}$

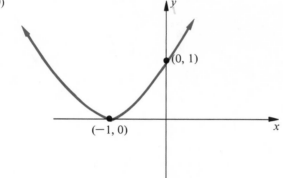

13. $x$-intercepts at $(2, 0)$ and $(-2, 0)$
vertex at $(0, 4)$
$y$-intercept at $(0, 4)$
range: $\{y \mid y \leq 4\}$

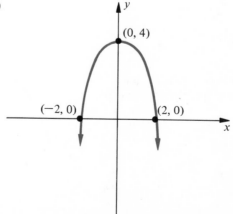

**15.** $x$-intercepts at $(5, 0)$ and $(-3, 0)$
vertex at $(1, 16)$
$y$-intercept at $(0, 15)$
range: $\{y \mid y \leq 16\}$

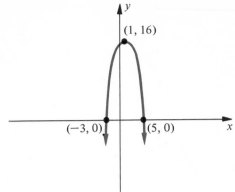

**17.** no $x$-intercepts
vertex at $(1, 2)$
$y$-intercept at $(0, 3)$
range: $\{y \mid y \geq 2\}$

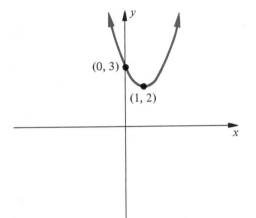

**19.** $x$-intercepts at $\left(1 \pm \dfrac{\sqrt{6}}{2}\right)$
vertex at $(1, 3)$
$y$-intercept at $(0, 1)$
range: $\{y \mid y \leq 3\}$

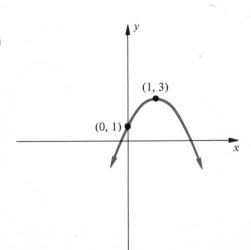

**21.** $f(x) = \frac{4}{3}x^2 - \frac{16}{3}x + 4$

**23.** (a)

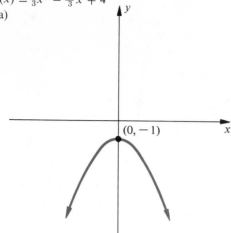

(b) none

(c) once

(d) twice

**25.** $5'' \times 5''$

**29.** 1,000 units

**31.** 30 ft $\times$ 60 ft

**33.** 40 sec; 25,600 ft

**Section 16, p. 72**

**1.** $\{(0, 0), (1, 1)\}$

**3.** $\{(2, -4), (1, -6)\}$

**5.** $\{(2, 1), (-1, 4)\}$

**7.** $\{(0, 0), (4, -2)\}$

**9.** $\{(1, 0), (-2, -3)\}$

**11.** $\left\{ \left( \frac{\sqrt{2}}{2}, \frac{\sqrt{2}}{2} \right), \left( -\frac{\sqrt{2}}{2}, -\frac{\sqrt{2}}{2} \right) \right\}$

**13.** $\{(1, 1)\}$

**15.** $\{(\frac{1}{4}, \frac{1}{2}), (0, 0)\}$

**17.** $\{(\sqrt{5}, 2), (-\sqrt{5}, 2), (\sqrt{5}, -2), (-\sqrt{5}, -2)\}$

**19.** $\{(4, 8), (4, -8)\}$

**21.** $8'' \times 15''$

**23.** 7, 9

**25.** Initial dimensions are $40 \times 75$.

**27.** 15 in $\times$ 20 in

**Section 17, p. 77**

**1.**

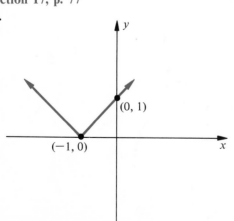

range is $\{y \mid y \geq 0\}$

**3.**

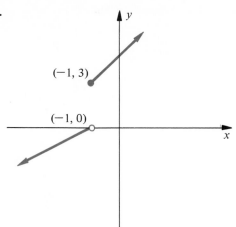

range is $\{y \mid y \geq 3 \text{ or } y < 0\}$

**5.**

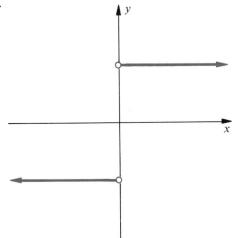

range is $\{3, -3\}$

**7.**

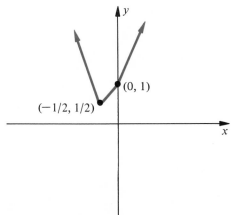

range is $\{y \mid y \geq \frac{1}{2}\}$

**9.**

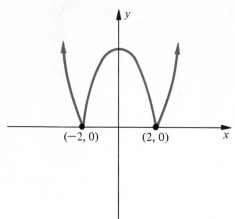

range is $\{y \mid y \geq 0\}$

**11.**

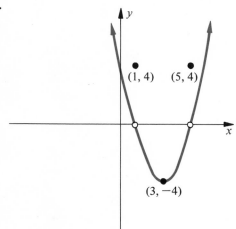

range is $\{y \mid y \geq -4,\, y \neq 0\}$

**13.**

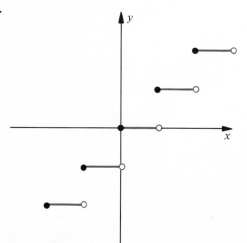

range is $\{-2, -1, 0, 1, 2\}$

**15.**

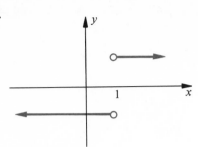

range is $\{-1, 1\}$

**17.** Same as 15
**19.** $h(x)$ is continuous at $x = 1$

**23.**

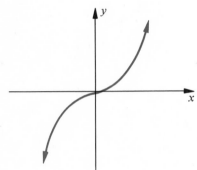

range is $\{y \mid y \in \mathbb{R}\}$

**25.**

range is $\{y \mid y \le 0 \text{ or } y > 1\}$

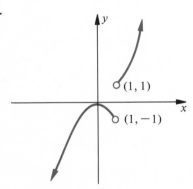

**27.**

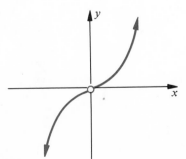

range is $\{y \mid y \in \mathbb{R}, y \ne 0\}$

**29.**

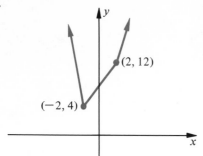

range is $\{y \mid y \geq 4\}$

**31.**

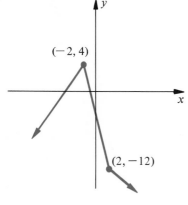

range is $\{y \mid y \leq 4\}$

**33.** $G(t) = \begin{cases} 2t, & [0, 6] \\ -3t + 30, & (6, 10] \\ 0, & (10, \infty) \end{cases}$

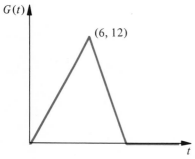

**35.** $C = 1.25 + 0.25[x]$

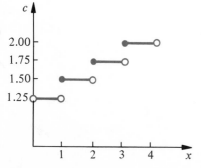

**Section 18, p. 82**

**1.**

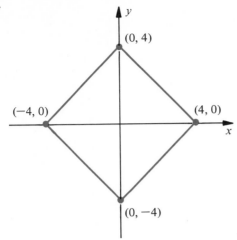

**3.**

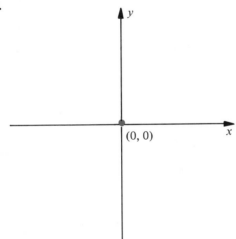

**5.**

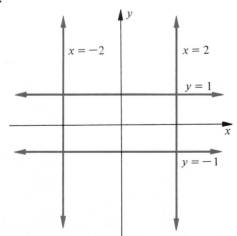

**7.**

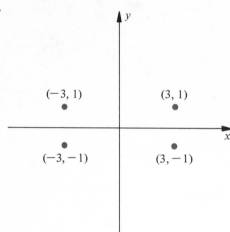

**9.**

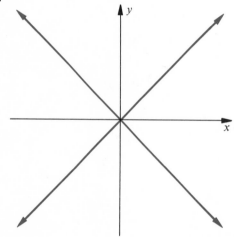

**11.**

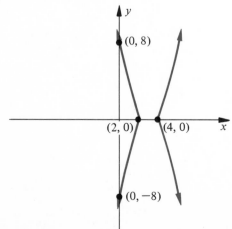

**13.**

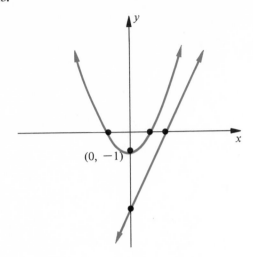

**15.** the origin

**17.**

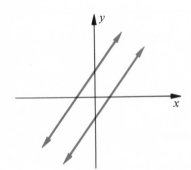

**19.**

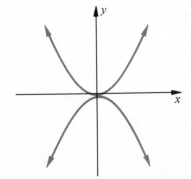

**21.**

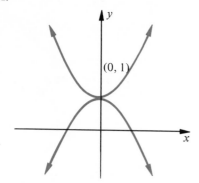

(0, 1)

## Section 19, p. 87

**1.** $r_1 = -\frac{1}{2}, r_2 = 3$
$2(x + \frac{1}{2})(x - 3)$

**3.** $r_1 = 6 + \sqrt{5}, r_2 = 6 - \sqrt{5}$
$[(x - 6) - \sqrt{5}][(x - 6) + \sqrt{5}]$

**5.** $r_1 = -\frac{1}{2} + \frac{1}{2}\sqrt{6}, r_2 = -\frac{1}{2} - \frac{1}{2}\sqrt{6}$
$4[(x + \frac{1}{2}) - \frac{1}{2}\sqrt{6}][(x + \frac{1}{2}) + \frac{1}{2}\sqrt{6}]$

**7.** $\mathbb{R}$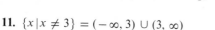
0

**9.** $\{x \mid x \leq -\frac{1}{2} \text{ or } x \geq 3\} = (-\infty, -\frac{1}{2}] \cup [3, \infty)$

$-1/2 \qquad 3$

**11.** $\{x \mid x \neq 3\} = (-\infty, 3) \cup (3, \infty)$

$3$

**13.** $\{x \mid x < \frac{3}{2} \text{ or } x > 2\} = (-\infty, \frac{3}{2}) \cup (2, \infty)$

$3/2 \qquad 2$

**15.** $\{x \mid x < -3 \text{ or } x > 3\} = \{-\infty, -3) \cup (3, \infty)$

$-3 \qquad 3$

**17.** $\mathbb{R}$
0

**19.** $\{x \mid -\frac{5}{2} < x < \frac{5}{2}\} = (-\frac{5}{2}, \frac{5}{2})$

$-5/2 \qquad 5/2$

**21.** $\{x \mid x \in \mathbb{R}, x \neq \frac{1}{2}\}$

$\frac{1}{2}$

**23.** $\{x \mid x \leq -1 \text{ or } x \geq 2\}$

$-1 \qquad 2$

**25.** $-\frac{2}{3}, \frac{1}{2}$

**27.** $l > \sqrt{41} - 1$

## Section 20, p. 93

**3.** $\{x \mid x \leq -6 \text{ or } -4 \leq x \leq -3 \text{ or } x \geq -2\}$

$-6 \qquad -4 \quad -3 \quad -2$

**5.** $\{x \mid -3 \leq x \leq 0 \text{ or } x \geq 3\}$

$-3 \qquad 0 \qquad 3$

**7.** $\{x \mid 0 < x < 1 \text{ or } x > 4\}$

$0 \quad 1 \qquad 4$

**9.** $\{x \mid x \leq -3 \text{ or } -2 < x < 2 \text{ or } x \geq 5\}$

$-3 \ -2 \qquad 2 \qquad 5$

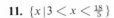

**11.** $\{x \mid 3 < x < \frac{18}{5}\}$

**13.** $\{x \mid x < 0 \text{ or } 1 < x \le 2\}$

**15.** $\{x \mid x < 0 \text{ or } 2 < x < 5\}$

**17.** $\{x \mid -3 < x < -1 \text{ or } 3 < x < 5\}$

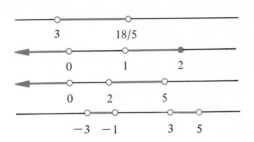

**19.** $D = \{x \mid -1 \le x \le 0 \text{ or } x \ge 1\}$
**21.** $D = \{x \mid -2 \le x < -1 \text{ or } 1 < x \le 2\}$
**25.** (c) and (d)
**27.** For $0 < e < 6$

## Section 21, p. 99

**1.**

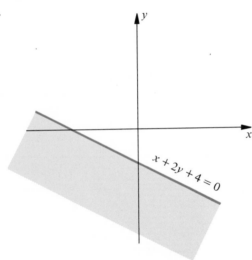

$x + 2y + 4 = 0$

**3.**

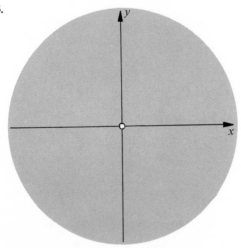

**5.**

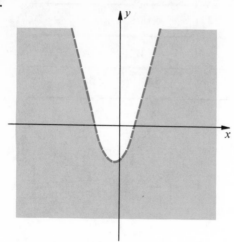

**7.**

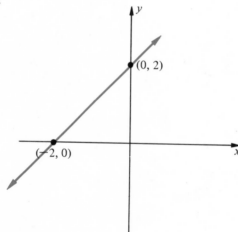

**9.**

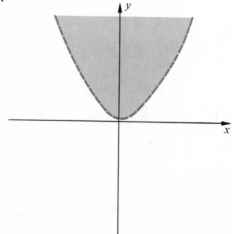

**11.**

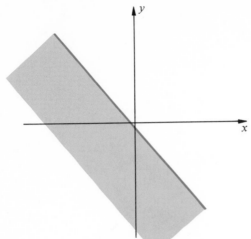

**13.**

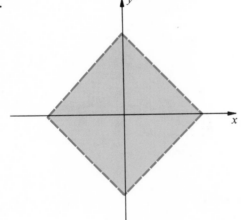

**15.**

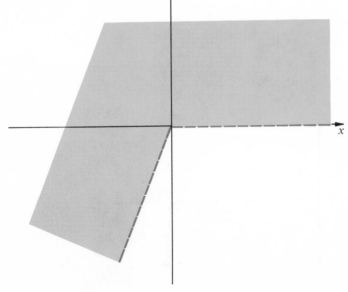

**17.**

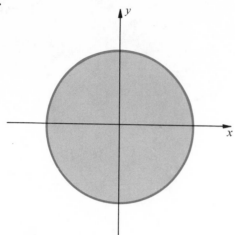

**19.**

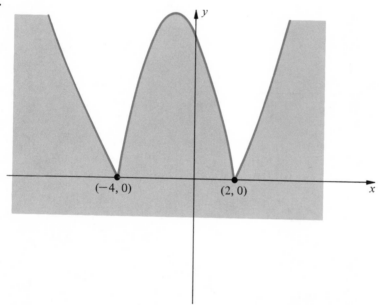

**21.**

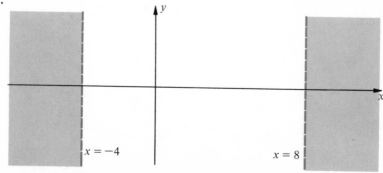

**23.**

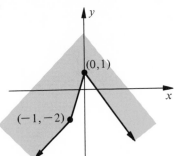

**29.**

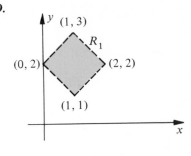

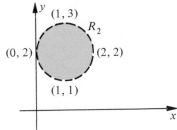

Examination of the graphs of $R_1$ and $R_2$ shows that each element of $R_1$ is contained in $R_2$.

The domain of $R_1$ *and* $R_2$ is

$$0 < x < 2.$$

The range of $R_1$ *and* $R_2$ is

$$1 < y < 3.$$

**25.**

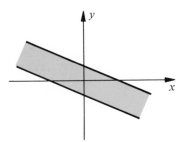

**Section 22, p. 104**

**1.**

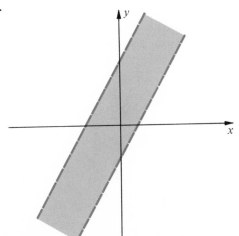

domain and range are all reals

**3.**

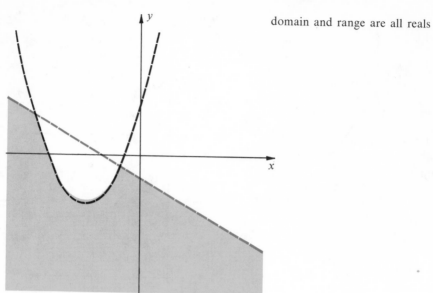

domain and range are all reals

**5.**

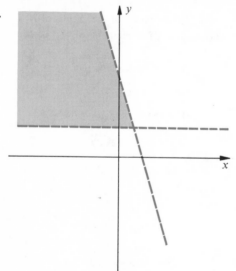

domain: $\{x \mid x < \frac{4}{3}\}$
range: $\{y \mid y > 2\}$

**7.** ∅

**9.**

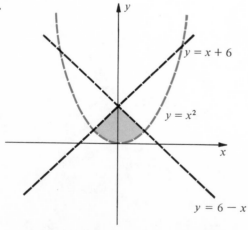

domain: $\{x \mid -2 < x < 2\}$
range: $\{y \mid 0 < y < 6\}$

**11.**

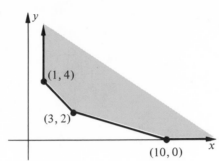

domain: $\{x \mid x \geq 1\}$
range: $\{y \mid y \geq 0\}$

**13.**

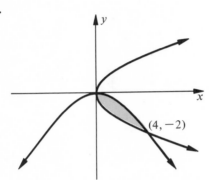

domain: $\{x \mid 0 \leq x \leq 4\}$
range: $\{y \mid -2 \leq y \leq 0\}$

**15.**

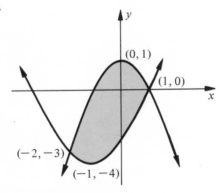

domain: $\{x \mid -2 \leq x \leq 1\}$
range: $\{y \mid -4 \leq y \leq 1\}$

**17.** 140 and 5
**19.** 116 and 94

**Section 23, p. 109**

**1.** third degree
**2.** fifth degree
**3.** not a polynomial
**4.** fifth degree
**5.** second degree in $x$, third degree in $y$, and fifth degree in $xy$
**6.** (a) 0      (c) 20      (e) 0      (g) $-2$      (i) 25
    (b) $-1$     (d) 5      (f) 0      (h) 55      (j) 15
**9.** $A = 6, B = -3$
**11.** $A = 5, B = -2$

**13.** (a) $m + n$      (c) $m$

     (b) $m$          (d) $m - n$

## Section 24, p. 114

**1.** (a) $q = 13, r = 18$      (c) $q = 3, r = 0$

     (b) $q = 25, r = 32$     (d) none

**2.** (a) $Q(x) = x + 3, R(x) = x + 2$

     (c) $Q(x) = \frac{2}{3}x^2 - \frac{7}{9}x - \frac{7}{27}, R(x) = -\frac{133}{27}x - \frac{13}{27}$

     (e) $Q(x) = x^5 + 6x + 3, R(x) = x - 1$

     (g) $Q(x) = x^2 + x + 1, R(x) = 0$

     (i) $Q(x) = x^3 + 3x^2 + 10x + 27, R(x) = 68x - 29$

**3.** $x^2 - x + 1$

**4.** $x - 1$

**5.** (a) $Q(x) = x^2 + 5x + 8, R(x) = 11$

       $x^3 + 3x^2 - 2x - 5 = (x - 2)(x^2 + 5x + 8) + 11$

     (c) $Q(x) = x^3 - 3x^2 - x + 3, R(x) = 0$

       $x^4 - 2x^3 - 4x^2 + 2x + 3 = (x + 1)(x^3 - 3x^2 - x + 3)$

     (e) $Q(x) = 2x^2 + 4x + 1, R(x) = 4$

       $4x^3 + 6x^2 - 2x + 3 = (2x - 1)(2x^2 + 4x + 1) + 4$

     (g) $Q(x) = x^3 + 2x^2 + 4x + 8, R(x) = 32$

       $x^4 + 16 = (x - 2)(x^3 + 2x^2 + 4x + 8) + 32$

     (i) $Q(x) = 2x^3 - 2x^2 - 6x + 4, R(x) = 6$

       $2x^4 - 3x^3 - 5x^2 + 7x + 4 = (x - \frac{1}{2})(2x^3 - 2x^2 - 6x + 4) + 6$

     (k) $Q(x) = 3x^2 + 6, R(x) = 0$

       $3x^3 + 4x^2 + 6x + 8 = (x + \frac{4}{3})(3x^2 + 6)$

**6.** (a) no

     (b) yes; $x^2 - 5 = (x + \sqrt{5})(x - \sqrt{5})$

**7.** (a) $(x + 2)(x - 2)(x^2 - 5)(x^2 + 9)$

     (b) $(x + 2)(x - 2)(x + \sqrt{5})(x - \sqrt{5})(x^2 + 9)$

**9.** (a) $(x + 2)^3$

     (b) $(x - 1)^3$

     (c) $(ax + b)^3$

**11.** (a) $Q(x) = x^2 - x, R(x) = 6$

     (c) $Q(x) = x^3 + x^2 - 2x + 1, R(x) = -3$

## Section 25, p. 118

**1.** (a) 14      (c) 0      (e) 11

     (b) 0      (d) 280      (f) $-5$

**2.** 3

**3.** 7

**4.** 37

**5.** (b) $a(x + 3)^2(x - 1)(x - 2) = 0, a \neq 0$

**6.** (b) $ax(x - 1)(x - 3) = 0, a \neq 0$

**7.** (b) $ax(x - 1)(2x - 1)(3x - 2) = 0, a \neq 0$

**8.** (a) and (c)

**9.** (b) and (c)

**11.** $x^3 - 3x^2 - x + 3$

**13.** (b) $-3, -2,$ and 2

**15.** $x^2 + x - 6 = 0; -3, 2$

**17.** $-22$

**Section 26, p. 124**

**1.** $-1, 1, 2$

**3.** $\frac{1}{2}$

**5.** $\frac{1}{2}, -3$

**7.** $1, 3, \frac{1}{3}, 9$

**9.** $-1, -1, -1, 3$

**11.** $-2, 1, 3$

**13.** $-3, -2, \frac{1}{2}, 2$

**15.** $(x^2 - 1)(x - 2)(x - 3)$

**17.** $-4(x - 1)(x - 2)(x - \frac{1}{2})$ or
$-2(x - 1)(x - 2)(2x - 1)$

**19.** $4, 0$

**21.** (a)

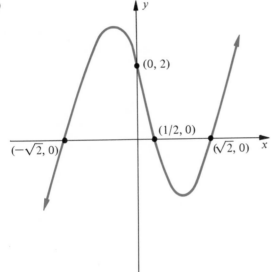

(c)

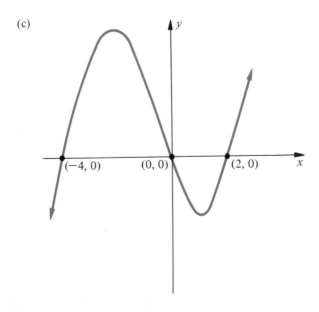

(e)

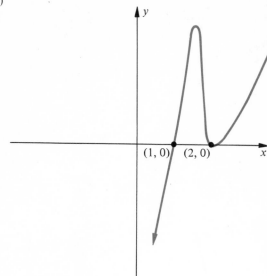

**22.** (a) $\{x \mid x \le -\sqrt{2} \text{ or } \frac{1}{2} \le x \le \sqrt{2}\}$
   (c) $\{x \mid x < -4 \text{ or } 0 < x < 2\}$
   (e) $\{x \mid x \ge 1\}$

**24.** (a) $x^4 - 6x^2 + 1$    **25.** (b) $\frac{1}{6}, \frac{1}{4}, -\frac{1}{4}$    **27.** $8, 9, 10$
   (c) $x^2 - 6x + 7$

**29.** (a)

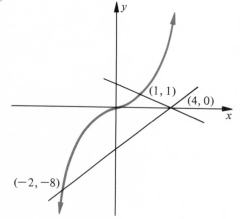

(c)

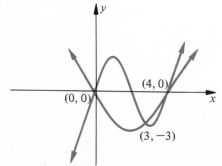

**Section 27, p. 130**

**1.** 1.5
**3.** 1.9
**5.** 1.6
**7.** $-2.2$
**9.** 4.6
**11.** (c) 1.41
**13.** (a) two
    (c) one
**15.** (a) 1        (b) 1.3        (c) 1.3        (d) 1.3        (e) 1.3
**21.** 3, $-2$
**23.** 4, $-1$
**25.** $(2, -2)$

**Section 28, p. 132**

**1.** (a)  Rational
    (c)  Rational
    (e)  Rational
    (g)  Not rational
    (i)  Not rational
    (k)  Rational
    (m) Not rational

**3.** (a)  Algebraic, domain: $x \geq 0$
    (c)  Algebraic, domain: $x > -2$
    (e)  Non-algebraic, domain: $\mathbb{R}$
    (g)  Algebraic, domain: $x \geq 0$
    (i)  Rational: domain: $x \neq \pm 2$

**Section 29, p. 145**

**1.** Domain: $\{x \mid x \neq 2\}$
    Range: $\{y \mid y \neq 0\}$
    Int.: $(0, -\frac{3}{2})$
    Asymptotes: $x = 2, y = 0$
    Symmetry: none

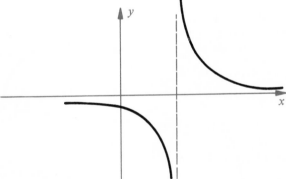

**3.** Domain: $\{x \mid x \neq 2\}$
    Range: $\{y \mid y \neq 3\}$
    Int.: $(0, 0)$
    Asymptotes: $x = 2, y = 3$
    Symmetry: none

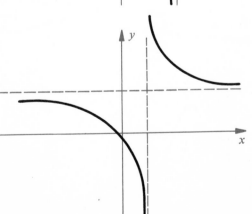

**5.** Domain: $\{x \mid x \in \mathbb{R}\}$
Range: $\{y \mid 0 < y \le 1\}$
Int.: $(0, 1)$
Asymptotes: $y = 0$
Symmetry: $y$-axis

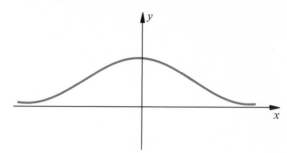

**7.** Note: $y = \dfrac{x(x - 2)}{(x - 2)}$
Domain: $\{x \mid x \ne 2\}$
Range: $\{y \mid y \ne 2\}$
Int.: $(0, 0)$
Asymptotes: none
Symmetry: none

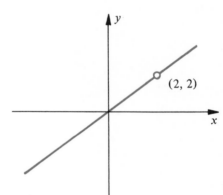

$(2, 2)$

**9.** Domain: $\{x \mid x \ne -3, x \ne 1\}$
Range: $\{y \mid y \in \mathbb{R}\}$
Int.: $(0, 0)$
Asymptotes: $x = 1$, $x = -3$, $y = 0$
Symmetry: none

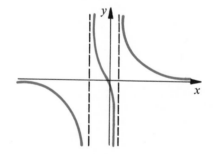

**11.** Domain: $\{x \mid x \in \mathbb{R}\}$
Range: $\left\{ y \,\middle|\, -\dfrac{\sqrt{2}}{4} \le y \le \dfrac{\sqrt{2}}{4} \right\}$
Int.: $(0, 0)$
Asymptotes: $y = 0$
Symmetry: origin

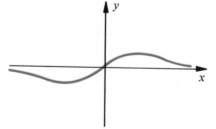

**13.** Domain: $\{x \mid x \in \mathbb{R}\}$
Range: $\{y \mid y \ge 0\}$
Int.: $(0, 0)$
Asymptotes: none
Symmetry: $y$-axis

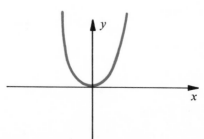

**15.** Domain: $\{x \mid x \neq -2, x \neq 1\}$
Range: $\{y \mid y \leq -\frac{4}{9} \text{ or } y > 0\}$
Int.: $(0, -\frac{1}{2})$
Asymptotes: $x = -2, x = 1, y = 0$
Symmetry: none

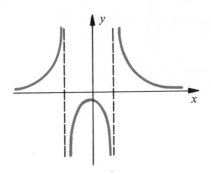

**17.** Domain: $\{x \mid x \in \mathbb{R}\}$.
Range: $\{y \mid 0 < y \leq 3\}$
Int.: $(0, 3)$
Asymptotes: $y = 0$
Symmetry: $y$-axis

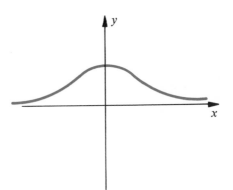

**19.** Domain: $\{x \mid x \neq \pm 1\}$
Range: $\{y \mid y \leq 0 \text{ or } y > 1\}$
Int.: $(0, 0)$
Asymptotes: $x = \pm 1, y = 1$
Symmetry: $y$-axis

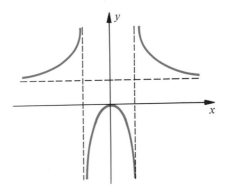

**21.** Domain: $\{x \mid x \neq 2 \text{ or } 6\}$
Range: $\{y \mid y \leq -6 \text{ or } y \geq 0\}$
Int.: $(0, 0)$
Asymptotes: $x = 2, x = 6, y = 2$
Symmetry: none

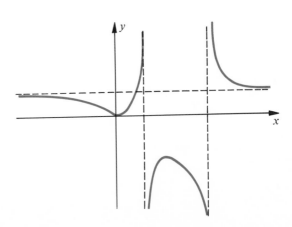

**23.** Domain: $\{x \mid x \neq -1\}$
   Range: $\{y \mid y \neq 0\}$
   Int.: $(0, -1)$
   Asymptotes: $x = -1$, $y = 0$
   Symmetry: none

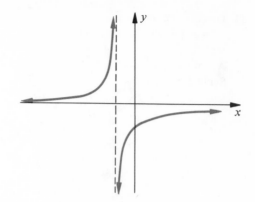

**27.** Yes
**31.** $x = 2$; none; $y = 3x + 8$
**33.** $x = \pm 2\sqrt{2}$; none; $y = x$
**35.** $x = \pm 3$; none; $y = 4x$
**37.** $x = 1$; none; $y = x + 1$

**Section 30, p. 153**
 **1.** Domain: $\{x \mid x \geq -4\}$
   Range: $\{y \mid y \geq 0\}$
   Int.: $(0, 2)$, $(-4, 0)$
   Asymptotes: none
   Symmetry: none

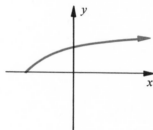

**3.** Domain: $\{x \mid x \in \mathbb{R}\}$
   Range: $\{y \mid y \geq 2\}$
   Int.: $(0, 2)$
   Asymptotes: none
   Symmetry: $y$-axis

**5.** Domain: $\{x \mid x > -4\}$
   Range: $\{y \mid y > 0\}$
   Int.: $(0, \frac{1}{2})$
   Asymptotes: $x = -4$, $y = 0$
   Symmetry: none

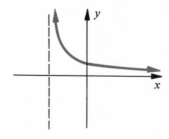

**7.** Domain: $\{x \mid x > 1\}$
Range: $\{y \mid y \geq 2\}$
Int.: none
Asymptotes: $x = 1$
Symmetry: none

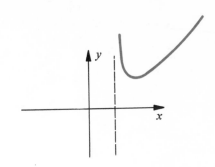

**9.** Domain: $\{x \mid -3 \leq x \leq 3\}$
Range: $\{y \mid 0 \leq y \leq 3\}$
Int.: $(0, 3)$, $(3, 0)$, $(-3, 0)$
Asymptotes: none
Symmetry: $y$-axis

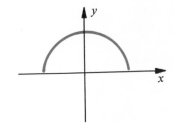

**11.** Domain: $\{x \mid x \neq -2\}$
Range: $\{y \mid y > 0\}$
Int.: $(0, \frac{1}{4})$
Asymptotes: $x = -2$, $y = 0$
Symmetry: none

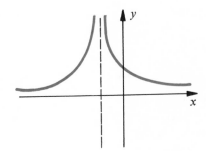

**13.** Domain: $\{x \mid x \in \mathbb{R}\}$
Range: $\{y \mid y \geq 0\}$
Int.: $(0, \sqrt[3]{4})$, $(2, 0))$
Asymptotes: none
Symmetry: none

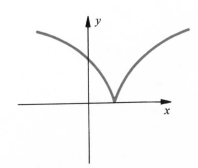

**15.** Domain: $\{x \mid x < -3 \text{ or } x > 3\}$
Range: $\{y \mid y < -1 \text{ or } y > 1\}$
Int.: none
Asymptotes: $x = \pm 3$, $y = \pm 1$
Symmetry: Origin

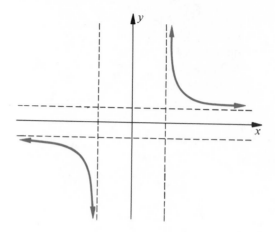

**Section 31, p. 159**

**1.** Domain: $\{x \mid -3 < x \le -1, 1 \le x < 3\}$
Range: $\{y \mid y \in \mathbb{R}\}$
Int.: $(1, 0)$, $(-1, 0)$
Asymptotes: $x = 3$, $x = -3$
Symmetry: $x$-axis, $y$-axis, origin

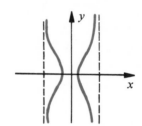

**3.** Domain: $\{x \mid x \le -1 \text{ or } x \ge 1\}$
Range: $\{y \mid y \in \mathbb{R}\}$
Int.: $(1, 0)$, $(-1, 0)$
Asymptotes: none
Symmetry: $x$-axis, $y$-axis, origin

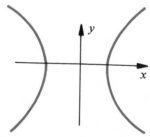

**5.** Domain: $\{x \mid x \le -\frac{1}{3} \text{ or } x \ge \frac{1}{3}\}$
Range: $\{y \mid y \in \mathbb{R}\}$
Int.: $(\frac{1}{3}, 0)$, $(-\frac{1}{3}, 0)$
Asymptotes: none
Symmetry: $x$-axis, $y$-axis, origin

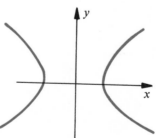

**7.** Domain: $\{x \mid -2 \le x \le 2\}$
Range: $\{y \mid -2 \le y \le 2\}$
Int.: $(0, 0)$, $(2, 0)$, $(-2, 0)$
Asymptotes: none
Symmetry: $x$-axis, $y$-axis, origin

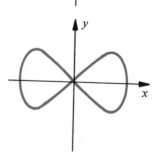

**9.** Domain: $\{x \mid -8 \leq x \leq 8\}$
Range: $\{y \mid -8 \leq y \leq 8\}$
Int.: $(0, 8)$, $(0, -8)$, $(8, 0)$, $(-8, 0)$
Asymptotes: none
Symmetry: $x$-axis, $y$-axis, origin

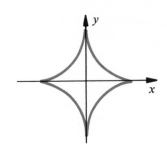

**11.** Domain: $\{x \mid x \geq 2\}$
Range: $\{y \mid y \in \mathbb{R}\}$
Int.: $(2, 0)$
Asymptotes: none
Symmetry: $x$-axis

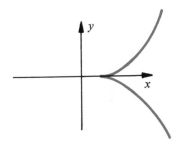

**13.** $(y^2 - x^2)(y - 2) = 0$
$(y - x)(y + x)(y - 2) = 0$

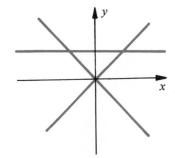

**15.** Domain: $\{x \mid x \leq 0 \text{ or } x > 2\}$
Range: $\{y \mid y \neq \pm 1\}$
Int.: $(0, 0)$
Asymptotes: $x = 2$, $y = \pm 1$
Symmetry: $x$-axis

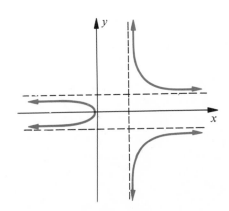

**17.** Domain: $\{x \mid -1 < x \le 1 \text{ or } x > 3\}$
Range: $\{y \mid y \in \mathbb{R}\}$

Int.: $(1, 0)$, $\left(0, \pm\dfrac{1}{\sqrt{3}}\right)$

Asymptotes: $x = -1$, $x = 3$, $y = 0$
Symmetry: $x$-axis

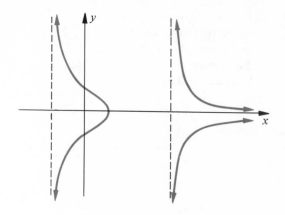

**19.** Domain: $\{x \mid -3 \le x \le 3\}$
Range: $\{y \mid -2 \le y \le 2\}$
Int.: $(0, \pm 2)$, $(\pm 3, 0)$
Asymptotes: none
Symmetry: $x$-axis, $y$-axis, origin.

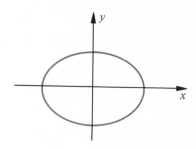

**20.**

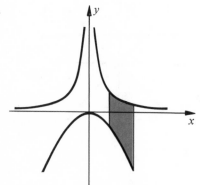

Domain: $\{x \mid 1 \le x \le 2\}$
Range: $\{y \mid -4 \le y \le 1\}$

**23.**

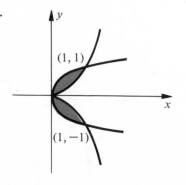

Domain: $\{x \mid 0 \le x \le 1\}$
Range: $\{y \mid -1 \le y \le 1\}$

**25.**

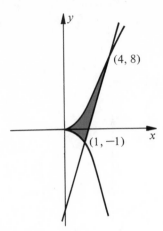

Domain: $\{x\,|\,0 \leq x \leq 4\}$
Range: $\{y\,|\,-1 \leq y \leq 8\}$

**Section 32, p. 162**

**1.** $f + g = x^2 + x$, domain: $\mathbb{R}$
$f - g = x - x^2$, domain: $\mathbb{R}$
$f \cdot g = x^3$, domain: $\mathbb{R}$
$\dfrac{f}{g} = \dfrac{1}{x}$, domain: $\mathbb{R}$, $x \neq 0$
$f \circ g = x^2$, domain: $\mathbb{R}$
$g \circ f = x^2$, domain: $\mathbb{R}$

**3.** $f + g = |x| + x^2 + 1$, domain: $\mathbb{R}$ for all
$f - g = |x| - (x^2 + 1)$
$f \cdot g = |x| \cdot (x^2 + 1)$
$\dfrac{f}{g} = \dfrac{|x|}{x^2 + 1}$
$f \circ g = |x^2 + 1| = x^2 + 1$
$g \circ f = x^2 + 1$

**5.** $f + g = x^3 + 4$, domain: $\mathbb{R}$ for all
$f - g = x^3 - 4$
$f \cdot g = 4x^3$
$\dfrac{f}{g} = \dfrac{x^3}{4}$
$f \circ g = 64$
$g \circ f = 4$

**7.** $f + g = x^2 + x^3$, domain: $\mathbb{R}$
$f - g = x^2 - x^3$, domain: $\mathbb{R}$
$f \cdot g = x^5$, domain: $\mathbb{R}$
$\dfrac{f}{g} = \dfrac{1}{x}$, domain: $\mathbb{R}$, $x \neq 0$
$f \circ g = x^6$, domain: $\mathbb{R}$
$g \circ f = x^6$, domain: $\mathbb{R}$

**9.** $f + g = 1 - x^2 + \sqrt{x}$, domain: $x \geq 0$
$f - g = 1 - x^2 - \sqrt{x}$, domain: $x \geq 0$
$f \cdot g = (1 - x^2) \cdot \sqrt{x}$, domain: $x \geq 0$
$\dfrac{f}{g} = \dfrac{(1 - x^2)}{\sqrt{x}}$, domain: $x > 0$
$f \circ g = 1 - (\sqrt{x})^2 = 1 - x$, domain: $x \geq 0$
$g \circ f = \sqrt{1 - x^2}$, domain: $-1 \leq x \leq 1$

**11.** $f[1 + x^2] = 1 + x^2$. Let $r = 1 + x^2$. Then $x = \sqrt{r - 1}$. $f(r) = 1 + (\sqrt{r - 1})^2 = 1 + r - 1$. $f(r) = r$. Thus $f(x) = x$.

**13.** $g(x) = \dfrac{x - 4}{3}$

**15.** Let $f(x) = \sqrt{x}$.        **17.** No; domain and range are not the same.

**19.**

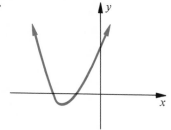

**21.**

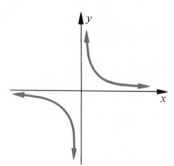

## Section 33, p. 168

**1.** $f(x) = 5x + 1$, domain: all reals, range: all reals.
  $f^{-1} = \frac{1}{5}(x - 1)$, domain: all reals, range: all reals.

**3.** no inverse

**5.** $f(x) = x^2 - 1$, domain: $x \leq 0$, range: $y \geq -1$
  $f^{-1}(x) = -\sqrt{x + 1}$, domain: $x \geq -1$, range: $y \leq 0$

**7.** $f(x) = (x - 2)^2$, domain: $x \geq 2$, range: $y \geq 0$
  $f^{-1}(x) = \sqrt{x} + 2$, domain: $x \geq 0$, range: $y \geq 2$

**9.** $f(x) = 4 - 3x$, domain: all reals, range: all reals.
  $f^{-1}(x) = \dfrac{4 - x}{3}$, domain: all reals, range: all reals.

**11.** $f(x) = -\sqrt{x + 1}$, domain: $x \geq -1$, range: $y \leq 0$
  $f^{-1}(x) = x^2 - 1$, domain: $x \leq 0$, range: $y \geq -1$

**13.** $f(x) = -\sqrt{x - 4}$, domain: $x \geq 4$, range: $y \leq 0$
  $f^{-1}(x) = x^2 + 4$, domain: $x \leq 0$, range: $y \geq 4$

**15.** $f(x) = x^2$, domain: $0 \leq x \leq 1$, range: $0 \leq y \leq 1$
  $f^{-1}(x) = \sqrt{x}$, domain: $0 \leq x \leq 1$, range: $0 \leq y \leq 1$

**17.** $f(x) = \dfrac{3}{x}$, domain: all reals, $x \neq 0$, range: all reals, $y \neq 0$

  $f^{-1}(x) = \dfrac{3}{x}[f(x)$ is its own inverse$]$

**19.** no inverse        **21.** no inverse        **23.** no inverse

**25.** $f(x) = \dfrac{x}{\sqrt{4 - x^2}}$, domain: $-2 < x < 2$, range: all reals

  $f^{-1}(x) = \dfrac{2x}{\sqrt{x^2 + 1}}$, domain: all reals; range: $-2 < y < 2$

**27.** (a), (b), (d), (f), (g), and (j) have inverses

**29.** $g(x) = x^2 - 6x + 11$

**Section 34, p. 175**

1. (a) 5
   (c) 2
   (e) b > 0, b ≠ 1
3. 16
5. (a)           7. (a)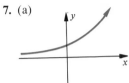

9. (a) 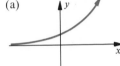          11. (a)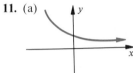

13. (a)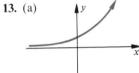

15. A line parallel to the x-axis and 1 unit above.
17. (a) 800,000
    (b) 1 hr, 2 hr
19. 51.56°C
21. (a)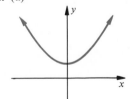

    (b)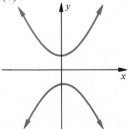

(c)

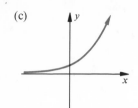

25.

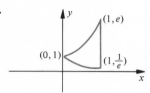

### Section 35A, p. 181

| | | | |
|---|---|---|---|
| **1.** $x = 3^y$ | **3.** $8 = 2^3$ | **5.** $100 = 10^2$ | **7.** $x = e$ |
| **9.** $N = a^y$ | **11.** $\log_2 32 = 5$ | **13.** $\log_b 6 = y$ | **15.** $\log_3 \left(\frac{1}{9}\right) = -2$ |
| **17.** $\ln 3 = x$ | **19.** $\log_2 y = x$ | **21.** 4 | **23.** 3 |
| **25.** $\frac{1}{2}$ | **27.** 2 | **29.** $-1$ | **31.** $-1$ |
| **33.** 0 | **35.** 9 | **37.** 4 | **39.** $-\frac{1}{2}$ |
| **41.** $\frac{1}{9}$ | **43.** $\frac{1}{27}$ | **45.** $\sqrt{e}$ | **47.** 6 |
| **49.** 2 | **51.** 7 | | |

53.

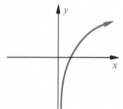

55.

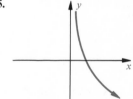

57.

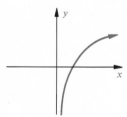

59.

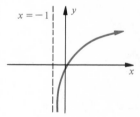

61.

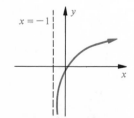

63.

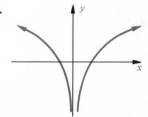

## Section 35B, p. 186

**7.** 3 **9.** 4 **11.** $\frac{5}{2}$ **13.** 3

**15.** $\log x^2 y^{1/2}$ **17.** $\ln \dfrac{x^2 y^3}{z^4}$ **19.** $\log_b \left( \dfrac{5z^3}{y^{2/3}} \right)$

**21.** $\ln \left( \dfrac{x^2 y^3 z^2}{81} \right)$ **23.** $\log_b 2 + \log_b x + \log_b y + \log_b z$

**25.** $3 \log_b y + \log_b z$ **27.** $\frac{2}{3} \log_b x$

**29.** $2 \log_b x + 3 \log_b y - 2 \log_b z$ **31.** 10

**33.** 7 **35.** 9 **37.** $-1, 4$ **39.** 10

**41.** $y = \dfrac{x^2}{10}$, $x > 0$ **43.** $y = \dfrac{6}{x}$, $x > 0$ **45.** $y = \dfrac{e}{8x^2}$, $x > 0$

**47.** $y = \sqrt{\dfrac{500}{x^3}}$, $x > 0$ **49.** $\{x \mid 0 < x < 1 \text{ or } x > 10\}$

**51.** $x = 1, e$ **53.** $\{x \mid 1 < x < e\}$

## Section 36, p. 192

**1.** (a) 1.3617 **2.** (a) $n = 304$ **3.** (a) $x = 2.3222$

(c) $0.1644 - 2$ (c) $n = 0.0517$ (c) $x = 1$

(e) 4.9200 (e) $n = 54.8$ (e) $x = 2.3026$

**5.** $x \le -3$ **7.** $x = 1, 2$ **9.** $x = 10^{100}$

**11.** $x = 10^{10^{10}}$ **13.** (a) $n = e^3$ **14.** (a) 2.9958

(c) $n = 10^{-2.3147}$ (c) 2.0796

(e) $n = 10^1$ (e) $-3.9120$

(g) $-1.6094$

**17.** (a) $k = 0.0578$ **19.** 7.8

(b) Approx. 19 yrs

## Section 37, p. 201

**1.** (a) $P(-\pi) = (-1, 0)$ (c) $P\left(\dfrac{7\pi}{2}\right) = (0, -1)$ (e) $P\left(-\dfrac{5\pi}{2}\right) = (0, -1)$

(b) $P(3\pi) = (-1, 0)$ (d) $P(6\pi) = (1, 0)$ (f) $P(0) = (1, 0)$

**3.** (a) $P\left(\dfrac{13\pi}{6}\right) = \left(\dfrac{\sqrt{3}}{2}, \dfrac{1}{2}\right)$ (c) $P\left(\dfrac{25\pi}{6}\right) = \left(\dfrac{\sqrt{3}}{2}, \dfrac{1}{2}\right)$

(b) $P\left(-\dfrac{5\pi}{3}\right) = \left(\dfrac{1}{2}, \dfrac{\sqrt{3}}{2}\right)$ (d) $P\left(-\dfrac{11\pi}{6}\right) = \left(\dfrac{\sqrt{3}}{2}, \dfrac{1}{2}\right)$

(e) $P\left(\dfrac{13\pi}{3}\right) = \left(\dfrac{1}{2}, \dfrac{\sqrt{3}}{2}\right)$

**5.** (a) second (c) third (e) fourth

(b) second (d) second

## Section 38, p. 209

**1.** 0 **3.** $\infty$ **5.** 1 **7.** 0

**9.** $-1$ **11.** $\sqrt{2}$ **13.** $\sqrt{2}$ **15.** $\dfrac{2}{\sqrt{3}}$

**17.** $\dfrac{\sqrt{3}}{2}$   **19.** 1   **21.** $\dfrac{-\sqrt{3}}{2}$   **23.** $\cos s = \frac{3}{5}$   $\sec s = \frac{5}{3}$

$\sin s = -\frac{4}{5}$   $\csc s = -\frac{5}{4}$

$\tan s = -\frac{4}{3}$   $\cot s = -\frac{3}{4}$

**25.** $b$, $c$, $d$, and $l$ are positive.

**26.** (a) $\dfrac{\pi}{3}, \dfrac{2\pi}{3}$   (c) $\dfrac{2\pi}{3}, \dfrac{4\pi}{3}$   (e) $\pi$   (g) $\dfrac{5\pi}{4}, \dfrac{7\pi}{4}$   (i) $\dfrac{\pi}{3}, \dfrac{5\pi}{3}$   (k) $\dfrac{\pi}{2}, \dfrac{3\pi}{2}$

**27.** (a) one and four   (c) four   (e) one and four
(b) four   (d) two   (f) one and three

**29.** $\dfrac{\sqrt{8}}{3}$   **31.** $\dfrac{-4}{5}$   **33.** $\pm\dfrac{5}{13}$   **35.** $\dfrac{1}{\sqrt{8}}$ or $\dfrac{\sqrt{2}}{4}$

**37.** (a) $-1$, 0, undefined   (e) 1, 0, undefined   (i) $\dfrac{\sqrt{3}}{2}, -\dfrac{1}{2}, -\sqrt{3}$

(c) $\dfrac{1}{2}, \dfrac{\sqrt{3}}{2}, \dfrac{1}{\sqrt{3}}$   (g) 0, 1, 0

**39.** $\sin s = \frac{15}{17}$, $\cos s = \pm\frac{8}{17}$, $\tan s = \pm\frac{15}{8}$

**41.** (a) $\frac{3}{4}$   (c) $\dfrac{2 + \sqrt{3}}{4}$   (e) 1   (g) $\frac{1}{2}$

**43.** (b) 0   (c) no   (d) no
**45.** $-\frac{3}{5}$   **46.** $-\frac{4}{5}$ and $-\frac{3}{5}$   **49.** $\frac{12}{5}$

## Section 39, p. 217

**1.** $\dfrac{\sqrt{6} + \sqrt{2}}{4}$   **3.** $\dfrac{\sqrt{6} + \sqrt{2}}{4}$   **5.** $\dfrac{\sqrt{3} + 1}{\sqrt{3} - 1}$   **7.** $\dfrac{\sqrt{2} - \sqrt{6}}{4}$

**9.** (a) $\frac{84}{85}, \frac{13}{85}$   (b) first
**25.** $\frac{1}{2}\sin 11 - \frac{1}{2}\sin 3$
**27.** $\frac{1}{2}\sin 13x - \frac{1}{2}\sin 3x$
**29.** $\frac{1}{2}\sin 7x - \frac{1}{2}\sin x$
**31.** $-2\cos\left(\frac{1}{2}\right)\sin\left(\frac{11}{2}\right)$
**33.** $2\sin\left(\frac{5}{2}\right)\sin\left(\frac{11}{2}\right)$
**35.** $2\cos\left(\dfrac{5x}{2}\right)\cos\left(\dfrac{x}{2}\right)$

**37.** $2\sin 4x \sin x$

## Section 40, p. 224

**1.** (a) $-\cos\dfrac{\pi}{4}$   (e) $-\cos 1.14$   (i) $-\tan 2.44 = \tan 0.7$

(c) $\cos\dfrac{2\pi}{5}$   (g) $\cos 2.44 = -\cos 0.7$   (k) $\sin\dfrac{\pi}{6}$

**2.** (a) 0.24740   (e) 0.84147   (i) $-0.17981$
(c) 3.8782   (g) $-2.1759$   (k) 0.96106

**3.** (a) $-\cos\dfrac{\pi}{6} = -\dfrac{\sqrt{3}}{2}$   (g) $-\sec\dfrac{\pi}{6} = -\dfrac{2}{\sqrt{3}}$

(c) same as (a)   (i) $-\sin\dfrac{3\pi}{4} = -\sin\dfrac{\pi}{4} = -\dfrac{\sqrt{2}}{2}$

(e) $\cos\dfrac{\pi}{3} = \dfrac{1}{2}$   (k) $-\sin\dfrac{\pi}{2} = -1$

**5.** (a) 0.91     (c) 1.299     (e) 1.46

    (b) 1.39     (d) 1.459     (f) 0.1

**7.** 1.39, 1.75

**9.** 1.46, 4.82

**11.** 4.53, 4.89

**13.** 1.68, 4.60

**15.** $\dfrac{3\pi}{4}, \dfrac{7\pi}{4}$

**17.** $\dfrac{\pi}{6}, \dfrac{5\pi}{6}$

**19.** $\dfrac{3\pi}{2}$

**20.** (b) (1) $\sin 0.57$     (3) $\cot 0.57$     (5) $-\cos 0.14$

       (2) $\cos 0.37$     (4) $\cos 0.43$     (6) $-\cot 0.43$

## Section 41, p. 228

**1.** (a) $\sin S = \dfrac{1}{\sqrt{5}}$, $\cos S = \dfrac{2}{\sqrt{5}}$, $\cot S = 2$, $\sec S = \dfrac{\sqrt{5}}{2}$, $\csc S = \sqrt{5}$

   (c) $\cos S = \dfrac{\sqrt{24}}{5}$, $\tan S = -\dfrac{1}{\sqrt{24}}$, $\cot S = -\sqrt{24}$, $\sec S = \dfrac{5}{\sqrt{24}}$, $\csc S = -5$

**2.** (a) $\sin\left(-\dfrac{9\pi}{2}\right) = -1$, $\cot\left(-\dfrac{9\pi}{2}\right) = 0$, $\tan\left(-\dfrac{9\pi}{2}\right)$ and $\sec\left(-\dfrac{9\pi}{2}\right)$ are undefined

      $\cos\left(-\dfrac{9\pi}{2}\right) = 0$, $\csc\left(-\dfrac{9\pi}{2}\right) = -1$

   (c) $\sin\dfrac{5\pi}{3} = \dfrac{-\sqrt{3}}{2}$, $\tan\dfrac{5\pi}{3} = -\sqrt{3}$, $\sec\dfrac{5\pi}{3} = 2$

      $\cos\dfrac{5\pi}{3} = \dfrac{1}{2}$, $\cot\dfrac{5\pi}{3} = -\dfrac{1}{\sqrt{3}}$, $\csc\dfrac{5\pi}{3} = -\dfrac{2}{\sqrt{3}}$

**3.** (a) $1 - \sin^2 S$       (b) $\dfrac{1}{\sin S}$, $1 + \cos S \neq 0$

   (c) $\dfrac{1}{1 - \sin^2 S}$       (d) $\pm\dfrac{\sin S}{\sqrt{1 - \sin^2 S}}$

**5.** $\sin S = \pm\sqrt{\dfrac{\tan^2 S}{1 + \tan^2 S}}$     $\cot S = \dfrac{1}{\tan S}$     $\csc S = \pm\sqrt{\dfrac{1 + \tan^2 S}{\tan^2 S}}$

   $\cos S = \pm\sqrt{\dfrac{1}{1 + \tan^2 S}}$     $\sec S = \pm\sqrt{1 + \tan^2 S}$

## Section 42, p. 233

**1.** (a) $-\dfrac{24}{25}$     (b) $\dfrac{7}{25}$     (c) $-\dfrac{24}{7}$

**3.** (a) $\dfrac{1}{\sqrt{17}}$     (b) $\dfrac{4}{\sqrt{17}}$     (c) $\dfrac{1}{4}$

**5.** (a) $\dfrac{\pm 2}{\sqrt{13}}$     (b) $\dfrac{\pm 3}{\sqrt{13}}$     (c) $\dfrac{\pm\sqrt{13}}{3}$

**7.** $\sin 2s = \pm 2y\sqrt{1 - y^2}$

   $\cos 2s = 2y^2 - 1$

   $\tan 2s = \pm\dfrac{2y\sqrt{1 - y^2}}{2y^2 - 1}$

**9.** $\dfrac{24}{25}, \dfrac{44}{125}$

**11.** (a) $\dfrac{\sqrt{3}}{2}$             (b) $\dfrac{\sqrt{3}}{2}$

     (c) $-\dfrac{\sqrt{3}}{2}$       (d) $\dfrac{\sqrt{2+\sqrt{3}}}{2}$

     (e) $\dfrac{\sqrt{2-\sqrt{3}}}{2}$      (f) $-\dfrac{1}{2}$

**13.** $\dfrac{3}{8} - \dfrac{\cos 2s}{2} + \dfrac{\cos 4s}{8}$

**15.** $\sin 6x$

**17.** $\sin x$

**19.** $\cos 6x$

**21.** $2 \cos 2x$

**23.** $-1$

**47.** (a) $\dfrac{z^2 + 1}{z^2 + 3}$      (c) $\dfrac{z^2 + 1}{3z^2 + 10z + 3}$

**Section 43, p. 241**

**1.** $P = 2\pi, A = 3$

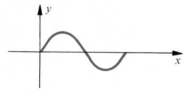

**3.** $P = 2\pi, A = 2$

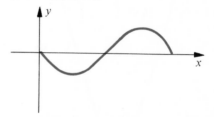

**5.** $P = 2\pi, A = 1$

**7.** $P = 8\pi, A = 1$

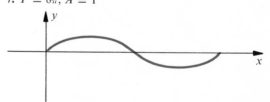

**9.** $P = 2\pi, A = \frac{1}{2}$

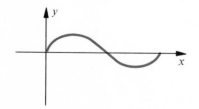

**11.** $P = 4\pi, A = 1$

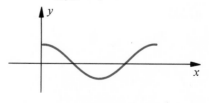

**13.** $P = \pi$, $A = 2$

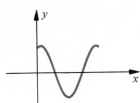

**17.** $P = 4\pi$, $A = \frac{3}{2}$, phase shift is $-1$

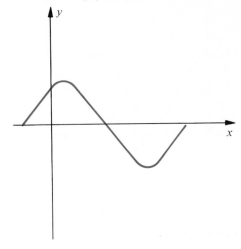

**19.** $P = \pi$, $A = 3$, phase shift is $\frac{\pi}{4}$   **21.** $P = \pi$, $A = 3$, phase shift is $\frac{\pi}{4}$

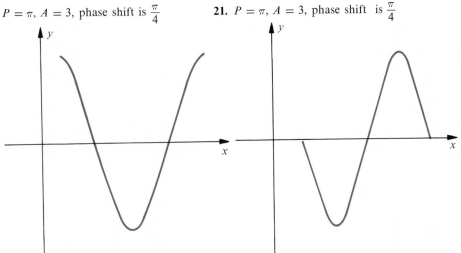

**25.** (a)

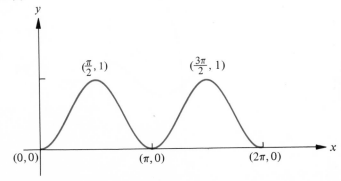

(b)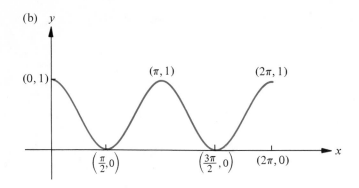

**Section 44, p. 245**

5.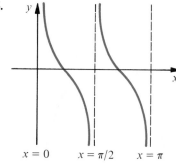

7.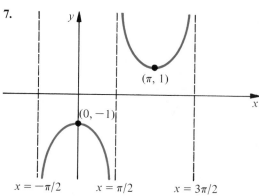

9.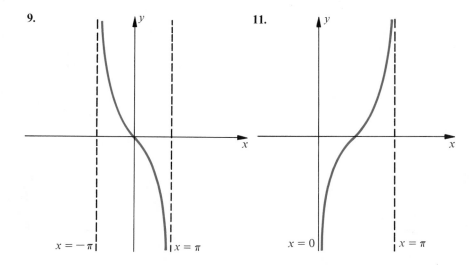

11.

**13.**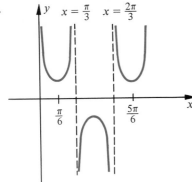

**15.**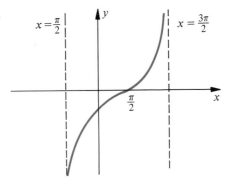

**17.** 0, 1, 2, 3

## Section 45, p. 250

**1.** $\dfrac{\pi}{3}$  **3.** 0  **5.** $\dfrac{\pi}{3}$  **7.** $\dfrac{3}{5}$  **9.** 1  **11.** $\dfrac{3 - \sqrt{8}}{3\sqrt{10}}$

**13.** $\dfrac{24}{25}$  **15.** $-0.2$  **17.** $-0.86$  **19.** $-\dfrac{\pi}{3}$

**21.** $\sqrt{1 - x^2}$  **23.** $\dfrac{1}{\sqrt{x^2 + 1}}$

**40.** (a) $x = \frac{1}{2}(\sin \frac{3}{2}y) + 1$  (c) $x = \sin(y - \pi)$  (e) $x = \tan(y) + 4$

**41.** $\dfrac{6x}{x^2 + 9} + \tan^{-1}\left(\dfrac{x}{3}\right)$

## Section 46, p. 255

**1.** $\dfrac{\pi}{6}, \dfrac{5\pi}{6}, \pi$

**3.** $\dfrac{7\pi}{6}, \dfrac{11\pi}{6}, \dfrac{3\pi}{2}$

**5.** $\pi$

**7.** $0, \pi, 2\pi$

**9.** $0, 2\pi$

**11.** $0, \dfrac{\pi}{4}, \dfrac{\pi}{2}, \dfrac{3\pi}{4}, \pi, \dfrac{5\pi}{4}, \dfrac{3\pi}{2}, \dfrac{7\pi}{4}, 2\pi, \dfrac{\pi}{12}, \dfrac{7\pi}{12}, \dfrac{13\pi}{12}, \dfrac{19\pi}{12}, \dfrac{5\pi}{12}, \dfrac{11\pi}{12}, \dfrac{17\pi}{12}, \dfrac{23\pi}{12}$

**13.** $\dfrac{\pi}{12}, \dfrac{11\pi}{12}, \dfrac{13\pi}{12}, \dfrac{23\pi}{12}$

**15.** $0, \pi, 2\pi, \dfrac{2\pi}{3}, \dfrac{4\pi}{3}$

**17.** $\dfrac{\pi}{2}, 2\pi - \cos^{-1}(\frac{4}{5})$

**19.** $2\pi + \sin^{-1}\left(-\dfrac{1}{\sqrt{10}}\right), \sin^{-1}\left(-\dfrac{1}{\sqrt{10}}\right) + \pi$

**21.** $\dfrac{\pi}{6}, \dfrac{5\pi}{6}, \pi$

**23.** $\dfrac{\pi}{8}, \dfrac{5\pi}{8}, \dfrac{9\pi}{8}, \dfrac{13\pi}{8}$

**25.** $\dfrac{\pi}{6}, \dfrac{7\pi}{6}, \dfrac{2\pi}{3}, \dfrac{5\pi}{3}$

**27.** $\dfrac{3\pi}{8}, \dfrac{11\pi}{8}, \dfrac{7\pi}{8}, \dfrac{15\pi}{8}$

**29.** $\dfrac{\pi}{3}, \dfrac{2\pi}{3}, \dfrac{4\pi}{3}, \dfrac{5\pi}{3}, \dfrac{\pi}{2}, \dfrac{3\pi}{2}$

**31.** $x = -\frac{1}{2}$

**33.** $x = \dfrac{\sqrt{2}}{2}$

**35.** $x = 1$

**37.** $x = 1$

**39.** $x = e^{k\pi}, k \in J$

**41.** $\varnothing$

**43.** (a) $y = \dfrac{x + 1}{x - 1}$

 (b) $x < 1$

**45.** $x = \dfrac{\pi}{24} + \dfrac{k\pi}{2}, k \in J$

 or

 $x = \dfrac{5\pi}{24} + \dfrac{k\pi}{2}, k \in J$

**47.** $x = \dfrac{\pi}{2} + k\pi, k \in J$

 or

 $x = \pi + 2k\pi, k \in J$

**Section 47, p. 261**

**1.** $150°$    **3.** $1080°$    **5.** $0°$    **7.** $570°$    **9.** $300°$

**11.** $\dfrac{2\pi}{3}$    **13.** $-\dfrac{4\pi}{9}$    **15.** $\dfrac{7\pi}{4}$    **17.** $\dfrac{\pi}{12}$    **19.** $\dfrac{3\pi}{2}$

**21.** $\theta_1 = 30° + k(360°), k \in J; \theta_1 = 240° + k(360°), k \in J$

**23.** $10\pi, \dfrac{40\pi}{3}$    **25.** $\theta = 0.29$

**Section 48, p. 265**

**1.** $\sin 30° = \dfrac{1}{2}$    $\tan 30° = \dfrac{1}{\sqrt{3}}$    $\sec 30° = \dfrac{2}{\sqrt{3}}$

 $\cos 30° = \dfrac{\sqrt{3}}{2}$    $\cot 30° = \sqrt{3}$    $\csc 30° = 2$

**2.** $\sin 60° = \dfrac{\sqrt{3}}{2}$    $\tan 60° = \sqrt{3}$    $\sec 60° = 2$

 $\cos 60° = \dfrac{1}{2}$    $\cot 60° = \dfrac{1}{\sqrt{3}}$    $\csc 60° = \dfrac{2}{\sqrt{3}}$

**3.** $\sin 90° = 1$, $\tan 90°$ is undefined, $\sec 90°$ is undefined $\cos 90° = 0$, $\cot 90° = 0$, $\csc 90° = 1$

**4.** $\sin 0° = 0$, $\tan 0° = 0$, $\sec 0° = 1$
$\cos 0° = 1$, $\cot 0°$ is undefined, $\csc 0°$ is undefined

**7.** (a) $\sin \theta = \dfrac{4}{5}$ $\quad \cos \theta = \dfrac{3}{5}$ $\quad\quad \tan \theta = \dfrac{4}{3}$

$\quad\quad \csc \theta = \dfrac{5}{4}$ $\quad \sec \theta = \dfrac{5}{3}$ $\quad\quad \cot \theta = \dfrac{3}{4}$

(c) $\sin \theta = \dfrac{4}{5}$ $\quad \cos \theta = -\dfrac{3}{5}$ $\quad \tan \theta = -\dfrac{4}{3}$

$\quad\quad \csc \theta = \dfrac{5}{4}$ $\quad \sec \theta = -\dfrac{5}{3}$ $\quad \cot \theta = -\dfrac{3}{4}$

(e) $\sin \theta = \dfrac{12}{13}$ $\quad \cos \theta = -\dfrac{5}{13}$ $\quad \tan \theta = -\dfrac{12}{5}$

$\quad\quad \csc \theta = \dfrac{13}{12}$ $\quad \sec \theta = -\dfrac{13}{5}$ $\quad \cot \theta = -\dfrac{5}{12}$

**9.** $\dfrac{\sqrt{2} + \sqrt{6}}{4}$ $\quad$ **11.** $\sqrt{\dfrac{2 - \sqrt{2}}{2 + \sqrt{2}}}$

**13.** $\dfrac{3 - \sqrt{3}}{3 + \sqrt{3}}$ $\quad$ **15.** $\dfrac{\sqrt{3}}{2}$

**17.** $\dfrac{\sqrt{2}}{2}$

### Section 49, p. 272

**1.** $a = 8 \sin 32° \approx 4.239$, $b = 8 \cos 32° \approx 6.784$, $B = 58°$

**3.** $c = \sqrt{801} \approx 28.3$, $A \approx 32°$, $B \approx 58°$

**5.** $A = 27°$, $b = 16 \sin 63° \approx 14.256$, $a = 16 \cos 63° \approx 7.264$

**7.** $B = 44°$, $c = 54 \csc 46°$, $b = 54 \cot 46°$

**9.** $\approx 78.54$

**11.** 17 ft; 36°

**15.** $\sin^{-1}\left(\dfrac{x}{2}\right) - \dfrac{x\sqrt{4 - x^2}}{4}$

**17.** 238.4 sq units

### Sections 50 and 51, p. 277

**1.** $b \approx 142.2$ $\quad\quad$ **3.** no solution $\quad\quad$ **5.** $A \approx 37°$ $\quad\quad$ **7.** $C \approx 57°$

**9.** $B \approx 62°$ or $118°$ $\quad\quad$ **11.** $A \approx 41°$, $B \approx 56°$, $C \approx 83°$ $\quad\quad$ **13.** $b \approx 6.7$

**15.** 335 mi

**17.** (b) 336 sq units

**19.** 30° and 150°

**23.** (b) $\approx 527$ gal

### Sections 52 and 53, p. 283

**2.** (a) $c \geq 0$ $\quad\quad$ (b) $\vec{A} = c\vec{B}$ $\quad\quad$ (c) $\vec{A} \perp \vec{B}$

### Section 54, p. 293

**1.** $-\vec{i} + 4\vec{j}$ $\quad\quad$ **3.** $2\vec{i}$ $\quad\quad$ **5.** $\vec{i} + 4\vec{j}$

**7.** $8\vec{i}$, $-4\vec{i} - 6\vec{j}$, $\sqrt{13}$, $3\sqrt{5}$, 8

**9.** $\vec{0}$, $2\vec{i} + 2\vec{j}$, $\sqrt{2}$, $\sqrt{2}$, 0

**11.** $\frac{4}{5}\vec{i} + \frac{3}{5}\vec{j}$    **13.** $(\frac{9}{2}, -\frac{1}{2})$

**15.** (a) $\dfrac{4}{\sqrt{2}}$, $2(\vec{i} + \vec{j})$    (c) $-\dfrac{2}{\sqrt{2}}$, $\vec{i} + \vec{j}$

**17.** $\dfrac{10}{\sqrt{5}}(-\vec{i} + 2\vec{j})$

**27.** (a) 6    (b) $\pm 4$    (c) $\pm 2\sqrt{3}$

**29.** 44 ft-lb

**Section 56, p. 298**

**1.** $6 - 2i$      **21.** $20 - 22i$      **41.** $x = -2$, $y = -\frac{1}{3}$

**3.** $-2 + 4i$      **23.** $(4 - 7i)/5$      **43.** $x = \dfrac{-1 \pm i\sqrt{3}}{2}$

**5.** $1 - 3i$      **25.** $-1 - i$      **45.** $x = 1$, $\dfrac{-1 \pm i\sqrt{3}}{2}$

**7.** $2 - i$      **27.** $-2 + 2i$      **47.** $x = 1$, $-1$, $i$, $-i$

**9.** $1 - 5i$      **29.** $(-3 - 2i)/4$      **49.** $x = (2 - i)/3$

**11.** $8 - 26i$      **31.** $-2i/3$      **51.** $x = 3i$, $-3i$

**13.** 3      **33.** $-i/2$      **59.** $a = 4$, $b = 4$

**15.** $21 + 6i$      **35.** $-\frac{1}{2}$

**17.** $3 - 4i$      **37.** $x = \frac{12}{5}$, $y = -\frac{9}{5}$

**19.** $-i$      **39.** $x = -\frac{13}{4}$, $y = 0$

**Section 57, p. 302**

**1.** (a) $(-1, \pi)$      **2.** (a) $(0, 0)$      **4.** (a) $\sqrt{2 - \sqrt{2}}$

(c) $\left(-2, -\dfrac{2\pi}{3}\right)$      (c) $(-1, -\sqrt{3})$      (c) $5\sqrt{2}$

(e) $(-6, 0)$      (e) $(-6, 0)$

**Section 58, p. 307**

**1.** $4 \operatorname{cis} 0$      **23.** $-2\sqrt{3} + 2i$

**3.** $3 \operatorname{cis} \dfrac{\pi}{2}$      **25.** $-4 + 0i$

**5.** $2 \operatorname{cis} \dfrac{\pi}{3}$      **27.** $4\sqrt{2} + 4\sqrt{2}i$

**7.** $2\sqrt{2} \operatorname{cis} 225°$      **29.** $\dfrac{1}{2} - \dfrac{\sqrt{3}}{2}i$

**9.** $16 \operatorname{cis} 150°$      **31.** $-\sqrt{2} + \sqrt{2}i$

**11.** $\operatorname{cis} \dfrac{5\pi}{4}$      **33.** $\dfrac{3\sqrt{2}}{2} - \dfrac{3\sqrt{2}}{2}i$

**13.** $16 \operatorname{cis} 300°$      **35.** $(\sqrt{3} - 2) + (1 + 2\sqrt{3})i$

**15.** $3 \operatorname{cis} \dfrac{\pi}{6}$      **37.** $-2$

**17.** $3 \operatorname{cis} \dfrac{7\pi}{6}$      **39.** $-1 - 2i$

**21.** 1      **41.** $2 \operatorname{cis} 30°$

**43.** $5 \operatorname{cis} \dfrac{\pi}{2}$          **49.** $2 \operatorname{cis} \pi$

**45.** $2 \operatorname{cis} \dfrac{4\pi}{3}$          **51.** $\operatorname{cis} 60°$

**47.** $\frac{3}{2} \operatorname{cis} 120°$          **53.** $4 \operatorname{cis} 0°$

### Section 59, p. 312

**1.** $32 \operatorname{cis} 300° = 16 - 16\sqrt{3}i$

**3.** $\dfrac{1}{8} \operatorname{cis} \dfrac{3\pi}{4} = \dfrac{-\sqrt{2}}{16} + \dfrac{\sqrt{2}}{16}i$

**5.** $-2 - 2i$

**7.** $1$

**9.** $-1$

**13.** $2 \operatorname{cis} 45° = \sqrt{2} + \sqrt{2}i$
$2 \operatorname{cis} 135° = -\sqrt{2} + \sqrt{2}i$
$2 \operatorname{cis} 225° = -\sqrt{2} - \sqrt{2}i$
$2 \operatorname{cis} 315° = \sqrt{2} - \sqrt{2}i$

**15.** $\operatorname{cis} 135° = -\dfrac{\sqrt{2}}{2} + \dfrac{\sqrt{2}}{2}i$

$\operatorname{cis} 315° = \dfrac{\sqrt{2}}{2} - \dfrac{\sqrt{2}}{2}i$

**17.** $\operatorname{cis} 45° = \dfrac{\sqrt{2}}{2} + \dfrac{\sqrt{2}}{2}i$

$\operatorname{cis} 225° = \dfrac{-\sqrt{2}}{2} - \dfrac{\sqrt{2}}{2}i$

**19.** $x = 2, \ -1 \pm \sqrt{3}i$

**21.** $r_1 = \operatorname{cis} 0° = 1$

$r_2 = \operatorname{cis} 60° = \dfrac{1}{2} + \dfrac{\sqrt{3}}{2}i$

$r_3 = \operatorname{cis} 120° = -\dfrac{1}{2} + \dfrac{\sqrt{3}}{2}i$

$r_4 = \operatorname{cis} 180° = -1$

$r_5 = \operatorname{cis} 240° = -\dfrac{1}{2} - \dfrac{\sqrt{3}}{2}i$

$r_6 = \operatorname{cis} 300° = \dfrac{1}{2} - \dfrac{\sqrt{3}}{2}i$

**23.** $r_1 = \sqrt{2} \operatorname{cis} 30° = \dfrac{\sqrt{6}}{2} + \dfrac{\sqrt{2}}{2}i$

$r_2 = \sqrt{2} \operatorname{cis} 210° = -\dfrac{\sqrt{6}}{2} - \dfrac{\sqrt{2}}{2}i$

**25.** $r_1 = \operatorname{cis} 90° = i$

$r_2 = \operatorname{cis} 210° = -\dfrac{\sqrt{3}}{2} - \dfrac{1}{2}i$

$r_3 = \operatorname{cis} 330° = \dfrac{\sqrt{3}}{2} - \dfrac{1}{2}i$

**Section 60, p. 325**

1. $-2$
3. $\frac{1}{4}$
5. $\frac{1}{3}$
7. $\frac{17}{6}$
9. (1) $2x + y - 7 = 0$
   (3) $x - 4y = 0$
   (5) $x - 3y + 2 = 0$
   (7) $17x - 6y - 14.9 = 0$
11. $x = 1$
13. $x + 2y - 9 = 0$
15. $5x - 2y + 10 = 0$
17. $x - y + 3 = 0$
19. $2x + y - 6 = 0$
    $3x + 5y - 9 = 0$
    $x - 3y + 11 = 0$
21. $3x - 2y + 5 = 0$
23. $3\sqrt{13} + \sqrt{65}$
25.

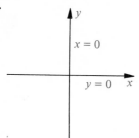

27. (a) $\dfrac{\sqrt{5}}{5}$
    (c) $2\sqrt{2}$
    (e) $2$

33. (a) $\cos\theta = \dfrac{4}{\sqrt{65}}$
    (c) $\cos\theta = \dfrac{\sqrt{2}}{2}$

35. $\dfrac{\sqrt{29}}{2}$, $2\sqrt{5}$, $\dfrac{\sqrt{101}}{2}$

**Section 62, p. 331**

1. $(x - 1)^2 + (y - 3)^2 = 4$
3. $x^2 + y^2 = 1$
5. $(x - \sqrt{2})^2 + (y - \sqrt{3})^2 = 25$
7. $(x - 2)^2 + y^2 = 9$
9. $(x - 1)^2 + (y - 2)^2 = 13$
11. $(x - 2)^2 + (y - 7)^2 = 36$
13. $(x + 2)^2 + (y - \frac{5}{2})^2 = \frac{49}{4}$
15. $r = \sqrt{35}$, $C(3, 4)$

**17.** $r = \sqrt{8}$, $C(-1, 2)$
**19.** $\varnothing$
**21.** $r = 4$, $C(-2, 0)$
**23.** $\varnothing$
**25.** $(-2, 0)$
**26.** (a) $(x + 4)^2 + (y - 3)^2 = 25$
    (c) $(x - 3)^2 + (y + 2)^2 = 25$
    (e) $(x - 2)^2 + (y - 2)^2 = 8$
**27.** $(x - 8)^2 + (y - 6)^2 = 100$ or $(x - 8)^2 + (y + 6)^2 = 100$
**29.** $(x - 1)^2 + (y - 2)^2 = 34$
**31.** $(x - 7)^2 + (y + 2)^2 = 4$
**33.** $(x - 5)^2 + (y - 2)^2 = 10$
**35.** (a) $(x - h)^2 + y^2 = 25$
    (c) $x^2 + y^2 = a^2$

**Section 63, p. 336**

**1.** $x' = x - 1$, $y' = y$; $(1, 0)$
**3.** $x' = x + 3$, $y' = y - 1$; $(-3, 1)$
**5.** $x' = x + 2$, $y' = y - 5$; $(-2, 5)$
**7.** $x' = x - 2$, $y' = y - 4$; $(2, 4)$
**9.** $x' = x - 1$, $y' = y - 2$; $(1, 2)$
**11.** $x'^2 + y'^2 = 25$
**13.** $x'^2 = 4y'$
**15.** $\dfrac{y'^2}{4} - \dfrac{x'^2}{1} = 1$
**17.** $2x'^2 = y'^3$
**19.** $x'y' - 2y'^2 = 1$; $(3, -1)$
**21.** $x'^3 = y'^2$; $(-1, -1)$
**23.** $x = x' + 3$, $y = y' - 1$
**25.** $y'^2 = 4x'$

**Section 64, p. 346**

| | Vertices | Focus | Eq. of Directrix |
|---|---|---|---|
| **1.** | $(0, 0)$ | $(-1, 0)$ | $x = 1$ |
| **3.** | $(0, 0)$ | $(0, 1)$ | $y = -1$ |
| **5.** | $(3, 2)$ | $(4, 2)$ | $x = 2$ |
| **7.** | $(-1, -1)$ | $(-2, -1)$ | $x = 0$ |
| **9.** | $(0, 2)$ | $(0, 3)$ | $y = 1$ |
| **11.** | $(-3, -3)$ | $(-3, -2)$ | $y = -4$ |
| **13.** | $(5, 1)$ | $(3, 1)$ | $x = 7$ |
| **15.** | $(-1, \frac{1}{2})$ | $(-\frac{3}{4}, \frac{1}{2})$ | $x = -\frac{5}{4}$ |

**17.** two straight lines: $y = -2 \pm 2\sqrt{2}$
**19.** one straight line: $x + y = 0$
**21.** $x^2 = 8y$
**23.** $(y - 1)^2 = 20(x + 5)$
**25.** $(x - 3)^2 = -16(y - 9)$
**27.** $y^2 = 16(x - 4)$
**29.** $(x + 2)^2 = -32(y - 4)$

**33.** 20 ft; 27 ft
**37.** $(y - 1)^2 = 2(x + 2)$
**39.** $(x - 3)^2 = y + 4$

## Section 65, p. 353

| *Vertices* | *Foci* |
|---|---|
| **1.** $(\pm 12, 0)$; $(0, \pm 9)$ | $(\pm 3\sqrt{7}, 0)$ |
| **3.** $(\pm 2, 0)$; $(0, \pm 5)$ | $(0, \pm\sqrt{21})$ |
| **5.** $(\pm 2, 0)$; $(0, \pm 3)$ | $(0, \pm\sqrt{5})$ |
| **7.** Circle, $r = 1$, $c(0, 0)$ | |
| **9.** $(\pm 1, 0)$; $(0, \pm 3)$ | $(0, \pm 2\sqrt{2})$ |
| **11.** $(-3, -2)$, $(5, -2)$; $(1, -4)$, $(1, 0)$ | $(1 \pm 2\sqrt{3}, -2)$ |
| **13.** $(-3, -5)$, $(1, -5)$; $(-1, -9)$, $(-1, -1)$ | $(-1, -5 \pm 2\sqrt{3})$ |
| **15.** $(-18, -2)$, $(8, -2)$; $(-5, -14)$, $(-5, 10)$ | $(-10, -2)$, $(0, -2)$ |
| **17.** $(-2, 0)$, $(-2, 6)$; $(-4, 3)$, $(0, 3)$ | $(-2, 3 \pm \sqrt{5})$ |
| **19.** $(1 \pm \sqrt{5}, 2)$, $(1, -3)$, $(1, -1)$ | $(3, -2)$, $(-1, -2)$ |

**21.** $\dfrac{x^2}{12} + \dfrac{y^2}{16} = 1$

**23.** $\dfrac{x^2}{144} + \dfrac{y^2}{169} = 1$

**25.** $\dfrac{x^2}{32} + \dfrac{(y - 2)^2}{36} = 1$

**27.** $\dfrac{x^2}{25} + \dfrac{y^2}{10} = 1$

**29.** $\dfrac{(x - 6)^2}{9} + \dfrac{(y + 4)^2}{4} = 1$

**31.** $\dfrac{(x - 1)^2}{4} + \dfrac{(y - 4)^2}{9} = 1$

**33.** $\dfrac{x^2}{100} + \dfrac{y^2}{84} = 1$

**35.** $\dfrac{(x + 1)^2}{36} + \dfrac{(y + 1)^2}{27} = 1$

**37.** $\dfrac{(x + 3)^2}{16} + \dfrac{(y - 2)^2}{9} = 1$

**39.** $y - 2 = \pm\dfrac{2\sqrt{5}}{5}(x - 6)$

**41.** $y^2 = 4(3 - \sqrt{5})(x + 3)$
$y^2 = 4(\sqrt{5} - 3)(x - 3)$

## Section 66, p. 361

| *Vertices* | *Foci* | *Asymptotes* |
|---|---|---|
| **1.** $(\pm 4, 0)$ | $(\pm 5, 0)$ | $y = \pm\frac{3}{4}x$ |
| **3.** $(\pm 15, 0)$ | $(\pm\sqrt{274}, 0)$ | $y = \pm\frac{7}{15}x$ |
| **5.** $(0, \pm 2)$ | $(0, \pm\sqrt{5})$ | $y = \pm 2x$ |
| **7.** $(0, \pm 1)$ | $\left(0, \pm 1\dfrac{\sqrt{13}}{2}\right)$ | $y = \pm\frac{2}{3}x$ |
| **9.** $(-2, 1)$, $(-2, 7)$ | $(-2, 4 \pm \sqrt{13})$ | $y - 4 = \pm\frac{3}{2}(x + 2)$ |
| **11.** $(5, -3)$, $(-3, -3)$ | $(6, -3)$, $(-4, -3)$ | $y + 3 = \pm\frac{3}{4}(x - 1)$ |
| **13.** $(-2, 2)$, $(-2, 6)$ | $(-2, -2 \pm 2\sqrt{5})$ | $y + 2 = \pm 2(x + 2)$ |
| **15.** $(0, 0)$, $(0, -\frac{4}{9})$ | $\left(0, -\dfrac{2}{9} \pm \dfrac{2\sqrt{10}}{9}\right)$ | $y + \dfrac{2}{9} = \pm\dfrac{x}{3}$ |
| **17.** $(-\frac{1}{2}, -\frac{7}{4})$, $(-\frac{1}{2}, \frac{9}{4})$ | $(-\frac{1}{2}, \frac{1}{4} \pm 2\sqrt{5})$ | $y - \frac{1}{4} = \pm\frac{1}{2}(x + \frac{1}{2})$ |
| **19.** $(-1, 2)$, $(1, 2)$ | $(\pm\sqrt{5}, 2)$ | $y - 2 = \pm 2x$ |

**21.** $\dfrac{x^2}{36} - \dfrac{y^2}{28} = 1$

**23.** $\dfrac{x^2}{4} - \dfrac{y^2}{12} = 1$

**25.** $\dfrac{y^2}{9} - \dfrac{x^2}{16} = 1$

**27.** $\dfrac{(y-5)^2}{4} - \dfrac{(x-2)^2}{21} = 1$

**29.** $\dfrac{(y-3)^2}{16} - \dfrac{(x+1)^2}{9} = 1$

**31.** $\dfrac{(x+4)^2}{36} - \dfrac{(y-1)^2}{16} = 1$

**35.** $\dfrac{x^2}{\frac{135}{4}} - \dfrac{y^2}{\frac{135}{9}} = 1$

**37.** $\dfrac{(x+2)^2}{9} - \dfrac{(y-1)^2}{4} = 1$

### Section 68, p. 365

**1.** circle

**3.** hyperbola

**7.** hyperbola

**9.** ellipse

### Section 69, p. 372

**1.** circle; $r = 5$, $C(-3, -2)$

**3.** hyperbola
vertices: $(-3 \pm \sqrt{2}, \frac{1}{2})$

foci: $\left(-3 \pm \dfrac{5\sqrt{2}}{2}, \dfrac{1}{2}\right)$

Asy.: $2y - 1 = \pm(x + 3)$

**5.** $\varnothing$

vertices: $\left(-\dfrac{1}{2} \pm \dfrac{\sqrt{3}}{2}, 3\right)$

foci: $\left(-\dfrac{1}{2} \pm \dfrac{\sqrt{15}}{2}, 3\right)$

Asy.: $y - 3 = \pm 2(x + \frac{1}{2})$

**9.** ellipse
vertices: $(-1, 0)$, $(5, 0)$, $(2, \pm\sqrt{3})$
foci: $(2 \pm \sqrt{6}, 0)$

**11.** ellipse: $\dfrac{x'^2}{1} + \dfrac{y'^2}{4} = 1$

**13.** hyperbola: $\dfrac{(x' - 3\sqrt{2})^2}{16} - \dfrac{(y' - \sqrt{2})^2}{16} = 1$

**15.** two straight lines $\dfrac{9x'^2}{2} - \dfrac{y'^2}{2} = 0$

**17.** ellipse: $\dfrac{\left(x' + \dfrac{\sqrt{2}}{2}\right)^2}{9} + \dfrac{\left(y' + \dfrac{5\sqrt{2}}{2}\right)^2}{4} = 1$

**19.** parabola: $y'^2 = 12x'$

**21.** (a) $35x^2 + 11y^2 - 10xy - 310x + 34y + 599 = 0$

**23.** (a) $7x^2 + 24xy - 144 = 0$

**29.** $\tan^{-1}(\frac{4}{3}) \approx 53°$

**Section 70, p. 384**

**1.**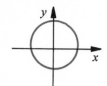

**3.**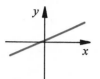

**5.**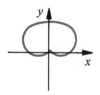

**7.**

**9.**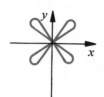

**13.**

**15.**

**17.**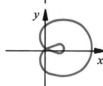

**19.**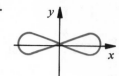

**21.**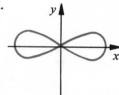

**23.**

**25.**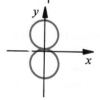

**26.** (a) $r = 2a \sin \theta$
    (c) $r^2 = a^2 \sin 2\theta$

**27.** (a) $x^2 + y^2 - x - y = 0$
    (c) $y = \sqrt{3}x$
    (e) $9x^2 + 25y^2 - 72x - 81 = 0$

**28.** (a) $\theta = 0, \dfrac{\pi}{6}, 5\dfrac{\pi}{6}, \dfrac{3\pi}{2}$

    (c) $\theta = 0, \dfrac{3\pi}{2}$

    (e) $\theta = -\dfrac{\pi}{3}, \dfrac{\pi}{3}$

    (g) $\left(3, -\dfrac{\pi}{3}\right), \left(3, \dfrac{\pi}{3}\right)$, pole

**Section 71, p. 389**

**1.** (a) $3, 5, 7, 9, 11; 25; 2k + 1; 2k + 3$

    (b) $\dfrac{1}{2}, \dfrac{1}{3}, \dfrac{1}{4}, \dfrac{1}{5}, \dfrac{1}{6}; \dfrac{1}{13}; \dfrac{1}{k+1}; \dfrac{1}{k+2}$

    (c) $\dfrac{1}{2}, \dfrac{1}{6}, \dfrac{1}{12}, \dfrac{1}{20}, \dfrac{1}{30}; \dfrac{1}{156}; \dfrac{1}{k(k+1)}; \dfrac{1}{(k+1)(k+2)}$

    (d) $-\dfrac{1}{5}, \dfrac{1}{25}, -\dfrac{1}{125}, \dfrac{1}{625}, -\dfrac{1}{3,125}; \left(-\dfrac{1}{5}\right)^{12}; \left(-\dfrac{1}{5}\right)^{k}; \left(-\dfrac{1}{5}\right)^{k+1}$

(e) $-1, \dfrac{1}{4}, -\dfrac{1}{9}, \dfrac{1}{16}, -\dfrac{1}{25}; \dfrac{1}{144}; \dfrac{(-1)^k}{k^2}; \dfrac{(-1)^{k+1}}{(k+1)^2}$

(f) $-\dfrac{1}{2}, \dfrac{1}{3}, -\dfrac{1}{4}, \dfrac{1}{5}, -\dfrac{1}{6}; \dfrac{1}{13}, \dfrac{(-1)^k}{k+1}; \dfrac{(-1)^{k+1}}{k+2}$

(g) 5, 8, 11, 14, 17; 38; $3k + 2$; $3k + 5$

(h) $\dfrac{3}{2}, \dfrac{9}{5}, \dfrac{27}{10}, \dfrac{81}{17}, \dfrac{243}{26}; \dfrac{3^{12}}{145}; \dfrac{3^k}{k^2+1}; \dfrac{3^{k+1}}{k^2+2k+2}$

(i) $-3, 9, -27, 81, -243; 3^{12}; (-1)^k 3^k; (-1)^{k+1} 3^{k+1}$

**3.** (a) 2, 3, 4, 5        (d) $5, \frac{5}{2}, \frac{5}{4}, \frac{5}{8}$

   (b) 2, 4, 16, 256     (e) 5, 19, 75, 299

   (c) 1, 2, 6, 24

**5.** value $= P(1 - r)^k$

## Section 72, p. 394

**1.** $\displaystyle\sum_{k=1}^{10} k$     **3.** $\displaystyle\sum_{k=2}^{n} k - 1$     **5.** $\displaystyle\sum_{k=1}^{100} 3k$     **9.** $\displaystyle\sum_{k=1}^{n} 2^k$

**11.** $\displaystyle\sum_{k=1}^{6} k^2$     **13.** $\displaystyle\sum_{k=1}^{n} k^2$     **15.** $2 + 3 + 4 + 5 + 6$

**17.** $2 \cdot 1 + 3 \cdot 2 + 4 \cdot 3 + 5 \cdot 4 + \cdots + 100 \cdot 99$

**19.** $2 \cdot 1 + 2 \cdot 2^2 + 2 \cdot 3^3 + 2 \cdot 4^4 + \cdots + 2 \cdot n^n$

**21.** $-\frac{1}{3} + \frac{1}{9} - \frac{1}{27} + \frac{1}{81}$

**23.** $\dfrac{1}{2 \cdot 1} + \dfrac{1}{2 \cdot 3} + \dfrac{1}{3 \cdot 4} + \cdots + \dfrac{1}{(n-1)n}$

**25.** $\dfrac{x^2}{2} + \dfrac{x^4}{4} + \dfrac{x^6}{6} + \cdots + \dfrac{x^{2n}}{2n}$

## Section 73, p. 396

**1.** $(3n - 1); S = 155$        **3.** $\dfrac{4n - 6}{3}; s = 130$

**5.** $x + y(n - 3); s = 20x + 150y$      **7.** $a = 7, d = 4$

**9.** $a = -33; S_n = -33$           **11.** $S_n = (217)(245)$

**13.** $3n - 2$

## Section 74, p. 399

**1.** $2 \cdot (\frac{1}{2})^{n-1}$ or $(\frac{1}{2})^{n-2}; s = \frac{127}{32}$

**3.** $\frac{1}{2}(-5)^{n-1}; s = \frac{521}{2}$

**5.** $2^{n-1}; s = 1{,}023$

**7.** $\left(\dfrac{a}{b}\right)\left(-\dfrac{b}{a}\right)^{n-1}; s = \dfrac{a^5 + b^5}{a^4 b + a^3 b^2}$

**9.** $b = \pm\frac{1}{3}$

**11.** $2(\frac{3}{2})^9$

**13.** (a) $s = \dfrac{31}{16}$      (c) $s = \dfrac{e^{10} - 1}{e^{10}(e - 1)}$

**15.** \$2,610.03

**Section 75, p. 402**

  **1.** $s = 2$      **3.** $s = 8$      **5.** $s = 9$      **7.** $s = \frac{20}{9}$

  **9.** $s = 4$      **11.** $|r| > 1$; no sum      **13.** $s = \dfrac{e}{e-1}$

**15.** $\frac{2}{33}$      **17.** 1      **19.** $\frac{19}{9}$      **21.** $\frac{1040}{333}$      **23.** 90 ft

**Section 77, p. 412**

  **1.** (a) 15

     (c) 1

     (e) 1

  **3.** $x^6 - 6x^5y + 15x^4y^2 - 20x^3y^3 + 15x^2y^4 - 6xy^5 + y^6$

  **5.** $\dfrac{x^4}{y^4} - \dfrac{4x^2}{y^2} + 6 - \dfrac{4y^2}{x^2} + \dfrac{y^4}{x^4}$

  **7.** $x^{15} - 5x^{12}y^3 + 10x^9y^6 - 10x^6y^9 + 5x^3y^{12} - y^{15}$

  **9.** $x^{4/3} - 4xy^{1/3} + 6x^{2/3}y^{2/3} - 4x^{1/3}y + y^{4/3}$

**11.** $x^2 + 4x + 6 + 4x^{-1} + x^{-2}$

**13.** 1.04060401

**15.** $5{,}376a^3b^6$

**17.** $101{,}376x^{15}y^7$

**19.** $\dfrac{28x^4y^6}{81}$

# INDEX

## A

Abscissa, 32
Absolute value
  equality, 21
  definition, 19, 21
  of a complex number, 304
  function, 73
  properties, 22
  relations, 79
Addition formula
  of cosine, 213
  of sine, 215
  of tangent, 216
Addition of two functions, 159
Algebra
  of functions, 159
  fundamental theorem, 116
Algebraic expression, 107
Ambiguous case in the solution of
   triangles, 273
Amplitude, 237
Angle
  coterminal, 258
  of depression, 271
  of elevation, 271
  initial side, 257
  measure, degree, radian, 258
  negative, 257
  quadrantal, 258

  sides, 257
  standard position, 258, 261
  terminal side, 257
  vertex, 257
Antilogarithm, 188
Arcosine function, 248
Arc length along a circle, 259
Arcsine function, 246
Arctangent function, 248
Area of a circle sector, 260
Area of a segment of a circle, 279
Argand diagram, 303
Argument of a complex number, 304
Associative laws of real numbers, 10
Asymptotes
  of cosecant functions, 244
  of cotangent functions, 243
  curvilinear, 146
  horizontal, 137
  of hyperbola, 356
  oblique, 144
  of secant functions, 243
  of tangent functions, 242
  vertical, 137
Axis, 32
  conjugate, 356
  imaginary, 303
  major, 349
  minor, 349
  real, 303